T0349049

BIOHYDROGEN

BIOHYDROGEN

Edited By

Ashok Pandey
CSIR–National Institute for Interdisciplinary Science and Technology,
Trivandrum, India

Jo-Shu Chang
National Cheng Kung University,
Taiwan

Patrick C. Hallenbeck
University of Montreal,
Montreal, Quebec, Canada

Christian Larroche
Universite Blaise Pascal,
Clermont-Ferrand, France

ELSEVIER   AMSTERDAM • BOSTON • HEIDELBERG • LONDON • NEW YORK • OXFORD
PARIS • SAN DIEGO • SAN FRANCISCO • SINGAPORE • SYDNEY • TOKYO

Elsevier
30 Corporate Drive, Suite 400, Burlington, MA 01803, USA
525 B Street, Suite 1800, San Diego, CA 92101-4495, USA

First edition **2013**

British Library Cataloguing in Publication Data
A catalogue record for this book is available from the British Library

Library of Congress Cataloging-in-Publication Data
A catalog record for this book is available from the Library of Congress

For information on all **Elsevier** publications
visit our web site at store.elsevier.com

Printed and bound in Poland
13 14 15 16 17   10 9 8 7 6 5 4 3 2 1

ISBN: 978-0-444-59555-3

Contents

Contributors

M. Kalim Akhtar University of Turku, Bioenergy Group, Tykistökatu 6A, 6krs, 20520 Turku, Finland

P. Suresh Babu Bioengineering and Environmental Centre, Council of Scientific and Industrial Research–Indian Institute of Chemical Technology, Hyderabad-500 007, India

G. Balachandar Department of Biotechnology, Indian Institute of Technology, Kharagpur, India

Bhavya Balagurumurthy Council of Scientific and Industrial Research–Indian Institute of Petroleum, Bio-Fuels Division, Dehradun 248005, India

Thallada Bhaskar Council of Scientific and Industrial Research–Indian Institute of Petroleum, Bio-Fuels Division, Dehradun 248005, India

Jaehwan Cha Clean Fuel Center, Korea Institute of Energy Research, Daejeon 305-343, Republic of Korea

K. Chandrasekhar Bioengineering and Environmental Centre, Council of Scientific and Industrial Research–Indian Institute of Chemical Technology, Hyderabad-500 007, India

Jo-Shu Chang Department of Chemical Engineering, National Cheng Kung University, Tainan, Taiwan

P. Chiranjeevi Bioengineering and Environmental Centre, Council of Scientific and Industrial Research–Indian Institute of Chemical Technology, Hyderabad-500 007, India

Chen-Yeon Chu, Green Energy Development Center, Feng Chia University, Taiwan; Master Program of Green Energy Science and Technology, Feng Chia University, Taiwan, and Department of Chemical Engineering, Feng Chia University, Taiwan

Philippe Constant Trace Gas Biogeochemistry Laboratory, INRS-Institut Armand-Frappier, Laval, Québec, Canada

Debabrata Das Department of Biotechnology, Indian Institute of Technology, Kharagpur, India

Patrick C. Hallenbeck University of Montreal, Montreal, Quebec, Canada

Patrik R. Jones University of Turku, Bioenergy Group, Tykistökatu 6A, 6krs, 20520 Turku, Finland

Gyoo Yeol Jung Department of Chemical Engineering & I-Bio Program, Pohang University of Science and Technology, Pohang 790-784, Korea

Namita Khanna Department of Biotechnology, Indian Institute of Technology, Kharagpur, India

Dong-Hoon Kim Clean Fuel Center, Korea Institute of Energy Research, Daejeon 305-343, Republic of Korea

Mi-Sun Kim Clean Fuel Center, Korea Institute of Energy Research, Daejeon 305-343, Republic of Korea

Man Kee Lam School of Chemical Engineering, Universiti Sains Malaysia, Engineering Campus, Seri Ampangan, 14300 Nibong Tebal, Pulau Pinang, Malaysia

Duu-Jong Lee Department of Chemical Engineering, National Taiwan University, Taipei 106, Taiwan

Keat Teong Lee School of Chemical Engineering, Universiti Sains Malaysia, Engineering Campus, Seri Ampangan, 14300 Nibong Tebal, Pulau Pinang, Malaysia

Chiu-Yue Lin Department of Environmental Engineering and Science, Feng Chia University, Taiwan; Green Energy Development Center, Feng Chia University, Taiwan, and Master Program of Green Energy Science and Technology, Feng Chia University, Taiwan

S. Venkata Mohan Bioengineering and Environmental Centre, Council of Scientific and Industrial Research–Indian Institute of Chemical Technology, Hyderabad-500 007, India

You-Kwan Oh Bioenergy Center, Korea Institute of Energy Research, Daejeon 305-343, Korea

Ashok Pandey Council of Scientific and Industrial Research–National Institute for Interdisciplinary Science and Technology, Trivandrum, India

Sunghoon Park Department of Chemical and Biomolecular Engineering, Pusan National University, Busan 609-735, Korea

Mukesh Kumar Poddar Council of Scientific and Industrial Research–Indian Institute of Petroleum, Bio-Fuels Division, Dehradun 248005, India

Subramanian Mohan Raj Center for Research and Development, PRIST University, Thanjavur, TN 613 403, India

Ganesh D. Saratale Department of Biochemistry and Department of Environmental Biotechnology, Shivaji University, Kolhapur-416 004, India

Rijuta G. Saratale Department of Biotechnology, Shivaji University, Kolhapur-416 004, India

Biswarup Sen Department of Environmental Engineering and Science, Feng Chia University, Taiwan

Kuan-Yeow Show Department of Environmental Engineering, Faculty of Engineering and Green Technology, University Tunku Abdul Rahman, Jalan University, Bandar Barat, 31900 Kampar, Perak, Malaysia

Rawel Singh Council of Scientific and Industrial Research–Indian Institute of Petroleum, Bio-Fuels Division, Dehradun 248005, India

Foreword

Realization that hydrogen can be produced by microorganisms goes back to 1896, when researchers at the Massachusetts Board of Health reported production of the gas by the cyanobacterium *Anabaena*. During the early part of the last century, others recognized that molecular hydrogen can be both a reactant as well as a product of metabolic (fermentative) reactions in heterotrophic and chemotrophic bacteria. The critical biological catalyst associated with hydrogen metabolism, hydrogenase, was in fact described and named by Marjory Stephenson and Leonard Stickland almost 80 years ago in 1931. In 1932, Cornelis van Niel recognized that hydrogen could act as an electron donor for the reduction of CO_2 in photosynthetic bacteria and thus directly related hydrogen metabolism with anaerobic photosynthesis. However, it was not until the late 1930s and early 1940s that Hans Gaffron recognized and characterized hydrogen metabolism in algae, though only under anaerobic conditions. All of this pioneering work and subsequent hydrogen metabolism studies from the 1940s to 1970s served as a foundation for more recent academic and applied interest in BioHydrogen. On the fundamental side, recent advances in genomics, systems biology and, more recently, synthetic biology are leading to a much more detailed understanding of the biology and chemistry of microbial hydrogen metabolism. In the applied arena, concerns about the relationship between the widespread use of fossil fuels on a worldwide basis, and its consequences, adverse environmental effects, and impending global change, have catalyzed a widespread interest in finding innovative ways of using microbes in future commercially viable biofuel- and Biohydrogen-producing systems to address these problems. In order to impact the current world order, more information at the basic level as well as practical studies at the applied level will be needed to overcome several challenges that currently prevent adoption of this vision. For biohydrogen, these include current limitations in hydrogen production efficiency, microbe bioengineering improvements, and substrate availability for non-photosynthetic processes. In the case of photosynthetic processes, light to hydrogen conversion efficiencies, photobioreactor designs (including optimal materials issues), and scale-up challenges associated with taking laboratory processes outdoors under conditions that cannot be completely controlled are most relevant. Of course, all of these challenges ultimately affect the cost of biohydrogen since in real life, new energy/fuel technologies will have to be cost competitive with older, mature processes.

"Biohydrogen," edited by Pandey, Chang, Hallenbeck, and Larroche, represents an important new look at this area, which, as one can see in the Table of Contents, is clearly "not a single technology." Most of the book's content emphasizes dark, fermentative processes, and rightly so since these processes have been demonstrated outside of the laboratory, government as well as privately funded scale-up facilities are currently being assessed in Asia and South America, they can be used to treat waste biomass and wastewater (conceivably solving two problems at once) and are closest to commercial application. Nevertheless, photo-fermentative bacterial and algal biohydrogen-production processes are also examined, as they are inherently more efficient than dark processes in terms of land usage and represent longer term technologies that substitute sunlight and water for biomass (product of previous photosynthesis) as the substrate.

In summary, we as a species must prepare for a world after fossil fuels (or after the point that we can continue to use them), and Biohydrogen presents a current review of the state-of-the-art of both potential nearer term and longer term biological hydrogen solutions.

Michael Seibert
National Renewable Energy Laboratory

Preface

There is a great deal of interest and ongoing effort in developing sustainable biofuels as one part of the puzzle of meeting the duel challenges of supplying increased future energy demands at the same time as reducing greenhouse gas emissions. Hydrogen could be a key future energy carrier due to its high energy content, its high efficiency of conversion to mechanical or electrical energy, and its intrinsic zero emissions. However, practical, renewable means for producing hydrogen are largely lacking. One possible approach is to use a microbial system to produce hydrogen using sunlight and water, or various waste streams, as substrates.

The present book, which is the second book in the series on BIOMASS being published by us, presents up-to-date state of the art information and knowledge by the internationally recognized experts and subject peers in various areas of biohydrogen. The topics being covered in the book range from the molecular level, the enzymes and metabolic pathways involved, to the level of bioprocess engineering and waste utilization, and include a treatment of scale-up and commercialization issues. Introductory chapters (Chapters 1 and 2) give a general background for this area as well as briefly outline the principles of the different microbial systems that are being investigated for hydrogen production. The following chapters (Chapters 3, 4, and 5) review the latest findings on hydrogenases, the enzymes responsible for proton reduction to hydrogen or for hydrogen oxidation, the thermodynamic limits to hydrogen yields imposed by the metabolic pathways, and attempts to increase biohydrogen yields through metabolic engineering. A series of chapters (Chapters 6–12) examine the details of the conversion of renewable resources, water or various waste streams to biohydrogen, highlighting the differences imposed by the substrate, or the advantages of particular microbial processes in dealing with a particular substrate. Engineering aspects of biohydrogen production, including bioprocess development and bioreactor design are also covered (Chapter 13) as well as initial attempts at scale-up and eventual commercialization (Chapter 14). Thus, the book covers the range of current research activities in biohydrogen production and should be of interest to anyone working in the general area of microbial bioprocesses, to researchers interested in biofuels in general, or to dedicated biohydrogen researchers. The text in all the chapters is supported by numerous clear, informative diagrams and tables. The book would be of special interest to the postgraduate students and researchers of applied biology, biotechnology, microbiology, biochemical, and chemical engineers working on biohydrogen. It is expected that the current discourse on biofuels R&D would go a long way in bringing out the exciting technological possibilities and ushering the readers toward the frontiers of knowledge in the area of biofuels, and this book will be helpful in achieving this discourse for biohydrogen.

We thank authors of all the articles for their cooperation and also for their preparedness in revising the manuscripts in a time-framed manner. We also acknowledge the help from the reviewers, who in spite of their busy professional activities, helped us by evaluating the manuscripts and gave their critical inputs to refine and improve the articles. We warmly thank Dr Marinakis Kostas and Dr Anita Koch and the team of Elsevier for their cooperation and efforts in producing this book.

Ashok Pandey
Jo-Shu Chang
Patrick C. Hallenbeck
Christian Larroche
Editors

Biohydrogen Production
An Introduction

S. Venkata Mohan*, Ashok Pandey[†]

*CSIR–Indian Institute of Chemical Technology, Hyderbad, India
[†]CSIR–National Institute for Interdisciplinary Science and Technology,
Trivandrum, India*

ESSENTIALS OF ENERGY

Increasing gaps between the energy requirement of the industrialized world and an inability to replenish needs from limited energy sources have resulted in a steep increase in fossil fuel utilization. This has not only put a severe strain on the depleting fossil fuels but also resulted in an alarming increase in pollution levels across the globe. An ever increasing level of greenhouse gases (GHGs) from the combustion of fossil fuels in turn aggravated the perils of global warming. Combustion of fossil fuels adds about 6 Gigatons (Gton $= 10^9$ tons) of carbon per year in the form of carbon dioxide to the atmosphere (IPCC, 2006). At present, the concentration of CO_2 is found to be exceeding 350 ppmv where it can potentially intensify the greenhouse effect by raising the global temperature. In the past few decades, human activities have released organic carbon, which is equivalent to that accumulated millions of years ago. The bulk emissions mainly come from motor vehicles, which alone account for more than 70% of the global carbon monoxide (CO) and 19% of CO_2 emissions (Goldemberg, 2008). According to the current consumption trends, oil (168.6 $\times 10^9$ tons), natural gas (177.4 $\times 10^{12}$ cubic meters), and coal (847.5 $\times 10^9$ tons) reserves may be depleted in another 133 years (BPSR, 2008). Petroleum products came in to existence to power everything when kerosene began to replace whale oil for lighting about 150 years ago (Grayson, 2011). The limited availability of global oil reserves and concerns about climate change from greenhouse gas emissions instigated marked interest in the development of clean and renewable energy alternatives to satisfy the growing energy demands. Therefore, diversification of energy resources is an essential requirement in the present-day energy scenario (Nouni, 2012).

Moreover, rapid development of alternative, renewable, carbon-neutral, and eco-friendly fuels is of paramount importance to fulfill the burgeoning energy demands.

HYDROGEN

Hydrogen (H_2) gas is an important and promising energy carrier that could play a significant role in the reduction of greenhouse gas emissions (Christopher and Dimitrios, 2012). On combustion, H_2 produces water and hence is as considered a clean and carbon-neutral energy carrier. With a high energy yield of 122 kJ/g, which is 2.75-fold greater than that of hydrocarbon fuels, H_2 is regarded as an ideal energy. Hydrogen gas is an attractive future energy carrier due to its potentially higher efficiency of conversion to usable power, low to nonexistent generation of pollutants, and high energy density. It is therefore considered a promising clean, renewable source of alternative energy in the realm of fossil fuel depletion and environmental pollution. In 1970, John Bockris proposed "hydrogen economy," which represents a system that delivers energy in the form of H_2. The main emphasis of this system is to promote H_2 as a potential fuel to solve some of the negative effects of using hydrocarbons. It is also believed that a H_2 fuel-based economy would be less polluting than a fossil fuel-based economy. In the context of energy systems, H_2 is best thought of an energy carrier, more akin to electricity than the fossil fuels (Lipman, 2011). Hydrogen is highly promising fuel for stationary and transportation uses. The high electrochemical reactivity of H_2 makes it ideal for fuel cells application in the presence of suitable catalysts. In fact, the sun's energy also results from the nuclear fusion of H_2.

The current merchant (purchased from H_2 producers) and captive (consumed by H_2 producer) market of H_2 is used mostly in oil refining, food production, metals treatment, and fertilizer manufacture (Lipman, 2011). Its applications in chemical processing, petroleum recovery and refining, metal production and fabrication, aerospace, and fuel cells are well established. Hydrogen is also used as a reducing and hydrogenating agent, shielding gas, food additive, rotor coolant, etc. At present, the largest demand for H_2 is observed in petroleum refinery and ammonia production (Markets and Markets, 2011). Hydrogen as an automotive fuel as an energy carrier for stationary power and transportation markets is an emerging sector with huge potential in the future. The market size of global H_2 production was estimated to be 53 million metric tons in 2010, in which 12% is shared by merchant H_2 and the rest with captive production (Markets and Markets, 2011). The hydrogen production market in terms of value was estimated to be \$82.6 billion in 2010 (Markets and Markets, 2011). The global H_2 production volume is forecasted to grow by a compound annual growth rate of 5.6% during 2011–2016 due to decreasing sulfur levels in petroleum products, lowering crude oil quality, and rising demand of H_2-operated fuel cell applications. The Asia and Oceania region is the largest market with 39% of global production share in 2010, accounting for a production of 21 million metric tons of hydrogen (Markets and Markets, 2011).

Even though H_2 is the most common element on Earth, it does not occur in elemental form. The industrial production of molecular H_2 is mainly from fossil sources through steam reforming of natural gas or methane supplementing marginally from the energy-intensive, water-splitting electrolysis process and as a by-product from some industrial processes. Global H_2 production currently exceeds 1 billion m^3/day, of which 48% is produced from natural gas, 30% from oil (often on-site in refineries), 18% from coal, and the remaining (4%) by water electrolysis (Jong, 2009). In conjunction to steam reforming, high-purity H$_2$ can also be produced by a water–gas shift reaction, which is one of the important industrial reactions used especially for ammonia synthesis. Partial oxidation of coal, heavy residual oils, and other low-value refinery products stands second in the production capacity after steam reforming of natural gas (Lipman, 2011; Markets and Markets, 2011). Autothermal reforming, catalytic partial oxidation, thermal decomposition, gasification, and pyrolysis were the other thermochemical routes used for H$_2$ production (Chen and Syu, 2010). The third largest production technology in terms of production capacity is steam gasification of coal (Markets and Markets, 2011). The production of H$_2$ from fossil fuels is accompanied by the production of greenhouse gases, namely, CO$_2$, CH$_4$, etc. (Jong, 2009). At present, H$_2$ synthesis from biomass/waste through biological routes has emerging interest due to its sustainable nature.

When traced by time, hydrogen gas was first produced (artificially) in the early 16th century by Robert Boyle through a reaction between iron and acid. In the 17th century, Henry Cavendish recognized H$_2$ gas as a discrete substance and named it as "flammable air" or "water-former." In 1783, Antoine Lavoisier gave the element the name "Hydrogen" (in Greek, *hydro* means water and *genes* meaning creator) (Stwertka, 1996). In 1939, Hans Gaffron noticed that *Chlamydomonas reinhardtii*, a typical green algae normally grown in pond scum, would occasionally change from producing oxygen to hydrogen, what is now called "biohydrogen."

TRANSITION TOWARD BIOENERGY

The looming energy crisis and climate change concerns, coupled with high oil prices and decreasing fossil fuel resources, have garnered significant global attention towards the development of alternative, renewable, carbon-neutral, and eco-friendly fuels to fulfill the burgeoning energy demands. Bioenergy presents an exciting and sustainable alternative for fossil fuels, which can defend the energy crisis and save the planet from the brink of environmental catastrophe. Bioenergy is deemed to have the potential to provide renewable and carbon-neutral energy through sustainable routes. It offers a strategy to diversify energy sources to reduce supply risks and to also help promote domestic rural economies (International Energy Agency (IEA), 2008). Bioenergy derived from microorganisms is of great interest in the present world's energy scenario due to its renewability. Bacteria have flexible and diverse metabolic machinery to convert/synthesize a variety of organics to various forms of bioenergy. Establishing a practicable link between terminal electron-acceptor-limited microorganism and an electron sink is the main basis for many of the bioenergy generation processes (Madsen, 2008). At present, extensive as well as intense research is being focused towards bioenergy generation from renewable resources throughout the world.

GENESIS OF BIOLOGICAL H$_2$ PRODUCTION

Biohydrogen is a natural and transitory by-product of various microbial-driven biochemical reactions. Generation of H$_2$ gas either by biological machinery or through thermochemical treatment of biomass can be defined as "biohydrogen" (Fig. 1). Incidentally,

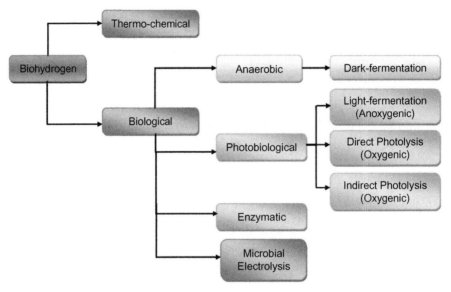

FIGURE 1   Schematic illustration representing various routes of biohydrogen production.

thermochemically produced H_2 is also being termed as biohydrogen due to the usage of biomass as a substrate/feedstock. On contrary, various biological routes are available for biohydrogen production pertaining to anaerobic/fermentation, photobiological, enzymatic, and electrogenic mechanisms. The research fraternity around the globe showed significant interest upon biological routes of H_2 production in the past two decades. The passing decade especially illustrated remarkable research on biohydrogen in both basic and applied fields.

Based on light dependency, the biological H_2 production processes can be further classified into light-independent (dark) fermentation and light-dependent photosynthetic processes. In another way, the photobiological process can again be classified into either a photosynthetic or a fermentation process depending on the carbon source and the biocatalyst used. Light-dependent processes can be through biophotolysis of water using green algae and cyanobacteria via direct and indirect biophotolysis or via photofermentation mediated by photosynthetic bacteria. Cyanobacteria and microalgae undergo direct and indirect biophotolysis to produce H_2 by utilizing inorganic CO_2 in the presence of sunlight and water, whereas photosynthetic bacteria (PSB) manifest H_2 production through photofermentation by consuming a wide variety of substrates ranging from inorganic to organic acids in the presence of light. However, the dark fermentation process is confined to anaerobic metabolism where, anaerobic microorganisms (mostly acidogenic bacteria) generate H_2 metabolically through an acetogenic process along with the generation of volatile fatty acids (VFA) and CO_2. H_2 production occurs in the absence of O_2 during fermentation. Both obligate and facultative bacteria are capable of producing H_2 under an oxygen-free environment. Microbial electrolysis is an *in situ* strategy where an external potential was applied to the microbial cells for enhancing biological H_2 production. The synthetic enzyme system mediates *in vitro* H$_2$ production, which is one of the fascinating routes envisaged by scientists. The biochemistry and metabolism involved with the biological routes vary significantly based on the function of biocatalyst used, operating conditions adapted, microenvironment employed, and substrate/feedstock used.

Photosynthetic Machinery of H$_2$ Production

Solar energy can be converted into biochemical energy with the help of a photosynthetic apparatus of hydrogenesis to molecular H$_2$ (Jordan et al., 2001). This is one of the promising approaches helpful for sustainable energy generation. A photosynthetic mechanism of H$_2$ production has been known since the early 1940s (Gaffron and Rubin, 1942). It is pertinent to note that H$_2$ production from the photosynthetic microbial community has contributed to the chemical evolution of the earth and could potentially also be a source of renewable H$_2$ in the future (Burow et al., 2012). However, the taxonomy of H$_2$-producing microorganisms, that is, hydrogenogens have been classified mainly into three physiologically distinct groups of microbes, which include prokaryotes and eukaryotes, varying from unicellular green algae (Gaffron and Rubin, 1942), cyanobacteria to photosynthetic bacteria (anoxygenic) (Gest and Kamen, 1949). Based on the photosynthetic organism, any of the two diverse photosynthetic machinery (anoxygenic or oxygenic) functions for H$_2$ production (Figs. 2–4). The mechanism of H$_2$ production is marginally different between these two processes (Beer et al., 2009; Kruse and Hankamer, 2010; Kruse et al., 2005). The light energy generates proton gradient and supply electrons (e$^-$) either from a water-splitting reaction (direct photolysis; Fig. 2) or from a parallel photosystem II (PSII)-independent process originating from the breakdown of starch molecules (indirect photolysis; Fig. 3). In green algae and cyanobacteria, light functions as a driving force for PSII, resulting in the production of oxidizing equivalents, which are used for the oxidation of water into electrons, protons (H$^+$), and O$_2$ through direct photolysis via oxygenic photosynthesis (Allakhverdiev et al., 2010; Doebbe, 2010; Krassen et al., 2009). In both the cases reduced ferredoxin serves as electron donor for [FeFe]-hydrogenases. The reducing powers are transferred to the chlorophyll α dimer (P700) residing in photosystem I

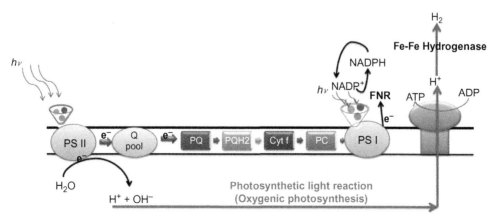

FIGURE 2 Mechanism of direct photolysis involved in oxygenic photosynthesis during H$_2$ production.

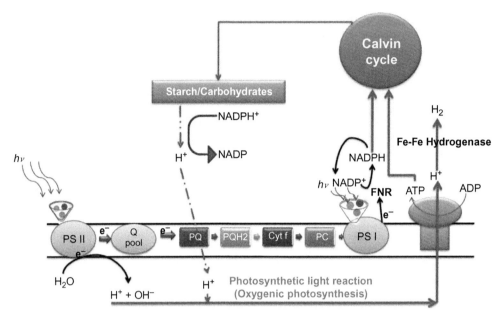

FIGURE 3   Mechanism of indirect photolysis involved in oxygenic photosynthesis during H_2 production.

(PSI). P700 gets excited by light absorption and yields e^- at a potential of -1.32 V (vs NHE) (Krabben et al., 2000). Finally, e^- are conducted through the internal electron transport chain to the iron–sulfur clusters located at the acceptor site of PSI at a potential of -0.45 V (Jordan et al., 1998). In both the photolysis processes, PSII driven water splitting to reducing equivalents and gets reduced to H_2 by the [FeFe]-hydrogenase (*HydA*). Hydrogenase acts as a H^+/e^- release

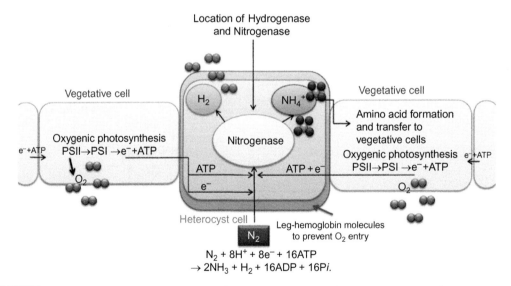

$$N_2 + 8H^+ + 8e^- + 16ATP \rightarrow 2NH_3 + H_2 + 16ADP + 16Pi.$$

FIGURE 4   Mechanism of cynobacterial H_2 production mediated through heterocyst (nitrogen fixation reaction).

valve by recombining H$^+$ (from the medium) and e$^-$ (from reduced ferredoxin) to produce H$_2$. Hydrogenase activity depends on the supply of reducing equivalents derived either directly by photosynthetic water splitting (driven by PSII) or indirectly from the degradation of organic molecules.

Microalgae are prominent for their extremely active [FeFe]-hydrogenase enzyme with the ability of high conversion (12–14%) of solar energy to molecular H$_2$ (Melis, 2009) from the oxidation of water molecules to generate H$_2$ in their chloroplast (Melis and Happe, 2001). However, during direct biophotolysis, oxygen (generated as a by-product of the function of PSII) is a dominant suppressor of the hydrogenase enzyme. Therefore, direct biophotolysis for H$_2$ production can be operated for short periods of time upon the start of illumination, before accumulating O$_2$, and inactivates the H$_2$ production process. In the indirect process, H$^+$ and e$^-$ generated from water are stored in the form of starch (synthesized from photophosphorylation) during photosynthesis, which fed into the plastoquinone pool and onto *HydA* via PSI under certain stress conditions (Melis 2007). The cyanobacteria and green algae accumulate reserve compounds through the Calvin cycle. When photosynthesis is shut off at night time, these reserve compounds provide energy to carry out the cellular metabolism (Krassen et al., 2009). The transition from aerobic to anaerobic conditions is often accompanied by cessation of a photosynthetic light reaction and generation of an excess reductant, which is finally turned into H$_2$ by hydrogenase. Under anaerobic conditions, two O$_2$-sensitive [FeFe]-hydrogenases (*HydA1* and *HydA2*) are induced to catalyze the reduction of H$^+$ to molecular H$_2$ (Doebbe, 2010). The processes of photosynthesis and H$_2$ evolution are coupled only by an indirect process and are time delayed (Krassen et al., 2009). Cyanobacteria have one additional mechanism for H$_2$ production, that is, via heterocyst during N$_2$ fixation. H$_2$ production can also be catalyzed by nitrogenase during nitrogen fixation, and the process is extremely O$_2$ sensitive. Cyanobacteria have a specific mechanism for protecting nitrogenase from O$_2$ through the localization of nitrogenase in the heterocyst of filamentous cyanobacteria (Vyas and Kumar, 1995), which is responsible for NH$_3$ and H$_2$ production (Fig. 4).

Under anaerobic conditions, photosynthetic bacteria use sunlight as a source of energy and produce H$_2$ and CO$_2$ by degrading the organic molecules (Fig. 5). Because, PSB does not require water as a source of e$^-$, it can easily circumvent the oxygen sensitivity issue that adversely affects the [FeFe]-hydrogenase. PSB can utilize both the visible (400–700 nm) and the near-infrared (700–950 nm) regions of the solar spectrum. Anoxygenic photosynthesis by PSB is advantageous for H$_2$ production due to the absence of oxygen, which is a scavenging molecule of reducing equivalents. PSB neither use water as an e$^-$ source nor produce O$_2$ photosynthetically (Blankenship et al., 1995). Light absorption by a dimer of bacteriochlorophyll (BChl) molecules instigates the reaction forming bacteriopheophytin (BPh) (Berg et al., 2002) (Fig. 4). The e$^-$ transfer proceeds from BPh to quinine pool (QA) and then to the cytochrome subunit of the reaction center generating a H$^+$ gradient, which drives the ATP generation and finally gets reduced to H$_2$ (Blankenship et al., 1995; Chandra and Venkata Mohan, 2011). The efficiency of light energy conversion to H$_2$ by PSB is much higher than that by cyanobacteria because of the less quantum of light energy requirement than water photolysis (Batyrova et al., 2012; Melis, 2012; Vyas and Kumar, 1995). The ability of PSB to trap energy over a wide range of the light spectrum without producing oxygen and its versatility in utilizing various substrates make the photofermentation process more feasible and viable.

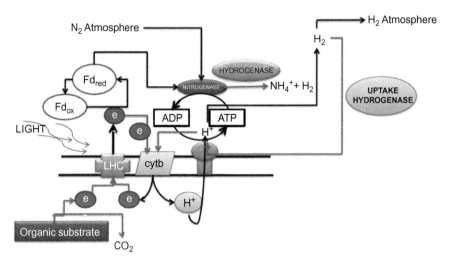

FIGURE 5   Anoxygenic H_2 production mechanism involved with photosynthetic bacteria.

In both oxygenic and anoxygenic photosynthesis, the reduced ferredoxin serves as an e^- donor for the [FeFe]-hydrogenases. Along with the hydrogenases, nitrogenase also plays a major role for H_2 production in all the photosynthetic systems. The main drawback of these two enzymes is that they get inhibited by the oxygen liberated from the photolysis of water. Holding back the oxygen feedback mechanism is essentially required to enhance H_2 production. A reduction in sulfate concentration during algal growth showed a decrement in photosynthesis, which reduced 90% oxygen production, sufficient to allow the hydrogenase enzyme to continue diverting e^- toward H^+ to yield H_2 for a longer period of time (Melis et al., 2000).

The molecular architecture of the photosynthetic membrane makes it possible to redirect photosynthetically generated reducing equivalents from PSI to H_2 production (Millsaps et al., 2001). PSI is a robust nanometer-scale molecular photovoltaic device (Lee et al., 2000) located on the nonappraised region of the thylakoid membrane. Platinum (Pt) can be precipitated on the stromal side of the photosynthetic thylakoid membrane at the site of e^- emergence from the PSI reaction center (Greenbaum, 1985). The platinized chloroplast thylakoids are capable of trapping this e^-, facilitating simultaneous photoevolution of H_2 and oxygen (Greenbaum, 1988a,b). Photosynthetic membranes have the capability to transform with metals other than Pt, such as osmium and ruthenium. Chemical platinization of PSI has no inhibitory effect on excitation transfer dynamics and/or the reaction center pigment (Lee et al., 1995). Functional nanoscale surface metallization at the reducing ends of isolated PSI was reported by substituting the negatively charged hexachloroplatinate ($[PtCl_6]^{2-}$) for negatively charged ferredoxin, the naturally occurring water-soluble electron carrier in photosynthesis (Millsaps et al., 2001). Visible light-induced enzymatic H_2 production with the platinum colloid using the photosensitization of Mg chlorophyll a (Mg Chl-a) was also reported (Saiki and Amao, 2004). Mg Chl-a acts as an effective photosensitizer with an absorption maximum at 670 nm. However, most of the photobiological process for H_2 production has major fundamental limitations, and practical engineering issues need to be resolved prior to practical development (Hallenbeck and Benemann, 2002).

Dark-Fermentative Biohydrogenesis

Microbial fermentation helps generate energy-rich reducing powers (NADH, FADH, etc.) from metabolism, which subsequently gets reoxidized during respiration with a simultaneous generation of biological energy molecules (ATP) in the presence of a terminal electron acceptor (TEA). Oxygen is a strong TEA existing in the biological system that helps in ATP generation during aerobic respiration by simultaneous regeneration of reducing powers. On the contrary, anaerobic respiration has the ability to utilize a wide range of compounds, namely, NO^{3-}, SO_4^{2-}, organic and inorganic compounds, and so on, as TEA with their simultaneous reduction and regeneration of reducing powers. However, ATP generation is not assured during these processes because the energy from the reducing equivalents [protons (H^+) and electrons (e^-)] will be utilized towards the completion of the terminal reduction reaction with TEA but not transferred to the bonding between ADP and inorganic phosphate (Pi) to generate ATP. Glycolysis is the primary metabolic pathway where a substrate gets converted to pyruvate, a central molecule of microbial fermentation. During aerobic fermentation, pyruvate transforms to CO_2 and H_2O with a simultaneous generation of reducing powers, which will further help in energy generation. On contrary, pyruvate has a diverse fate under anaerobic fermentation based on operating conditions. Pyruvate enters the acidogenic pathway and generates VFA, namely, acetic acid, propionic acid, butyric acid, malic acid, and so on, in association with the generation of H$_2$ [Eqs. (1–4)].

$$C_6H_{12}O_6 + 2H_20 \rightarrow 2CH_3 \cdot COOH + 2CO_2 + 4H_2 (\text{acetic acid}) \tag{1}$$

$$C_6H_{12}O_6 \rightarrow CH_3 \cdot CH_2 \cdot CH_2 \cdot COOH + 2CO_2 + 2H_2 (\text{butyric acid}) \tag{2}$$

$$C_6H_{12}O_6 + 2H_2 \rightarrow 2CH_3 \cdot CH_2 \cdot COOH + 2H_2O \ (\text{propionic acid}) \tag{3}$$

$$C_6H_{12}O_6 + 2H_2 \rightarrow COOH \cdot CH_2 \cdot CH_2O \cdot COOH + CO_2 (\text{malic acid}) \tag{4}$$

$$C_6H_{12}O_6 \rightarrow CH_3 \cdot CH_2OH + CO_2 (\text{ethanol}) \tag{5}$$

Both obligate and facultative acidogenic bacteria (AB) can catalyze H$_2$ production from organic substrates (Dinopoulou et al., 1988; Hallenbeck and Benemann, 2002; Klein et al., 2005; Lalit Babu et al., 2009; Vardar-Schara et al., 2008; Venkata Mohan, 2009, 2010; Venkata Mohan et al., 2011a,b, 2012). Facultative anaerobes convert pyruvate to acetyl-CoA and formate by the action of pyruvate formate lyase and further H$_2$ is produced by formate hydrogen lyase (Vardar-Schara et al., 2008). Figure 6 illustrates the biochemical route of the dark fermentative pathway for H$_2$ production. While obligate anaerobes convert pyruvate to acetyl-CoA and CO_2 through pyruvate ferredoxin oxidoreductase, this oxidation process requires the reduction of ferredoxin (Fd) (Kraemer and Bagley, 2007; Vardar-Schara et al., 2008). The proton-reducing reactions facilitate the generation of H$_2$, a common fermentation by-product during electron acceptor-limited microbial processes (Madsen, 2008). Interconversion of metabolites takes place during substrate degradation in anaerobic fermentation, which increases the availability of reducing equivalents in the cell. The H^+ from redox mediators (NADH/FADH) gets detached in the presence of the NADH-dehydrogenase enzyme and gets reduced to H$_2$ in the presence of the hydrogenase enzyme with the help of the e^- donated by the

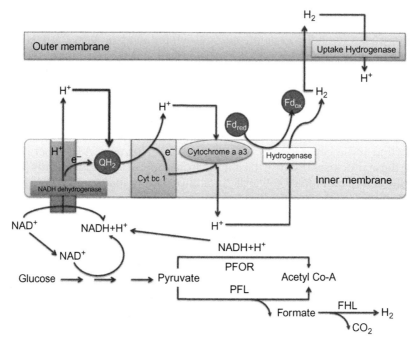

FIGURE 6 Schematic illustration of substrate conversion and H_2 production mechanism during dark fermentation.

oxidized ferredoxin (cofactor), while membrane-bound protein complexes (NADH dehydrogenase and cytochrome b–c_1) and mobile carrier proteins (quinine and cytochrome C) facilitate the e^- transport through the quinone (Q) pool (Fig. 6). The continuous interconversions of Q and H^+ (from the cytosol) to QH_2 and QH_2 to Q and H^+ facilitate the e^- transfer to the cytochrome b–c_1 complex (Cyt bc1) and further to the cytochrome aa_3. Finally, from the cytochrome aa_3, the e^- gets transferred to the iron-containing protein Fd. This reduced Fd donates e^- to the active site component of the hydrogenase enzyme, which reduces the H^+ with this e^- producing H_2 (Vardar-Schara et al., 2008).

Hydrogenase and nitrogenase are the two important enzymes involved in fermentative H_2 production by catalyzing the reversible reduction of H^+ to H_2 (Hallenbeck and Benemann, 2002). Both enzymes contain the complex metal clusters at their active site with diverse subunits. [FeFe]-hydrogenase and [NiFe]-hydrogenase are the two important enzymes involved in microbial H_2 production. [FeFe]-hydrogenase removes the excess reducing equivalents. [NiFe]-hydrogenase, also called uptake hydrogenase, is involved in the quinine pool, where e^- are used directly or indirectly for NAD/NADP reduction (Hallenbeck and Benemann, 2002). Hydrogenases catalyze the reduction of protons to H_2 by oxidizing a suitably strong reductant, including the natural electron carrier proteins ferredoxin and/or flavodoxin, which have redox potentials near that of the H_2 electrode (-420 mV) (Hallenbeck and Benemann, 2009). NADPH is too positive (-320 mV) to serve as a direct reductant of hydrogenase, except in hyperthermophiles, where the H_2 redox potential is near this potential.

The nitrogenase enzyme contains two component protein systems, Mo–Fe protein and Fe protein, which are involved in the H$_2$ production process. Phylum Firmicutes (*Clostridia* and *Bacilli*) dominated in the microbial community of long-term operated bioreactors with various wastewaters (Goud et al., 2012). Generally, Firmicutes are widespread in the environment, mostly with diverse metabolic activity, and drive H$_2$ production through iron-dependent hydrogenase (Venkata Mohan et al., 2011a). Nitrogenases use Mg-ATP and e$^-$ to reduce a variety of substrates during H$_2$ production. Dehydrogenase is another important enzyme involved in the interconversion of metabolites and the transfer of H$^+$ between metabolic intermediates through redox reactions using several mediators (NAD$^+$, FAD$^+$, etc.) (Srikanth et al., 2010). Both dehydrogenase and hydrogenase functions are important in maintaining H$^+$ equilibrium in the cell and reducing them to H$_2$.

Fermentative conversion of an organic substrate to its end products involves a series of interrelated biochemical reactions, namely, hydrolysis, acidogenesis, acetogenesis, and methanogenesis, manifested by five physiologically distinct groups of microorganisms. The complex organic compounds get degraded to monomers during hydrolysis by hydrolytic microorganisms. Further, these monomers will be fermented by acidogenic bacteria to generate a mixture of low molecular weight organic acids associated with H$_2$ during acidogenesis [Eqs. (1–5)]. The reversible interconversion of acetate production from H$_2$ and CO$_2$ by acetogens and homoacetogens can also be considered for H$_2$ production. Finally, the acetoclastic methanogens convert these organic acids to CH$_4$ and CO$_2$ through methanogenesis (Angenent et al., 2004; Venkata Mohan, 2009). Acetogens grow in syntrophic association with the hydrogenotrophic methanogens (H$_2$ consuming) and keep H$_2$ partial pressure low enough to allow acidogenesis to become thermodynamically favorable by interspecies H$_2$ transfer. Henceforth, the methanogenic activity needs to be suppressed to make H$_2$ a sole metabolic by-product.

In Vitro Hydrogenesis

In vitro H$_2$ production using a synthetic enzymatic pathway for directing electrons to the hydrogenase is one of the interesting and alternative routes for the synthesis of H$_2$. This approach facilitates the improvement of typically low yields encountered with the *in vivo* process where the microorganism has its own metabolic requirements that need to be satisfied, along with the accumulation of inhibitory fermentative by-products. *In vitro* H$_2$ production helps achieve higher yields as the cellular metabolic needs are eliminated (Smith et al., 2012). Synthetic biology infers the engineering and biological entities by assembling the interchangeable parts of the natural biology into the systems that function unnaturally (Benner and Sismour, 2005; Endy, 2005). Woodward and co-workers (2000) used the enzymes of the oxidative pentose phosphate pathway (PPP) and coupled hydrogenase (purified from the bacterium *Pyrococcus furiosus*) to use NADP$^+$ as the electron carrier for producing 11.6 mol H$_2$ per mol glucose-6-phosphate (G6P). Biohydrogen is the major product of this pathway, unlike that produced by intermediate metabolic pathways of bacterial fermentation (Woodward et al., 2000). In extension to this, Zhang and co-workers (2007) used a synthetic pathway consisting of 13 purified enzymes from the PPP coupled with the [NiFe]-hydrogenase (*Pyrococcus furiosus*) and achieved 8.35 mol H$_2$ per mole of G6P.

Alternatively, when all the enzymes of the oxidative branch of the PPP cycle, apart from 6-phosphogluconolactonase, were used with the hydrogenase, $NADP^+$, and G6P about 97% of the maximum stoichiometric yield of H_2 from G6P was attained (Zhang et al., 2007). The yield of biohydrogen from the oxidative PPP represents the reaction almost to completion as a result of the hydrogenase oxidizing NADPH as soon as it is formed and swept out. The type of synthetic enzymes may vary with the type of substrate used for H_2 production. H_2 production from starch and water was also reported employing 13 different synthetic enzymes (Zhang et al., 2007). Smith and co-workers (2012) demonstrated a new enzymatic approach for the conversion of reducing equivalents to H_2 by using ferredoxin-NADPH-reductase to transfer electrons from NADPH to ferredoxin, which after oxidation delivers the electrons to a [FeFe]-hydrogenase, resulting in H_2 production. This alternative *in vitro* pathway enables utilization of the fastest known [FeFe]-hydrogenases and activates electron delivery by the native electron donor. Even though higher conversion yields are achievable with the existing *in vitro* methodologies compared to fermentative processes, using multiple purified enzymes for a full-scale industrial process would be prohibitively expensive (Smith et al., 2012).

Electrically Driven Biohydrogenesis

Microbial electrolysis cell (MEC) represents an alternative electrically driven H_2 production process, which facilitates the conversion of electron equivalents in organic compounds to H_2 gas by combining microbial metabolism with bioelectrochemical reactions (Venkata Mohan et al., 2008; Liu et al., 2005; Rozendal et al., 2006). MEC, more or less resembles a microbial fuel cell where the basic difference exists with the requirement of a small input of external potential to facilitate the conversion of biodegradable material into H_2. Protons transferred to the cathode were reduced to form H_2 in the presence of electrons coming from the anode (as an electron sink) under the applied voltage, which is essentially required to cross the endothermic barrier to form H_2 gas. The standard redox potential for the reduction of protons to H_2 is -0.414 V. A potential greater than 0.11 V in addition to that generated by bacteria (-0.3 V) facilitates good H_2 production at the cathode (Logan and Grot, 2005; Cheng and Logan, 2007). This approach provides a route for extending H_2 production to pass through the endothermic barrier imposed by the microbial formation of fermentation end products (Cheng and Logan, 2007; Logan et al., 2008), and the potential required is relatively low compared to the theoretically applied voltage of 1.23 V for water electrolysis (Rozendal et al., 2007). Acetate (-0.279 V) can be converted to H_2 (-0.414 V) in a cathodic reaction against the thermodynamic gradient with the application of a relatively small voltage (-0.135 V) (Hong et al., 2012). In practice, a relatively higher voltage than this is required due to over potentials created by physicochemical and microbial factors (Hallenbeck, 2011). Application of an external voltage also results in the selective growth of electrochemically active microbes on anodes, which can effectively sink electrons (Srikanth et al., 2010; Venkata Mohan and Lenin Babu, 2011; Wang et al., 2009). MEC documented more than 90% of H_2 recovery as against 33% with the dark fermentation process (Cheng and Logan, 2007). MEC is also termed with different nomenclatures as bioelectrochemically assisted microbial reactor or electrohydrogenesis or biocatalyzed electrolysis cell. MEC showed a capability of converting a wide variety of soluble organic matter to H$_2$ (Call and Logan, 2008) or methane production (Clauwaert and Verstraete, 2009) with simultaneous waste treatment. The application of MEC can also be seen with the usage of acidogenic effluents rich in fatty acids as the primary substrate for additional H$_2$ production associated with simultaneous treatment of wastewater (Lalaurette et al. 2009; Venkata Mohan and Lenin Babu, 2011). Low-energy consumption compared to conventional water electrolysis, high product (H$_2$) recovery, and substrate degradation than the dark fermentation process are some of the potential benefits that make MEC an alternate process.

The nature of the biocatalyst, electrode materials, membrane, applied potential, and nature of the substrate and its loading rate and configuration play vital roles in the performance of MEC. MEC was initially operated in a dual chamber and later shifted to a single chamber. The dual-chamber system allows separate capture of H$_2$ (cathode) and CO$_2$ (anode) and prevents fouling of the cathode by anodic bacteria. On the contrary, separation leads to inhibitory pH changes, as there is potential acidification of the anode chamber due to the production of protons and basification of the cathode chamber from proton consumption (Hallenbeck, 2011). Lowering the applied potential and eliminating the membrane creates a single-chamber MEC (Hu et al., 2008). Eliminating the membrane attenuated pH energy loss and ohmic energy loss (Clauwaert et al., 2008; Lee et al., 2009), which were significant for a dual-chamber MEC (Rozendal et al., 2007). More recently, single-chamber MEC were operated to eliminate some of the inherent disadvantages (Hu et al., 2008; Liang et al., 2011), which significantly reduce the internal resistance (Hallenbeck, 2011). Much research is being focused on working with wastewater as a substrate for operating MEC (Dictor et al., 2010; Escapa et al., 2009; Venkata Mohan and Lenin Babu, 2011; Wagner et al., 2009). Integrating MEC with other processes is also gaining much attention (Cusick et al., 2010; Lalaurette et al., 2009; Wang et al., 2011). A number of challenges need to be addressed before MEC can be applied on a practical level. Little applied voltage with an elevated current density will be an essential challenge for moving bench-scale MEC to commercial applications (Lee and Rittmann, 2010). Integrating the wastewater treatment with MEC to produce economically feasible biohydrogen generation is recent interest among the research fraternity.

Thermochemical Process

Thermochemical treatment of biomass is the only nonbiological process pertaining to biohydrogen production. Thermochemical processes involve either gasification or pyrolysis (heating biomass in the absence of oxygen) to produce a H$_2$-rich stream of gas known as "syngas" (a blend of hydrogen and carbon monoxide) (Lipman, 2011). Thermochemical processes for H$_2$ production involve thermally assisted chemical reactions that release the hydrogen from hydrocarbons or water (Yildiz and Kazimi, 2006). The thermochemical process can utilize a broad range of feedstock. Gasification of biomass at temperatures above 1000 K in the presence of oxygen and/or steam undergoes partial oxidation and/or steam reforming reactions yielding a gas and char product (Navarro et al., 2009). The char is subsequently reduced to form H$_2$, CO, CO$_2$, and CH$_4$. This process is more favorable for H$_2$ production than pyrolysis. Gasification by partial oxidation has been around for over 150 years (Jong, 2009). Low temperature (<1000 °C) gasification yields a significant amount of hydrocarbon, whereas at a higher temperature the yield of syngas contains virtually no hydrocarbons (Navarro et al.,

2009). Pyrolysis facilitates thermal decomposition of biomass at a temperature of 650–800 K (1–5 bar) in the absence of air to yield liquid oils, solid charcoal, and gaseous compounds (Navarro et al., 2009). Hydrogen can be produced directly through fast or flash pyrolysis if both high temperature and sufficient volatile phase residence time are provided (Navarro et al., 2009). Pyrolysis followed by reforming of bio-oil and gasification has received significant amount of interest as these provide an improved quality fuel product (Saxena et al., 2008). Gasification followed by reforming of the syngas and pyrolysis (fast) followed by reforming of the carbohydrate fraction of the bio-oil are the two major thermochemical routes used for H_2 production (Balat, 2010). A water–gas shift is used along with pyrolysis and gasification processes to convert the reformed gas into H_2, and pressure swing adsorption is used to purify the product (Balat, 2010). A supercritical water condition (pressures >221 bar; temperatures >647 K) in the absence of oxygen can convert biomass into fuel gases, which can be separated easily from the water phase by cooling to ambient temperature (Navarro et al., 2009). The cost of H_2 production from supercritical water gasification of wet biomass was several times higher than the current price of H_2 from steam methane reforming (Balat, 2010).

Today, the most widely used and least expensive technology used to produce H_2 is catalytic steam reforming of natural gas (Momirlan and Veziroglu, 2002). Steam reforming converts hydrocarbons to CO and H_2 driven by steam addition either through thermal reforming at high temperatures (>1100°C) or catalytic reforming (>650°C) over nickel-based catalysts (Blackader and Rensfelt, 1984). The large-scale production of H_2 from natural gas and other available hydrocarbons through catalytic reforming processes remains the least expensive source of H_2 (Navarro et al., 2009). Gasification is still the most mature technology for large-scale H_2 production, but it needs integrated demonstration plants at sufficiently large scale, including catalytic gas upgrading (Jong, 2009). Pyrolysis processes for H_2 production are at present in the smaller scale demonstration phase, while supercritical water gasification is still in the early stage of development and research. Hydrogen production involving reforming technologies produces large amounts of CO_2, which has an impact on global warming (Navarro et al., 2009). One way to reduce CO_2 emissions is to apply reforming methods to alternative renewable precursors.

SCIENTOMETRIC EVALUATION OF THE RESEARCH ON BIOHYDROGEN

Scientometric analysis was performed employing ISI Web of Knowledge (Thomson Reuters) using SCI-expanded (since 1987), science (CPCI-S, since 1990), and social science and humanities (CPCI-SSH, since 1990) data to characterize the literature published on biohydrogen during the last three decades (1986–2012) to enumerate the trends. From 2004, a significant increase in biohydrogen publications was observed, accounting for 1479 records with a concomitant increase in citations (21,983) and H-index (68). With reference to time, literature related to biohydrogen research showed a steep rise after 2004 and reached maximum records of 250 in the year 2008 (total citations: 2500), which remained more or less the same in 2009 with increased citations (3500) followed by a steep increment in 2010 (records, 300; citations, 4250). In 2011, about 340 publications were documented with citations of more than 6800. Average citation per year also showed an increasing trend year by year

(2009, 11.73; 2010, 12.10; 2011, 14.2; 2012, 19.120). Out of the total records, 62.8% relates to the fermentation process (total citations: 14146), 10.2% relates exclusively to photosynthesis process (total citations: 3211), 4% relates to application of microbial electrolysis (total citations: 1025), and 1.3% relates to pyrolysis process (total citations: 930). The enzymatic process showed only a few records in the search. Based on the process, the fermentation process showed a H-index of 55 (average citation per year, 15.12; citations in 2011, 4500), followed by the photosynthetic process (H-index, 26; average citation per year, 21.3; citations in 2011, 1000), microbial electrolysis (H-index, 16; average citation per year, 16.8; citations in 2011, 350), and pyrolysis (H-index, 9; average citation per year, 46.5; citations in 2011, 350). When maximum records were considered, application of fermentation for H_2 production documented 220 publications in 2011, photosynthesis recorded 35 (2010), microbial electrolysis recorded 20 (2010), and pyrolysis recorded 5 (2009). Most of the records are in the form of articles (84.9%), followed by conference proceeding papers (18.3%) and reviews (6.7%). When waste (accounting for 30% of wastewater) is considered as a substrate, about 58% records relate to waste usage for biohydrogen production. The application of waste usage as a substrate gained prominence since 2004 and steadily reached a maximum in the year 2011 (records >180; citations, 4000; H-index, 53). When records were segregated according to the biocatalyst usage, about 23.5% of records showed an application of mixed culture as the biocatalyst for biohydrogen production. On the basis of journals, the *International Journal of Hydrogen Energy* published 40.4% of the biohydrogen research literature, followed by *Bioresource Technology* (12.1%), *Biotechnology and Bioengineering* (2.3%), *Advanced Materials Research* (2.0%), and *Biomass Bioenergy* (1.4%). Country wise, China is first in biohydrogen-based publications with 343 records (23.9%), followed by the United States (186 records; 13.7%), Taiwan (160 records; 11.2%), India (109 records; 7.7%), and Korea (101 records; 7.1%). When biohydrogen research was integrated with wastewater, India is second (44 records, 19.2%) after China (61 records, 26.7%), followed by the United States (29 records, 12.7%), Taiwan (22 records, 9.6%), and Canada/Thailand (12 records, 5.2%).

WASTE AS RENEWABLE FEEDSTOCK/SUBSTRATE FOR BIOHYDROGEN PRODUCTION

Rejected materials from natural and anthropogenic activities are recently being considered as a potential feedstock/substrate for harnessing renewable bioenergy. Reducing the treatment cost of waste/wastewater and finding ways to produce useful/value-added products from treatment have been gaining importance due to its sustainable nature. In the contemporary energy scenario, environmental scientists are gradually shifting their focus from "pollution control" to "resource exploitation from waste." Biological processes are generally preferred to treat wastewater as they are technically feasible, simple, economical, and eco-friendly. Biological approaches also facilitate the conversion of negative-valued organic waste to useful forms of energy while simultaneously achieving the objective of pollution control. An enormous quantity of waste/wastewater is available, which is composed of a reasonably good biodegradable carbon fraction associated with inherent net-positive energy. The regulatory need for their treatment prior to disposal makes them an ideal commodity to produce H_2 from the anaerobic treatment. Utilizing wastes as a potential source for

H_2 generation through biological routes has instigated considerable interest due to its sustainable nature and further opening up a new avenue for the utilization of renewable and inexhaustible energy sources (Li and Fang, 2007; Van Ginkel et al., 2005; Venkata Mohan, 2009, 2010; Zhang et al., 2007). Biohydrogen generation from renewable wastewater associated through treatment simultaneously reduces the overall effluent treatment cost by creating additional revenue and makes the whole process environmentally sustainable (Venkata Mohan et al., 2007, 2012). Municipal and industrial wastewater, along with the waste generated from agriculture and food-processing industries, contains enough organic load, which can be tapped beneficially if utilized appropriately. Integration of biohydrogen production with an existing effluent treatment plant is the futuristic goal envisaged with the usage of wastewater as the primary feedstock.

Biologically derived organic material and their residues, such as agricultural crops and their waste by-products, wood and wood waste, food processing waste, aquatic plants, and algae, constitute a large source of biomass, which can also be used as fermentable substrates (Saratale et al., 2008). Biomass precursors derived from plant crops, agricultural residues, woody biomass, and so on are being used as feedstock for generating H_2 by both thermochemical and biological routes (Nath and Das, 2003). Different types of lignocellulosic/agricultural residues, such as sugarcane and sweet sorghum bagasse, cornstalks and stover, fodder maize, and wheat straw, and forestry residues, such as wood trimmings, have been studied as potential renewable feedstocks for dark-fermentative biohydrogen production (Ntaikou et al., 2010). Unlike wastewater, cellulosic material or solid waste requires an initial pretreatment step to make the organic fraction soluble and bioavailable to the microorganism for metabolic reaction. The highly crystalline and water-insoluble nature makes cellulose recalcitrant to the hydrolysis (Saratale et al., 2008). Direct microbial assimilation of cellulosic materials facilitates low H_2 yields. Development of novel and effective cellulase enzymes and optimization and improvement of cellulase system, as well as engineering approaches on cellulose pretreatment and saccharification, are gaining increasing interest to overcome this limitation (Saratale et al., 2008). Combinations of chemical, mechanical, and enzymatic pretreatment methods were used to pretreat cellulose-based feedstock to usable carbohydrates. Techniques, namely, high temperature, high or low pH, hydrolytic enzymes, microwaves, ultrasound, radiation, and pulsed electric fields, were applied for this purpose (Rittmann, 2008; Venkata Mohan et al., 2008b). On the contrary, wastewater represents a readily available carbon source.

Much of the work reported on biohydrogen production concerning wastewater utility was specifically found with the dark fermentation process. Fermentative H_2 production is relatively less energy intensive and more environmentally sustainable due to the utilization of waste and wastewater as the substrate and to its operational feasibility at ambient temperature and pressure. Exploitation of wastewater as a substrate for H_2 production with simultaneous wastewater treatment will lead to a new avenue for the utilization of renewable and inexhaustible energy sources. In conjunction with the wastewater treatment, this process is capable of solving two issues: reduction of pollutants in waste and generation of a clean alternative fuel (Cuetos et al., 2007; Gomez et al., 2006). Certain inherent limitations—low substrate conversion efficiency, accumulation of carbon-rich acid intermediates, drop in system pH, etc.—still exist with the process, which needs considerable attention prior to process upscaling. At present, basic and applied research is on the way to gaining more insight into the process of understanding and establishing optimized conditions. Especially after initiating the usage of wastewater as a substrate, a great deal of attention has been paid to the application of mixed consortia as a biocatalyst for the acidogenic fermentation process of H_2 production.

BIOCATALYST FOR BIOHYDROGEN PRODUCTION

Selection of an appropriate biocatalyst or inoculum significantly influences the metabolic end-product formation, which is also true with H_2 evolution. Diverse groups of microorganisms—anaerobic, photosynthetic (heterotrophic and autotrophic), and microalgae—are capable of producing H_2 by taking advantage of their specific metabolic route under defined conditions. Bacteria capable of producing H_2 are reported to widely exist in natural environments. Obligate anaerobes, thermophiles, methanogens, and a few facultative anaerobes are involved in metabolic H_2 production. Initial research on biohydrogen was mostly confined towards the usage of pure cultures as a biocatalyst with a defined substrate. It is believed that the application of mixed consortia as a biocatalyst is one of the promising and practical options for the scaling up of the biohydrogen technology, especially when wastewater is used as the substrate. Also mixed cultures are usually preferred because of operational flexibility, diverse biochemical functions, stability, and the possibility of using a broad range of substrates, as well as restricting the requirement of sterile conditions (Angenent et al., 2004; Venkata Mohan, 2010; Wang and Wan, 2009). From an engineering point of view, producing H_2 by mixed culture offers a lower operational cost and ease of control in concurrence to the possibility of using waste as a feedstock (Venkata Mohan, 2010). The feasibility of H_2 production with typical anaerobic-mixed consortia is limited, as it gets consumed rapidly by methanogens (Venkata Mohan et al., 2008a, 2012). Shifting or regulating the metabolic pathway towards acidogenesis and inhibiting methanogenesis facilitate higher H_2 yields (Venkata Mohan et al., 2012). Pretreatment of the biocatalyst plays a vital role in the selective enrichment of mixed consortia to shifting the metabolic function towards acidogenesis (Venkata Mohan, 2008, 2010; Venkata Mohan et al., 2008a, 2009; Zhu and Beland, 2006). Application of pretreatment to parent inoculum facilitates the selective enrichment of acidogenic bacteria capable of producing H_2 as the end product with the simultaneous prevention of hydrogenotrophic methanogens (Goud and Venkata Mohan, 2012; Venkata Mohan et al., 2008a). Physiological differences between H_2-producing AB and H_2 uptake bacteria (methanogens) form a fundamental basis for the methods used for the preparation of H_2-producing inoculum (Srikanth et al., 2010; Venkata Mohan et al., 2008a; Zhu and Beland, 2006). Pretreatment also prevents competitive growth and coexistence of other H_2-consuming bacteria.

PROSPECTS OF BIOHYDROGEN DOMAIN

Use of bioenergy reduces GHG emissions (International Energy Agency (IEA), 2004), although the extent of reduction depends on the production technology (Farrell et al., 2006). At present, bioenergy production is relatively expensive compared to fossil fuels (International Energy Agency (IEA), 2004), and introduction of more efficient techniques for their production may help in the overall cost reduction. Governments around the world are encouraging the bioenergy sector to commercialize bioenergy production in the form of subsidies. Both basic and applied research on biohydrogen production is presently in the developing phase. Initial interest in biohydrogen research was much visible with the photobiological route using specific strains and a defined medium. Low rates of H_2 production and the inhibitory effect of O_2 on the hydrogenase and nitrogenase enzymes (Orskav et al., 1968; Rittmann, 2008) are some of the inherent disadvantages linked with the photobiological process. The O_2 sensitivity of H_2 production, combined with the competition between hydrogenases and NADPH-dependent carbon dioxide fixation, is the main limitation for the commercialization of photosynthetic water splitting coupled to hydrogenase-catalyzed H_2 production (Yacoby et al., 2011). High activation energy needed to drive hydrogenase and the low solar conversion efficiencies are also as considered major limitations. Key technical challenges include overcoming the oxygen sensitivity of hydrogenase enzymes, outcompeting the other metabolic pathways for photosynthetic reductants, and ensuring adequate efficiency when capturing and converting solar energy (Melis and Happe, 2001). At present, much attention is being focused on eliminating O_2 sensitivity, which is a key issue retarding the functional advantage of the photosynthetic process, especially with the evolution of H_2.

Light-independent dark fermentation that can be operated at ambient temperatures and pressures is gaining importance as a practically viable method among the biological routes. Significant progress was witnessed in the dark fermentation process in the last decade due to its feasibility of utilizing a broad range of substrates, including waste/wastewater with mixed cultures as biocatalysts. The dark fermentation process is relatively less energy intensive, technically much simpler, requires low operating costs, and is more stable and robust (Bhaskar et al., 2008; Gustavo et al., 2008; Hallenbeck and Benemann, 2002; Idania et al., 2005; Kraemer and Bagley, 2007; Venkata Mohan et al., 2009, 2010). The process simplicity, efficiency, and fewer footprints are some of the striking features of the dark fermentation process, which makes it practically more feasible for the mass production of H_2 (Venkata Mohan et al., 2012). Despite the striking advantages, low yields, and production rates, low substrate conversion efficiency and fatty acid-rich wastewater generated from the acidogenic process are major barriers to the practical implementation of this technology. The accumulation of acidogenic by-products causes a sharp drop in the pH, resulting in inhibition of the fermentation process. The undissociated soluble metabolites can permeate through the cell membrane of H_2-producing bacteria and then dissociate in the cell, leading to a physiological imbalance, resulting in cell lysis, especially at higher concentrations (Wang and Wan, 2009). H_2 yield is lower when more reduced organic compounds, such as lactic acid, propionic acid, and ethanol, are produced as fermentation products, as these represent the end products of metabolic pathways that bypass the major H_2-producing reaction (Venkata Mohan, 2010). In practice, yields are lower, as NADH oxidation by NADH-ferredoxin oxidoreductase is inhibited under standard conditions and proceeds only at very low partial pressures of H_2 (Angenent et al., 2004). However, biological limitations, such as H_2 end product inhibition and acid or solvent accumulation, limit the molar yield. Even under optimum conditions, about 60–70% of the original organic matter remains as residues in the wastewater. Apart from lower conversion efficiency, the nonutilized organic fraction usually remaining as a soluble fermentation product from the acidogenic process is another major concern that needs to be resolved. Environmental and economic concerns suggest that it is advisable to use the residual carbon fraction of the acidogenic outlet for additional energy generation in the process of its treatment (Mohanakrishna et al., 2010b). Integrated approaches were studied extensively to overcome some of the persistent limitations by utilizing fatty acid wastewater as the primary substrate. Integration with the secondary treatment processes—methanogenesis (Venkata Mohan et al., 2008c), acidogenic fermentation (Mohanakrishna et al., 2010c), photobiological process (Chandra and Venkata Mohan, 2011; Laurinavichene et al., 2012; Srikanth et al., 2009), microbial electrolysis (Mohanakrishna et al., 2010a; Venkata Mohan and Lenin Babu, 2011), bioplastics production (Reddy and Venkata Mohan, 2012), and lipid accumulation using microalgae (Venkata Mohan and Prathima Devi, 2012)—was evaluated with a diverse degree of success. These integration approaches facilitate a reduction in wastewater load with the advantage of value addition in the form of product recovery. Photosynthetic bacteria can readily utilize the organic acids generated from the dark fermentation process to produce additional H_2 (Chandra and Venkata Mohan, 2011; Srikanth et al., 2009).

Fundamental understanding of the potential limiting factors is necessary to overcome some limitations in the direction of enhancing process efficiency. Optimization of process parameters is essential for upscaling of the technology. Intensive research on biohydrogen is underway, and several novel approaches have been studied to surpass some of the persistent drawbacks. Several metabolic engineering approaches, namely, providing metabolic energy to overcome thermodynamic barriers, expression of heterologous proteins, including hydrogenases and rerouting metabolism to achieve more complete substrate degradation, and increased electron flux for proton reduction, are being developed for higher H_2 yields (Hallenbeck et al., 2012). To establish an environmentally sustainable biohydrogen technology, a multidisciplinary research approach is vital (Venkata Mohan, 2009, 2010). Process engineering and optimization of operational factors govern the performance of any biological system and also have a considerable influence on fermentative H_2 production. With the documented improvements in the performance of this technology since the early 2000s, it can be presumed that a level for practical application can be achieved in a relatively short period of time.

Acknowledgments

SVM wishes to thank Director, CSIR-IICT for his kind encouragement and support. SVM also acknowledges the financial support from CSIR in the form of XII five year network project on 'Sustainable Waste Management Technologies for Chemical and Allied Industries-SETCA' and Ministry of New and Renewable Energy (MNRE, Project No. 103/131/2008-NT) on "Biohydrogen".

References

Allakhverdiev, S.I., Thavasi, V., Kreslavski, V.D., Zharmukhamedov, S.K., Klimov, V.V., Ramakrishna, S., et al., 2010. Photosynthetic hydrogen production. J. Photochem. Photobiol. 11, 87–99.
Angenent, L.T., Karim, K., Dahhan, M.H., Wrenn, B.A., Espinosa, R., 2004. Production of bioenergy and biochemicals from industrial and agricultural wastewater. Trends Biotechnol. 22, 477–485.
Balat, M., 2010. Thermochemical routes for biomass-based hydrogen production. Energy Sources A. Recovery Utilization Environ. Effects 32 (15), 1388–1398.

Batyrova, K.A., Tsygankov, A.A., Kosourov, S.N., 2012. Sustained hydrogen photoproduction by phosphorus-deprived *Chlamydomonas reinhardtii* cultures. Int. J. Hydrogen Energy 37, 8834–8839.

Beer, L.L., Boyd, E.S., Peters, J.W., Posewitz, M.C., 2009. Engineering algae for biohydrogen and biofuel production. Curr. Opin. Biotechnol. 20, 264–271.

Benner, S.A., Sismour, A.M., 2005. Synthetic biology. Nat. Rev. Genet. 6, 533–543.

Berg, J.M., Tymoczko, J.L., Stryer, L., 2002. Biochemistry, fifth ed. W.H. Freeman & Co., New York.

Bhaskar, Y.V., Venkata Mohan, S., Sarma, P.N., 2008. Effect of substrate loading rate of chemical wastewater on fermentative biohydrogen production in biofilm configured sequencing batch reactor. Bioresour. Technol. 99, 6941–6948.

Blackader, W., Rensfelt, E., 1984. Synthesis gas from wood and peat: The mino process. In: Bridgwater, A.V. (Ed.), Thermochemical Processing of Biomass. Butterworth, London, pp. 137–149.

Blankenship, R.E., Medigan, M.T., Bauer, C.E. (Eds.), 1995. Anoxygenic Photosynthetic Bacteria. Kluwer Academic, Dordresht, The Netherlands, pp. 1005–1028.

BPSR, 2008. http://www.ief.org/whatsnew/Pages/BPStatisticalReview.aspx.

Burow, L.C., Woebken, D., Bebout, B.M., McMurdie, P.J., Singer, S.W., Pett-Ridge, J., et al., 2012. Hydrogen production in photosynthetic microbial mats in the Elkhorn Slough estuary, Monterey Bay. ISME J. 6 (4), 863–874.

Call, D., Logan, B.E., 2008. Hydrogen production in a single chamber microbial electrolysis cell lacking a membrane. Environ. Sci. Technol. 42 (9), 3401–3406.

Chandra, R., Venkata Mohan, S., 2011. Microalgal community and their growth conditions influence biohydrogen production during integration of dark-fermentation and photo-fermentation processes. Int. J. Hydrogen Energy 36, 12211–12219.

Chen, W., Syu, Y., 2010. Hydrogen production from water gas shift reaction in a high gravity (Higee) environment using a rotating packed bed. Int. J. Hydrogen Energy 35 (19), 10179–10189.

Cheng, S., Logan, B.E., 2007. Sustainable and efficient biohydrogen production via electrohydrogenesis. PNAS 104, 18871–18873.

Christopher, K., Dimitrios, R., 2012. A review on exergy comparison of hydrogen production methods from renewable energy sources. Energy Environ. Sci. 5, 6640–6651.

Clauwaert, P., Verstraete, W., 2009. Methanogenesis in membraneless microbial electrolysis cell. Appl. Microbiol. Biotechnol. 82 (5), 829–836.

Clauwaert, P., Toledo, R., Ha, D.V.D., Crab, R., Verstraete, W., Hu, H., et al., 2008. Combining biocatalyzed electrolysis with anaerobic digestion. Water Sci. Technol. 57, 575–579.

Cuetos, M.J., Gomez, X., Escapa, A., Moran, A., 2007. Evaluation and simultaneous optimization of bio-hydrogen production using 32 factorial design and the desirability function. Power Sources 169, 131–139.

Cusick, R.D., Kiely, P.D., Logan, B.E., 2010. A monetary comparison of energy recovered from microbial fuel cells and microbial electrolysis cells fed winery or domestic wastewaters. International Journal of Hydrogen Energy 35 (17), 8855–8861.

Dictor, M.C., Joulian, C., Touze, S., Ignatiadis, I., 2010. Electro-stimulated biological production of hydrogen from municipal solid waste Dominique guyonnet. Int. J. Hydrogen Energy 35, 10682–10692.

Dinopoulou, G., Sterritt, R.M., Lester, J.N., 1988. Anaerobic acidogenesis of a complex wastewater kinetics of growth, inhibition, and product formation. Biotechnol. Bioeng. 31, 969–978.

Doebbe, A., 2010. The interplay of proton, electron, and metabolite supply for photosynthetic H_2 production in *Chlamydomonas reinhardtii*. J. Biol. Chem. 285, 30247–30260.

Endy, D., 2005. Foundations for engineering biology. Nature 438, 449–453.

Escapa, A., Manuel, M.F., Moran, A., Gomez, X., Guiot, S.R., Tartakovsky, B., 2009. Hydrogen production from glycerol in a membraneless microbial electrolysis cell. Energy Fuels 23, 4612–4618.

Farrell, A.E., Plevin, R.J., Turner, B.T., Jones, A.D., Hare, M.O., Kammen, D.M., 2006. Ethanol can contribute to energy and environmental goals. Science 311, 506–508.

Gaffron, H., Rubin, J., 1942. Fermentative and photochemical production of hydrogen in algae. J. Gen. Physiol. 26, 219–240.

Gest, H., Kamen, M.D., 1949. Studies on the metabolism of photosynthetic bacteria. IV. Photochemical production of molecular hydrogen by growing cultures of photosynthetic bacteria. J. Bacteriol. 58, 239–245.

Ginkel, S.V., Logan, B., 2005. Increased biological hydrogen production with reduced organic loading. Water Res. 39, 3819–3826.

Goldemberg, J., 2008. Environmental and ecological dimensions of biofuels. In: Proceedings of the Conference on the Ecological Dimensions of Biofuels. Washington, DC.

Gomez, X., Moran, A., Cuetos, M.J., Sanchez, M.E., 2006. The production of hydrogen by dark fermentation of municipal solid wastes and slaughterhouse waste: A two-phase process. J. Power Sources 157, 727–732.

Goud, R.K., Venkata Mohan, S., 2012. Acidic and alkaline shock pretreatment to enrich acidogenic biohydrogen producing mixed culture: Long term synergetic evaluation of microbial inventory, dehydrogenase activity and bio-electro kinetics. RSC Advances 2 (15), 6336–6353.

Goud, R.K., Raghavulu, S.V., Mohanakrishna, G., Naresh, K., Venkata Mohan, S., 2012. Predominance of Bacilli and Clostridia in microbial community of biohydrogen producing biofilm sustained under diverse acidogenic operating conditions. Int. J. Hydrogen Energy 37, 4068–4076.

Grayson, M., 2011. Nature outlook: Biofuels. Nature 474, S1.

Greenbaum, E., 1985. Platinized chloroplasts: A novel photocatalytic material. Science 230, 1373–1375.

Greenbaum, E., 1988a. Interfacial photoreactions at the photosynthetic membrane interface: An upper limit for the number of platinum atoms required to form a hydrogen-evolving platinum metal catalyst. J. Phys. Chem. 92, 4571–4574.

Greenbaum, E., 1988b. Energetic efficiency of hydrogen photoevolution by algal water-splitting. Biophys. J. 54, 365–368.

Gustavo, D.V., Felipe, A.M., Antonio de, L.R., Elías, R.F., 2008. Fermentative hydrogen production in batch experiments using lactose, cheese whey and glucose: Influence of initial substrate concentration and pH. Int. J. Hydrogen Energy 33, 4989–4997.

Hallenbeck, P.C., 2011. Microbial paths to renewable hydrogen production. Biofuels 2, 285–302.

Hallenbeck, P.C., Benemann, J.R., 2002. Biological hydrogen production: Fundamentals and limiting processes. Int. J. Hydrogen Energy 27, 1185–1193.

Hallenbeck, P.C., Benemann, J.R., 2009. Biohydrogen: The microbiological production of hydrogen fuel. In: Doelle, H.W., DaSilva, E.J. (Eds.), Biotechnology. In Encyclopedia of Life Support Systems (EOLSS), UNESCO, Eolss Publishers, Oxford, UK.

Hallenbeck, P.C., Abo-Hashesh, M., Ghosh, D., 2012. Strategies for improving biological hydrogen production. Bioresour. Technol. 110, 1–9.

Hong, Y.J., Ryu, J.H., Kim, C.H., Lee, W.K., Tran, T.V.T., Lee, H.L., et al., 2012. Enhancing hydrogen production efficiency in microbial electrolysis cell with membrane electrode assembly cathode Original Research Article. Journal of Industrial and Engineering Chemistry 18, 715–719.

Hu, H., Fan, Y., Liu, H., 2008. Hydrogen production using single chamber membrane-free microbial electrolysis cells. Water Res. 42 (15), 4172–4178.

Idania, V.V., Sparling, R., Risbey, D., Noemi, R.S., Hec, M., Poggi Varaldo, H.M., 2005. Hydrogen generation via anaerobic fermentation of paper mill wastes. Bioresour. Technol. 96, 1907–1913.

International Energy Agency (IEA), 2004. Biofuels for transport: An International Perspective.

International Energy Agency (IEA), 2008. Key World Energy Statistics 2008. OECD/IEA, Paris.

IPCC, 2006. IPCC guidelines for national greenhouse gas inventories. Prepared by the National Greenhouse Gas Inventories Programme (Eggleston, H.S., Buendia, L., Miwa, K., Ngara, T., Tanabe, K., Eds.). IGES, Japan.

Jong, W., 2009. Sustainable hydrogen production by thermochemical biomass processing. In: Gupta, R.B. (Ed.), Hydrogen Fuel: Production, Transport, and Storage. Taylor & Francis Group, LLC.

Jordan, R., Nessau, U., Schlodder, E., 1998. Charge Recombination between the Iron_Sulphur Clusters and P700. In: Garab, G. (Ed.), Photosynthesis: Mechanisms and Effects. Kluwer Academic Publishers, Dordrecht, The Netherlands, pp. 663–666.

Jordan, P., Fromme, P., Witt, H.T., Klukas, O., Saenger, W., Krauss, N., 2001. Three dimensional structure of cyanobacterial photosytstem I at 2.5 Å resolution. Nature 411, 909–917.

Klein, D.W., Prescott, L.M., Harley, J., 2005. Microbiology. McGraw-Hill, New York.

Kraemer, J.T., Bagley, D.M., 2007. Improving the yield from fermentative hydrogen production. Biotechnol. Lett. 29, 685–695.

Krabben, L., Schlodder, E., Jordan, R., Carbonera, D., Giacometti, G., Lee, H., et al., 2000. Influence of the axial ligands on the spectral properties of P700 of photosystem I: A study of site-directed mutants. Biochemistry 39, 13012–13025.

Krassen, H., Schwarze, A.A., Friedrich, B., Ataka, B., Lenz, O., Heberle, J., 2009. Photosynthetic hydrogen production by a hybrid complex of photosystem I and [NiFe]-hydrogenase. ACS Nano. 3 (12), 4055–4061.

Kruse, O., Hankamer, B., 2010. Microalgal hydrogen production. Curr. Opin. Biotechnol. 21 (3), 238–243.

Kruse, O., Rupprecht, J., Bader, K.P., Hall, T.S., Schenk, P.M., Finazzi, G., et al., 2005. Improved photobiological H_2 production in engineered green algal cells. J. Biol. Chem. 280, 34170–34177.

Lalaurette, E., Thammannagowda, S., Mohagheghi, A., Maness, P.C., Logan, B.E., 2009. Hydrogen production from cellulose in a twostage process combining fermentation and electrohydrogenesis. Int. J. Hydrogen Energy 34 (15), 6201–6210.

Lalit Babu, V., Venkata Mohan, S., Sarma, P.N., 2009. Influence of reactor configuration on fermentative hydrogen production during wastewater treatment. Int. J. Hydrogen Energy 34, 3305–3312.

Laurinavichene, T.V., Belokopytov, B.F., Laurinavichius, K.S., Khusnutdinova, A.N., Seibert, M., Tsygankov, A.A., 2012. Towards the integration of dark- and photo-fermentative waste treatment. Int. J. Hydrogen Energy 37, 8800–8810.

Lee, H.S., Rittmann, B.E., 2010. Significance of biological hydrogen oxidation in a continuous single-chamber microbial electrolysis cell. Environ. Sci. Technol. 44, 948–954.

Lee, J.W., Lee, I., Laible, P.D., Owens, T.G., Greenbaum, E., 1995. Chemical platinization and its effect on excitation transfer dynamics and P700 photooxidation kinetics in isolated photosystem I. Biophys. J. 69, 652–659.

Lee, I., Lee, J.W., Stubna, A., Greenbaum, E., 2000. Measurement of electrostatic potentials above oriented photosystem I reaction centers. J. Phys. Chem. B 104, 2439–2443.

Lee, H.S., Torres, C.I., Parameswaran, P., Rittmann, B.E., 2009. Fate of H_2 in an upflow single-chamber microbial electrolysis cell using a metal-catalyst-free cathode. Environ. Sci. Technol. 43, 7971–7976.

Li, C., Fang, H.H.P., 2007. Fermentative hydrogen production from wastewater and solid wastes by mixed cultures. Critical Rev. Environ. Sci. Technol. 37, 1–39.

Liang, P., Wang, H.Y., Xia, X., Huang, X., Mo, Y.H., Cao, X.X., et al., 2011. Carbon nanotube powders as electrode modifier to enhance the activity of anodic biofilm in microbial fuel cells. Biosens. Bioelectron. 26, 3000–3004.

Lipman, T., 2011. An overview of hydrogen production and storage systems with renewable hydrogen case studies. In: Energy Efficiency and Renewable Energy Fuel Cell Technologies Program (US DOE Grant DE-FC3608GO18111 A000).

Liu, H., Grot, S., Logan, B.E., 2005. Electrically assisted microbial production of hydrogen from acetate. Environ. Sci. Technol. 39, 4317–4320.

Logan, B.E., Grot, S., 2005. A bioelectrochemically assisted microbial reactor (BEAMR) that generates hydrogen gas. U.S. Patent Application. 60/588,022.

Logan, B.E., Call, D., Cheng, S., Hamelers, H.V.M., Sleutels, T.J.A., Jeremiasse, A.W., et al., 2008. Microbial electrolysis cells for high yield hydrogen gas production from organic matter. Environ. Sci. Technol. 42 (23), 8630–8640.

Madsen, E.L., 2008. Environmental Microbiology: From Genomes to Biogeochemistry. Blackwell Publishing, Malden, MA.

Markets and Markets, 2011. Hydrogen Generation Market-by Merchant & Captive Type, Distributed & Centralized Generation, Application & Technology-Trends & Global Forecasts (2011–2016). Report code: EP1708. http://www.marketsandmarkets.com/Market-Reports/hydrogen-generation-market-494.html.

Melis, A., 2007. Photosynthetic H_2 metabolism in *Chlamydomonas reinhardtii* (unicellular green algae). Planta 226, 1075–1086.

Melis, A., 2009. Solar energy conversion efficiencies in photosynthesis: Minimizing the chlorophyll antennae to maximize efficiency. Plant Sci. 177, 272–280.

Melis, A., 2012. Photosynthesis-to-fuels: From sunlight to hydrogen, isoprene, and botryococcene production. Energy Environ. Sci. 5, 5531–5539.

Melis, A., Happe, T., 2001. Hydrogen production: Green algae as a source of energy. Plant Physiol. 127 (3), 740–748.

Melis, A., Zhang, L., Forestier, M., Ghirardi, M.L., Seibert, M., 2000. Sustained photobiological hydrogen gas production upon reversible inactivation of oxygen evolution in the green alga *Chlamydomonas reinhardtii*. Plant Physiol. 122, 127–133.

Millsaps, J.F., Bruce, B.D., Lee, J.W., Greenbaum, E., 2001. Nanoscale photosynthesis: Photocatalytic production of hydrogen by platinized photosystem I reaction centers. Photochem. Photobiol. 73 (6), 630–635.

Mohanakrishna, G., Goud, R.K., Venkata Mohan, S., Sarma, P.N., 2010a. Enhancing biohydrogen production through sewage supplementation of composite vegetable based market waste. Int. J. Hydrogen Energy 35 (2), 533–541.

Mohanakrishna, G., Venkata Mohan, S., Sarma, P.N., 2010b. Utilizing acid-rich effluents of fermentative hydrogen production process as substrate for harnessing bioelectricity: An integrative approach. Int. J. Hydrogen Energy 35, 3440–3449.

Mohanakrishna, G., Venkata Mohan, S., Sarma, P.N., 2010c. Bio-electrochemical treatment of distillery wastewater in microbial fuel cell facilitating decolorization and desalination along with power generation. J. Hazard Mater. 177, 487–494.

Momirlan, M., Veziroglu, T.N., 2002. Current status of hydrogen energy. Energy Edu. Sci. Technol. 6, 141–179.

Nath, K., Das, D., 2003. Hydrogen from biomass. Curr. Sci. 85 (30), 265–271.

Navarro, R.M., Sanchez-Sanchez, M.C., Alvarez-Galvan, F., Valle, D., Fierro, J.L.G., 2009. Hydrogen production from renewable sources: Biomass and photocatalytic opportunities. Energy Environ. Sci. 2, 35–54.

Nouni, M.R., 2012. Hydrogen energy and fuel cell technology: Recent developments and future prospects in India. Renewable Energy-Akshay Urja 5 (5), 10–14.

Ntaikou, I., Antonopoulou, G., Lyberatos, G., 2010. Biohydrogen production from biomass and wastes via dark fermentation: A review. Waste Biomass Valor 1, 21–39.

Orskav, E.R., Flatt, W.P., Moe, P.W., 1968. Fermentation balance approach to estimate extent of fermentation and efficiency of volatile fatty acid formation in ruminants. J. Dairy Sci. 51, 1429–1435.

Reddy, M.V., Venkata Mohan, S., 2012. Effect of substrate load and nutrients concentration on the polyhydroxyalkanoates (PHA) production using mixed consortia through wastewater treatment. Bioresour. Technol. 114, 573–582.

Rittmann, B.E., 2008. Opportunities for renewable bioenergy using microorganisms. Biotechnol. Bioeng. 100, 203–212.

Rozendal, R., Hamelers, H.V.M., Euverink, G.J.W., Metz, S.J., Buisman, C.J.N., 2006. Principle and perspectives of hydrogen production through biocatalyzed electrolysis. Int. J. Hydrogen Energy 31, 1632–1640.

Rozendal, R.A., Hamelers, H.V.M., Molenkmp, R.J., Buisman, J.N., 2007. Performance of single chamber biocatalyzed electrolysis with different types of ion exchange membranes. Water Res. 41 (9), 1984–1994.

Saiki, Y., Amao, Y., 2004. Visible light-induced enzymatic hydrogen production from oligosaccharides using Mg chlorophyll-a and platinum colloid conjugate system. Int. J. Hydrogen Energy 29, 695–699.

Saratale, G.D., Chen, S., Lo, Y., Saratale, J.L.G., Chang, J., 2008. Outlook of biohydrogen production from lignocellulosic feedstock using dark fermentation: A review. J. Sci. Ind. Res. 67, 962–979.

Saxena, R.C., Seal, D., Kumar, S., Goyal, H.B., 2008. Thermo-chemical routes for hydrogen rich gas from biomass: A review. Renew. Sustain. Energy Rev. 12, 1909–1927.

Smith, P.R., Bingham, A.S., Swartz, J.R., 2012. Generation of hydrogen from NADPH using an [FeFc] hydrogenase. Int. J. Hydrogen Energy 37, 2977–2983.

Srikanth, S., Venkata Mohan, S., Prathima Devi, M., Lenin Babu, M., Sarma, P.N., 2009. Effluents with soluble metabolites generated from acidogenic and methanogenic processes as substrate for additional hydrogen production through photo-biological process. Int. J. Hydrogen Energy 34, 1771–1779.

Srikanth, S., Venkata Mohan, S., Lalit Babu, V., Sarma, P.N., 2010. Metabolic shift and electron discharge pattern of anaerobic consortia as a function of pretreatment method applied during fermentative hydrogen production. Int. J. Hydrogen Energy 35, 10693–10700.

Stwertka, A., 1996. A Guide to the Elements. Oxford University Press, pp. 16–21.

Van Ginkel, S., Oh, S., Logan, B.E., 2005. Biohydrogen gas production from food processing and domestic wastewaters. Int. J. Hydrogen Energy 30, 1535–1542.

Vardar-Schara, G., Maeda, T., Wood, T.K., 2008. Metabolically engineered bacteria for producing hydrogen via fermentation. Microbial Biotechnol. 1, 107–125.

Venkata Mohan, S., 2008. Fermentative hydrogen production with simultaneous wastewater treatment: Influence of pretreatment and system operating conditions. J. Sci. Ind. Res. 67 (11), 950–961.

Venkata Mohan, S., 2009. Harnessing of biohydrogen from wastewater treatment using mixed fermentative consortia: Process evaluation towards optimization. Int. J. Hydrogen Energy 34, 7460–7474.

Venkata Mohan, S., 2010. Waste to renewable energy: A sustainable and green approach towards production of biohydrogen by acidogenic fermentation. In: Singh, O.V., Harvey, S.P. (Eds.), Sustainable Biotechnology: Sources of Renewable Energy. Springer, The Netherlands, pp. 129–164.

Venkata Mohan, S., Lenin Babu, M., 2011. Dehydrogenase activity in association with poised potential during biohydrogen production in single chamber microbial electrolysis cell. Bioresour. Technol. 102, 8457–8465.

Venkata Mohan, S., Lalit Babu, V., Sarma, P.N., 2007. Anaerobic biohydrogen production from dairy wastewater treatment in sequencing batch reactor (AnSBR): Effect of organic loading rate. Enz. Microbial Technol. 41 (4), 506–515.

Venkata Mohan, S., Lalit Babu, V., Sarma, P.N., 2008a. Effect of various pretreatment methods on anaerobic mixed microflora to enhance biohydrogen production utilizing dairy wastewater as substrate. Bioresour. Technol. 99, 59–67.

Venkata Mohan, S., Lalit Babu, V., Srikanth, S., Sarma, P.N., 2008b. Bio-electrochemical evaluation of fermentative hydrogen production process with the function of feeding pH. Int. J. Hydrogen Energy 33 (17), 4533–4546.

Venkata Mohan, S., Mohankrishna, G., Sarma, P.N., 2008c. Integration of acidogenic and methanogenic processes for simultaneous production of biohydrogen and methane from wastewater treatment. Int. J. Hydrogen Energy 33, 2156–2166.

Venkata Mohan, S., Veer Raghuvulu, S., Mohanakrishna, G., Srikanth, S., Sarma, P.N., 2009. Optimization and evaluation of fermentative hydrogen production and wastewater treatment processes using data enveloping analysis (DEA) and Taguchi design of experimental (DOE) methodology. Int. J. Hydrogen Energy 34, 216–226.

Venkata Mohan, S., Srikanth, S., Lenin Babu, M., Sarma, P.N., 2010. Insight into the dehydrogenase catalyzed redox reactions and electron discharge pattern during fermentative hydrogen production. Bioresour. Technol. 101, 1826–1833.

Venkata Mohan, S., Agarwal, L., Mohanakrishna, G., Srikanth, S., Kapley, A., Purohit, H.J., et al., 2011a. Firmicutes with iron dependent hydrogenase drive hydrogen production in anaerobic bioreactor using distillery wastewater. Int. J. Hydrogen Energy 36, 8234–8242.

Venkata Mohan, S., Mohanakrishna, G., Srikanth, S., 2011b. Biohydrogen production from industrial effluents. In: Pandey, A., Larroche, C., Ricke, S.C., Dussap, C.G., Gnansounou, E. (Eds.), Biofuels: Alternative Feedstocks and Conversion Processes. Academic Press, Burlington, MA, pp. 499–524.

Venkata Mohan, S., Goud, R.K., 2012. Pretreatment of biocatalyst as viable option for sustained production of biohydrogen from wastewater treatment. In: Mudhoo, A. (Ed.), Biogas Production: Pretreatment Methods in Anaerobic Digestion. John Wiley & Sons, Inc., Hoboken, NJ.

Venkata Mohan, S., Prathima Devi, M., 2012. Fatty acid rich effluent from acidogenic biohydrogen reactor as substrate for lipid accumulation in heterotrophic microalgae with simultaneous treatment. Bioresour. Technol. 123, 627–635.

Venkata Mohan, S., Chiranjeevi, P., Mohanakrishna, G., 2012. A rapid and simple protocol for evaluating biohydrogen production potential (BHP) of wastewater with simultaneous process optimization. Int. J. Hydrogen Energy 37, 3130–3141.

Vyas, D., Kumar, H.D., 1995. Nitrogen fixation and hydrogen uptake in four cyanobacteria. Int. J. Hydrogen Energy 20 (2), 163–168.

Wagner, R.C., Regan, J.M., Oh, S.E., Zuo, Y., Logan, B.E., 2009. Hydrogen and methane production from swine wastewater using microbial electrolysis cells. Water Res. 43 (5), 1480–1488.

Wang, J., Wan, W., 2009. Factors influencing fermentative hydrogen production: A review. Int. J. Hydrogen Energy 34, 799–811.

Wang, B., Wan, W., Wang, J., 2009. Effect of ammonia concentration on fermentative hydrogen production by mixed cultures. Bioresour. Technol. 100, 1211–1213.

Wang, A., Sun, D., Cao, G., Wang, H., Ren, N., Wu, W.M., et al., 2011. Integrated hydrogen production process from cellulose by combining dark fermentation, microbial fuel cells, and a microbial electrolysis cell Original Research Article. Bioresour. Technol. 102, 4137–4143.

Woodward, J., Orr, M., Cordray, K., Greenbaum, E., 2000. Enzymatic production of biohydrogen. Nature 405 (29), 1014–1015.

Yacoby, I., Pochekailov, S., Toporik, H., Ghirardi, M.L., King, P.W., Zhang, S., 2011. Photosynthetic electron partitioning between [FeFe]-hydrogenase and ferredoxin:NADP$^+$-oxidoreductase (FNR) enzymes in vitro. Proc. Natl. Acad. Sci. USA 108 (23), 9396–9401.

Yildiz, B., Kazimi, M.S., 2006. Efficiency of hydrogen production systems using alternative nuclear energy technologies. Int. J. Hydrogen Energy 31, 77–92.

Zhang, Y.H.P., Evans, B.R., Mielenz, J.R., Hopkins, R.C., Adams, M.W.W., 2007. High-yield hydrogen production from starch and water by a synthetic enzymatic pathway. PLoS ONE 2 (5), e456.

Zhu, H., Beland, M., 2006. Evaluation of alternative methods of preparing hydrogen producing seeds from digested wastewater sludge. Int. J. Hydrogen Energy 31, 1980–1988.

Fundamentals of Biohydrogen

Patrick C. Hallenbeck

University of Montreal, Montreal, Quebec, Canada

INTRODUCTION

A great deal of research is presently being carried out on the biological production of hydrogen that is driven by the need for renewable hydrogen for a possible future hydrogen economy. Hydrogen is being investigated extensively as a future fuel, with many demonstration projects and prototype hydrogen-powered vehicles on display throughout the world (Tollefson, 2010). Because more than 85% of the hydrogen presently on the market is made by steam reforming of methane, there is a real need for a sustainable means of hydrogen production. Hydrogen has many characteristics that make it a more desirable biofuel than others currently being developed, including a high gravimetric density, lower emissions, and higher efficiency of energy conversion, as it can be converted in fuel cells rather than combusted. However, the many technical problems associated with hydrogen storage and conversion to motive power, coupled with the fact that it cannot be blended into existing motor fuels like many liquid biofuels, means that widespread use of hydrogen as a fuel is a long-term prospect.

Perhaps surprisingly there are a variety of microbial paths to hydrogen production (Hallenbeck, 2011) and a number of possible substrates. This chapter presents an overview of the metabolic basis of the hydrogen-producing systems, along with a survey of possible substrates that do not compete with the food supply. Each possible scenario has certain inherent advantages, but each also has technical challenges that must be overcome before they can be implemented on a practical level. These are discussed, along with strategies for overcoming them. Finally, a number of tools and approaches that are being applied to increasing biohydrogen production rates and yields are outlined.

ENZYMES

Knowledge about the synthesis and mode of action of the different enzymes catalyzing hydrogen evolution is now well advanced, with much recent progress in understanding many of the molecular details (see Chapter 4). Not only does this now permit the heterologous expression of functional enzymes, perhaps laying the way for improvements in biohydrogen production through metabolic engineering, it also allows the rational design of changes in these catalysts, either to possibly improve catalysis directly or to create hydrogenases with desirable properties that do not necessarily occur in nature, that is, oxygen stability.

Even though proton reduction to molecular hydrogen is perhaps the simplest chemical reaction, several different enzymatic systems have evolved the capability to carry this out (Hallenbeck, 2012b; Kim and Kim, 2011). Two enzymatic systems, [NiFe]- and [FeFe]-hydrogenase, are dedicated to either proton reduction or hydrogen oxidation. In general, most [NiFe]-hydrogenases carry out hydrogen oxidation, but some exceptions are known and form the basis for some well-known hydrogen-evolving bacterial metabolic processes, such as the production of hydrogen from formate by hydrogenase 3 in *Escherchia coli* (see later, dark fermentation section). This class of enzyme is of particular potential interest, as some oxygen-tolerant members are known, all unfortunately carrying out hydrogen oxidation. In the future, once the molecular details of the oxygen tolerance are known, it might be possible to transfer these to a [NiFe]-hydrogenase more apt for hydrogen production. The enzymes of this class that do make hydrogen are mostly membrane bound, but a few exceptions are known, such as the cytoplasmic "reversible" hydrogenase from cyanobacteria (Carrieri et al., 2011; Hallenbeck and Benemann, 1978; Horch et al., 2012; McIntosh et al., 2011). A great deal of research has gone into understanding hydrogen metabolism in cyanobacteria (McIntosh et al., 2011), as they are a candidate for carrying out biophotolysis (see later) (Hallenbeck, 2012a).

The [FeFe]-hydrogenases are, in general, soluble and most are thought to be involved in proton reduction to hydrogen as a way to dispose of excess electrons during fermentations. Their turnover number in this direction is much higher than that of most of the [NiFe]-hydrogenases, making them the preferred hydrogenase in biohydrogen production schemes. With the relatively recent discovery of the [FeFe]-hydrogenase maturation system and subsequent elucidation of many of the details of the biosynthesis of the active site H cluster (Chapter 4; Nicolet and Fontecilla-Camps, 2012; Peters and Broderick, 2012), it is now possible to attempt to express this hydrogenase heterologously in the hope of bringing in a highly active hydrogen-producing catalyst. To date, the best-studied [FeFe]-hydrogenase and the only ones so far to be expressed heterologously successfully belong to the single subunit ferredoxin-dependent class. However, even complete pathways coupling this enzyme with pyruvate oxidation have been introduced and expressed successfully in *E. coli*, an organism lacking both [FeFe]-hydrogenase and its maturation system (Akhtar and Jones, 2009; Veit et al., 2008). However, hydrogen yields were low, indicating a need for further metabolic engineering.

Finally, in some cases, hydrogen evolution is catalyzed by nitrogenase, whose natural function of course is to reduce N_2 to ammonia, and whose active site is completely different from that of hydrogenases (Hu and Ribbe, 2011). These include hydrogen production by heterocystous cyanobacteria (Bothe et al., 2010; Hallenbeck, 2012a) and photoferementation by photosynthetic bacteria (Adessi and Philippis, 2012; Keskin et al., 2011; see Chapter 7). During the catalytic cycle, 1 mol of H_2 is released for every mole of N_2 fixed (Seefeldt et al., 2012). However, much more hydrogen is obtained in the absence of N_2 as nitrogenase turnover continues, with all the electron flux going to reducing protons to hydrogen. The enzyme mechanism of nitrogenase is completely different from that of hydrogenases, and even when supplied with high-energy electrons at a low enough redox potential to theoretically drive proton reduction thermodynamically, energy in the form of ATP is required. Because two electrons are required for every electron transferred from the Fe protein of nitrogenase to the Mo–Fe protein, four ATP are necessary to produce one H_2.

$$4ATP + 2e^- + 2H^+ \rightarrow 2H_2 + 4ADP$$

On the one hand, this represents an additional energy drain on metabolism to produce hydrogen that might be otherwise produced by a standard hydrogenase with no ATP hydrolysis. Thus, in some cases, replacing nitrogenase with a hydrogenase might be thought to produce a system with higher efficiency. This has yet to be reported. On the other hand, due to this requirement for ATP hydrolysis, hydrogen production by nitrogenase is essentially irreversible. Consequently, there is no feedback inhibition by high hydrogen partial pressures in the gas phase, and therefore some systems using nitrogenase can essentially drive substrate conversion to hydrogen to completion (see photofermentation, Chapter 7).

OVERVIEW OF HYDROGEN-PRODUCING SYSTEMS

The variety of microbial processes that can be used to produce hydrogen is perhaps at first surprising, but in reality this is a reflection of the great utility in hydrogen as either a metabolic end product loaded with electrons or a high-energy substrate for a large number of different natural microbial processes. The focus on what follows is on hydrogen-evolving systems, and four major types can be distinguished—those dependent on water-splitting photosynthesis, photofermentation by photosynthetic bacteria, dark fermentation, and microbial electrolysis cells.

Direct and Indirect Biophotolysis

The most inherently attractive biohydrogen system is the use of organisms that carry out oxygenic photosynthesis, capturing solar energy and coupling water splitting to proton reduction, a process that has been called biophotolysis. This system has the tremendous theoretical advantages of an essentially limitless supply of substrate and, potentially, the availability of tremendous energies with a total irradiance of 1.74×10^{17} W. Of course, as discussed later and elsewhere, there are a number of serious obstacles to successfully harnessing this capacity and making a practical system. Biophotolytic processes can be thought of as two types—direct and indirect. In direct biophotolysis, electrons derived from water splitting are boosted in energy through photosystem II (PSII) and photosystem I (PSI) to ferredoxin and are used to reduce a hydrogen-producing enzyme directly. In indirect biophotolysis, photosynthesis is used to capture solar energy and convert it to some form of carbohydrate, which is stored and then later used to produce hydrogen.

Two different systems can carry out biophotolysis and have been studied extensively: hydrogen production by heterocystous cyanobacteria and hydrogen production by green algae, principally sulfur-deprived cultures of the green alga *Chlamydomonas* (Hallenbeck, 2012a; Jones and Mayfieldt, 2012; Kruse and Hankamer, 2010) (Fig. 1). Hydrogen evolution by heterocystous cyanobacteria is driven by nitrogenase, whereas that of green algae is driven by a [FeFe]-hydrogenase. Because the hydrogen-producing reaction occurs in the heterocyst and oxygenic photosynthesis coupled with CO_2 fixation in the vegetative cell, this is actually a microscopic-indirect biophotolysis process and the hydrogen-evolving catalyst, nitrogenase, is naturally protected from oxygen inactivation (Fig. 1A). In contrast, the [FeFe]-hydrogenase of green algae is inactivated rapidly by photosynthetically produced oxygen unless the cultures are incubated under special conditions. In practice, sustained hydrogen evolution has been achieved by sulfur deprivation of acetate grown cells. Sulfur deprivation causes a drastic ($\sim$90%) reduction in photosynthetic capacity. The greatly decreased rates of oxygen production, coupled with enhanced respiration due to the presence of acetate, result in the culture going anaerobic and producing hydrogen with some electrons coming from the residual water-splitting reaction (true biophotolysis) and some coming from stored carbon reserves (indirect biophotolysis) (Fig. 1B) (Eroglu and Melis, 2011; Tsygankov, 2012).

Recent advances in this area include the first direct demonstration of an indirect biophotolysis process using cyanobacteria where the nonheterocystous *Plectonema boryanum* was cycled multiple times through an aerobic, nitrogen-limited stage, allowing for glycogen accumulation, and a second anaerobic, hydrogen-producing stage, catalyzed by nitrogenase (Huesemann et al., 2010). As well, sustained hydrogen production by single-cell, nonheterocystous cyanobacterium, *Cyanothece*, has been reported, although this was either when the cultures had been supplemented with glycerol, allowing for respiratory protection (Bandyopadhyay et al., 2010), or when photosynthetically evolved oxygen is removed by Ar sparging (Melnicki et al., 2012).

Thus although amply demonstrated, hydrogen production by these systems will require substantial research and development (R&D) before the major obstacles to practical hydrogen production by this route are eliminated. Possible solutions include developing an oxygen-tolerant hydrogenase (green algae) or replacing nitrogenase with a hydrogenase (cyanobacteria, Table 1). A number of other possible strategies for improving hydrogen production have been proposed and are currently under investigation, including reducing the antenna size (Kosourov et al., 2011), direct downregulation or mutation of PSII proteins (Scoma et al., 2012; Surzycki et al., 2007), variations in operational parameters (Esquivel et al., 2011; Grossman et al., 2011), or even the heterologous expression of hydrogenase and ferredoxin (Ducat et al., 2011).

Photofermentative Hydrogen Production by Photosynthetic Bacteria

In another hydrogen-producing process involving the capture and conversion of light energy, photosynthetic bacteria degrade a variety of substrates to hydrogen and carbon dioxide,

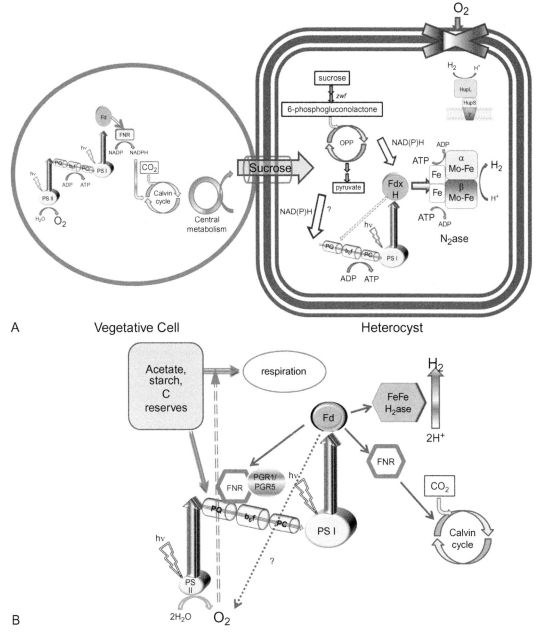

FIGURE 1 Different biophotolytic systems. (A) Hydrogen production by heterocystous cyanobacteria. The heterocyst, a specialized cell, is induced in some filamentous cyanobacteria by nitrogen deprivation. The heterocyst undergoes many morphological and metabolic changes, which enable it to provide an anaerobic environment for nitrogenase. Even though the cells are in an environment supersaturated with oxygen, no oxygen is evolved from photosynthesis since photosystem II (PSII) is absent, the specialized cell envelope impedes the inward diffusion of oxygen, and an active respiratory chain consumes any traces of oxygen rapidly. Reduced carbon, sucrose, necessary for hydrogen production, must be imported from the vegetative cells, which carry out CO_2 fixation through oxygenic photosynthesis. Reductant, reduced fdxH, generated by the oxidative pentose pathway (OPP) and ATP, formed by cyclic photophosphorylation around photosystem I (PSI), drive hydrogen production by nitrogenase. (B) Hydrogen production by sulfur-deprived green algae. Hydrogen production by a sulfur-deprived green alga (*Chlamydomonas reinhardtii*) is shown. A reduction in PSII caused by sulfur deprivation and any oxygen evolved by residual PSII activity is respired, consuming the added acetate or fixed carbon reserves, thus providing an anaerobic environment where the oxygen-sensitive [FeFe]-hydrogenase can function. About half of the evolved hydrogen comes from residual water-splitting activity and the rest is produced with high-energy electrons made from fixed carbon by PSI activity. Several pathways lead to reduction of ferredoxin (Fd), the electron donor for hydrogenase.

TABLE 1   Different Biohydrogen Processes and Their Advantages and Disadvantages

Process/ organisms	Substrate	Advantages	Barriers	Possible improvements
Biophotolysis (green algae, cyanobacteria)	Water	Very abundant, inexpensive substrate Solar energy capture, high solar flux	Oxygen evolved by photosynthesis incompatible with hydrogen-evolving catalyst	Create O_2-resistant hydrogenase (algae) Replace N_2ase with H_2ase (cyanobacteria)
Photofermentation (purple nonsulfur photosynthetic bacteria)	Organic acids Sugars Glycerol	Nearly complete substrate conversion Able to use various waste streams or effluents from dark fermenters	Low-light conversion efficiencies Low volumetric rates of production	Elimination of competing pathways Substitution of hydrogenase for nitrogenase Reduction in photosynthetic antenna size
Dark fermentation (strict anaerobes, facultative anaerobes)	Sugars, many different carbohydrate-rich materials	Simple reactor technology Capable of using a wide variety of substrates High volumetric rates	Low yields (<4 H_2/ glucose) Many side products	Improve yields through metabolic engineering Use a second stage to derive additional energy from side products
Microbial electrolysis (wide variety of exoelectrogens, *Geobacter*, *Shewanella*)	Very wide range of organic substrates: sugars, organic acids, proteins	Complete substrate degradation possible H_2 from wide range of compounds H_2 from organic acids	Excess voltage needed Low current densities Expensive cathodes (platinum)	Develop biocathodes Ni-doped or stainless-steel electrodes Improved microbes?

often nearly stoichiometrically, in a process that has been called photofermentation. Because this process involves hydrogen production by nitrogenase, consequently both high-energy electrons coming from the substrate and ATP, generated via cyclic bacterial photosynthesis, must be supplied (Fig. 2).

   A great deal of research has gone into this area in the 60 years since it was first described with studies on a variety of organisms, the use of many different substrates, and examination of the effects of a plethora of factors, including light intensity, nutrient regime, and cell immobilization (Adessi and Philippis, 2012; Keskin et al., 2011; Li and Fang, 2009). Most commonly, various organic acids are used as substrate, and nearly stoichiometric conversion to hydrogen is achieved in many cases. However, other substrates can be used.

   A relatively high-yield conversion of sugars to hydrogen has been shown (Abo-Hashesh et al., 2011a; Ghosh et al., 2012a; Keskin and Hallenbeck, 2012a,b), and the completely stoichiometric conversion of waste glycerol to hydrogen has been achieved (Ghosh et al., 2012b,c; Sabourin-Provost and Hallenbeck, 2009). These are all light-dependent processes

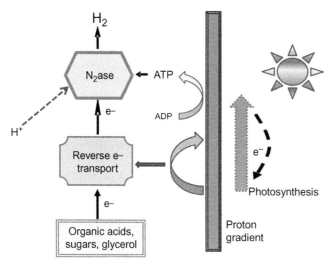

FIGURE 2 Photofermentative hydrogen production by photosynthetic bacteria. Hydrogen production pathways in nonsulfur photosynthetic bacteria are depicted. Organic compounds are assimilated and used for biomass biosynthesis. If the substrates are more reduced than the redox state of the biomass, excess reducing equivalents (NADH) must be disposed for growth to continue. This is accomplished by nitrogenase, which, expressed under nitrogen-limiting conditions, reduces protons to hydrogen. Bacterial photosynthesis supports the requirements for nitrogenase activity in two ways by converting the captured light energy into a proton gradient. Both ATP synthesis and reduced ferredoxin, generated by reversed electron flow, needed for nitrogenase activity are produced from this proton gradient (from Keskin et al., 2011).

where captured light energy drives electron flow through the photosynthetic apparatus, generating a proton gradient. This gradient is then used to satisfy both requirements for nitrogenase activity: ATP, through ATP synthase, and high-energy electrons, through reverse electron flow.

The need for redox balancing of metabolism is what drives hydrogen production by nitrogenase under these conditions. Reoxidation of NADH is required to dissipate excess electrons in the supplied carbon substrate, a need satisfied under nitrogen-replete conditions by the seemingly paradoxical fixation of carbon dioxide by RuBisCO (McKinlay and Harwood, 2010).

Even though the substrate is converted to hydrogen at relatively high yields, there are a number of barriers to practical application of this system at present, including low volumetric rates of hydrogen production and low-light conversion efficiencies (Table 1), which, for one thing, translate into a requirement for inordinately large surface areas for any photobioreactors. Different strategies could be put to use in attempts to increase rates, yields, or photosynthetic efficiencies (Table 1). These are discussed in more detail in Chapter 7. Finally, photofermentation, as discussed in Chapter 7, is often discussed as a possible second stage to extract more hydrogen from the by-products of dark fermentative hydrogen production.

Dark Fermentative Production of Hydrogen

By far the largest numbers of studies on biological hydrogen production have involved the use of dark fermentation with a variety of different bacteria. Fermentative hydrogen production is inherently attractive for a number of reasons (Table 1) (Abo-Hashesh and Hallenbeck, 2012; Hallenbeck, 2009; Hallenbeck and Ghosh, 2009). For one thing, the reactor technology can be simple, with perhaps some concern over the maintenance of anaerobic conditions when the fermentation is being carried out by strict anaerobes, *Clostridium*, for example. This is less of a consideration of course when facultative anaerobic bacteria, such as *E. coli*, are used. Dark fermentation can be carried out using either pure cultures or mixed cultures that have been enriched for spore formers (strict anaerobes) by heat treatment. The advantage of mixed cultures is that they usually possess a wide range of hydrolytic activities so a variety of substrates and complex polymeric substrates can be used. In addition, at least in some bioreactor configurations, high volumetric rates of hydrogen production can be achieved. The highest volumetric rates of hydrogen production are observed with some type of immobilized culture, usually in the form of self-immobilized granules, which form in the reactor over time when the reactor is seeded with mixed-culture inoculums, such as obtained by heat treatment of sludge from an anaerobic digestor.

A number of metabolic pathways potentially lead from the sugar that is catabolized to hydrogen with pyruvate as the key intermediate (Fig. 3). Given the variety of hydrogenases, especially in the Firmicutes (Calusinska et al., 2010), a variety of means of reducing hydrogenase appear to be available. Some routes are relatively firmly established by the evidence currently in hand, whereas others are based on properties inferred from genetic signatures and thus are more speculative and will require further verification.

Two major possibilities for the anaerobic degradation of pyruvate leading to hydrogen are known. The pyruvate formate lyase pathway is used by some organisms to degrade all or part of the pyruvate to formate, a redox-neutral process. Formate can accumulate as a fermentation by-product, but in some organisms, such as *E. coli*, hydrogen is produced from formate via a special class of [Ni-Fe]-hydrogenase, the Ech hydrogenase, which is part of a formate: hydrogen lyase complex, which is induced upon media acidification, as it mainly serves to reduce acidity by removing formic acid. Thus, increasing the activity of this complex can be a means of increasing hydrogen production (Abo-Hashesh et al., 2011b; Hallenbeck et al., 2012; Seol et al., 2012). In other organisms, including some Firmicutes, a trimeric [FeFe]-hydrogenase may be present and, perhaps anchored to the cytoplasmic membrane by an associated membrane protein, couple formate oxidation to proton reduction. In an alternate pathway, pyruvate can be converted to acetyl-CoA and reduced ferredoxin by pyruvate ferredoxin oxidoreductase (PFOR). The reduced ferredoxin can then drive hydrogen production by the well-known monomeric *Clostridium pasteurianum*-type [FeFe]-hydrogenase.

During substrate oxidation to pyruvate, NADH is released (Fig. 3). Although some organisms, for example, enterobacteria, such as *E. coli*, cannot derive additional hydrogen from this source, others, such as Firmicutes, are thought to possess a variety of mechanisms for doing so. Additional reduced ferredoxin can perhaps be generated through NADH oxidation by NADH:ferredoxin oxidoreductase (NFO). Soluble NFO activity was described several decades ago; however, this enzyme has never been purified nor its encoding gene identified. Perhaps ferredoxin can be reduced using NADH through the more recently described membrane complex of the Rnf family. In this scenario, membrane potential is used to provide the energy needed for the thermodynamically unfavorable reduction of ferredoxin by NADH.

Some multimeric [FeFe]-hydrogenases that have been identified in genome sequences, similar to Hnd from the sulfate reducers, could possibly evolve hydrogen directly from NAD(P)H at very low P_{H_2S}. Finally, a timeric bifurcating [FeFe]-hydrogenase has been described that could use NADH and reduced ferredoxin simultaneously to evolve hydrogen with the excess energy available from ferredoxin oxidation driving the unfavorable oxidation

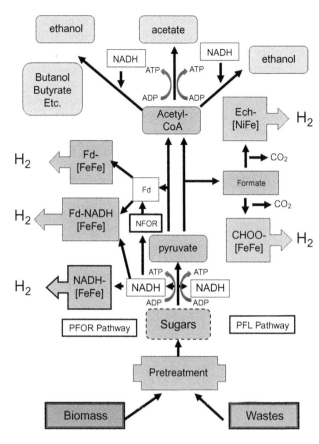

FIGURE 3 Metabolic pathways leading to hydrogen production in dark fermentation. Shown are the various possibilities for connection of different hydrogenases to glycolysis and pyruvate degradation suggested by the present evidence. Some aspects are speculative and will require further verification. Some organisms can degrade at least part of the pyruvate to formate through the pyruvate formate lyase (PFL) pathway. In organisms such as *Escherichia coli*, hydrogen is produced from formate via a special class of [NiFe]-hydrogenase, the Ech hydrogenase. In other organisms, including some Firmicutes, a trimeric hydrogenase may, if present and perhaps anchored to the cytoplasmic membrane by an associated membrane protein, couple formate oxidation to proton reduction. Alternately, pyruvate can be converted to acetyl-CoA and reduced ferredoxin by PFOR. The reduced ferredoxin can then drive hydrogen production by the well-known monomeric *C. pasteurianum*-type hydrogenase. Substrate oxidation to pyruvate releases NADH. Although some organisms, for example, enterobacteria, such as *E. coli*, cannot derive more hydrogen from this source, others, the Firmicutes, are thought to possess a variety of mechanisms for doing so. Additional reduced ferredoxin can perhaps be produced through NADH oxidation by NADH:ferredoxin oxidoreductase (Nfo). Nfo could be the soluble activity described several decades ago or the more recently described membrane complex of the Rnf family. Some multimeric hydrogenases [TE (M2)], similar to Hnd from the sulfate reducers, could possibly evolve hydrogen directly from NAD(P)H at very low P_{H_2S}. Finally, the newly described timeric bifurcating hydrogenase [TR (M3)] could simultaneously use NADH and reduced ferredoxin to evolve hydrogen with the excess energy available from ferredoxin oxidation driving the unfavorable oxidation of NADH.

of NADH. Genome analysis suggests that, in fact, this enzyme may be fairly widespread, suggesting that hydrogen production using the NADH generated during glycolysis might actually be more favorable than previously thought.

In practice, a number of considerations have to be taken into account in carrying out dark hydrogen fermentations. As alluded to earlier, bioreactor design and operation are important factors in maximizing hydrogen volumetric productivity and yields (Jung et al., 2011; Show et al., 2011). The choice of a pure organism (Lee et al., 2011) or the use of mixed cultures containing synergistic activities (Hung et al., 2011) may also be an important consideration. In general, mixed cultures that are immobilized, either on inert supports or by formation of self-aggregates, offer the highest productivities, with granular aggregates showing superior performance over biofilm type systems (Chojnacka et al., 2011). In fact, this type of system has been scaled up successfully to the pilot scale (Lin et al., 2011) and is at the center of an industrial biological hydrogen effort (the Brazilian company Ergostech). Although mixed flora in general are to be preferred due to their inherently wider substrate range, it should be pointed out that some single organisms, such as *E. coli*, are naturally endowed with the capacity to use a wide range of sugars and sugar derivatives (Ghosh and Hallenbeck, 2009). Additionally, many studies have shown that much is to be gained by the optimization of operational parameters, whether with pure cultures (Ghosh and Hallenbeck, 2010) or with mixed cultures (Won and Lau 2011), as hydrogen production, especially yields, can often be increased by carefully controlling factors such as nutrient limitation (Bisaillon et al., 2006; Turcot et al., 2008).

Hydrogen Production in Microbial Electrolysis Cells

A novel technique for producing hydrogen from a variety of substrates, using what are called microbial electrolysis cells (MECs) (Fig. 4), has seen very rapid development over the last few years (Geelhoed et al., 2010; Kiely et al., 2011; Pant et al., 2012). These are basically modified microbial fuel cells (MFCs), which have been under study for decades, with much recent research and patent activity on these devices (Yang et al., 2011). In fact, the anodic reactions are the same in the two devices. At this electrode, electrogenic microbes completely degrade various organic substrates to CO_2 and electrons in what is essentially an anaerobic respiration with the electrode (anode) serving as an electron sink. Organisms that couple with the anode do so through a variety of mechanisms, including the excretion of extracellular mediators and direct physical contact through extracellular structures such as "nanowires," and can carry out complete substrate degradation to CO_2, protons, and electrons. The current generated flows to the cathode where reduction of oxygen takes place (MFC). These basic cells can be modified to produce variants carrying out different types of bioelectrochemistry

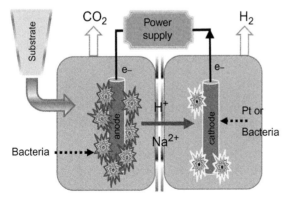

FIGURE 4   Hydrogen production by a microbial electrolysis cell. A microbial electrolysis cell is conceptually an extension of the concept of a microbial fuel cell. Electrogenic microbes at the anode degrade organic matter to CO_2 and protons using the anode as an electron sink. A variety of mechanisms, including excreted soluble electron carriers and conductive biological nanowires, assure good electrical conductivity between the organism's metabolism and the physical hardware. Application of additional voltage drives hydrogen evolution at the cathode, which usually contains platinum. In one variation, a biocathode, electrons are fed to microbes, which then use the delivered power to reduce protons to hydrogen.

with several types of useful reactions from electromethanogenesis to microbial desalination (Cao et al., 2009; Cheng et al., 2009). Of course, here we are interested in how this type of cell can be coupled to hydrogen production in a device that has been called microbial electrolysis cell. A schema of a typical MEC is shown in Figure 4.

A MEC produces hydrogen by supplementing the voltage generated at the anode with voltage from an external power source, driving hydrogen evolution at the cathode (Fig. 4). For one thing, this allows hydrogen production from substrates whose redox potential would normally not allow it. Thus, acetate (-0.279 V) can theoretically be converted to hydrogen (-0.414 V) with the application of a relatively small voltage (-0.135 V) [29]. In practice, additional voltage is required due to overpotentials created by physical–chemical and biological factors.

Research and development on MECs has produced a variety of configurations. Initial studies used two-chamber devices with ion-permeable membranes separating the anodic and cathodic chambers. This geometry allows easy capture of hydrogen at the cathode and carbon dioxide at the anode, prevents fouling of the cathode by anodic bacteria, and allows the development of separate microbial communities in the two compartments if a biocathode is used to catalyze hydrogen production. Problems encountered with this configuration include pH changes with acidification of the anode chamber due to the production of protons and alkalinization of the cathode chamber due to proton consumption and high resistance caused by the membrane, thus increasing the voltage requirement for hydrogen evolution. These problems are avoided by single-chamber MECs, which can generate higher current densities than dual-chamber configurations and thus give significantly higher hydrogen production rates. These in turn, however, have their potential drawbacks, which could decrease hydrogen yields or coulombic efficiencies. For one, methane production at the expense of produced hydrogen may increase with this configuration, thus decreasing yields. Another significant problem could be the development of a biological "short circuit" with microbes at the anode consuming the produced hydrogen using the anode as an electron sink, thus creating a futile cycle and drastically increasing the amount of current needed to generate a given amount of net hydrogen. As much as two-thirds of the current density seen in one of these devices may be due to hydrogen recycling (Lee and Rittmann, 2010).

There have been rapid, significant improvements in the performance of these devices since 2007, and partial progress toward overcoming these barriers has been made. Improved electrode materials have been developed with better anodic configurations offering higher surface-to-volume properties and thus higher current densities (graphite brushes, graphite felts, carbon papers). At the cathode, success has been reported for the replacement of costly and rare platinum with stainless steel or various metal alloys. It may even be possible to produce an effective biocathode where hydrogen-producing microbes are immobilized on the electrode, and success has been reported in initial studies, although current densities were quite low. Thus further improvement in cell geometries and other physical chemical parameters could help reduce internal resistances, thus decreasing overpotentials and the excess applied voltage requirement. Further R&D might well bring this technology to a level where it could be applied on a practical level in a relatively short period of time, representing a useful technology for the conversion of various waste streams to hydrogen.

Use of Hybrid Systems

Some studies have examined the use of hybrid two-stage systems to increase overall energy extraction from some substrates (Luo et al., 2011). In this scenario, carbohydrate-rich substrates are first partially converted to hydrogen in a typical optimized dark fermentation reactor. The organic acid by-products are then treated in a second stage to recover additional energy. There are three possible alternatives currently in various stages of development. In one scenario, the dark hydrogen fermenter is coupled with an anaerobic digestor, for which organic acids are ideal substrates. The result, now proven in several pilot scale studies, is a stream of hydrogen and methane along with a significant total reduction in the chemical oxygen demand. Although it is probably more desirable to produce a single gaseous product, hydrogen and hydrogen/methane mixtures have the distinct advantage of burning much cleaner than methane alone, and treating the complex substrate matrix with a dark fermentation reactor appears to provide overall greater substrate decomposition than methane digestion alone.

Two other possibilities for a second stage are under study that do produce additional hydrogen—the use of photofermentation (Adessi et al., 2012) or MEC. Both of these are eminently suited for using the organic acid effluent and, with the introduction of additional energy, either sunlight (photofermentation) or electricity (MEC), converting it almost entirely to hydrogen (and carbon dioxide). Both these processes have already been discussed earlier and, as noted previously, will require further R&D before they can be used on a practical level. Thus, for example, the use of photofermentation presents a number of technical barriers, including sensitivity to fixed nitrogen, low-light conversion efficiencies, inability to use high-light intensities, and the need for low-cost, transparent, hydrogen-impermeable photobioreactors. Similarly, although MECs offer the possibility of extracting additional hydrogen from dark fermentation effluents, they also require more R&D before they can be employed on a practical level.

TOOLS

In addition to techniques that can increase hydrogen production through bioprocess engineering (Chen et al., 2011), a variety of tools are now available either for a detailed molecular analysis of various biohydrogen processes or for bringing about changes on a molecular level to increase hydrogen production rates or yields. Different techniques can be used to study the composition of the microbial population present in mixed cultures involved in the dark fermentative production of hydrogen and also allow the study of population changes over time (Li et al., 2011). These are useful in understanding what makes an effective mixed culture fermentation and have brought about an appreciation of the fact that, in some cases, nonhydrogen-producing bacteria may have a role to play through their metabolic interactions with the hydrogen-producing bacteria present (Hung et al., 2011).

A variety of strategies can be proposed for increasing biohydrogen production and many of these depend on the use of metabolic or protein engineering to achieve the desired goals (Hallenbeck and Ghosh 2012; Hallenbeck et al., 2011, 2012). However, metabolic engineering should be guided intelligently to be successful. For relatively simple modifications, the changes to be made are usually obvious and straightforward, but even so, unplanned metabolic alterations may take place due to activation of alternative pathways, metabolic shifts brought about by changes in metabolite pools, or unexpected activities of known enzymes. Therefore, efforts to make major changes in existing pathways or to introduce novel pathways should be based on knowledge obtained about existing metabolic fluxes and metabolomics (Baran et al., 2009; Cai et al., 2011; Zamboni and Sauer, 2009). When combined with the available genomics information, this will allow the construction of relatively accurate metabolic models to guide the rational design of changes to organisms to introduce the desired characteristics (Tyo et al., 2010). For example, software has been described (OptKnock, OptStrain, and OptReg) to find the optimal strategies for the overproduction of fuels and chemicals (Pharkya and Maranas, 2006).

Metabolic engineering opens up a variety of avenues for potentially improving biohydrogen production, and several different metabolic engineering strategies for increasing biohydrogen production under dark fermentative as well as phototrophic conditions have been described (Hallenbeck and Ghosh, 2012; Hallenbeck et al., 2011, 2012; Kontur et al., 2012; Oh et al., 2011; Srirangan et al., 2011; Work et al., 2012). Metabolic engineering allows the modification of metabolic pathways to either increase the production of natural products or produce novel products through the use of systematic and quantitative analysis of pathways, and molecular biology approaches to remodel the genome. Metabolic engineering could be used to overcome limiting factors for biohydrogen production in various systems in a variety of ways, including increasing substrate utilization, increasing the flow of electrons to hydrogen-producing pathways, and engineering more efficient and/or oxygen-resistant hydrogen-evolving enzymes. Metabolic engineering could be used at several different levels for process improvement in dark fermentation (Fig. 5) (Hallenbeck et al., 2011). In terms of substrate utilization, metabolic engineering could be used to extend the range of usable substrates, necessary in many cases if abundant lignocellulosic substrates are to be used as feedstock. Thus, an organism could be given either the capacity for direct lignocellulosics degradation or the ability to use the mixture of pentoses and hexoses available after enzymatic deconstruction of this feedstock. Finally, once the sugars are converted to pyruvate, the key intermediate, metabolic engineering could be used to increase the rates and/or yields of hydrogen either through the modification of existing pathways or by the introduction of novel hydrogen-producing pathways. A variety of tools for achieving these types of modifications are now available.

The metabolic engineering horizons have expanded greatly with recently introduced new techniques that allow—in principle—an almost limitless series of markerless knockouts to be made in a single strain, as well as the introduction of very large amounts of foreign DNA into the chromosome. While most of this work has been carried out with E. coli, many of these can be, and have been, extended to other Gram-negative and Gram-positive bacteria. The use of phage-encoded, site-specific recombination systems—FLP/FRT, Cre/Lox, or λ Xis/Int—is the key to making strains with multiple specific deletions. Introduction of a specific short recombination site (FRT, Lox, Int) allows subsequent excision when the specific excisase is expressed. A useful additional tool, the use of special recombinases permits transformation with polymerase chain reaction products that have an antibiotic resistance marker flanked by recognition sites and as little as 40 to 50 bp of genomic

Roles for Metabolic Engineering in Hydrogen Production

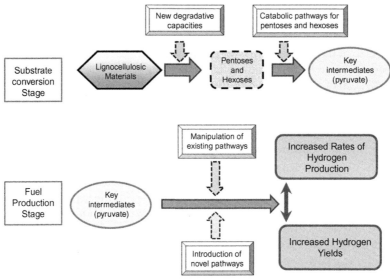

FIGURE 5 Potential roles for metabolic engineering in biohydrogen production. Biofuel production from fixed carbon substrates can be divided conceptually into two levels: acquisition and conversion of complex substrates to key metabolic intermediates (pyruvate) and conversion of key metabolic intermediates to the desired biofuel. Metabolic engineering can play a role in several different ways: to (1) add pathways to an organism, enabling it to directly use a wider range of complex substrates, (2) add pathways permitting the conversion of a wider range of monomers to key metabolic intermediates, (3) boost production of a biofuel that is produced naturally by the organism, or (4) add pathways leading to the production of a novel biofuel (from Abo-Hashesh et al., 2012).

homology at each end. This simple and powerful method has been called recombineering and allows insertional inactivation without intermediate cloning steps. This system has been well developed for *E. coli* and closely related organisms (Sharan et al., 2009) and allows the systematic analysis of unknown gene function and gene regulatory networks, as well as genome-wide testing of mutational effects in a common background. Many different mutational studies have been carried out with *E. coli* and have been reviewed elsewhere (Abo-Hashesh et al., 2011b; Hallenbeck et al., 2011, 2012).

There is a great deal of interest in the Gram-positive Firmicutes, well known for their powers of fermentation, but until relatively recently most have been difficult to work with genetically so very few studies have been reported on the metabolic engineering of biohydrogen production. Now, a set of tools, the ClosTron system and pMTL80000 modular shuttle vector system, is available and has been used successfully in *Clostridium difficile*, *Clostridium acetobutylicum*, *Clostridium sporogenes*, *Clostridium botulinum*, *Clostridium beijerinckii*, and *Clostridium perfringens* (Heap et al., 2010). This system is based on a bacterial group II intron that can insert into a specific target site via an RNA-mediated "retrohoming" mechanism and can be reprogrammed to the appropriate target site by alteration of the DNA sequence of the appropriate part of the intron. These newly developed tools will allow the future engineering of a variety of pathways to increase hydrogen production.

CONCLUSION

Hydrogen as an alternative fuel has a number of advantages and is the focus of a great deal of R&D from the engineering of suitable storage to the development of prototype cars. Biological hydrogen production could provide a renewable hydrogen stream that is presently lacking in the development of a future hydrogen economy. A number of possible means for the microbial production of hydrogen are possible and are currently under intensive investigation. Recent research has brought about many advances in knowledge in this area, ranging from fundamental understanding of the structure and function of hydrogenases to pilot scale conversion of wastes to hydrogen.

The field is sufficiently developed now to be able to identify and target the limiting factors that impede the full development of a microbial process for renewable hydrogen production to power a future hydrogen economy. Since details of the synthesis and molecular mechanisms of the different hydrogenases are now well understood, a solid basis for the rational design of improved catalysts has been established. A number of challenges remain to making biophotolysis, where solar energy is directly converted to hydrogen, practical, including low conversion efficiencies, and the oxygen sensitivity of hydrogenase. Likewise, photofermentation suffers from low-light conversion efficiencies. Possible improvements here are possible through a number of strategies, including the substitution of hydrogenase for the energy-demanding nitrogenase. A variety of organisms with different metabolic pathways have been shown to be capable of producing hydrogen through the breakdown of various carbohydrates or waste streams with the achievement of high volumetric rates of hydrogen production, but low yields and inefficient waste treatment remain problems to be overcome. Improvements in hydrogen yields may be possible through metabolic engineering or the development of hybrid two-stage systems employing second stages of methane digestion, photofermentation or MECs. The use of MECs is promising, and research is leading to rapid improvement, but challenges remain in finding inexpensive, efficient cathode materials and a way to increase current densities. Given the relatively large R&D effort in biohydrogen research, and the number of options available, a breakthrough in one of the various approaches is possible, enabling its practical application within the next one or two decades.

Acknowledgments

I thank the many students who have worked in my laboratory over the years on biofuel projects and NSERC and FQRNT for research support.

References

Abo-Hashesh, M., Hallenbeck, P.C., 2012. Fermentative hydrogen production. In: Hallenbeck, P.C. (Ed.), Microbial Technologies in Advanced Biofuels Production. Springer, New York, pp. 77–92.

Abo-Hashesh, M., Ghosh, D., Tourigny, A., Taous, A., Hallenbeck, P.C., 2011a. Single stage photofermentative hydrogen production from glucose: An attractive alternative to two stage photofermentation or co-culture approaches. Int. J. Hydrogen Energy 36 (21), 13889–13895.

Abo-Hashesh, M., Wang, R., Hallenbeck, P.C., 2011b. Metabolic engineering in dark fermentative hydrogen production; theory and practice. Bioresour. Technol. 102 (18), 8414–8422.

Adessi, A., Philippis, R., 2012. Hydrogen production: Photofermentation. In: Hallenbeck, P.C. (Ed.), Microbial Technologies in Advanced Biofuels Production. Springer, New York, pp. 53–75.

Adessi, A., De Philippis, R., Hallenbeck, P.C., 2012. Combined systems for maximum substrate conversion. In: Hallenbeck, P.C. (Ed.), Microbial Technologies in Advanced Biofuels Production. Springer, New York, pp. 107–126.

Akhtar, M.K., Jones, P.R., 2009. Construction of a synthetic YdbK-dependent pyruvate:H-2 pathway in *Escherichia coli* BL21(DE3). Metab. Eng. 11 (3), 139–147.

Bandyopadhyay, A., Stöckel, J., Min, H., Sherman, L.A., Pakrasi, H.B., 2010. High rates of photobiological H2 production by a cyanobacterium under aerobic conditions. Nat. Commun. 1, 139.

Baran, R., Reindl, W., Northen, T.R., 2009. Mass spectrometry based metabolomics and enzymatic assays for functional genomics. Curr. Opin. Microbiol. 12, 547–552.

Bisaillon, A., Turcot, J., Hallenbeck, P.C., 2006. The effect of nutrient limitation on hydrogen production by batch cultures of *Escherichia coli*. Int. J. Hydrogen Energy 31 (11), 1504–1508.

Bothe, H., Schmitz, O., Yates, M.G., Newton, W.E., 2010. Nitrogen fixation and hydrogen metabolism in cyanobacteria. Microbiol. Mol. Biol. Rev. 74 (4), 529–551.

Cai, G.Q., Jin, B., Monis, P., Saint, C., 2011. Metabolic flux network and analysis of fermentative hydrogen production. Biotechnol. Adv. 29 (4), 375–387.

Calusinska, M., Happe, T., Joris, B., Wilmotte, A., 2010. The surprising diversity of clostridial hydrogenases: A comparative genomic perspective. Microbiology 156, 1575–1588.

Cao, X.X., Huang, X., Liang, P., Xiao, K., Zhou, Y., Zhang, X., et al., 2009. A new method for water desalination using microbial desalination cells. Environ. Sci. Technol. 43 (18), 7148–7152.

Carrieri, D., Wawrousek, K., Eckert, C., Yu, J., Maness, P.C., 2011. The role of the bidirectional hydrogenase in cyanobacteria. Bioresour. Technol. 102 (18), 8368–8377.

Chen, C.Y., Liu, C.H., Lo, Y.C., Chang, J.S., 2011. Perspectives on cultivation strategies and photobioreactor designs for photo-fermentative hydrogen production. Bioresour. Technol. 102 (18), 8484–8492.

Cheng, S.A., Xing, D.F., Call, D.F., Logan, B.E., 2009. Direct biological conversion of electrical current into methane by electromethanogenesis. Environ. Sci. Technol. 43 (10), 3953–3958.

Chojnacka, A., Błaszczyk, M.K., Szczęsny, P., Nowak, K., Sumińska, M., Tomczyk-Żak, K., et al., 2011. Comparative analysis of hydrogen-producing bacterial biofilms and granular sludge formed in continuous cultures of fermentative bacteria. Bioresour. Technol. 102 (21), 10057–10064.

Ducat, D.C., Sachdeva, G., Silver, P.A., 2011. Rewiring hydrogenase-dependent redox circuits in cyanobacteria. Proc. Natl. Acad. Sci. USA 108 (10), 3941–3946.

Eroglu, E., Melis, A., 2011. Photobiological hydrogen production: Recent advances and state of the art. Bioresour. Technol. 102 (18), 8403–8413.

Esquivel, M.G., Amaro, H.M., Pinto, T.S., Fevereiro, P.S., Malcata, F.X., 2011. Efficient H-2 production via *Chlamydomonas reinhardtii*. Trends Biotechnol. 29 (12), 595–600.

Geelhoed, J.S., Hamelers, H.V.M., Stams, A.J.M., 2010. Electricity-mediated biological hydrogen production. Curr. Opin. Microbiol. 13 (3), 307–315.

Ghosh, D., Hallenbeck, P.C., 2009. Fermentative hydrogen yields from different sugars by batch cultures of metabolically engineered *Escherichia coli* DJT135. Int. J. Hydrogen Energy 34 (19), 7979–7982.

Ghosh, D., Hallenbeck, P.C., 2010. Response surface methodology for process parameter optimization of hydrogen yield by the metabolically engineered strain *Escherichia coli* DJT135. Bioresour. Technol. 101 (6), 1820–1825.

Ghosh, D., Sobro, I.F., Hallenbeck, P.C., 2012a. Optimization of the hydrogen yield from single-stage photofermentation of glucose by *Rhodobacter capsulatus* JP91 using response surface methodology. Bioresour. Technol. 123, 199–206.

Ghosh, D., Sobro, I.F., Hallenbeck, P.C., 2012b. Stoichiometric conversion of biodiesel derived crude glycerol to hydrogen: Response surface methodology study of the effects of light intensity and crude glycerol and glutamate concentration. Bioresour. Technol. 106, 154–160.

Ghosh, D., Tourigny, A., Hallenbeck, P.C., 2012c. Near stoichiometric reforming of biodiesel derived crude glycerol to hydrogen by photofermentation. Int. J. Hydrogen Energy 37 (3), 2273–2277.

Grossman, A.R., Catalanotti, C., Yang, W., Dubini, A., Magneschi, L., Subramanian, V., et al., 2011. Multiple facets of anoxic metabolism and hydrogen production in the unicellular green alga *Chlamydomonas reinhardtii*. New Phytol. 190 (2), 279–288.

Hallenbeck, P.C., 2009. Fermentative hydrogen production: Principles, progress, and prognosis. Int. J. Hydrogen Energy 34 (17), 7379–7389.

Hallenbeck, P.C., 2011. Microbial paths to renewable hydrogen production. Biofuels 2, 285–302.

Hallenbeck, P.C., 2012a. Hydrogen production by cyanobacteria. In: Microbial Technologies in Advanced Biofuels Production, Part 2. Springer, New York, pp. 15–28.

Hallenbeck, P.C., 2012b. Fundamentals of dark hydrogen fermentations; multiple pathways and enzymes. In: Azbar, Levin (Ed.), State of the Art and Progress in Production of Biohydrogen. Bentham Science Publishing, pp. 94–111.

Hallenbeck, P.C., Benemann, J.R., 1978. Characterization and partial purification of reversible hydrogenase from *Anabaena cylindrica*. FEBS Lett. 94 (2), 261–264.

Hallenbeck, P.C., Ghosh, D., 2009. Advances in fermentative biohydrogen production: The way forward? Trends Biotechnol. 27 (5), 287–297.

Hallenbeck, P.C., Ghosh, D., 2012. Improvements in fermentative biological hydrogen production through metabolic engineering. J. Environ. Manag. 95, S360–S364.

Hallenbeck, P.C., Ghosh, D., Abo-Hashesh, M., Wang, R., 2011. Metabolic engineering for enhanced biofuels production with emphasis on the biological production of hydrogen. In: Talyor, J.C. (Ed.), Advances in Chemistry Research, vol. 6. Nova Science, pp. 125–154.

Hallenbeck, P.C., Abo-Hashesh, M., Ghosh, D., 2012. Strategies for improving biological hydrogen production. Bioresour. Technol. 110, 1–9.

Heap, J.T., Kuehne, S.A., Ehsaan, M., Cartman, S.T., Cooksley, C.M., Scott, J.C., et al., 2010. The ClosTron: Mutagenesis in Clostridium refined and streamlined. J. Microbiol. Methods 80 (1), 49–55.

Horch, M., Lauterbach, L., Lenz, O., Hildebrandt, P., Zebger, I., 2012. NAD(H)-coupled hydrogen cycling: Structure-function relationships of bidirectional NiFe hydrogenases. FEBS Lett. 586 (5), 545–556.

Hu, Y.L., Ribbe, M.W., 2011. Biosynthesis of the metalloclusters of molybdenum nitrogenase. Microbiol. Mol. Biol. Rev. 75 (4), 664.

Huesemann, M.H., Hausmann, T.S., Carter, B.M., Gerschler, J.J., Benemann, J.R., 2010. Hydrogen generation through indirect biophotolysis in batch cultures of the nonheterocystous nitrogen-fixing cyanobacterium *Plectonema boryanum*. Appl. Biochem. Biotechnol. 162 (1), 208–220.

Hung, C.II., Chang, Y.T., Chang, Y.J., 2011. Roles of microorganisms other than Clostridium and Enterobacter in anaerobic fermentative biohydrogen production systems: A review. Bioresour. Technol. 102 (18), 8437–8444.

Jones, C.S., Mayfieldt, S.P., 2012. Algae biofuels: Versatility for the future of bioenergy. Curr. Opin. Biotechnol. 23 (3), 346–351.

Jung, K.W., Kim, D.H., Kim, S.H., Shin, H.S., 2011. Bioreactor design for continuous dark fermentative hydrogen production. Bioresour. Technol. 102 (18), 8612–8620.

Keskin, T., Hallenbeck, P.C., 2012a. Hydrogen production from sugar industry wastes using single-stage photofermentation. Bioresour. Technol. 112, 131–136.

Keskin, T., Hallenbeck, P.C., 2012b. Enhancement of biohydrogen production by two-stage systems: Dark and photofermentation. In: Baskar, C., Shikha Baskar, S., Dhillon, R.S. (Eds.), Biomass Conversion. Springer-Verlag, Berlin, pp. 313–340.

Keskin, T., Abo-Hashesh, M., Hallenbeck, P.C., 2011. Photofermentative hydrogen production from wastes. Bioresour. Technol. 102 (18), 8557–8568.

Kiely, P.D., Regan, J.M., Logan, B.E., 2011. The electric picnic: Synergistic requirements for exoelectrogenic microbial communities. Curr. Opin. Biotechnol. 22 (3), 378–385.

Kim, D.H., Kim, M.S., 2011. Hydrogenases for biological hydrogen production. Bioresour. Technol. 102 (18), 8423–8431.

Kontur, W.S., Noguera, D.R., Donohue, T.J., 2012. Maximizing reductant flow into microbial H-2 production. Curr. Opin. Biotechnol. 23 (3), 382–389.

Kosourov, S.N., Ghirardi, M.L., Seibert, M., 2011. A truncated antenna mutant of *Chlamydomonas reinhardtii* can produce more hydrogen than the parental strain. Int. J. Hydrogen Energy 36 (3), 2044–2048.

Kruse, O., Hankamer, B., 2010. Microalgal hydrogen production. Curr. Opin. Biotechnol. 21 (3), 238–243.

Lee, H.S., Rittmann, B.E., 2010. Significance of biological hydrogen oxidation in a continuous single-chamber microbial electrolysis cell. Environ. Sci. Technol. 44 (3), 948–954.

Lee, D.J., Show, K.Y., Su, A., 2011. Dark fermentation on biohydrogen production: Pure culture. Bioresour. Technol. 102 (18), 8393–8402.

Li, R.Y., Fang, H.H.P., 2009. Heterotrophic photofermentative hydrogen production. Crit. Rev. Environ. Sci. Technol. 39 (12), 1081–1108.

Li, R.Y., Zhang, T., Fang, H.H.P., 2011. Application of molecular techniques on heterotrophic hydrogen production research. Bioresour. Technol. 102 (18), 8445–8456.

Lin, C.Y., Wu, S.Y., Lin, P.J., Chang, J.S., Hung, C.H., Lee, K.S., et al., 2011. A pilot-scale high-rate biohydrogen production system with mixed microflora. Int. J. Hydrogen Energy 36 (14), 8758–8764.

Luo, G., Xie, L., Zhou, Q., Angelidaki, I., 2011. Enhancement of bioenergy production from organic wastes by two-stage anaerobic hydrogen and methane production process. Bioresour. Technol. 102 (18), 8700–8706.

McIntosh, C.L., Germer, F., Schulz, R., Appel, J., Jones, A.K., 2011. The NiFe -hydrogenase of the Cyanobacterium *Synechocystis* sp PCC 6803 works bidirectionally with a bias to H-2 production. J. Am. Chem. Soc. 133 (29), 11308–11319.

McKinlay, J.B., Harwood, C.S., 2010. Carbon dioxide fixation as a central redox cofactor recycling mechanism in bacteria. Proc. Natl. Acad. Sci. USA 107 (26), 11669–11675.

Melnicki, M.R., Pinchuk, G.E., Hill, E.A., Kucek, L.A., Fredrickson, J.K., Konopka, A., et al., 2012. Sustained H_2 production driven by photosynthetic water splitting in a unicellular cyanobacterium. mBio 3, 1–7.

Nicolet, Y., Fontecilla-Camps, J.C., 2012. Structure-function relationships in FeFe -hydrogenase active site maturation. J. Biol. Chem. 287 (17), 13532–13540.

Oh, Y.K., Raj, S.M., Jung, G.Y., Park, S., 2011. Current status of the metabolic engineering of microorganisms for biohydrogen production. Bioresour. Technol. 102 (18), 8357–8367.

Pant, D., Singh, A., Van Bogaert, G., Olsen, S.I., Nigam, P.S., Diels, L., et al., 2012. Bioelectrochemical systems (BES) for sustainable energy production and product recovery from organic wastes and industrial wastewaters. RSC Adv. 2 (4), 1248–1263.

Peters, J.W., Broderick, J.B., 2012. Emerging paradigms for complex iron-sulfur cofactor assembly and insertion. Annu. Rev. Biochem. 81, 429–450.

Pharkya, P., Maranas, C.D., 2006. An optimization framework for identifying reaction activation/inhibition or elimination candidates for overproduction in microbial systems. Metab. Eng. 8, 1–13.

Sabourin-Provost, G., Hallenbeck, P.C., 2009. High yield conversion of a crude glycerol fraction from biodiesel production to hydrogen by photofermentation. Bioresour. Technol. 100 (14), 3513–3517.

Scoma, A., Krawietz, D., Faraloni, C., Giannelli, L., Happe, T., Torzillo, G., 2012. Sustained H-2 production in a *Chlamydomonas reinhardtii* D1 protein mutant. J. Biotechnol. 157 (4), 613–619.

Seefeldt, L.C., Hoffman, B.M., Dean, D.R., 2012. Electron transfer in nitrogenase catalysis. Curr. Opin. Chem. Biol. 16 (1–2), 19–25.

Seol, E., Jang, Y., Kim, S., Oh, Y.K., Park, S., 2012. Engineering of formate-hydrogen lyase gene cluster for improved hydrogen production in *Escherichia coli*. Int. J. Hydrogen Energy 37 (20), 15045–15051.

Sharan, S.K., Thomason, L.C., Kuznetsov, S.G., Court, D.L., 2009. Recombineering: A homologous recombination-based method of genetic engineering. Nat. Protoc. 4, 206–223.

Show, K.Y., Lee, D.J., Chang, J.S., 2011. Bioreactor and process design for biohydrogen production. Bioresour. Technol. 102 (18), 8524–8533.

Srirangan, K., Pyne, M.E., Perry Chou, C., 2011. Biochemical and genetic engineering strategies to enhance hydrogen production in photosynthetic algae and cyanobacteria. Bioresour. Technol. 102 (18), 8589–8604.

Surzycki, R., Cournac, L., Peltier, G., Rochaix, J.D., 2007. Potential for hydrogen production with inducible chloroplast gene expression in Chlamydomonas. Proc. Natl. Acad. Sci. USA 104 (44), 17548–17553.

Tollefson, J., 2010. Hydrogen vehicles: Fuel of the future? Nature 464, 1262–1264.

Tsygankov, A.A., 2012. Hydrogen production: Light-driven processes. In: Hallenbeck, P.C. (Ed.), Microbial Technologies in Advanced Biofuels Production. Springer, New York, pp. 29–51.

Turcot, J., Bisaillon, A., Hallenbeck, P.C., 2008. Hydrogen production by continuous cultures of *Escherchia coli* under different nutrient regimes. Int. J. Hydrogen Energy 33 (5), 1465–1470.

Tyo, K.E.J., Kocharin, K., Nielsen, J., 2010. Toward design-based engineering of industrial microbes. Curr. Opin. Microbiol. 13, 255–262.

Veit, A., Akhtar, M.K., Mizutani, T., Jones, P.R., 2008. Constructing and testing the thermodynamic limits of synthetic NAD(P)H:H-2 pathways. Microbial Biotechnol. 1 (5), 382–394.

Won, S.G., Lau, A.K., 2011. Effects of key operational parameters on biohydrogen production via anaerobic fermentation in a sequencing batch reactor. Bioresour. Technol. 102 (13), 6876–6883.

Work, V.H., D'Adamo, S., Radakovits, R., Jinkerson, R.E., Posewitz, M.C., 2012. Improving photosynthesis and metabolic networks for the competitive production of phototroph-derived biofuels. Curr. Opin. Biotechnol. 23 (3), 290–297.

Yang, Y., Guoping Sun, G., Xu, M., 2011. Microbial fuel cells come of age. J. Chem. Technol. Biotechnol. 86, 625–632.

Zamboni, N., Sauer, U., 2009. Novel biological insights through metabolomics and C-13-flux analysis. Curr. Opin. Microbiol. 12, 553–558.

Metabolic Engineering of Microorganisms for Biohydrogen Production

You-Kwan Oh[*][1], *Subramanian Mohan Raj*[†][1],
Gyoo Yeol Jung[‡], *Sunghoon Park*[§]

[*]Clean Fuel Department, Korea Institute of Energy Research, Daejeon, Korea
[†]Center for Research and Development, PRIST University, Thanjavur, India
[‡]Department of Chemical Engineering & I-Bio Program, Pohang University of Science and Technology, Pohang, Korea
[§]Department of Chemical and Biomolecular Engineering, Pusan National University, Busan, Korea

INTRODUCTION

Hydrogen (H_2) is a nonpolluting renewable fuel. Biologically, H_2 can be produced via biophotolysis, photofermentation, and dark fermentation. The biophotolysis of water, carried out by green microalgae and cyanobacteria, is the most desirable, as it only requires water and sunlight (Kruse et al., 2005; Prince and Kheshgi, 2005). However, low photochemical efficiency and serious by-product (oxygen, O_2) inhibition on hydrogenase (H_2ase) (Brentner et al., 2010) are critical problems. In photofermentation, carried out by anoxygenic photosynthetic bacteria, H_2 is produced by nitrogenase in the absence of ammonium ions (NH_4^+). In this case, O_2 is not released and therefore O_2 inhibition is not an issue (Brentner et al., 2010; Harwood, 2008). However, the low H_2 production activity of nitrogenase, repression of the expression of nitrogenase by NH_4^+, and low photochemical efficiency are major drawbacks (Harwood, 2008; Koku et al., 2002). Dark fermentation by facultative and obligate anaerobic

[1]These authors contributed equally to this work.

bacteria is another method of producing H_2. It proceeds at a higher rate compared to photofermentation and photolysis (Kivisto et al., 2010; Levin et al., 2004; Ntaikou et al., 2009). However, the low H_2 yield on organic substrates, caused by the excessive formation of various by-products, is considered the major drawback (Brentner et al., 2010; Hallenbeck and Ghosh, 2009; Oh et al., 2002, 2003).

The purpose of this chapter is to update recent developments in metabolic engineering to improve microbial biohydrogen production. Major challenges of the H_2 production methods and the efforts made to deal with these challenges are covered in three separate sections: biophotolysis, photofermentation, and dark fermentation. The stability, efficiency, yield, and production rate of H_2-producing biocatalysts in the three methods are discussed.

BIOPHOTOLYSIS

General Overview

Biophotolysis is a water-splitting process occurring in biological systems. Molecular O_2 and H_2 are produced, with light as the energy source. Biophotolysis proceeds in two distinctive ways—directly and indirectly (Brentner et al., 2010). Direct biophotolysis has been best studied in the microalgae *Chlamydomonas reinhardtii*. It relies on photosystems (both PSI and PSII) and hydrogenase (Fig. 1). Absorption of light in the form of photons by PSII (680 nm) and/or PSI (700 nm) generates a strong oxidant that can oxidize water into protons, electrons/reducing equivalents, and O_2. The electrons reduce protons to form H_2, according to Equation (1) (Brentner et al., 2010).

$$2\,H_2O \underset{PS\ \&\ H_2ase}{\overset{h\nu}{\rightarrow}} 2\,H_2 + O_2 \tag{1}$$

Indirect biophotolysis is found in many cyanobacteria. Here, the electrons or reducing equivalents from water are stored as endogenous reserve carbohydrates before being used to form H_2. This means that, in indirect biophotolysis, photosynthesis for carbohydrate accumulation is coupled with dark fermentation for H_2 production [Eqs. (2a) and (2b)] (Antal and Lindblad, 2005)

$$6\,CO_2 + 6\,H_2O \underset{PS}{\overset{h\nu}{\rightarrow}} C_6H_{12}O_6 + 6\,O_2 \tag{2a}$$

$$C_6H_{12}O_6 + 6H_2O \underset{H_2ase}{\rightarrow} 12\,H_2 + 6\,CO_2 \tag{2b}$$

Improvement of Biophotolysis

Despite the advantages of just requiring light and water for H_2 production, biophotolysis seriously suffers from two major challenges: (i) high sensitivity of hydrogenase to O_2 (Brentner et al., 2010) and (ii) low-light conversion efficiency, less than 2% (Lindblad et al., 2002; Melis, 2002). Oxygen sensitivity of hydrogenase can be addressed by either engineering hydrogenase to be more resistant to O_2 or controlling the O_2 level low in the neighborhood of

[Indirect biophotolysis]

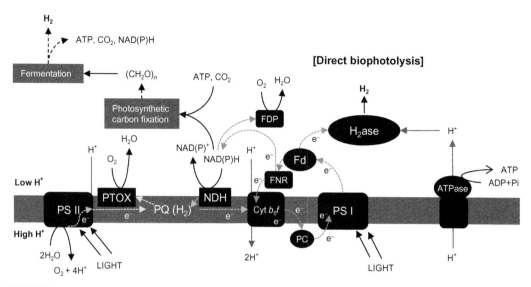

[Direct biophotolysis]

FIGURE 1 Direct and indirect biophotolysis processes of photosynthetic microorganisms. This figure is modified from Akkerman et al. (2003) and Peltier et al. (2010). The gray bar represents the lipid bilayer of the thylakoid membrane. The electron transfer reactions and directions are drawn as gray dotted arrows. Black broken lines indicate the multiple reactions. ATPase, ATP synthase; Cyt b_6f, cytochrome b_6f complex; Fd, ferredoxin; FDP, flavodiiron protein; FNR, ferredoxin NAD(P)$^+$ reductase; H$_2$ase, hydrogenase; NDH, NAD(P)H dehydrogenase; PC, plastocyanin; PQ, plastoquinones; PS, photosystem; PTOX, plastid terminal oxidase.

hydrogenase during the H_2 production reaction. The first approach, that is, making hydrogenases O_2 resistant, is a better solution and several studies have been conducted to this end (Liebgott et al., 2011; Wu et al., 2011). However, engineering of algal hydrogenases is difficult. Hydrogenases are very complex in structure, and genetic toolboxes in microalgae have not been well developed yet. The other approach, that is, controlling the O_2 level low, has been studied more extensively and some successful results have been reported (Berberoglu et al., 2008; Esper et al., 2006; Govindjee, 2002; Melis et al., 2000; Mussgnug et al., 2007; Polle et al., 2002; Surzycki et al., 2007). For example, decoupling of the hydrogenase reaction from O_2-evolving photosynthesis has been attempted by temporarily interrupting the photosynthetic processes using sulfur deprivation (Melis et al., 2000). The cultivation of *Chlamydomonas* under sulfur deprivation inhibited the synthesis of sulfur-containing polypeptides of the PSII reaction center, which eventually decreased O_2 formation. However, the algal cells did not survive for more than a few days when the sulfur became depleted (Esper et al., 2006). In another study, Surzycki et al. (2007) developed an inducible system for chloroplast gene expression in *Chlamydomonas*. This system enabled the PSII activity to be turned off and thereby anaerobic conditions could be created when needed. Under the dark cycle, O_2, which accumulated during photosynthesis, could be consumed by respiration. Furthermore, PSII could be reactivated by replenishing a minimum amount of sulfur. By repeating the dark activation–light production cycles, H_2 production in the mutant *Chlamydomonas* was prolonged by up to 15 days.

In an attempt to improve the light conversion efficiency in biophotolysis, a reduction in the size of a light-harvesting complex (LHC) has been studied (Berberoglu et al., 2008; Govindjee, 2002; Mussgnug et al., 2007; Polle et al., 2002). Polle et al. (2002) generated a mutant *C. reinhardtii* strain that had a truncated antenna size of PSII by random mutagenesis. This mutant strain showed an improved photosynthetic productivity (by 6- to 7-fold) and H_2 production yield (by 3-fold) when compared to its parental strain. The RNA interference technique has also been used to downregulate LHC in *C. reinhardtii* (Mussgnug et al., 2007). This increased the photosynthetic quantum yield (ϕPSII) from 0.256 to 0.464 (1.8-fold improvement) and decreased the sensitivity of PS to photoinhibition. Consequently, the overall photochemical efficiency improved under excess light. The specific disruption of the regulatory gene *tla1* of LHC has been shown to decrease the expression of the light-harvesting antenna and improved the photochemical efficiency in many green microalgae, including *C. reinhardtii*, *Scenedesmus obliquus*, and *Chlorella vulgaris* (Melis and Mitra, 2010).

The H_2 production efficiency in biophotolysis has also been improved by engineering the electron transport system. The electrons in biophotolysis are metabolized for the reduction of O_2 or cyclic electron transport. In one study, the electron flux to the O_2 reduction pathway in green microalgae was decreased by introducing the hydrogenase-programmed polypeptide proton channel to the thylakoid membrane (Lee and Greenbaum, 2003). This increased the proton gradient across the membrane and H_2 production efficiency as well. In another approach, the *Chlamydomonas* mutant *Stm6* was developed in which cyclic electron transfer around PSI was inhibited (Kruse et al., 2005). The H_2 production rate in the mutant strain was fivefold higher than that of its wild-type counterpart.

PHOTOFERMENTATION

General Overview

Hydrogen production by photofermentation, also called photoheterotrophic H_2 production, differs from biophotolysis in that it utilizes carbon substrates such as organic acids as electron donors. The mechanism of photofermentation has been studied extensively in purple nonsulfur bacteria (PNSB) (Brentner et al., 2010; Hallenbeck and Benemann, 2002; Harwood, 2008; Koku et al., 2002; McKinlay and Harwood, 2010). Under an anaerobic condition, the electrons scavenged from organic acids are transferred to oxidized ferredoxin (Fd_{ox}) through a series of membrane-bound, electron transport carrier molecules (McKinlay and Harwood, 2010; Fig. 2). The electrons in reduced ferredoxin (Fd_{rd}) are used primarily to reduce molecular dinitrogen (N_2) to ammonia (NH_3) by the action of nitrogenase (N_2ase) [Eq. (3)]. However, in the absence of N_2, nitrogenase catalyzes the reduction of protons to produce H_2 [Eq. (4)] (Brentner et al., 2010; Harwood, 2008). The electrochemical gradient generated during the electron transfer is used for ATP synthesis.

$$N_2 + 8\,H^+ + 8\,e^- + 16\,ATP \xrightarrow[N_2ase]{h\nu} 2\,NH_3 + H_2 + 16\,ADP + 16\,P_i \tag{3}$$

$$8\,H^+ + 8\,e^- + 16\,ATP \xrightarrow[N_2ase]{h\nu} 4\,H_2 + 16\,ADP + 16\,P_i \tag{4}$$

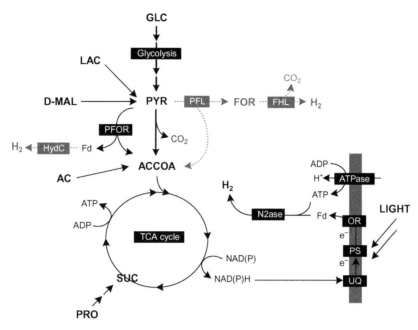

FIGURE 2 Photoheterotrophic H_2 production metabolic pathway from organic acids and glucose in purple nonsulfur bacteria. Glucose uptake is dependent on species. Diagram was modified from Koku et al. (2002) and McKinlay and Harwood (2010). New pathways introduced for production of H_2 under a dark fermentative condition are represented by dotted gray lines (Kim et al., 2008a). AC, acetate; ACCOA, acetyl-CoA; D-MAL, D-malate; Fd, ferredoxin; FHL, formate:hydrogen lyase; FOR, formate; GLC, glucose; HydC, ferredoxin dependent hydrogenase; LAC, lactate; N2ase, nitrogenase; OR, oxidoreductase; PFL, pyruvate:formate lyase; PFOR, pyruvate:ferredoxin oxidoreductase; PRO, propionate; PS, photosystem; PYR, pyruvate; SUC, succinate; UQ, ubiquinone.

The reduction of protons or N_2 requires a large amount of ATP to overcome the high activation energy (Fe-protein, $E_o'r$-250 mV; Fe-protein-MgATP, $E_o'r$-400 mV). Also, two ATPs are required to transfer every electron from the Fe-protein to nitrogenase (Harwood, 2008). The metabolic network through membrane-bound enzymes and electron carriers has not been fully identified. To understand and engineer the metabolism of photofermentation, extensive studies including x-omics, biochemical, and genetic characterization are required.

There are several issues with photofermentative H_2 production. First, the rate of H_2 production is very low, as the catalytic turnover of typical nitrogenase is only ~ 5 s^{-1} (Hallenbeck and Benemann, 2002; Harwood, 2008). Reoxidation of the produced H_2 by uptake hydrogenase is another problem. Furthermore, the requirement of a high cellular ATP level for efficient H_2 production is considered another important problem. This section summarizes metabolic engineering approaches for photofermentation with an emphasis on three key aspects: photosystem, enzymes, and carbon metabolism.

Photosystem

Hydrogen production requires a large amount of ATP. Purple nonsulfur bacteria synthesize ATP by anoxygenic cyclic photophosphorylation, which involves a photosystem and electron transferring proteins (Fig. 2). Light is collected by two light-harvesting complexes referred to as "core" and "peripheral" complexes, channeled into the reaction complex, which initiates a cyclic electron flow through several carriers. During the photosynthetic process, a proton gradient is generated that is utilized for ATP synthesis (Harwood, 2008; Koku et al., 2002; McKinlay and Harwood, 2010). The rate of ATP-dependent H_2 production by nitrogenase is directly proportional to light intensity up to a certain level that does not inhibit cell growth (Koku et al., 2002). In many cases, the self-shading of cells due to their light-harvesting complexes reduces light penetration and therefore lowers H_2 production (Kim et al., 2006a).

Many researchers have attempted to improve the light-harvesting efficiency of PNSB. For example, Kondo et al. (2002) generated a mutant of *Rhodobacter sphaeroides* RV using ultraviolet irradiation, which absorbs less light at a wide range of wavelengths (350–1000 nm). The growth of the mutant strain was similar to that of the wild type, but the H_2 production rate was improved by 1.5-fold when cultivated in a plate-type photobioreactor. Similarly, the *pucBA*-deleted mutant of *R. sphaeroides* KCTC 12085, lacking the B800-850 light-harvesting complex, exhibited increased H_2 production by approximately 2-fold (Kim et al., 2006a). Eltsova et al. (2010) developed a recombinant strain of *R. sphaeroides* pRK pur DD13, lacking a peripheral light-harvesting antenna complex, and confirmed that the culture of the mutant showed a higher volumetric H_2 production rate. Ozturk et al. (2006) reported H_2 production from various *Rhodobacter* species harboring the genetically modified electron carrier cytochromes or lacking cyt cbb_3 oxidase or quinol oxidase. Mutant strains with modified cytochromes exhibited a decreased H_2 production yield by ~4-fold. The H_2 production rate with the cbb_3 deletion mutant increased by 1.8-fold compared to that of the wild-type strain. The terminal electron acceptor cytochrome cbb_3 oxidase of *Rhodobacter* species serves as a redox signal to the RegB/RegA global regulatory system. This regulatory system indirectly activates the synthesis of nitrogenase by activating expression of the *nifA2* gene, which encodes one of the two functional copies of the NifA transcriptional activator for nitrogenase structural genes. In addition, the RegA protein directly represses the structural gene expression of uptake hydrogenase. It was suggested that inactivation of the cyt cbb_3 oxidase eliminated its inhibitory effects on the RegB/RegA system, which consequently induced nitrogenase but inhibited uptake hydrogenase (Ozturk et al., 2006).

Enzymes

Two enzymes—nitrogenase and hydrogenase—play key roles in the H_2 metabolism of PNSB. Nitrogenase produces H_2 under nitrogen-deficient conditions [Eq. (4)], whereas hydrogenase oxidizes H_2 to recycle electrons, protons, and ATP for energy metabolism (Brentner et al., 2010; Koku et al., 2002; Mathews and Wang, 2009). Hydrogen production in PNSB is often limited by low nitrogenase activity (Hallenbeck and Benemann, 2002; Harwood, 2008). To improve cellular nitrogenase activity, this enzyme can be engineered to have a higher catalytic turnover number or enhancement of the cellular expression level of the enzyme can be attempted. Most of the studies thus far have focused on improving the expression level of nitrogenase rather than engineering it. The expression of nitrogenase is strictly controlled at the transcriptional and post-transcriptional levels in response to the availability of a nitrogen source, particularly NH_4^+. In the presence of NH_4^+, two PII-like proteins, GlnB and GlnK, activate DraT, which inhibits nitrogenase by ADP-ribosylation. The two PII-like proteins also control the activity of NifA, a transcriptional activator of nitrogenase structural genes. Furthermore, under nitrogen-replete conditions, GlnB inhibits a sensor kinase, NtrB, which is involved in dephosphorylation of the transcriptional activator NtrC and subsequently inactivates the transcription of *nifA* (Kim et al., 2008b). Even at 20 μM of NH_4^+, nitrogenase is fully repressed and H_2 production becomes negligible (Koku et al., 2002).

Many efforts have been directed toward preventing a repression of the expression of nitrogenase. For example, Drepper et al. (2003) reported that in a *glnB-glnK* double mutant of *R. capsulatus* the synthesis of nitrogenase, and thus its enzymatic activity, is enhanced greatly during cultivation in the presence of 200 μM NH_4^+. H_2 production of the mutant strain was 1.5-fold higher than that of the wild-type strain. Kim et al. (2008b) reported that in another PNSB, *R. sphaeroides* KCTC 12085, the deletion of both *glnB* and *glnK* genes alleviated the repression of nitrogenase genes by NH_4^+ and hence improved H_2 production. The effect of the deletion of *glnA1*, encoding glutamine synthase (a key enzyme in the assimilation pathways of NH_4^+), on nitrogenase activity and H_2 production was also studied in *R. sphaeroides* 6016 in the presence of glutamine and NH_4^+ as nitrogen sources (Li et al., 2010). The mutant *R. sphaeroides* produced a substantial amount of H_2 at relatively high NH_4^+ concentrations (15 to 40 mM), as the elimination of glutamine synthase prevented NH_4^+ assimilation from culture media *in vivo*.

Hydrogen production in PNSB can also be improved by deleting uptake hydrogenases (Hup). According to Rey et al. (2006), H_2 production in Hup-negative *Rhodopseudomonas palustris* was improved greatly under nitrogen-fixing conditions with malate as the carbon source. Two mutants, HupV$^-$ and HupS$^-$, produced H_2 at 120 and 110 μmol H_2/mg protein, respectively, while the wild type only produced less than 1 μmol H_2/mg protein. Other *Rhodobacter* species, such as *R. capsulatus* MT1131 (Ozturk et al., 2006) and *R. sphaeroides* RV (Franchi et al., 2004), also showed an improved rate of H_2 production, by $\approx$36%, when Hup was disrupted. Liu et al. (2010) confirmed a high H_2 production yield in two mutant strains of *R. sphaeroides* 6016—one without uptake hydrogenase HupSL and the other lacking its positive regulator protein HupR. Kars et al. (2009) reported a similar result with the *hupSL$^-$* mutant of *R. sphaeroides* O.U. 001. In another study, Kim et al. (2006b) tested the double-mutant *R. sphaeroides* KD131, which lacks both uptake hydrogenase (Hup) and poly-β-hydroxybutyrate (PHB) synthase (Phb). This mutant strain showed a much higher H_2 production of 3.34 ml H_2/mg cell (on a dry weight basis) than that of the wild-type parent strain (1.32 ml H_2/mg cell). Poly-β-hydroxybutyrate synthase, which is responsible for the accumulation of carbon storage material PHB in many PNSBs, is a major competitor to nitrogenase for electrons (Harwood, 2008; Koku et al., 2002). It was suggested that the reduction of PHB synthesis is an efficient method for increasing H_2 yield in PHB-synthesizing PNSB.

Carbon Metabolism and Metabolic-Flux Analysis

Purple nonsulfur bacteria use a wide variety of carbon substrates, including organic acids and glucose (Fig. 2), but they do not produce H_2 from the fermentation of these carbon substrates without light (Harwood, 2008; Koku et al., 2002; McKinlay and Harwood, 2010).

The incorporation in PNSB of fermentative H_2 production pathways, which are naturally present in most fermentative microorganisms, is beneficial because it enables the recombinant PNSB to produce H_2 under both light and dark conditions. Introduction of such a fermentative H_2 production pathway into the PNSB *R. sphaeroides* KCTC 12085 has been reported. Kim et al. (2008a) cloned and successfully expressed several genes from *Rhodospirillum rubrum* encoding pyruvate:formate lyase (PFL), formate:hydrogen lyase (FHL), a transcriptional activator for FHL, ferredoxin (Fd)-dependent hydrogenase (HydC), and other maturation proteins (see Fig. 2). The recombinant produced up to 0.36 mol H_2/mol glucose when cultured under dark anaerobic conditions. Interestingly, under photoheterotrophic conditions, the H_2 production yield of this recombinant was also elevated to 4 mol H_2/mol glucose, which is twofold higher than that of its parental wild-type strain (2 mol H_2/mol glucose). They claimed that with the recombinant strain, 2 mol H_2/mol glucose was produced by the nitrogenase and the additional 2 mol H_2/mol glucose was achieved by the newly synthesized H_2 pathways via FHL (1 mol H_2/mol glucose) and HydC (1 mol H_2/mol glucose). This study demonstrates that expression of fermentative H_2 production pathways can improve the H_2 production capability of PNSB.

In silico metabolic pathway models for PNSB have been developed. In one study, Golomysova et al. (2010) developed a model that consisted of 314 biochemical reactions and observed that the prediction power of the model was carbon source dependent. With lactate as the carbon substrate, the model well predicted the H_2 production rate (predicted 5.5 mmol H_2/g cell/h vs measured 5.3 mmol H_2/g cell/h). However, with malate (1.42 mmol H_2/g cell/h vs 1.11 mmol H_2/g cell/h) and acetate (1.25 mmol H_2/g cell/h vs 0.63 mmol H_2/g cell/h) as substrates, a substantial discrepancy was observed. Based on this model, Golomysova et al. (2011) also attempted mutation studies to get *in silico* mutant strains that can produce H_2 with a higher rate. They found that mutation in such genes as the ones encoding ribulose-1,5-bisphosphate carboxylase/oxygenase (RuBisCO), acetyl-CoA carboxylase, threonine:ammonia lyase, PFL, and pyruvate dehydrogenase was beneficial in improving the H_2 production rate. In another study, Hadicke et al. (2011) developed a model with 143 biochemical reactions. They reported that an *in silico* mutant devoid of the RuBisCO enzyme showed higher H_2 yields for several substrates, such as succinate, malate, and fructose. Although several models have been studied, they seem to have limitations in simulating the behavior of PNSB. More elaborate models and iterative refinement of the model(s) are required. To this end, various x-omics, biochemical, and physiological data are needed.

DARK FERMENTATION

General Overview

Dark fermentative H_2 production is considered the most practical among the various biological H_2 production methods at present (Brentner et al., 2010; Levin et al., 2004). It differs from the other processes in that it utilizes organic substrates as the sole source of energy and electrons. Three types of biochemical reactions are currently known to generate H_2 in dark fermentation. The first one, which is typical in *Escherichia coli* and Enterobacteriaceae (Hallenbeck and Benemann, 2002; Hallenbeck and Ghosh, 2009; Kim et al., 2008c, 2009),

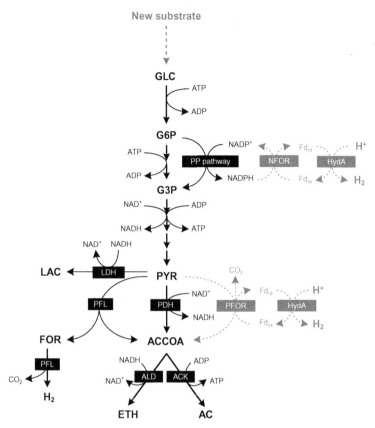

FIGURE 3 Fermentative H_2 production metabolic pathway from glucose in *E. coli* and enterobacteria. New pathways introduced for improving H_2 production are represented by dotted gray lines (Akhtar and Jones, 2009; Veit et al., 2008). AC, acetate; ACCOA, acetyl-CoA; ACK, acetate kinase; ALD, alcohol dehydrogenase; ETH, ethanol; Fd_{ox}, oxidized ferredoxin; Fd_{rd}, reduced ferredoxin; FOR, formate; GLC, glucose; G3P, glyceraldehyde-3-phosphate; G6P, glucose-6-phosphate; HydA, ferredoxin-dependent hydrogenase; LAC, lactate; LDH, lactate dehydrogenase; PP pathway, pentose phosphate pathway; NFOR, NAD(P)H:ferredoxin oxidoreductase; PDH, pyruvate dehydrogenase; PFL, pyruvate:formate lyase; PFOR, pyruvate:ferredoxin oxidoreductase; PYR, pyruvate.

employs two major enzymes: (1) pyruvate:formate lyase and (2) formate:hydrogen lyase (Fig. 3). Pyruvate formed via the Embden–Meyerhof–Parnas (EMP) pathway is split into acetyl-CoA and formate by PFL under anaerobic conditions. Formate is then cleaved to H_2 and CO_2 by FHL [Eqs. (5a) and (5b)] (Hallenbeck, 2005).

$$\text{Pyruvate} + \text{CoA} \underset{\text{PEL}}{\rightarrow} \text{Acetyl} - \text{CoA} + \text{Formate} \qquad (5a)$$

$$\text{Formate} \underset{\text{FHL}}{\rightarrow} H_2 + CO_2 \qquad (5b)$$

The second type of H_2-producing reaction involves pyruvate:ferredoxin oxidoreductase (PFOR) and Fd-dependent hydrogenase (HydA) (Hallenbeck, 2005). Pyruvate:ferredoxin oxidoreductase catalyzes the oxidative decarboxylation of pyruvate to form acetyl-CoA and

CO_2 under anaerobic conditions. In this reaction, the electrons are first transferred to Fd_{ox}, an electron acceptor with a highly negative potential (E_o' −420 mV) (Chabriere et al., 2001). The electrons in Fd_{rd} are then transferred to protons to form H_2 by HydA, according to Equations (6a) and (6b). This type of H_2-producing reaction is typical for *Clostridium* species (Mathews and Wang, 2009).

$$\text{Pyruvate} + \text{CoA} + 2\,Fd_{ox} \underset{\text{PFOR}}{\leftrightarrow} \text{Acetyl} - \text{CoA} + CO_2 + 2\,Fd_{rd} \qquad (6a)$$

$$2\,Fd_{rd} + 2\,H^+ \underset{\text{HydA}}{\rightarrow} 2\,Fd_{ox} + H_2 \qquad (6b)$$

The third type of reaction utilizes NAD(P)H to evolve H_2. This reaction is catalyzed by two major enzymes—NAD(P)H:ferredoxin oxidoreductase (NFOR) and HydA (Wang et al., 2010). In this reaction, Fd_{ox} is reduced by NAD(P)H, which is formed during carbon metabolism. The electrons in Fd_{rd} are then transferred to protons to form H_2 by HydA [Eqs. (7a)–(7c)]. This H_2-producing reaction is reported to exist in many thermophilic bacteria and some *Clostridium* species (Silva et al., 2000).

$$\text{Glucose} + 2\,NAD^+ \underset{\text{EMP}}{\rightarrow} 2\,\text{pyruvate} + 2\,\text{NADH} \qquad (7a)$$

$$2\,\text{NADH} + 4\,Fd_{ox} \underset{\text{NFOR}}{\rightarrow} 2\,NAD^+ + 4\,Fd_{rd} \qquad (7b)$$

$$4\,Fd_{rd} + 4\,H^+ \underset{\text{HydA}}{\rightarrow} 4\,Fd_{ox} + 2\,H_2 \qquad (7c)$$

Dark fermentative H_2 production is often compared with fermentative methane (CH_4) production. Methane is produced in the anaerobic digestion of organic wastes, and its production processes have been well established commercially. Compared to CH_4, H_2 is more valuable as an energy source and chemical feedstock. Nevertheless, the H_2 production processes by dark fermentation are less well developed than CH_4 production processes due to a number of drawbacks (Brentner et al., 2010; Hallenbeck and Ghosh, 2009). Apart from the difficulties in operating the reactor system stably, H_2 production gives a low yield on sugars. According to the stoichiometry of glucose oxidation, 12 mol of H_2 can be generated from 1 mol of glucose. However, the maximum yield in dark fermentation is 4 mol H_2/mol glucose, that is, only 33% of the stoichiometric maximum. The low H_2 yield is linked to microbial metabolism. The H_2 is to be produced during carbon conversion to various alcohols and organic acids, not from the simple oxidation reactions of glucose to CO_2 and water (Levin et al., 2004). Accumulation of the alcohols and acids beyond a certain level inhibits cell growth, which further decreases the H_2 yield. In order to improve the H_2 production yield in dark fermentation, the production of H_2 from NAD(P)H directly or indirectly via Fd has been suggested (Akhtar and Jones, 2009; Oh et al., 2008; Veit et al., 2008). However, this route is significantly affected by several factors, such as H^+ concentration, NADH/NAD^+ ratio, H_2 partial pressure, and temperature. Moreover, because NAD(P)H- or Fd-dependent hydrogenases are reversible, the forward reaction leading to H_2 formation is inhibited by H_2 accumulation. Compared to CH_4 fermentation, limited substrate availability is another problem encountered with dark H_2 fermentation. For example, lignocellulosic biomass, the largely available plant biomass, and starch are not used readily for H_2 production by most H_2-producing bacteria. This section discusses the current metabolic engineering efforts to improve dark H_2 fermentation.

Metabolic Engineering for Extending Substrate Utilization

The extension of substrate utilization is an important topic in metabolic engineering for H_2-producing microorganisms. Compared to other H_2 production processes, the dark H_2 production process can utilize a wide variety of substrates, including wastes. However, except for glucose, the use of other biomasses, particularly cellulosic biomass, has not been studied extensively so far. Cellulose and hemicellulose need to be hydrolyzed to fermentable sugars before they can be utilized by most H_2 producers. A separate hydrolysis is costly and often results in the generation of inhibitory compounds such as furfurals and/or hydroxyl methyl furfurals (Abril and Abril, 2009). It is beneficial if the H_2-producing microorganisms can directly utilize the cellulosic biomass without a pretreatment for hydrolysis. In the case of starch, which is often found in agricultural products or wastes, hydrolysis is relatively easy compared to that of cellulosic biomass. However, the incorporation of amylase activity in H_2-producing microorganisms can improve the economic feasibility of fermentative H_2 production processes.

There are several extreme thermophilic bacteria, clostridium species, and fibrolytic bacteria that can produce H_2 directly from cellulose, hemicellulose, and/or starch (Geng et al., 2010; Kongjan and Angelidaki, 2010; Ntaikou et al., 2009; Verhaart et al., 2010). All of these bacteria share the same feature in that cellulolytic and/or starch hydrolytic enzymes are expressed along with various hydrogenases when the polymeric biomass is present in growth media. However, the use of these bacteria often suffers from low H_2 production rates and poor bacterial growth (Geng et al., 2010). The limited availability of a genetic toolbox is the major problem to address these issues with these native H_2 producers.

The extension of substrate utilization has been studied actively with *E. coli*. In one study, a recombinant *E. coli* expressing *amyE*, encoding an α amylase, was developed and the direct production of H_2 from soluble starch was demonstrated (Akhtar and Jones, 2009). In another study, a cell-free system was investigated that was composed of 13 enzymes comprising the "synthetic starch hydrolytic pathway" (Zhang et al., 2007): starch phosphorylase yielding glucose-1-phosphate (G-1-P) from starch; phosphoglucomutase, which converts G-1-P to glucose-6-phosphate; 10 enzymes of the pentose phosphate pathway; and the enzyme(s) required for H_2 production from NADPH. This cell-free system could produce 5.19 mol H_2/mol glucose equivalent of the starch consumed. The same group also reported the production of H_2 from cellobiose using the cell-free enzymes (Ye et al., 2009). They produced 11.2 mol H_2 and 5.64 mol CO_2 from 1 mol of an anhydroglucose unit of cellobiose. However, the cell-free system is not practical, as the cost of the enzymes is high and the reaction is slow. The reconstitution of similar enzyme systems in *E. coli*, along with extracellular hydrolyzing enzymes, will be very interesting.

Metabolic Engineering of H_2-Producing Native Pathways in Dark Fermentation

The metabolic engineering of native H_2-producing pathways focuses mainly on the improvement of H_2 production yields on carbon substrates. This includes the overexpression of several enzymes and redirection of the carbon flux by eliminating competitive reactions in H_2 production pathways. With facultative anaerobic bacteria that utilize PFL and FHL, the maximum achievable theoretical yield is 2 mol H_2/mol glucose. In strict anaerobic microorganisms such as *Clostridium* sp., where H_2 production depends on NFOR and PFOR, the theoretical maximum is 4 mol H_2/mol glucose. The maximum yield in strict anaerobes is achieved when acetate is produced as the only by-product. When other by-products, such as butyrate, propionate, lactate, and alcohols, are formed extensively, the H_2 yield decreases below 2 mol H_2/mol glucose (Levin et al., 2004). Therefore, the metabolic engineering of these native H_2 production pathways targets the achievement of yields of 2 or 4 mol H_2/mol glucose.

Several extreme thermophilic bacteria can naturally produce H_2 at a high yield, from 3.3 to 4.0 mol H_2/mol glucose (Raj et al., 2012; Verhaart et al., 2010). These bacteria use both NFOR and PFOR for H_2 production. From a thermodynamic perspective, the H_2 production from NAD(P)H is unfavorable, but the high H_2 production yield near the theoretical maximum of 4 mol H_2/mol glucose indicates that NFOR and HydA function efficiently in some thermophilic bacteria at elevated temperatures (Kanai et al., 2005; Munro et al., 2009). It was thus suggested that a high temperature is beneficial for obtaining a high H_2 production yield (Raj et al. 2012; Verhaart et al., 2010). However, with most extreme thermophilic bacteria, the volumetric productivity is very low, as the cell density is low compared to mesophilic bacteria (Hallenbeck and Ghosh, 2009). The high partial pressure of H_2 often reduces H_2 production yields in these extremophiles by shifting the metabolic pathway toward the formation of other reduced by-products such as lactate, ethanol, acetone, and butanol (Levin et al. 2004). The energy cost for maintaining such high temperatures should also be justified for the economic viability of the thermophilic process (Oh et al., 2004; Raj et al., 2012). Although metabolic engineering for improving H_2 production yields with these extremophiles is difficult, Sato and co-workers (2005) developed and demonstrated methods for gene cloning and knockout in *Thermococcus kodakaraensis* KOD1. The KOD1 could produce H_2 at 3.3 mol H_2/ mol glucose equivalent during the continuous culture in starch medium (Kanai et al., 2005).

Some metabolic engineering studies have been reported for *Clostridium* sp., such as *C. paraputrificum* and *C. tyrobutyricum*. For example, a [FeFe]-hydrogenase (HydA) was overexpressed in *C. paraputrificum* M-21 (Morimoto et al., 2005), which improved the H_2 yield to 2.4 mol from 1.4 mol/mol glucose. Jo et al. (2010) also reported that *C. tyrobutyricum* JM1 with overexpressed hydrogenase (HydA) showed an improved H_2 yield (1.8 mol H_2/mol glucose) compared to the parental strain (1.2 mol H_2/mol glucose). Liu et al. (2006) inactivated *ack*, which encodes the acetate kinase of *C. tyrobutyricum*, and increased the H_2 production yield by 1.5-fold compared with the wild-type strain. A universal gene knockout system (ClosTron) for *Clostridium* sp. such as *C. acetobutylicum*, *C. difficile*, *C. botulinum*, and *C. sporogenes* was developed by Minton's group (Heap et al., 2007, 2010). Combination of this ClosTron and other gene expression systems in *Clostridium* sp. is expected to improve our understanding and engineering on the carbon and H_2 production metabolic pathways.

With *E. coli*, several research groups have engineered the FHL complex and carbon metabolic pathways (Table 1). Yoshida et al. (2005) genetically modified the *E. coli* strain by inactivating the FHL repressor (*hycA*), the negative regulator for FHL. This *hycA* deletion mutant was further engineered by overexpressing FHL and its activator genes (*fhlA*). This strain showed the highest volumetric productivity of 300 liter H_2/liter/h in a high cell density culture (93 g dry weight/liter) with formate as the substrate. This study seems to be promising, provided that inexpensive formate becomes available. In a consecutive study, Yoshida et al. (2006) improved the H_2 yield in *E. coli* to 1.82 mol from 1.08 mol/mol of glucose by disrupting *ldhA* encoding lactate dehydrogenase and *frdBC* encoding fumarate dehydrogenase. In another report, Kim et al. (2009) showed the sequential improvement of the H_2 yield by

TABLE 1   Metabolic engineering of *E. coli* for improvement of H_2 production

E. coli strain	Metabolic engineering approach	H$_2$ production from glucose		References
		Yield (mol H$_2$/mol glucose)	Rate (mmol H$_2$/g cell/h)	
W3110	Wild-type		*ca.* 90 [a]	Yoshida
W3110 $\Delta hycA/fhlA$	Inactivate FHL repressor and overexpress FHL		*ca.* 260 [a]	et al., 2005
W3110	Wild-type	1.08	9.5	Yoshida
W3110 $\Delta ldhA$ $\Delta frdBC$	Inactivate lactate dehydrogenase and fumarate dehydrogenase	1.82	13.4	et al., 2006
BW25113	Wild-type	1.20	30.6 [a]	Kim
SH1 (BW25113 $\Delta hycA$)	Inactivate negative regulator for FHL	1.17	58.2 [a]	et al., 2009
SH2 (SH1 $\Delta hyaAB$)	Inactivate uptake hydrogenase 1	1.37	46.2 [a]	
SH3 (SH2 $\Delta hybBC$)	Inactivate uptake hydrogenase 2	1.48	49.2 [a]	
SH4 (SH3 $\Delta ldhA$)	Inactivate lactate dehydrogenase	1.61	45.6 [a]	
SH5 (SH4 $\Delta frdAB$)	Inactivate fumarate reductase	1.80 / 2.11 [b]	52.2 [a]	
BW25113/pCA24N	Wild-type containing empty vector (pCA24N)	0.47	3.7 [c]	Maeda et al., 2008
BW25113 $\Delta hyaB$ $\Delta hybC$ $\Delta hycA$ $\Delta fdoG$/pCA24N-FhlA	Inactivate uptake hydrogenases and overexpress FHL complex	0.70	12 [c]	
BL21(DE3) pA pYdbK	Wild-type harboring *E. coli* ydbK encoding a putative pyruvate:flavodoxin/ferredoxin-oxidoreductase	0.67	-	Akhtar and Jones, 2009
BL21(DE3) $\Delta iscR$ pAF pYdbK	Inactivate negative transcriptional regulator of the *isc* operon and express Fd-dependent hydrogenase (*hydA*) of *C. acetobutylicum*, maturation factors of *hydA*, and [4Fe-4S]-ferredoxin of *C. pasteurianum*	1.88	-	

- Not determined.
[a] Measured using formate as a substrate.
[b] Measured under low H$_2$ partial pressure.
[c] Measured based on protein (mmol H$_2$/g protein/h).

engineering of the FHL complex and carbon metabolism. By disrupting the gene *hycA*, a two-fold improvement in FHL activity was observed. The subsequent deletion of two uptake hydrogenases (1, *hya*; and 2, *hyb*) improved the H$_2$ production yield from 1.2 to 1.48 mol/mol of glucose, indicating that the activity of uptake hydrogenases is critically important under anaerobic, glucose fermentation conditions. After disrupting the *ldhA* and *frdAB* genes in addition to *hycA*, *hya*, and *hyb*, the H$_2$ yield was improved to 1.8 mol H$_2$/mol glucose under high H$_2$ partial pressure. When the same multiple mutated strain was subjected to glucose fermentation under low H$_2$ partial pressure, a yield of 2.11 mol H$_2$/mol glucose was obtained. This is slightly higher than the theoretical maximum achievable with *E. coli* strains relying on the formate-dependent pathway. In a similar approach, Maeda et al. (2008) improved the H$_2$ production yield in the *E. coli* strain BW25113 $\Delta hyaB$ $\Delta hybC$ $\Delta hycA$ $\Delta fdoG$/pCA24N-FhlA by overexpressing the FHL complex and inactivating the uptake hydrogenases. Seol et al. (2010) showed a high volumetric productivity of 2.4 liter H_2/liter/h using immobilized cells of the genetically engineered recombinant *E. coli* SH5, which has multiple deletion mutations in uptake hydrogenases (*hya* and *hyb*), *hycA*, and the carbon metabolic pathways (*ldhA* and *frdAB*). Also, they demonstrated that whole-cell FHL activity could be improved more than threefold by increasing the expression of formate dehydrogenase (FdhH) and disrupting *iscR*, a negative regulator of the iron–sulfur cluster in the recombinant *E. coli* SH5 (Seol et al., 2012). The increase in FHL activity was accompanied by an increase in the activity of its member enzymes, such as HycE (hydrogenase 3) and/or FdhH.

Inhibition of hydrogenases by O_2 is a critical issue in H_2 production, and it is known that [NiFe]-hydrogenases are more resistant to O_2 than [FeFe]-hydrogenases (Frey, 2002). Although the physiological function of [NiFe]-hydrogenase 1 has been considered an uptake hydrogenase, Kim et al. (2010) suggested that [NiFe]-hydrogenase 1 of an *E. coli* K strain can function as an evolving hydrogenase when expressed in *E. coli* BL21, which is devoid of H_2-producing capability. They observed that the recombinant *E. coli* BL21 expressing [NiFe]-hydrogenase 1 produced H_2 at 12.5 ml H_2/liter/h under microaerobic conditions. Furthermore, they reported that, even under the atmospheric condition, the purified recombinant [NiFe]-hydrogenase 1 displayed an active H_2 production at $\sim$12 nmol H_2/mg protein/min. This O_2-resistant property of hydrogenase 1 is expected to have a significant impact in H_2 production research (Kim et al., 2010).

In most Enterobacteriaceae members, H_2 is produced via a formate-dependent pathway that involves both PFL and FHL (Hallenbeck and Benemann, 2002; Hallenbeck and Ghosh, 2009; Kim et al., 2008c, 2009). However, in *Enterobacter aerogenes* sp., an additional NADH-dependent pathway via putative, membrane-bound hydrogenase has been proposed to be functional (Lu et al., 2010; Nakashimada et al., 2002; Zhang et al., 2005, 2009). Nakashimada et al. (2002) showed NADH-dependent H_2 formation in the cell-free extract (especially membrane fraction) of *E. aerogenes* AY-2 and suggested the presence of such a NADH-dependent H_2 production pathway. Zhao et al. (2009) developed recombinant *E. aerogenes* IAM1183 strains by disrupting *hycA*, encoding FHL repressor, and/or *hybO*, encoding the small subunit of uptake hydrogenase, and observed that the mutant strains had improved H_2 production yields (*ΔhycA*, 0.73; *ΔhybO*, 0.78; *ΔhycA ΔhybO*, 0.83 mol H_2/mol glucose) compared to the parental strain (0.65 mol H_2/mol glucose). Zhang et al. (2009) suggested that the exogenous addition of NADH or NAD^+ could change the intracellular redox state of *E. aerogenes* and, subsequently, the H_2 production through formate and NADH pathways. In a consecutive study, Lu et al. (2010) improved the H_2 yield in *E. aerogenes* IAM 1183 to 1.70 from 0.91 mol H_2/mol glucose by inactivating NADH-consuming lactate dehydrogenase and overexpressing NAD^+-dependent formate dehydrogenase. However, detailed genetic and biochemical characterizations for NADH-dependent H_2 production machineries in *E. aerogenes* sp. such as membrane-bound hydrogenase(s) and electron carrier protein(s) have not been reported and wait further investigation.

Incorporation of Nonnative Pathways for H_2 Production in Dark Fermentation

The incorporation of nonnative pathways has been studied to improve the yield of H_2 production in *E. coli* (see Fig. 3). Here, the target of metabolic engineering is the achievement of yields of more than $2\,mol\,H_2/mol$ glucose by utilizing Fd-dependent or NAD(P)H-dependent hydrogenases. In order to demonstrate H_2 production by Fd-dependent hydrogenase in *E. coli*, Akhtar and Jones (2009) constructed an *E. coli* BL21(DE3) $\Delta iscR$ recombinant strain that expresses an Fd-dependent hydrogenase of *C. acetobutylicum* ([FeFe]-hydrogenase, *hydA*), maturation factors of *hydA* (*hydF*, *hydE*, and *hydG*), [4Fe-4S]-ferredoxin of *C. pasteurianum*, and a putative pyruvate:flavodoxin/ferredoxin-oxidoreductase (*ydbK* of *E. coli*). *E. coli* BL21(DE3) was selected as the host strain as it is devoid of any H_2 production ability (Veit et al., 2008). Akhtar and Jones (2009) successfully expressed various genes related to the Fd-dependent H_2 production in *E. coli* BL21(DE3) and demonstrated the production of H_2 by the recombinant at a yield of $1.88\,mol\,H_2/mol$ glucose (Table 1). Deletion of the *iscR* gene, encoding a negative transcriptional regulator of the *isc* operon (encoding genes responsible for Fe-S protein assembly), enhanced hydrogenase activity and the H_2 production rate (Akhtar and Jones, 2008). In another study, Veit et al. (2008) constructed an NAD(P)H-dependent H_2-producing pathway in *E. coli* BL21(DE3) (Fig. 3). The recombinant expressed a ferredoxin-dependent hydrogenase and [4Fe-4S]-ferredoxin along with two different NFORs—one from *Bacillus subtilis* and the other from *Chlorobium tepidum*. The reconstructed pathway was functional, but NAD(P)H-dependent H_2 production was severely inhibited by H_2. Furthermore, the yield of H_2 production was as low as $0.04\,mol\,H_2/mol$ glucose when assuming that 2 mol of NAD(P)H is produced per mole of glucose. Veit et al. (2008) suggested that both of the NFORs used in their studies were not efficient as they might not be involved with H_2 metabolism in the original microorganisms (Seo et al., 2004). It was also suggested that the low NFOR-dependent H_2 yield could be partially attributed to the low NAD(P)H/NAD(P)$^+$ ratio (Veit et al., 2008).

Agapakis et al. (2010) reported that H_2 production via Fd-dependent hydrogenase can be improved by manipulating the interaction between hydrogenase and Fd via protein surface engineering in *E. coli*. Various synthetic H_2-producing electron circuits containing Fd-dependent hydrogenases (from *C. acetobutylicum*, *C. saccharobutylicum*, *C. reinhardtii*, and *Shewanella oneidensis*), ferredoxins (from *C. acetobutylicum*, *Spinacia olearcea*, and *Zea mays*), and PFORs (from *C. acetobutylicum*, *Desulfovibrio africanus*, and *E. coli*) were constructed in *E. coli* BL21(DE3) that had multiple deletions in uptake hydrogenases (*ΔhycE*, *ΔhyaB*, and *ΔhybC*) and competing carbon pathways (*Δfpr*, *ΔydbK*, *Δhcr*, *Δhcr*, *ΔyeaX*, *ΔhcaD*, or *ΔfrdB* single deletion). In order to improve the interaction between hydrogenase and Fd, synthetic scaffolds were built from modular scaffold domains from eukaryotic signal transduction (PDZ, SH3, and GBD domains) and imported into *E. coli*. The production of H_2 via the newly synthesized circuits was affected significantly by the scaffold configurations, such as the position of Fd, the length of the flexible linker connecting Fd to the scaffold, the modulation ratio of Fd to hydrogenase on the scaffold, and the direct fusion ratio of Fd to hydrogenase. There was some improvement, but the resulting H_2 production yield was still as low as $\approx 0.4\,mol\,H_2/mol$ glucose (Agapakis et al., 2010).

Metabolic Reconstruction and *In Silico* Modeling

In silico metabolic models are useful for analyzing metabolic fluxes from experimental results and/or estimating the effects of modifications of metabolic reactions on cell growth and

H_2 yield (Oh et al., 2007). Cai et al. (2010) developed a metabolic network for *C. butyricum* W5 with 34 reactions. The simple model could well predict the experimental H_2 production yields and specific H_2 production rates in two different culture media. Manish et al. (2007) developed a central metabolic pathway model of *E. coli* for anaerobic glucose fermentation based on 27 biochemical reactions. This model successfully predicted the experimental H_2 production yields for the wild type (0.17 mol H_2/mol glucose) and a mutant strain lacking lactate dehydrogenase (0.23 mol H_2/mol glucose). Oh et al. (2008) constructed a metabolic pathway model for *Citrobater amalonaticus* Y19 with 81 biochemical reactions. They suggested that the high H_2 production yield of ~ 8.7 mol H_2/mol glucose is possible if H_2 is produced from NAD(P)H by nonnative NAD(P)H-linked hydrogenase [EC 1.12.1.2(3)] where NAD(P)H is supplied at a high rate through the PP pathway. Nogales et al. (2012) developed a genome-scale metabolic model for the hyperthermophilic archeon *Thermotoga maritima* with 562 intracellular and 83 extracellular biochemical reactions. They reported that the *in silico* mutant devoid of fructose bisphosphate aldolase and 2-dehydro-3-deoxy-phosphogluconate aldolase while having nonnative NADPH:ferredoxin reductase showed the high H_2 yield of 7.6 mole per glucose. In this mutant, glucose flux was directed exclusively toward the PP pathway, and H_2 production was conducted by a bifurcating hydrogenase enzyme utilizing both Fd and NADH. However, Oh et al. (2008) indicated that H_2 production via the NAD(P) H-linked hydrogenase is thermodynamically unfavorable in the presence of H_2. Thus far, several metabolic-flux models for fermentative H_2 production have been developed. However, as indicated earlier, the prediction power of these models is rather limited. More elaborate genome-wide models and their refinements with the aid of various x-omics, biochemical, and physiological data are needed.

PERSPECTIVES AND FUTURE DIRECTIONS

Recent progress in metabolic engineering has increased our understanding of biological H_2 production pathways and has significantly improved the performance of various H_2-producing microorganisms. However, many critical issues still remain for the implementation of practical biohydrogen production processes. In direct biophotolysis, studies to alleviate O_2 inhibition of hydrogenase are still required. Because O_2 suppresses the expression of hydrogenase at the transcriptional level, rational modifications of transcriptional regulatory proteins of hydrogenase structural genes can be attempted. In addition, PSII of LHC can be confined in a compartment with a less efficient oxidase. The effective bifurcation of electron circuits to a "PSI-Fd-hydrogenase compartment" and O_2 is expected to donate electrons for the reduction of protons and O_2. In this decoupling process, a certain number of electrons (price) need to be spent on O_2 detoxification in order to obtain sustained H_2 production.

Photoheterotrophic H_2 production is largely dependent on three factors: the activity of nitrogenase, the flow of carbon and electrons, and the supply of ATP via photophosphorylation. These three factors are closely interconnected and tightly regulated under most physiological conditions. However, most studies thus far have been limited to the individual biochemical components (in particular, nitrogenase and/or uptake hydrogenases) or to subsets of the overall cellular networks (H_2 production from pyruvate, PHB synthesis, or photophosphorylation), without considering the complex and network-wide interactions. Improvements in photoheterotrophic H_2 production require integrated metabolic engineering approaches in the context of the global metabolism of microorganisms. They could include the combinational engineering of currently existing strategies mentioned in this chapter, the incorporation of nonnative carbon pathways to extend substrate utilization (especially complex sugars and lignocellulosic biomass), protein engineering of nitrogenase to improve catalytic turnover rates and/or interactions with Fd, and the development of more elaborate *in silico* metabolic models involving photosystems and regulatory mechanisms.

In dark fermentative H_2 production, improving the H_2 production yield on sugars is the major topic of investigation. Without incorporating NAD(P)H-dependent H_2 production pathways, the yield cannot go above $4 \, mol \, H_2/mol$ glucose. Theoretical analyses have suggested that a yield of $\sim 9 \, mol \, H_2/mol$ glucose is possible if H_2 is produced through NAD(P)H-linked hydrogenase (Oh et al., 2008). NAD(P)H-dependent H_2 production is affected by several parameters, such as the $NAD(P)H/NAD(P)^+$ ratio, H_2 partial pressure, temperature, and H^+ concentration. However, studies on these parameters thus far were rather qualitative and fragmented. For example, the effect of the $NAD(P)H/NAD(P)^+$ ratio on H_2 production was not examined carefully. In order to utilize the NAD(P)H-dependent pathway, the parameters affecting this pathway and their interactions should be investigated in a quantitative manner.

In particular, in order to overcome the thermodynamic limitations in NAD(P)H-dependent H_2 production reactions, increasing the H^+ concentration and/or decreasing H_2 pressure will be important. In this context, thermoacidophilic bacteria or the reconstitution of NAD(P)H-dependent H_2 production reactions in bacterial periplasm where a higher H^+ concentration is maintained might be studied. Another important issue is extension of the substrate utilization for H_2 production. The economic feasibility of the dark H_2 fermentation process may be largely dependent on the availability of inexpensive carbon sources, such as cellulosic materials and organic wastes. Improvements in the H_2 production rate and yield from these recalcitrant substrates are currently in the early stages of research and could also be important subjects for metabolic engineering. Although huge challenges exist in biological H_2 production processes for commercialization, the potential of H_2 as a "clean energy" will further motivate efforts in the metabolic engineering of biological H_2 producers.

Acknowledgments

This study was supported financially by the Korean Ministry of Education, Science and Technology through the Advanced Biomass R&D Center (ABC; Grant No. 2010-0029799), KAIST, Korea. This study was also supported by the Brain Korea 21 program at Pusan National University. Dr. S. M. Raj is grateful to PRIST University, TN, India for their financial assistance.

References

Abril, D., Abril, A., 2009. Ethanol from lignocellulosic biomass. Cien. Inv. Agr. 36, 177–190.

Agapakis, C.M., Ducat, D.C., Boyle, P.M., Wintermute, E.H., Way, J.C., Silver, P.A., 2010. Insulation of a synthetic hydrogen metabolism circuit in bacteria. J. Biol. Eng. 4, 3.

Akhtar, M.K., Jones, P.R., 2008. Deletion of *iscR* stimulates recombinant clostridial Fe-Fe hydrogenase activity and H_2-accumulation in *Escherichia coli* BL21(DE3). Appl. Microbiol. Biotechnol. 78, 853–862.

Akhtar, M.K., Jones, P.R., 2009. Construction of a synthetic YdbK-dependent pyruvate:H$_2$ pathway in *Escherichia coli* BL21(DE3). Metab. Eng. 11, 139–147.

Akkerman, M.J., Rocha, J.M.S., Reith, J.H., Wijffels, R.H., 2003. Photobiological hydrogen production: Photochemical efficiency and bioreactor design. In: Reith, J.H., Wijffels, R.H., Barten, H. (Eds.), Bio-methane and Bio-hydrogen. Dutch Biological Hydrogen Foundation, The Netherlands, pp. 124–145.

Antal, T.K., Lindblad, P., 2005. Production of H$_2$ by sulphur-deprived cells of the unicellular cyanobacteria *Gloeocapsa alpicola* and *Synechocystis* sp. PCC 6803 during dark incubation with methane or at various extracellular pH. J. Appl. Microbiol. 98, 114–120.

Berberoglu, H., Jay, J., Pilon, L., 2008. Effect of nutrient media on photobiological hydrogen production by *Anabaena variabilis* ATCC 29413. Int. J. Hydrogen Energy 33, 1172–1184.

Brentner, L.B., Jordan, P.A., Zimmerman, J.B., 2010. Challenges in developing biohydrogen as a sustainable energy source: Implications for a research agenda. Environ. Sci. Technol. 44, 2243–2254.

Cai, G., Jin, B., Saint, C., Monis, P., 2010. Metabolic flux analysis of hydrogen production network by *Clostridium butyricum* W5: Effect of pH and glucose concentrations. Int. J. Hydrogen Energy 35, 6681–6690.

Chabriere, E., Vernede, X., Guigliarelli, B., Charon, M.H., Hatchikian, E.C., Fontecilla-Camps, J.C., 2001. Crystal structure of the free radical intermediate of pyruvate:ferredoxin oxidoreductase. Science 294, 2559–2563.

Drepper, T., Gross, S., Yakunin, A.F., Hallenbeck, P.C., Masepohl, B., Klipp, W., 2003. Role of GlnB and GlnK in ammonium control of both nitrogenase systems in the phototrophic bacterium *Rhodobacter capsulatus*. Microbiology-Sgm 149, 2203–2212.

Eltsova, Z.A., Vasilieva, L.G., Tsygankov, A.A., 2010. Hydrogen production by recombinant strains of *Rhodobacter sphaeroides* using a modified photosynthetic apparatus. Appl. Biochem. Microbiol. 46, 487–497.

Esper, B., Badura, A., Rogner, M., 2006. Photosynthesis as a power supply for (bio-) hydrogen production. Trends Plant Sci. 11, 543–549.

Franchi, E., Tosi, C., Scolla, G., Della Penna, G., Rodriguez, F., Pedroni, P.M., 2004. Metabolically engineered *Rhodobacter sphaeroides* RV strains for improved biohydrogen photoproduction combined with disposal of food wastes. Marine Biotechnol. (NY) 6, 552–565.

Frey, M., 2002. Hydrogenases: Hydrogen-activating enzymes. Chembiochem 3, 153–160.

Geng, A., He, Y., Qian, C., Yan, X., Zhou, Z., 2010. Effect of key factors on hydrogen production from cellulose in a co-culture of *Clostridium thermocellum* and *Clostridium thermopalmarium*. Bioresour. Technol. 101, 4029–4033.

Golomysova, A.N., Ivanov, P.S., 2011. Investigation of the anaerobic metabolism of *Rhodobacter capsulatus* by means of a flux model. Biophysics 56, 74–85.

Golomysova, A., Gomelsky, M., Ivanov, P.S., 2010. Flux balance analysis of photoheterotrophic growth of purple nonsulfur bacteria relevant to biohydrogen production. Int. J. Hydrogen Energy 35, 12751–12760.

Govindjee, 2002. A role for a light-harvesting antenna complex of photosystem II in photoprotection. Plant Cell 14, 1663–1668.

Hadicke, O., Grammel, H., Klamt, S., 2011. Metabolic network modelling of redox balancing and biohydrogen production in purple nonsulfur bacteria. BMC Syst. Biol. 5, 150.

Hallenbeck, P., 2005. Fundamentals of the fermentative production of hydrogen. Water Sci. Technol. 52, 21–29.

Hallenbeck, P.C., Benemann, J.R., 2002. Biological hydrogen production; fundamentals and limiting processes. Int. J. Hydrogen Energy 27, 1185–1193.

Hallenbeck, P.C., Ghosh, D., 2009. Advances in fermentative biohydrogen production: The way forward? Trends Biotechnol 27, 287–297.

Harwood, C.S., 2008. Nitrogenase-catalyzed hydrogen production by purple non-sulfur photosynthetic bacteria. In: Wall, J.D., Harwood, C.S., Demain, A. (Eds.), Bioenergy. ASM Press, Washington, DC, pp. 259–271.

Heap, J.T., Pennington, O.J., Cartman, S.T., Carter, G.P., Minton, N.P., 2007. The ClosTron: A universal gene knockout system for the genus *Clostridium*. J. Microbiol. Methods 70, 452–464.

Heap, J.T., Kuehne, S.A., Ehsaan, M., Cartman, S.T., Cooksley, C.M., Scott, J.C., et al., 2010. The ClosTron: Mutagenesis in *Clostridium* refined and streamline. J. Microbiol. Methods 80, 49–55.

Jo, J.H., Jeon, C.O., Lee, S.Y., Lee, D.S., Park, J.M., 2010. Molecular characterization and homologous overexpression of [FeFe]-hydrogenase in *Clostridium tyrobutyricum* JM1. Int. J. Hydrogen Energy 35, 1065–1073.

Kanai, T., Imanaka, H., Nakajima, A., Uwamori, K., Omori, Y., Fukui, T., et al., 2005. Continuous hydrogen production by the hyperthermophilic archaeon, *Thermococcus kodakaraensis* KOD1. J. Biotechnol. 116, 271–282.

Kars, G., Gunduz, U., Yucel, M., Rakhely, G., Kovacs, K.L., Eroglu, I., 2009. Evaluation of hydrogen production by *Rhodobacter sphaeroides* O.U.001 and its *hupSL* deficient mutant using acetate and malate as carbon sources. Int. J. Hydrogen Energy 34, 2184–2190.

Kim, E.J., Kim, J.S., Kim, M.S., Lee, J.K., 2006a. Effect of changes in the level of light harvesting complexes of *Rhodobacter sphaeroides* on the photoheterotrophic production of hydrogen. Int. J. Hydrogen Energy 31, 121–127.

Kim, M.S., Baek, J.S., Lee, J.K., 2006b. Comparison of H_2 accumulation by *Rhodobacter sphaeroides* KD131 and its uptake hydrogenase and PHB synthase deficient mutant. Int. J. Hydrogen Energy 31, 121–127.

Kim, E.J., Kim, M.S., Lee, J.K., 2008a. Hydrogen evolution under photoheterotrophic and dark fermentative conditions by recombinant *Rhodobacter sphaeroides* containing the genes for fermentative pyruvate metabolism of *Rhodospirillum rubrum*. Int. J. Hydrogen Energy 33, 5131–5136.

Kim, E.J., Lee, M.K., Kim, M.S., Lee, J.K., 2008b. Molecular hydrogen production by nitrogenase of *Rhodobacter sphaeroides* and by Fe-only hydrogenase of *Rhodospirillum* rubrum. Int. J. Hydrogen Energy 33, 1516–1521.

Kim, S., Seol, E., Raj, S.M., Park, S., Oh, Y.K., Ryu, D.D.Y., 2008c. Various hydrogenases and formate-dependent hydrogen production in *Citrobacter amalonaticus* Y19. Int. J. Hydrogen Energy 33, 1509–1515.

Kim, S., Seol, E., Oh, Y.K., Wang, G.Y., Park, S., 2009. Hydrogen production and metabolic flux analysis of metabolically engineered *Escherichia* coli strains. Int. J. Hydrogen Energy 34, 7417–7427.

Kim, J.Y.H., Jo, B.H., Cha, H.J., 2010. Production of biohydrogen by recombinant expression of [NiFe]-hydrogenase 1 in *Escherichia coli*. Microbial Cell Factories 9, art. No. 54.

Kivisto, A., Santala, V., Karp, M., 2010. Hydrogen production from glycerol using halophilic fermentative bacteria. Bioresour. Technol. 101, 8671–8677.

Koku, H., Eroglu, I., Gunduz, U., Yucel, M., Turker, L., 2002. Aspects of the metabolism of hydrogen production by *Rhodobacter sphaeroides*. Int. J. Hydrogen Energy 27, 1315–1329.

Kondo, T., Arakawa, M., Hirai, T., Wakayama, T., Hara, M., Miyake, J., 2002. Enhancement of hydrogen production by a photosynthetic bacterium mutant with reduced pigment. J. Biosci. Bioeng. 93, 145–150.

Kongjan, P., Angelidaki, I., 2010. Extreme thermophilic biohydrogen production from wheat straw hydrolysate using mixed culture fermentation: Effect of reactor configuration. Bioresour. Technol. 101, 7789–7796.

Kruse, O., Rupprecht, J., Bader, K.P., Thomas-Hall, S., Schenk, P.M., Finazzi, G., et al., 2005. Improved photobiological H_2 production in engineered green algal cells. J. Biol. Chem. 280, 34170–34177.

Lee, J.W., Greenbaum, E., 2003. A new oxygen sensitivity and its potential application in photosynthetic H_2 production. Appl. Biochem. Biotechnol. 105–108, 303–313.

Levin, D.B., Pitt, L., Love, M., 2004. Biohydrogen production: Prospects and limitations to practical application. Int. J. Hydrogen Energy 29, 173–189.

Li, X., Liu, T., Wu, Y., Zhao, G., Zhou, Z., 2010. Derepressive effect of NH_4^+ on hydrogen production by deleting the *glnA1* gene in *Rhodobacter sphaeroides*. Biotechnol. Bioeng. 106, 564–572.

Liebgott, P.P., de Lacey, A.L., Burlat, B., Cournac, L., Richaud, P., Brugna, M., et al., 2011. Original design of an oxygen-tolerant [NiFe] hydrogenase: Major effect of a valine-to-cysteine mutation near the active site. J. Am. Chem. Soc. 133 (4), 986–997.

Lindblad, P., Christensson, K., Lindberg, P., Pinto, F., Tsygankov, A., 2002. Photoproduction of H_2 by wild type *Anabaena* PCC 7120 and a hydrogen uptake deficient mutant: From laboratory experiments to outdoor culture. Int. J. Hydrogen Energy 27, 1271–1281.

Liu, X., Zhu, Y., Yang, S.T., 2006. Construction and characterization of *ack* deleted mutant of *Clostridium tyrobutyricum* for enhanced butyric acid and hydrogen production. Biotechnol. Prog. 22, 1265–1275.

Liu, T., Li, X., Zhou, Z., 2010. Improvement of hydrogen yield by *hupR* gene knock-out and *nifA* gene overexpression in *Rhodobacter sphaeroides* 6016. Int. J. Hydrogen Energy 35, 9603–9610.

Lu, Y., Zhao, H., Zhang, C., Lai, Q., Wu, X., Xing, X.H., 2010. Alteration of hydrogen metabolisms of *ldh*-deleted *Enterobacter aerogenes* by overexpression of NAD(+)-dependent formate dehydrogenase. Appl. Microbiol. Biotechnol. 86, 255–262.

Maeda, T., Sanchez-Torres, V., Wood, T.K., 2008. Metabolic engineering to enhance bacterial hydrogen production. Microbial Biotechnol. 1, 30–39.

Manish, S., Venkatesh, K.V., Banerjee, R., 2007. Metabolic flux analysis of biological hydrogen production by *Escherichia coli*. Int. J. Hydrogen Energy 32, 3820–3830.

Mathews, J., Wang, G., 2009. Metabolic pathway engineering for enhanced biohydrogen production. Int. J. Hydrogen Energy 34, 7404–7416.

McKinlay, J.B., Harwood, C.S., 2010. Photobiological production of hydrogen gas as a biofuel. Curr. Opin. Biotechnol. 21, 244–251.

Melis, A., 2002. Green alga hydrogen production: Progress, challenges and prospects. Int. J. Hydrogen Energy 27, 1217–1228.

Melis, A., Mitra, M., 2010. Suppression of Tla1 gene expression for improved solar energy conversion efficiency and photosynthetic productivity in plants and algae. United States Patent 7, 745,696.

Melis, A., Zhang, L., Forestier, M., Ghirardi, M.L., Seibert, M., 2000. Sustained photobiological hydrogen gas production upon reversible inactivation of oxygen evolution in the green alga *Chlamydomonas reinhardtii*. Plant Physiol. 122, 127–136.

Morimoto, K., Kimura, T., Sakka, K., Ohmiya, K., 2005. Overexpression of a hydrogenase gene in *Clostridium paraputrificum* to enhance hydrogen gas production. FEMS Microbiol. Lett. 246, 229–234.

Munro, S.A., Zinder, S.H., Walker, L.P., 2009. The fermentation stoichiometry of *Thermotoga neapolitana* and influence of temperature, oxygen, and pH on hydrogen production. Biotechnol. Prog. 25, 1035–1042.

Mussgnug, J.H., Thomas-Hall, S., Rupprecht, J., Foo, A., Klassen, V., McDowall, A., et al., 2007. Engineering photosynthetic light capture: impacts on improved solar energy to biomass conversion. Plant Biotechnol. J. 5, 802–814.

Nakashimada, Y., Rachman, M.A., Kakizono, T., Nishio, N., 2002. Hydrogen production of *Enterobacter aerogenes* altered by extracellular and intracellular redox states. Int. J. Hydrogen Energy 27, 1399–1405.

Nogales, J., Gudmundsson, S., Thiele, I., 2012. An in silico re-design of the metabolism in *Thermotoga maritima* for increase biohydrogen production. Int. J. Hydrogen Energy 37, 12205–12218.

Ntaikou, I., Koutros, E., Kornaros, M., 2009. Valorisation of wastepaper using the fibrolytic/hydrogen producing bacterium *Ruminococcus albus*. Bioresour. Technol. 100, 5928–5933.

Oh, Y.K., Seol, E.H., Lee, E.Y., Park, S., 2002. Fermentative hydrogen production by a new chemoheterotrophic bacterium *Rhodopseudomonas palustris* P4. Int. J. Hydrogen Energy 27, 1373–1379.

Oh, Y.K., Seol, E.H., Kim, J.R., Park, S., 2003. Fermentative biohydrogen production by a new chemoheterotrophic bacterium *Citrobacter* sp. Y19. Int. J. Hydrogen Energy 28, 1353–1359.

Oh, Y.K., Kim, S.H., Kim, M.S., Park, S., 2004. Thermophilic biohydrogen production from glucose with trickling biofilter. Biotechnol. Bioeng. 88, 690–698.

Oh, Y.K., Palsson, B.O., Park, S.M., Schilling, C.H., Mahadevan, R., 2007. Genome-scale reconstruction of metabolic network in *Bacillus subtilis* based on high-throughput phenotyping and gene essentiality data. J. Biol. Chem. 282, 28791–28799.

Oh, Y.K., Kim, H.J., Park, S., Kim, M.S., Ryu, D.D.Y., 2008. Metabolic-flux analysis of hydrogen production pathway in *Citrobacter amalonaticus* Y19. Int. J. Hydrogen Energy 33, 1471–1482.

Ozturk, Y., Yucel, M., Daldal, F., Mandaci, S., Gunduz, U., Turker, L., et al., 2006. Hydrogen production by using *Rhodobacter capsulatus* mutants with genetically modified electron transfer chains. Int. J. Hydrogen Energy 31, 1545–1552.

Peltier, G., Tolleter, D., Billon, E., Cournac, L., 2010. Auxiliary electron transport pathways in chloroplasts of microalgae. Photosynth. Res. 106, 19–31.

Polle, J.E.W., Kanakagiri, S., Jin, E., Masuda, T., Melis, A., 2002. Truncated chlorophyll antenna size of the photosystems: A practical method to improve microalgal productivity and hydrogen production in mass culture. Int. J. Hydrogen Energy 27, 1257–1264.

Prince, R.C., Kheshgi, H.S., 2005. The photobiological production of hydrogen: potential efficiency and effectiveness as a renewable fuel. Crit. Rev. Microbiol. 31, 19–31.

Raj, S.M., Talluri, S., Christopher, L.P., 2012. Thermophilic hydrogen production from renewable resources: Current status and future perspectives. Bioenergy Res. 1–17.

Rey, F.E., Oda, Y., Harwood, C.S., 2006. Regulation of uptake hydrogenase and effects of hydrogen utilization on gene expression in *Rhodopseudomonas palustris*. J. Bacteriol. 188, 6143–6152.

Sato, T., Fukui, T., Atomi, H., Imanaka, T., 2005. Improved and versatile transformation system allowing multiple genetic manipulations of the hyperthermophilic archaeon *Thermococcus kodakaraensis*. Appl. Environ. Microbiol. 71, 3889–3899.

Seo, D., Kamino, K., Inoue, K., Sakurai, H., 2004. Purification and characterization of ferredoxin-NADP$^+$ reductase encoded by *Bacillus subtilis yumC*. Arch. Microbiol. 182, 80–89.

Seol, E., Manimaran, A., Jang, Y., Kim, S., Oh, Y.K., Park, S., 2010. Sustained hydrogen production from formate using immobilized recombinant *Escherichia coli* SH5. Int. J. Hydrogen Energy 36, 8681–8686.

Seol, E., Jang, Y., Kim, S., Oh, Y.K., Park, S., 2012. Engineering of formate-hydrogen lyase gene cluster for improved hydrogen production in *Escherichia coli*. Int. J. Hydrogen Energy 37, 15045–15051.

Silva, P.J., Ban, E.C., Wassink, H., Haaker, H., Castro, D.B., Robb, F.T., et al., 2000. Enzymes of hydrogen metabolism in *Pyrococcus furiosus*. Eur. J. Biochem. 267, 6541–6551.

Surzycki, R., Cournac, L., Peltier, G., Rochaix, J.D., 2007. Potential for hydrogen production with inducible chloroplast gene expression in *Chlamydomonas*. Proc. Natl. Acad. Sci. USA 104, 17548–17553.

Veit, A., Akhtar, K., Mizutani, T., Jones, P.R., 2008. Constructing and testing the thermodynamic limits of synthetic NAD(P)H:H$_2$ pathways. Microbial Biotechnol. 1, 382–394.

Verhaart, M.R.A., Bielen, A.A.M., Oost, J.V.D., Stams, A.J.M., Kengen, S.W.M., 2010. Hydrogen production by hyperthermophilic and extremely thermophilic bacteria and archaea: Mechanisms for reductant disposal. Environ. Technol. 31, 993–1003.

Wang, S., Huang, H., Moll, J., Thauer, R.K., 2010. NADP$^+$ reduction with reduced ferredoxin and NADP$^+$ reduction with NADH are coupled via an electron-bifurcating enzyme complex in *Clostridium kluyveri*. J. Bacteriol. 192, 5115–5123.

Wu, X., Liang, Y., Li, Q., Zhou, J., Long, M., 2011. Characterization and cloning of oxygen-tolerant hydrogenase from *Klebsiella oxytoca* HP1. Res. Microbiol. 162 (3), 330–336.

Ye, X., Wang, Y., Hopkins, R.C., Adams, M.W.W., Evans, B.R., Mielenz, J.R., et al., 2009. Spontaneous high-yield production of hydrogen from cellulosic materials and water catalyzed by enzyme cocktails. Chem. Sus. Chem. 2, 149–152.

Yoshida, A., Nishimura, T., Kawaguchi, H., Inui, M., Yukawa, H., 2005. Enhanced hydrogen production from formic acid by formate hydrogen lyase-overexpressing *Escherichia* coli strains. Appl. Environ. Microbiol. 71, 6762–6768.

Yoshida, A., Nishimura, T., Kawaguchi, H., Inui, M., Yukawa, H., 2006. Enhanced hydrogen production from glucose using *ldh*- and *frd*-inactivated *Escherichia* coli strains. Appl. Microbiol. Biotechnol. 73, 67–72.

Zhang, C., Xing, X.H., Lou, K., 2005. Rapid detection of a gfp-marked *Enterobacter aerogenes* under anaerobic conditions by aerobic fluorescence recovery. FEMS Microbiol. Lett. 249, 211–218.

Zhang, Y.H.P., Evans, B.R., Mielenz, J.R., Hopkins, R.C., Adams, M.W.W., 2007. High-yield hydrogen production from starch and water by a synthetic enzymatic pathway. PLoS ONE 5, 456.

Zhang, C., Ma, K., Xing, X.H., 2009. Regulation of hydrogen production by *Enterobacter aerogenes* by external NADH and NAD$^+$. Int. J. Hydrogen Energy 34, 1226–1232.

Zhao, H., Ma, K., Lu, Y., Zhang, C., Wang, L., Xing, X.H., 2009. Cloning and knock-out of formate hydrogen lyase and H$_2$-uptake hydrogenase genes in *Enterobacter aerogenes* for enhanced hydrogen production. Int. J. Hydrogen Energy 34, 186–194.

Insurmountable Hurdles for Fermentative H$_2$ Production?

Patrik R. Jones and M. Kalim Akhtar

University of Turku, Bioenergy Group, Turku, Finland

The switch from nonrenewable to "renewable" energy consumption is one grand societal challenge that requires the continued development of novel and/or improved methods and tools to convert solar energy into consumer-friendly forms on a massive scale. Such a switch would possibly also require a change in the form of the dominant energy vector(s) that is used, in turn potentially forcing a revision of the entire infrastructure for energy storage, distribution, and utilization.

Dihydrogen is still regarded as a "future" fuel, despite the century-old idea of a global H$_2$-based society. As an energy vector, H$_2$ has both advantages and disadvantages; however, its facile link to electricity and innocuous environmental impact from combustion remains strongly in its favor. The availability of a large number of approaches (e.g., electrolytic, photobiological, fermentative) for the conversion of diverse energy forms into H$_2$ will be important for easy accessibility and availability of energy that is independent of geographical location.

In this chapter, one such approach, the fermentative conversion of glucose, produced and stored by photobiological organisms, to H$_2$ is discussed. The topic has been covered comprehensively in several up-to-date reviews (Hallenbeck and Ghosh, 2012; Rittmann and Herwig, 2012). The present account focuses on two seemingly insurmountable hurdles to enhance fermentative H$_2$ production, along with a brief discussion on measures that could be implemented to overcome such hurdles.

THE FIRST HURDLE IS THE THERMODYNAMIC LIMITATION

Interpretation of the metabolic capabilities of native H$_2$ producers is made difficult by the presence of multiple H$_2$-metabolizing pathways, differential regulation of H$_2$-related genes, varied sensitivity of hydrogenases toward O$_2$, and incomplete knowledge regarding the *in vivo* substrate specificities of H_2-metabolizing systems. In light of this, the engineering of synthetic microbial H_2 systems can shed some insights into some of the metabolic aspects of H_2 synthesis, including the biochemical properties of the H_2 pathway components. Based on the targeted engineering of *Escherichia coli* BL21(DE3), which is void of H_2 metabolism due to multiple metabolic deficiencies (Pinske et al., 2011), the difference in thermodynamic limitations between ferredoxin and NADPH-dependent H_2 pathways was clearly demonstrated (Akhtar and Jones, 2009; Veit et al., 2008). A comparison among the synthetic pathways for NAD(P)H:H_2, NADPH:H_2, and pyruvate:ferredoxin:H_2 in a closed and controlled environment suggested that the partial H_2 pressure at which H_2 reactions reach an equilibrium in minimal glucose media is increased in the order of NADH ($<100\,Pa$), NADPH ($600–1000\,Pa$), and pyruvate ($>10,000\,Pa$) (Akhtar and Jones, 2009; Veit et al., 2008). Such differences in thermodynamic outcome can be partly accounted for by the choice of redox cofactors given the difference in free energy between the reduced and the oxidized states of ferredoxin and $NAD(P)^+/NAD(P)H$.

Although it is tempting to simply rely on reaction equilibrium constants for various metabolic intermediates to draw conclusions regarding thermodynamic limitations of pathways, several uncertainties need to be considered. For example, there is a lack of knowledge regarding the metabolic homeostasis of cofactor couples that may result in potentially large errors in such estimates. Whole cell analysis, after all, indicates a wide range of feasible ratios between oxidized and reduced nicotinamide cofactors (Henry et al., 2007), and it is therefore quite likely that the ratio of reduced to oxidized cofactors is not constant under batch conditions. This issue of redox homeostasis is further compounded by the fact that, in addition to the free form, nicotinamide cofactors exist to varying degrees in protein-associated states (Patterson et al., 2000). Considering that electron-carrying cofactors exist in equilibra with a multitude of cellular reactions, each of which in turn is coupled to the oxidation of the initial carbon source (e.g. glucose), the biochemically effective intracellular ratio of reduced to oxidized cofactors is therefore difficult to estimate based on metabolite quantification alone, particularly given the notorious lability of NADP(H) (Pollak et al., 2007). Further study under controlled conditions thus is required to identify the actual thermodynamic limitations of metabolic pathways *in vivo*.

The uncertainties described earlier may possibly explain why van Neil and colleagues have found varying thermodynamic limitations with *Caldicellulosiruptor* species across multiple studies (van Niel et al., 2003; Willquist et al., 2011), exceeding the estimated theoretical equilibrium point up to a factor of 5 (Willquist et al., 2011). As Willquist et al. (2011) suggested, if the end product could be harvested in a continuous fashion and at a rate exceeding its production, the thermodyamic hurdle could possibly be overcome. This may be a logical direction toward mass scale commercial consumption, as H_2 would need to be stored, distributed, and used in the form of compressed gas in order for it to be of any practical benefit. However, energy-intensive methods such as compression and/or gas upgrading will incur additional costs and, for this reason, biological systems capable of generating higher, as opposed to lower, partial pressures would be far more desirable. In one study, gas upgrading by stripping; which involves separation of H_2 from CO_2; was identified as the greatest capital cost in a H_2 fermentation process model (Ljunggren and Zacchi, 2010). Given that hydrogen separation and compression will be crucial for distribution and transportation, the cost of such postproduction processing therefore warrants further investigation.

The thermodynamic challenge is fundamentally interesting as it places a constraint on the evolution of organisms that rely on protons as the main electron acceptor to produce H_2. Nature's response to this thermodynamic dilemma can be observed within both socio-biological and biochemical contexts. A classic example of the sociobiological solution is the symbiosis between anaerobic ciliates and methanogens in which hydrogen is removed continuously and rapidly to allow reduction of CO_2 to methane (Nowack and Melkonian, 2010).

More recently in 2009, a biochemical solution was uncovered by Schut and Adams (2009) with the discovery of a trimeric hydrogenase which catalyzes H_2 synthesis by parallel utilization of both NADH and reduced ferredoxin as electron donors. This enzyme allows the transfer of electrons from NADH to protons, despite an unfavorable reaction equilibrium [at standard conditions, the Gibbs free energy change for NADH:H_2 is predicted to be 34.9 kJ/mol (Table 1A)], and if the measured concentration of the cofactor ratio is considered, the equilibrium point is reached at a very low partial H_2 pressure (Veit et al., 2008). By coupling this reaction with the more favorable transfer of electrons from reduced ferredoxin to protons (21.3 kJ/mol, Table 1B), the authors argued that the equilibrium point is shifted to higher partial H_2 pressure. It is difficult to estimate how much higher the *in vivo* equilibrium point of the trimeric hydrogenase reaction can be raised other than by experimentally testing such a pathway in a synthetic H_2-neutral model system, as was done earlier for synthetic NADP(H):H_2 pathways (Veit et al., 2008). Under standard conditions, the predicted Gibbs free energy change of the trimeric hydrogenase reaction (21.3 kJ/mol, Table 1C) does not differ much from the simpler variant, NADH:H_2. The relative ratio of reduced to oxidized ferredoxin and redox potential of the relevant ferredoxin(s), however, would also need to be considered; the maintenance of a low NADH/NAD^+ ratio is one of the main issues with the NADH:H_2 pathway. This discovery, nevertheless, provides a potential biochemical explanation for the observation of the high partial H_2 pressure that has been reported in closed vessels of fermentative species that use NADH as a main electron shuttle in glycolysis and where the carbon-containing fermentation products clearly do not carry all electrons derived from glucose oxidation (Soboh et al., 2004; Willquist et al., 2011). The potential utility of trimeric hydrogenases in engineered systems will undoubtedly be of great interest to test. A synthetic

TABLE 1   Predicted Gibbs Free Energy Change under Standard Conditions (0.1 M, pH 7.0)[a]

	Stoichiometry	kJ/mol (eQuilibrator)	kJ/mol (per H_2)
A	$NADH + 2H^+ \rightarrow H_2 + NAD^+ + H^+$	37.3	37.3
B	2 ferredoxin(red) $+ 2H^+ \rightarrow$ 2 ferredoxin(ox) $+ H_2$	21.3	21.3
C	$NADH + 2$ ferredoxin(red) $+ 3 H^+ \rightarrow 2 H_2 + NAD^+ + 2$ ferredoxin(ox)	58.7	29.3
D	Glucose $+ 12 H_2O \rightarrow 6 CO_2$(total) $+ 12 H_2$	203.8	17.0
E	Glucose $+ 8 H_2O \rightarrow$ acetate $+ 4 CO_2$(total) $+ 8 H_2$	36.1	4.5
F	Glucose $+ 6 H_2O \rightarrow 1.5$ acetate $+ 3 CO_2$(total) $+ 6 H_2$	−50.9	−8.5
G	Glucose $+ 4 H_2O \rightarrow 2$ acetate $+ 2 CO_2$(total) $+ 4 H_2$	−142.4	−35.6

[a] *The Gibbs free energy changes are predicted using the online tool eQuilibrator (Flamholz et al., 2012).*

system, however, may not necessarily reflect the ratio of reduced to oxidized ferredoxin that exists in organisms where ferredoxin is a general electron acceptor/donor, such as *Clostridium* spp. and cyanobacteria.

Until now, the most successful H$_2$-producing systems have been demonstrated using thermophilic species (van Niel et al., 2003; Willquist et al., 2011; Zeidan et al., 2010). The thermodynamic equilibrium of H$_2$ pathways is shifted slightly by the rise in temperature under thermophilic conditions, in favor of H$_2$ synthesis. Also, the solubility of gases will be influenced, once again in favor of H$_2$ production. However, the shift in temperature is unlikely to greatly alter the thermodynamic limitation of H$_2$ pathways. More importantly, thermophilic species have most likely evolved under greater pressure to utilize H$_2$ as one of its central electron sinks. As a consequence, important enzymatic components that help minimize the impact of H$_2$ accumulation on H$_2$ pathways are found most commonly in thermophiles, such as glyceraldehyde-3-phosphate (GAP):ferredoxin oxidoreductase (GAPOR) (van der Oost et al., 1998) and the trimeric hydrogenase (Schut and Adams, 2009). Rather than engineering synthetic systems in well-known biotechnological hosts, an attractive alternative would be to engineer the best available production strain [e.g., *Caldicellulosiruptor* spp. (Willquist et al., 2011)] with the aim of maximizing the potential productivity with a minimal number of modifications. The main drawback with such an approach is often the lack of appropriate methodologies to genetically engineer thermophilic species though recently, there have been several reports of the successful development of tools for genetic engineering of thermophilic microorganisms, including *Clostridium thermocellum* (Olson and Lynd, 2012) and *Pyrococcus furiosus* (Chandrayan et al., 2012).

THE SECOND HURDLE IS INCOMPLETE OXIDATION OF SUBSTRATE

Glucose is oxidized in two steps of classical fermentative metabolism: (1) glyceraldehyde 3-phosphate to 1, 3 -biphosphoglycerate and (2) pyruvate to acetyl-CoA. The engineering of high-yielding fermentative H$_2$ pathways has so far only been reported for the pyruvate step (Akhtar and Jones, 2009; Maeda et al., 2008; Yoshida et al., 2006). The lack of success or interest for GAP-dependent H$_2$ synthesis is likely due to the dependence on NAD$^+$ which would be considered as the poorest choice of electron acceptor given its low ratio of reduced to oxidized cofactor, as discussed above (see also Veit et al., 2008). Native species, such as *Caldicellulosiruptor saccharolyticus*, are nevertheless capable of channeling electrons from GAP to H$_2$ and most likely employ a dual substrate-specific hydrogenase similar to that discovered by Schut and Adams (2009). Theoretically, a direct route from GAP to H$_2$ via ferredoxin could also be constructed using a GAPOR. Such enzymes have been identified in both thermophilic (van der Oost et al., 1998) and mesophilic organisms (Park et al., 2007). However, no GAPOR-dependent H$_2$ pathway has been demonstrated to date in either native species or engineered systems.

Even if all electrons released from both the GAP and PYR steps could be channeled to H$_2$, the majority of electrons would eventually be lost in the form of excreted acetate, arising from the breakdown of acetyl-CoA. Further substrate oxidation would therefore be necessary to enable higher yields of fermentative H_2. In stoichiometric modeling, a change of the objective function from biomass to H_2, together with the addition of appropriate H_2 pathways, results in H_2 yields exceeding the so-called "fermentative limit." Such theoretical solutions are typically dependent on imaginative and likely impossible loops in the stoichiometric models and/or single or circular flux through the oxidative pentose phosphate pathway (Jones, 2008; Nogales et al., 2012). In practice, however, a 100% glycolytic flux through the pentose phosphate pathway is lethal under anoxic conditions, even in the presence of a functional NADPH:H_2 pathway (Veit and Jones, unpublished observations).

Alternative solutions to maximize substrate oxidation include external processing of acetate by either electrically assisted fuel cells (Geelhoed and Stams, 2011) or photobiological organisms (Wang et al., 2012). Still within the fermentative host, it could well be possible to reengineer fermentative metabolism to enhance the degree of substrate oxidation in order to minimize the loss of electrons through excretion. The evolved design of glycolytic metabolism of *E. coli* was studied comprehensively from a range of different criteria (Bar-Even et al., 2012), including intermediate toxicity, intermediate stability, intermediate loss, enzyme oxygen sensitivity, and reaction energetics. Although natural evolved variants of glycolysis exist, it was concluded that the most common core glycolytic pathway sequence has evolved for very good biochemical reasons. Not all of the considered criteria are important for stoichiometric maximization of biomass, although all in one way or another may influence the actual biomass formation rate, the objective function of most stoichiometric analyses. However, in a biotechnological H_2 production system, the objective function would be accumulation of H_2, not biomass. Obviously, some compromise would be necessary in order to synthesize and maintain the activities of the biological catalyst(s), thus stoichiometric efficiency, as well as nonstoichiometric criteria, such as those reviewed by Bar-Even et al. (2012), still need some consideration. From a simple thermodynamic inspection (Table 1D–1G), complete oxidation would not be possible in the absence of O_2, even without considering any demand for biomass synthesis, unless the product (H_2) can be removed more rapidly than it is synthesized. With the assumption that H_2 removal at low partial H_2 pressure is more costly than the value of the product itself, it is unclear if complete oxidation would ever be a commercially viable option. An energetic and financial evaluation of H_2 removal would be valuable in determining this question, but as far as we are aware, there has not been any such published study.

ENGINEERING H₂ PATHWAYS WITH MAXIMUM CAPABILITY

Once these hurdles are overcome, if ever, electrons will need to be directed to H_2. The recombinant synthesis of both [FeFe]-hydrogenase (Posewitz et al., 2004) and [NiFe]-hydrogenase (Weyman et al., 2011), along with biochemical characterization of numerous hydrogenases, has already been well documented (Kim and Kim, 2011; Redwood et al., 2008; Soboh et al., 2004). The substrate specificity of H_2 pathway components, particularly hydrogenases and ferredoxin-dependent oxidoreductases, will ultimately limit the scope for novel solutions that aim to enhance substrate oxidation in glycolytic catabolism. The choice of electron acceptor/donor will also need to be considered carefully in order to maximize metabolic flux. The importance of choosing appropriate ferredoxins in synthetic H_2 pathways, for example, is known to influence total pathway capability (Agapakis and Silver, 2010; Veit et al., 2008). This may be caused by a difference in biochemical properties relating to substrate specificities and/or redox potentials (Cammack et al., 1977). Although *in vitro* studies suggest that no relationship exists between ferredoxin redox potential and H$_2$ pathway flux (Fitzgerald et al., 1980), the impact of a redox potential on H$_2$ pathways *in vivo* remains to be studied. The stimulation of secondary pathways such as Fe–S cluster metabolism, required for active synthesis of H$_2$ pathway components, has also been demonstrated to enhance the metabolic competitivity of complex H$_2$ pathways (Akhtar and Jones, 2008, 2009), illustrating the importance of cofactor supply in metabolic engineering.

CONCLUSION

Because individual reactions operate within a highly complex network of coupled reactions, it is difficult to predict the metabolic outcome of engineered pathways based solely on thermodynamic calculations of individual reactions. Barring the thermodynamic limitations imposed by fermentative metabolites and redox components, engineering steps could still be implemented to favor H$_2$ yields. This could be accomplished by metabolic means, via modification of the reduced to oxidized redox cofactor ratio, or through physical means, via reduction of the partial pressure by removal of the hydrogen end product. Additional engineering of the H$_2$ pathway components (e.g., altering of ferredoxin redox potentials, rational engineering of oxygen-labile hydrogenases) to remove some of the thermodynamic limitations also remains a viable but challenging prospect. As it is also likely to be difficult, if not impossible, to reengineer evolved fermentative metabolism in order to enhance substrate oxidation, the outlook is bleak for fermentative H$_2$ yields ever to be improved beyond what the best native species are capable of. A combination of different technologies will therefore need to be implemented in order to maximize the efficiency of the entire photobiological fermentative conversion chain in which sunlight and water are converted into H$_2$.

References

Agapakis, C.M., Silver, P.A., 2010. Modular electron transfer circuits for synthetic biology: insulation of an engineered biohydrogen pathway. Bioeng. Bugs 1, 413–418.

Akhtar, M., Jones, P., 2008. Deletion of iscR stimulates recombinant clostridial Fe-Fe hydrogenase activity and H2-accumulation in *Escherichia coli* BL21(DE3). Appl. Microbiol. Biotechnol. 78, 853–862.

Akhtar, M., Jones, P., 2009. Construction of a synthetic YdbK-dependent pyruvate:H2 pathway in *Escherichia coli* BL21 (DE3). Metab. Eng. 11, 139–147.

Bar-Even, A., Flamholz, A., Noor, E., Milo, R., 2012. Rethinking glycolysis: On the biochemical logic of metabolic pathways. Nat. Chem. Biol. 8, 509–517.

Cammack, R., Rao, K.K., Bargeron, C.P., Hutson, K.G., Andrew, P.W., Rogers, L.J., 1977. Midpoint redox potentials of plant and algal ferredoxins. Biochem. J. 168, 205–209.

Chandrayan, S.K., McTernan, P.M., Hopkins, R.C., Sun, J., Jenney, F.E., Adams, M.W., 2012. Engineering hyperthermophilic archaeon *Pyrococcus furiosus* to overproduce its cytoplasmic [NiFe]-hydrogenase. J. Biol. Chem. 287, 3257–3264.

Fitzgerald, M.P., Rogers, L.J., Rao, K.K., Hall, D.O., 1980. Efficiency of ferredoxins and flavodoxins as mediators in systems for hydrogen evolution. Biochem. J. 192, 665–672.

Flamholz, A., Noor, E., Bar-Even, A., Milo, R., 2012. eQuilibrator: The biochemical thermodynamics calculator. Nucleic Acids Res. 40, D770–D775.

Geelhoed, J.S., Stams, A.J., 2011. Electricity-assisted biological hydrogen production from acetate by Geobacter sulfurreducens. Environ. Sci. Technol. 45, 815–820.

Hallenbeck, P.C., Ghosh, D., 2012. Improvements in fermentative biological hydrogen production through metabolic engineering. J. Environ. Manag. 95 (Suppl.), S360–S364.

Henry, C., Broadbelt, L., Hatzimanikatis, V., 2007. Thermodynamics-based metabolic flux analysis. Biophys. J. 92, 1792–1805.

Jones, P., 2008. Improving fermentative biomass-derived H_2-production by engineering microbial metabolism. Int. J. Hydrogen Energy 33, 5122–5130.

Kim, D.H., Kim, M.S., 2011. Hydrogenases for biological hydrogen production. Bioresour. Technol. 102, 8423–8431.

Ljunggren, M., Zacchi, G., 2010. Techno-economic evaluation of a two-step biological process for hydrogen production. Biotechnol. Prog. 26, 496–504.

Maeda, T., Sanchez-Torres, V., Wood, T.K., 2008. Metabolic engineering to enhance bacterial hydrogen production. Microbial Biotechnol. 1, 30–39.

Nogales, J., Gudmundsson, S., Thiele, I., 2012. An in silico re-design of the metabolism in *Thermotoga maritima* for increased biohydrogen production. Int. J. Hydrogen Energy 37, 12205–12218.

Nowack, E.C., Melkonian, M., 2010. Endosymbiotic associations within protists. Philos. Trans. R. Soc. Lond. B Biol. Sci. 365, 699–712.

Olson, D.G., Lynd, L.R., 2012. Transformation of *Clostridium thermocellum* by electroporation. Methods Enzymol. 510, 317–330.

Park, M., Mizutani, T., Jones, P., 2007. Glyceraldehyde-3-phosphate ferredoxin oxidoreductase from *Methanococcus maripaludis*. J. Bacteriol. 189, 7281–7289.

Patterson, G.H., Knobel, S.M., Arkhammar, P., Thastrup, O., Piston, D.W., 2000. Separation of the glucose-stimulated cytoplasmic and mitochondrial NAD(P)H responses in pancreatic islet beta cells. Proc. Natl. Acad. Sci. USA 97, 5203–5207.

Pinske, C., Bönn, M., Krüger, S., Lindenstrauss, U., Sawers, R.G., 2011. Metabolic deficiences revealed in the biotechnologically important model bacterium *Escherichia coli* BL21(DE3). PLoS ONE 6, e22830.

Pollak, N., Dölle, C., Ziegler, M., 2007. The power to reduce: Pyridine nucleotides—Small molecules with a multitude of functions. Biochem. J. 402, 205–218.

Posewitz, M.C., King, P.W., Smolinski, S.L., Zhang, L., Seibert, M., Ghirardi, M.L., 2004. Discovery of two novel radical S-adenosylmethionine proteins required for the assembly of an active [Fe] hydrogenase. J. Biol. Chem. 279, 25711–25720.

Redwood, M.D., Mikheenko, I.P., Sargent, F., Macaskie, L.E., 2008. Dissecting the roles of *Escherichia coli* hydrogenases in biohydrogen production. FEMS Microbiol. Lett. 278, 48–55.

Rittmann, S., Herwig, C., 2012. A comprehensive and quantitative review of dark fermentative biohydrogen production. Microb. Cell Fact. 11, 115.

Schut, G., Adams, M., 2009. The iron-hydrogenase of *Thermotoga maritima* utilizes ferredoxin and NADH synergistically: A new perspective on anaerobic hydrogen production. J. Bacteriol. 191, 4451–4457.

Soboh, B., Linder, D., Hedderich, R., 2004. A multisubunit membrane-bound [NiFe] hydrogenase and an NADH-dependent Fe-only hydrogenase in the fermenting bacterium *Thermoanaerobacter tengcongensis*. Microbiology 150, 2451–2463.

van der Oost, J., Schut, G., Kengen, S.W., Hagen, W.R., Thomm, M., de Vos, W.M., 1998. The ferredoxin-dependent conversion of glyceraldehyde-3-phosphate in the hyperthermophilic archaeon *Pyrococcus furiosus* represents a novel site of glycolytic regulation. J. Biol. Chem. 273, 28149–28154.

van Niel, E.W., Claassen, P.A., Stams, A.J., 2003. Substrate and product inhibition of hydrogen production by the extreme thermophile, *Caldicellulosiruptor saccharolyticus*. Biotechnol. Bioeng. 81, 255–262.

Veit, A., Akhtar, M., Mizutani, T., Jones, P., 2008. Constructing and testing the thermodynamic limits of synthetic NAD(P)H:H2 pathways. Microbial Biotechnol. 1, 382–394.

Wang, R.Y., Shi, Z.Y., Chen, J.C., Wu, Q., Chen, G.Q., 2012. Enhanced co-production of hydrogen and poly-(R)-3-hydroxybutyrate by recombinant PHB producing *E. coli* over-expressing hydrogenase 3 and acetyl-CoA synthetase. Metab. Eng. 14, 496–503.

Weyman, P.D., Vargas, W.A., Chuang, R.Y., Chang, Y., Smith, H.O., Xu, Q., 2011. Heterologous expression of *Alteromonas macleodii* and *Thiocapsa roseopersicina* [NiFe] hydrogenases in *Escherichia coli*. Microbiology 157, 1363–1374.

Willquist, K., Pawar, S.S., Van Niel, E.W., 2011. Reassessment of hydrogen tolerance in *Caldicellulosiruptor saccharolyticus*. Microb. Cell Fact. 10, 111.

Yoshida, A., Nishimura, T., Kawaguchi, H., Inui, M., Yukawa, H., 2006. Enhanced hydrogen production from glucose using ldh- and frd-inactivated *Escherichia coli* strains. Appl. Microbiol. Biotechnol. 73, 67–72.

Zeidan, A.A., Rådström, P., van Niel, E.W., 2010. Stable coexistence of two *Caldicellulosiruptor* species in a de novo constructed hydrogen-producing co-culture. Microb. Cell Fact. 9, 102.

Hydrogenase

Philippe Constant*, *Patrick C. Hallenbeck[†]*

*Trace Gas Biogeochemistry Laboratory, INRS-Institut Armand-Frappier, Laval, Québec, Canada
[†]University of Montreal, Montreal, Quebec, Canada

INTRODUCTION

The discovery of H_2 has been reported in a seminal publication of the English scientist Henry Cavendish, who meticulously examined the inflammability of gases produced from the dissolution of metals in acids and fermentation (Cavendish, 1766). Two decades later, production of water following ignition of the inflammable air described by Cavendish inspired the French chemist Antoine Lavoisier to name the inflammable air hydrogen from the Greek *"hydro"* and *"genes"* meaning "water" and "born of." The high gravimetric energy content combined to combustion free of carbon and ozone precursor emissions made H_2 a promising alternative to fossil fuel, especially in the context of the energy crisis and climate change mitigation. However, exploitation of H_2 energy potentials comes at a price, as, in nature, most of the hydrogen is bound in molecular form into organic and water molecules. A great deal of research is underway to extract H_2 out of these elements and harness the energy potential of H_2. Today, more than 95% of H_2 is produced from hydrocarbons by stream reforming or partial oxidation. These methods are energy consuming and are still dependent on fossil fuels, generating CO_2, black carbon particles, and climate-relevant reactive gases as by-products. Establishment of a H_2-based economy remains a challenging task and there is much we can learn from nature. Indeed, recent progress and technological advances achieved in genetic and enzymatic bioengineering demonstrate that investigation of microbial H_2 metabolism offers promising perspectives for the sustainable production of biohydrogen. The sophisticated H_2 metabolic pathways of the microbial world are the outcome of an extensive evolution—probably originating from the emergence of early life on Earth, according to certain evolutionary models. For instance, examination of geochemical properties of hydrothermal vents formed in deep oceans inspired the hypothesis that H_2 was the electron donor and CO_2 the acceptor for the generation of an ion gradient for the synthesis of ATP through chemiosmosis in the "Last Universal Common Ancestor" (Lane et al., 2010). Methanogenic archaea and bacteria producing H_2 as a by-product of fermentation probably coexisted in anaerobic environments exposed to geological CO_2 and H_2. Symbiotic association of the strict autotrophic archaea with the bacteria, based on H_2 dependence, would have enabled the archaeal host to thrive in H_2-depleted ecological niches, leading to the origin of the first eukaryote (Martin and Muller, 1998; Moreira and López-García, 1998).

Nowadays, H_2 is still distributed ubiquitously in the environment, albeit at a lower level than found in hydrothermal vents and in the early atmosphere. H_2 is an obligatory by-product of nitrogenase and organic matter fermentation, supporting microbial symbiotic interactions in soil (Golding and Dong, 2010). In marine and freshwater, H_2 is usually in equilibrium with the atmosphere, with a concentration of 0.40 nM. In the atmosphere, H_2 is present at a trace level, with a mixing ratio of 0.530 ppmv. In contrast to other climate-relevant trace gases, the atmospheric level of H_2 has remained stable over the last decades, even though anthropogenic activity generates increasing emissions of H_2 and precursors, such as methane and nonmethane hydrocarbons (Novelli et al., 1999). This stability of atmospheric burden is explained by the high turnover of H_2—microbes scavenge H_2 directly from the atmosphere and consume H_2 produced in soil and water, limiting net emissions to the atmosphere (Constant et al., 2009).

Hydrogenase is the enzyme catalyzing the interconversion of H_2 into protons and electrons ($H_2 \leftrightarrow 2H^+ + 2e^-$) in bacteria, archaea, and eukaryotes. Even though several microorganisms using H_2 as an energy source attracted the attention of scientists in the 1800's, Stephenson and Stickland (1931) were the first to propose the existence of hydrogenases and to report the kinetic properties as well as the O_2 sensitivity of these enzymes. More recently, several crystal structures of hydrogenase helped unveil the geometry and mode of action of their active site (Fontecilla-Camps et al., 2007), and extensive phylogenetic analyses revealed that microorganisms harboring genes encoding hydrogenases are taxonomically diverse and ubiquitous in the environment (Vignais and Billoud, 2007). These enzymes are utilized to generate energy, disperse reducing equivalents produced during fermentation, or generate reduced cofactors involved in several reactions of cellular metabolism.

THREE DISTINCT CLASSES OF HYDROGENASES

Hydrogenases are discriminated into three distinct classes based on the metal content of their active site, namely, [NiFe]-hydrogenase, [FeFe]-hydrogenase, and [Fe]-hydrogenase. [NiFe]-hydrogenases are only found in archaea and bacteria, catalyzing either H_2 oxidation or production *in vivo*, while [FeFe]-hydrogenases have been detected in bacteria and eukaryotes where they exclusively catalyze H_2 production. However, [Fe]-hydrogenase is distributed unevenly in hydrogenotrophic methanogenic archaea and uses H_2 to provide reducing equivalents for intermediate steps of CO_2 reduction to methane. The difference in the occurrence of hydrogenases belonging to the three different classes, combined with the common features of their active site's ligands, indicates that hydrogenases have emerged from convergent evolution. All of these enzymes contain complex metal centers that are very sensitive to O_2 inactivation. [FeFe]-hydrogenases are destroyed quickly and irreversibly by exposure to O_2, while [NiFe]-hydrogenases react with O_2 to give intermediates that can be reactivated by reduction and [Fe]-hydrogenases remain active in the presence of O_2 (Fontecilla-Camps et al., 2007; Lyon et al., 2004). O_2 tolerance is an important issue for the utilization of hydrogenases in biotechnological applications. For biohydrogen production applications, a compromise must be made between O_2 tolerance and H_2 production capacity of the selected catalyst (Friedrich et al., 2011).

[NiFe]-HYDROGENASES

[NiFe]-Hydrogenases comprise minimally two structural subunits. The small β subunit ($\approx$30 kDa) comprises up to three iron–sulfur clusters acting as a relay of electrons between the active site and the physiological electron mediator of the hydrogenase. The large α subunit ($\approx$60 kDa) comprises four conserved cysteine residues coordinating the NiFe active site (Fig. 1). The NiFe active site comprises two cyanide (CN^-) and one carbon monoxide (CO) ligands, essential for H_2 activation. Several microorganisms possess the genetic capacity to express multiple [NiFe]-hydrogenases, for example, *Ralstonia eutropha* and *Escherichia coli*, whose genome encodes four different [NiFe]-hydrogenases, conferring highly beneficial metabolic flexibility. However, commonly there are multiple operons encoding the structural genes and some accessory genes, with a common set of hydrogenase pleiotropic genes (so-called hyp genes), being used for core generic functions required for [NiFe]-hydrogenase maturation. The *hypABCDEF* genes coordinate the NiFe cofactor and assembly of the hydrogenase. HypC, HypD, HypE, and HypF proteins participate in the synthesis and transfer of the Fe-$(CN^-)_2$-(CO) moiety to the precursor of the large subunit of [NiFe]-hydrogenase, while HypA and HypB are involved in nickel storage and insertion (Böck et al., 2006; Lenz et al., 2010; Swanson et al., 2011; Vignais and Colbeau, 2004). The large subunit is synthesized in a precursor form, which is processed by proteolytic cleavage at the C-terminal end, after nickel incorporation. Because of the incompatibility of the maturation apparatus, heterologous expression of hydrogenase often requires simultaneous transfer of structural and accessory genes in heterologous hosts already harboring hydrogenases. Nevertheless, heterologous expression of hydrogenase in closely related species has been reported; furthermore, functional [NiFe]-hydrogenase of *Thiocapsa roseopersinina* BBS has been expressed in *E. coli* (Palágyi-Mészáros et al., 2009). Even more striking, expression of hydrogenase from the archaea *Pyrococcus furiosus* in *E. coli* was made possible only if the archaeal endopeptidase

FIGURE 1   Active site structure of the [NiFe]-hydrogenase from *Allochromatium vinosum* (pbd 3MYR) in the oxidized state as determined by X-ray crystallography (Ogata et al., 2010). The figure was extracted from the pdb entry using the Jmol viewer. Shown are the Ni and Fe atoms, as well as the CN^- and CO.

TABLE 1 Consensus L1 and L2 Signatures in the Large Subunit of the Hydrogenase Generated from the Alignment of Sequences Comprising [NiFe]-Hydrogenase (Constant et al., 2011; Vignais and Billoud, 2007)[a]

Group	Function	L1 and L2 signatures in the large subunit
1	Membrane-bound uptake hydrogenase	L1 [EGMQS]Rx**C**[GR][IV]**C**xxx[HT]xxx[AGS]x[VANQD] L2 [AFGIKLMV][HMR]xx[HR][AS][AFLY][DN]P**C**[FILMV]x**C**[AGS]xH
2a	Cyanobacterial uptake hydrogenase	L1 PR[AIV]**C**GI**C**xHxLxx[AST] L2 Vx[KR]S[FHY]Dx**C**x**VC**[ST][TV][HK]
2b	H$_2$-sensing hydrogenase	L1 PR[IV]**C**GI**C**S[IV][AS]Q[GS]xA L2 H[IV]VRSFDP**C**MV**C**T[AV]H
3a	F$_{420}$-reducing hydrogenase	L1 R[FIV]**C**G[ILV]**C**[PQ]x[APT]H[ACGT]x[AS][AGS] L2 R[ACS]YD[IP]**C**[AILV][AS]**C**xHx[ILMV]
3b	Bifunctional (NADP) hydrogenase	L1 R[IV]**C**[AGS][FIL]**C**xxx[HY]xx[AST][ANS]xx[AS][AILV] L2 R[ANS][FHY]DP**C**IS**C**[AS][ATV]H
3c	Methyl viologen-reducing hydrogenase	L1 Px[FILV][TV][ADPST]x[IV]**C**G[IV]**C**xxxHxx[AC][AS]xxA L2 E[FMV][AGLV][FIV]Rx[FY]DP**C**x[AS]**C**[AS][ST]Hx[AILV]
3d	Bidirectional NAD(P)-linked hydrogenase	L1 Ex[APV]xxxxRx**C**G[IL]**C**xx[AS]Hx[IL][ACS][AGS][AGNSV][KR][ATV]XD L2 DP**C**[IL]S**C**[AS][AST]H[ASTV]x[AG]xx[APV]
4	Membrane-bound H$_2$ evolving hydrogenase	L1 **C**[GS][ILV]**C**[AGNS]xxH L2 [DE][PL]**C**x[AGST]**C**x[DE][RL]
5	High-affinity hydrogenase	L1 SRIx**C**GI**C**GDNHATxCS[VCI] L2 MR[TA]VRSFDPx**C**LP**C**GVH

[a] *Bold, underlined C's represent cysteine residues coordinating the NiFe cofactor in the large subunit of the hydrogenases.*

was coexpressed with the structural subunits of the enzyme (Sun et al., 2010). The possibility of expressing recombinant [NiFe]-hydrogenases led to sophisticated innovations, including the bioengineered hybrid of photosystem I of cyanobacteria with [NiFe]-hydrogenase for light-driven H$_2$ production (Ihara et al., 2006) and synthetic pathways for the production of biohydrogen (Zhang et al., 2007), in addition to providing fundamental information on the physiological role of the enzyme.

Phylogenetic analysis of the gene encoding for the small or the large structural subunits of [NiFe]-hydrogenase revealed the occurrence of five phylogenetically distinct groups, whose definition is supported by the physiological role of the isoenzymes. Examination of the large subunit coding sequences unveiled the presence of group-specific conserved amino acid motifs in the regions of the cysteine residues coordinating the NiFe cofactor (Table 1). These signatures can be utilized to determine from which phylogenetic group hydrogenase sequences derived from public databases or environmental samples belong.

Group 1, Membrane-Bound [NiFe]-Hydrogenase (MBH)

Membrane-bound [NiFe]-hydrogenase allows microorganisms to use H$_2$ as an energy source by coupling H$_2$ oxidation to the reduction of carbon dioxide, sulfate, fumarate, nitrate, and iron under anaerobic conditions, or oxygen in aerobic conditions. The enzyme is a key component of the electron transport chain, channeling electrons to the quinone pool via the membrane-integral diheme cytochrome b. This association is ensured by the interaction between the hydrophobic C-terminal region of the small subunit of the MBH and the cytochrome. H_2 is oxidized at the periplasmic side of the cytoplasmic membrane, and electrons derived from the reaction are transferred to cytochrome b, coupling electron transfer with transmembrane proton translocation and thus energy respiration. Protons released in the periplasmic space by the MBH, instead of redox reactions of menaquinone in the electron transport chain, contribute to the proton motive force (Gross et al., 1998; Kröger et al., 2002). Maturation of the hydrogenase is followed by translocation of the assembled structural subunit of the enzyme, ensured by the conserved twin-arginine translocation motif (i.e., RRxFxK) in the N terminus of the small subunit (Gross et al., 1999; Wu et al., 2000). Notwithstanding these common operational features, MBHs are particularly diverse in terms of architecture and physiology. This heterogeneity was indeed highlighted in an extensive phylogenetic analysis of group 1 MBH categorized into several clades, well supported by the organization of the MBH gene cluster, the occurrence of additional subunits and the structure of the iron–sulfur clusters in the small subunit, and, overall, the degree of O_2 tolerance of the enzyme (Pandelia et al., 2012). O_2-sensitive MBHs are the ancestral form, found in archaea and bacteria. This ancestral clade comprises the well-known periplasmic hydrogenase from sulfate-reducing bacteria, including *Desulfovibrio gigas* utilized to report the first three-dimensional structure of a hydrogenase (Volbeda et al., 1995). In these bacteria, transmembrane high molecular mass multihemecytochrome c links H_2 oxidation in the periplasm to the sulfate reduction taking place in the cytoplasm (Matias et al., 2005). Another prototype of this clade is the MBH of the methanogenic archaea *Methanosarcina mazei* Göl. In total, methanogenic archaea possess up to four different hydrogenases for the generation of energy and reducing equivalents for the reduction of CO_2 to CH_4 (Fig. 2). The F_{420}-nonreducing hydrogenase is channeling electrons to cytochrome and methanophenazine, resulting in the electrochemical proton gradient used for ATP generation (Ide et al., 1999). Phylogenetic analysis of group 1 hydrogenase suggests that MBHs displaying low O_2 tolerance were the second clade to emerge during the course of evolution. Enzymes belonging to this clade form two main clusters. First, the Isp cluster consists of tetraheterodimeric hydrogenases possessing the subunits Isp1 and Isp2, sharing homology with cytochrome b and heterosulfite reductase, respectively. The enzyme is found in obligate and facultative autototrophic bacteria capable of using H_2 as the sole energy source. The enzyme has been purified from *Aquifex aeolicus* (Brugna-Guiral et al., 2003) and *Hydrogenobacter thermophilus* TK-6 (Ishii et al., 2000)—two hyperthermophilic bacteria belonging to the Acquificales, which represent the earliest branching order of the bacteria domain. These obligate autotrophic bacteria use the reductive tricarboxylic acid cycle to fix CO_2 and H_2 or elemental sulfur as an energy source. The electron transport chain of these bacteria is composed of two Isp-MBHs for which the exact physiological role remains elusive. It has been proposed that the first Isp-MBH delivers electrons to the quinone pool and ultimately reduces molecular O_2 under microaerophilic conditions, while electrons generated by the second enzyme are channeled to the sulfur reductase via quinines (Guiral et al., 2005). Similar MBH, coupling H_2 oxidation to sulfur reduction has been purified from the photosynthetic purple bacterium *Thiocapsa roseoperscina* BBS (Palágyi-Mészáros et al., 2009; Rakhely et al., 1998) and the thermophilic archaea *Acidianus*

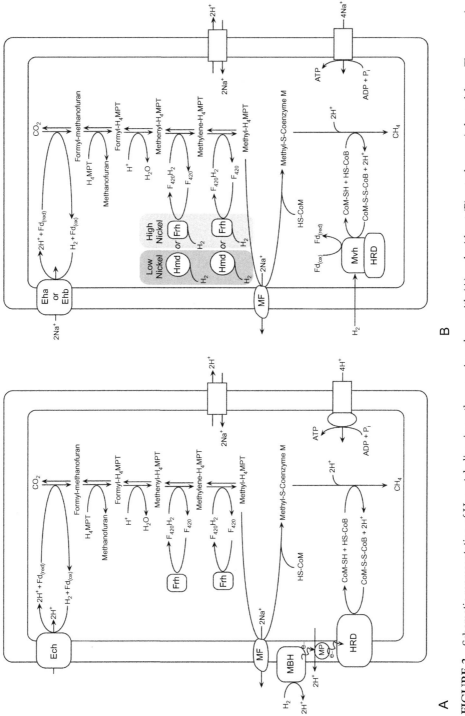

FIGURE 2 Schematic representation of H_2 metabolism in methanogenic archaea with (A) and without (B) cytochrome (adapted from Thauer et al., 2010). Only members of the Methanosarcinales contain cytochrome, which is essential for acetogenic methanogenesis. Ech, energy-converting hydrogenase (group 4); Frh, F_{420}-reducing hydrogenase (group 3A); MF, methyl transferase; MBH, membrane-bound hydrogenase (group 1); HRD, heterodisulfide reductase; Eha/Ehb, ferredoxin-reducing, energy-converting hydrogenases (group 4); Hmd, [Fe]-hydrogenase; Mvh, methylviologen reducing hydrogenase (group 3C).

ambivalens (Laska et al., 2003). Second, MBH displaying low O_2 tolerance comprises the subclade HypA, with the hydrogenase-2 of *E. coli* as the prototype. This is a heterotetrameric complex composed of the core structural subunits (HybOC) and two proteins (HybAB) essential to shuttle electrons to the quinone pool (Dubini et al., 2002). The emergence of oxygenic photosynthesis 2.7 billion years ago may have led to the development of O_2-tolerant MBH, namely, the clade 6C according to the nomenclature proposed by Pandelia et al. (2012). MBHs belonging to this group comprise an unusual [4Fe-3S] proximal cluster coordinated by six cysteine residues in contrast to the canonical four cysteine residues coordinating the [4Fe-4S] cluster in other [NiFe]-hydrogenases (Fritsch et al., 2011; Shomura et al., 2011; Volbeda et al., 2012). This unusual proximal cluster was shown to confer O_2 tolerance to the active site of the enzyme—a prerequisite for knallgas bacteria, which are facultative chemoautotrophic bacteria oxidizing H_2 under aerobic conditions. The so-called 6C signature was observed in the small subunit of Bacteriodes, Chlorobi, and α-/β-/γ-Proteobacteria (Pandelia et al., 2012), including the hydrogenase-1 in *E. coli*, which is coupled to terminal electron acceptors of higher redox potential than hydrogenase-2 (Dubini et al. 2002). The biotechnological potential of O_2-tolerant hydrogenase was demonstrated by implementing the MBH of *R. eutropha* into a fuel cell generating electric current from H_2, even in the presence of O_2 (Vincent et al., 2005). Indeed, MBH is not susceptible to CO poisoning and achieves the same current density as a platinum catalyst, a feature that argues for their integration into biofuel cells and biohydrogen production systems (Jones et al., 2002).

Group 2, Soluble Uptake [NiFe]-Hydrogenase

[NiFe]-hydrogenases belonging to group 2 are cytoplasmic and can be discriminated into two main clades. Clade 2A includes uptake hydrogenases found in nitrogen-fixing cyanobacteria and bacteria (Bothe et al., 2010; Tamagnini et al., 2007). This enzyme plays dual roles in protection of the nitrogenase against O_2 and H_2 by recycling the H_2 generated as an obligate by-product of nitrogen fixation (Peterson and Burris 1978). This pathway has been utilized for photobiological H_2 production, and obtaining uptake hydrogenase-deficient strains rapidly became a standard to improve the yield of these biotechnological applications (Lopes Pinto et al., 2002). The uptake hydrogenase is located in the heterocysts of cyanobacteria, a differentiated structure specialized for nitrogen fixation into which cell division and photosynthesis are arrested, providing a microaerobic environment for O_2-sensitive nitrogenase and hydrogenase. Expression of the hydrogenase structural gene is often orchestrated by the *hupL* gene rearrangement in the heterocyst (Tamagnini et al., 2000). In vegetative cells of *Anabaena* sp. PCC 7120, for instance, the *hupL* gene encoding for the large subunit of the hydrogenase is interrupted by a 9.5-kb DNA fragment specifying a recombinase (*xisC*). In heterocysts, XisC ensures excision of the DNA fragment by site-specific recombination, allowing expression of the H_2 uptake hydrogenase (Carrasco et al., 2005). In other cyanobacteria lacking this regulatory excision element, hydrogenase genes expression is under the control of the transcription regulator NtcA, which is a key component of nitrogen control and heterocyst development in cyanobacteria (Weyman et al., 2008). Even though physiological and biochemical data suggest that cyanobacterial uptake hydrogenase is a membrane-bound enzyme, the link between the enzyme and the respiratory chain has not been demonstrated clearly. Experimental evidence suggests that the core enzyme is associated with a membrane-bound protein channeling electrons to the plastoquinone pool for ATP generation. The importance of the enzyme was demonstrated with *Anabaena* sp. PCC 7120 for which an uptake hydrogenase-deficient mutant showed higher H_2 emissions and lower growth yield than the wild-type strain (Lindblad et al., 2002).

In aerobic environments, H_2 is typically present at the trace level, with the exception of H_2 pulses found in soil surrounding N_2-fixing root nodules (Golding and Dong, 2010). Knallgas bacteria have developed an efficient strategy to optimize the expression of their hydrogenase apparatus, when a sufficient level of H_2 is present in their environment. They developed H_2-sensing regulatory hydrogenase (RH), belonging to clade 2B. RH controls transcription of the genes specifying structural and auxiliary components of MBH (group 1) and SH (group 3) in knallgas bacteria. Existence of a hydrogenase complex, regulating the expression of hydrogenase, was first proposed following the characterization of *hupUV* genes in *Bradyrhizobium japonicum* and *Rhodobacter capsulatus*. Deletion of *hupUV* genes, showing high homology to the [NiFe]-hydrogenase small and large subunits, resulted in a derepression of hydrogenase activity in both knallgas bacteria (Black et al., 1994; Elsen et al., 1996). The molecular mechanisms behind this regulation have been investigated extensively in *R. eutropha* (Lenz et al., 2002). This knallgas bacteria is harboring the energy-generating MBH (group 1) and SH (group 3D) enzymes (Fig. 3). In addition to carbon limitation inducing the expression of hydrogenases through a pleiotropic sigma factor (Römermann et al., 1989; Schwartz et al., 1998), heterologous complementation studies revealed for the first time that hydrogenase gene control under the RH was dependent on H_2, stimulating hydrogenase genes transcription (Lenz et al., 1997). The essential components for H_2-dependent transcriptional control are the proteins HoxB, HoxC, HoxJ, and HoxA (Lenz and Friedrich, 1998). HoxB and HoxC encode for the small and large subunits of the RH, interacting directly or indirectly with the histidine protein kinase HoxJ. In the absence of H_2, HoxJ uses ATP as a phosphate donor for phosphorylation of the Ntrc-like response regulator HoxA, which is then incapable of activating MBH and SH operons (Fig. 3). The RH activity is not coupled to an energy-generating electron transport process, and the *R. eutropha* strain lacking both MBH and SH activity while overexpressing the RH was shown unable to grow autotrophically with H_2 and CO_2 (Kleihues et al., 2000). As knallgas bacteria are thriving in aerobic environments, O_2 has exerted a strong selection pressure on RH, resulting in development of an atypical hydrophobic gas channel blocking O_2 access to the active site of the enzyme, while allowing the transport of H_2 (Buhrke et al., 2005). This inspired modifications of the gas channel bridging the active site to the surface of the enzyme in other hydrogenases, especially in the context of the design of O_2-tolerant biocatalysts for biohydrogen production. For instance, substitution of two hydrophobic residues by two methionines in the gas channel of the periplasmic hydrogenase of *Desulfovibrio fructosovorans* (group 1 MBH) conferred O_2 tolerance by restricting O_2 access to the active site, while facilitating reactivation of the active site exposed to oxidizing conditions (Dementin et al., 2009).

Group 3, Bidirectional Heteromultimeric Cytoplasmic [NiFe]-Hydrogenase

Structural subunits of hydrogenases belonging to group 3 are associated with modules able to bind soluble cofactors such as NAD/NADP and cofactor 420 (F_{420}). Their activity is

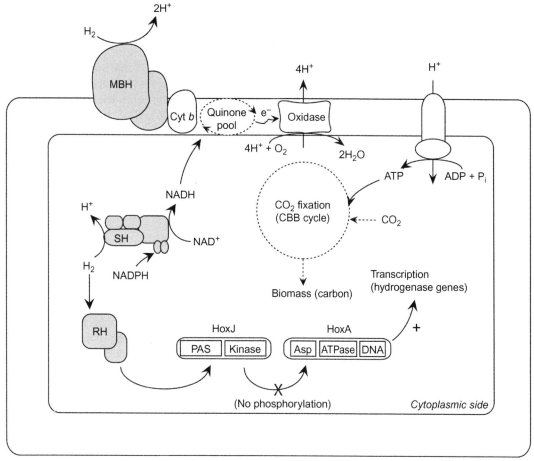

FIGURE 3 Schematic representation of H_2 metabolism in *Ralstonia eutropha*. Auxiliary and structural genes specifying the membrane-bound uptake hydrogenase (MBH) and soluble hydrogenase (SH) are under control of the H_2-sensing hydrogenase (RH). Although the megaplasmid harboring all the genetic requirements for lithoautotrophic growth possess genes specifying the auxiliary and structural components of the putative high-affinity hydrogenase (group 5), the physiological role of the enzyme, if any, remains obscure. PAS, Per–Arnt–Sim domain necessary for signal transduction; Kinase, histidine kinase domain; Asp, aspartate residues phosphorylated in the absence of H_2; DNA, DNA-binding site.

bidirectional *in vivo*, and they are involved in electron transport energy and ensure redox balance in the cells. These enzymes are found in bacteria and archaea, and are defined by four different phylogenetic clades. Hydrogenases belonging to clade 3A comprise the F_{420}-reducing hydrogenase (Frh) found in methanogenic archaea (Fig. 2). The enzyme was first purified under aerobic conditions from *Methanothermobacter marburgensis* and required reductive anaerobic activation (Fox et al., 1987; Livingston et al., 1987). This is a heterotrimeric hydrogenase comprising the small subunit (FrhG) accommodating three [4Fe-4S] clusters, the large subunit (FrhA) accommodating the NiFe center, and the third subunit (FrhB) with one [4Fe-4S] cluster and one FAD acting as an electron relay for reduction of the F_{420} coenzyme. F_{420} is a two-electron carrier essential for two reduction steps in hydrogenotrophic methanogenesis (Fig. 2), as well as other anabolic reactions (Thauer et al., 2010).

Hydrogenases belonging to clade 3B include the tetrameric bifunctional sulfhydrogenases of hyperthermophile, such as the anaerobic archaea *Pyrococcus furiosis*. This archaea possesses two bifunctional sulfhydrogenases demonstrating different kinetic properties and for which the exact physiological role remains elusive (Ma et al., 2000). The term bifunctional sulfhydrogenase refers to the ability of the enzyme to reduce S^0 to H_2S *in vitro* and to use NADPH as an electron donor (Ma et al., 1993, 1994). It was first proposed that sulfhydrogenase acts as an electron valve, generating H_2 to disperse of reducing equivalents derived from fermentation and producing NADPH for a biosynthetic purpose from H_2 (Silva et al., 2000; van Haaster et al., 2008). However, studies have demonstrated that dispersion of a reducing equivalent through H_2 production mediated by the bifunctional sulfhydrogenases is insufficient to ensure growth in the absence of elemental sulfur. Reducing equivalents generated in sugar metabolism are disposed of as H_2 by a membrane-bound hydrogenase belonging to group 4 (Schut et al., 2012).

Hydrogenases belonging to clade 3C comprise the F_{420}-nonreducing hydrogenase (Mvh; methyl viologen-reducing hydrogenase) found in methanogenic archaea without cytochromes (Fig. 2). This cytoplasmic hydrogenase comprises the small and large structural subunits (MvhGA) and an additional subunit (MvhD) mediating electron transfer between the hydrogenase and the heterodisulfide reductase (Stojanowic et al., 2003). Considering the fact that ferredoxin enhances the specific rate of CoM-S-S-CoB reduction, and also that Mvh gene operons contain gene-specifying polyferredoxin that copurify with the complex, it has been proposed that the complex couples the endergonic reduction of ferredoxin to the exergonic reduction of CoM-S-S-CoB (Thauer et al., 2010). In methanogens with cytochrome, these reactions are coupled by the generation of chemiosmotic gradients involving the group 1 MBH and a group 4 Ech [NiFe]-hydrogenase (Fig. 2).

Hydrogenases belonging to clade 3D use the NAD(P) cofactor. The prototype enzyme of this subgroup found in bacteria was first characterized in the knallgas bacterium *R. eutropha*. The enzyme is composed of two different catalytic subunits: a heterodimeric hydrogenase (HoxHY) and a NADH-dehydrogenase module (HoxUF). Two HoxI proteins are attached to the diaphorase moiety and serve as the NADPH-binding site. The enzyme couples the reversible H_2 oxidation to NAD^+ reduction, balancing the $NAD^+/NADH$ pool in the cell. Indeed, reducing equivalents in the form of NADH are necessary for several cell functions, including energy metabolism (Fig. 3). During the transition to anaerobic conditions, the enzyme clears the cell of excess reducing equivalents. As the hydrogenase is not coupling H_2 oxidation to the reduction of $NADP^+$, it was proposed that HoxI proteins play a role in activation of the active site of the enzyme (Burgdorf et al., 2005). O_2 tolerance of the enzyme has been attributed to the incorporation of additional cyanide ligands to the NiFe active site during maturation of the enzyme (Van der Linden et al., 2004), and it has been proposed that this conformational change was partly attributed to HypX, responsible for the incorporation of additional nickel-bound cyanide during maturation of the hydrogenase (Bleijlevens et al., 2004). This paradigm shifted when the active site of the hydrogenase was examined in whole cells instead of using purified enzyme preparations. These analyses demonstrated that architecture of the active site corresponded to canonical $CN_{(2)}^- - CO$ ligands (Horch et al., 2010). Investigations are in progress to unveil how the nonstandard cofactors (flavin mononucleotide and [4Fe-4S] cluster) of the SH preserve H_2 oxidation activity under aerobic conditions (Lauterbach et al., 2011). In terms of biotechnological application, diaphorase subunits of the soluble hydrogenase from *R. eutropha* were coupled to the structural subunits of hydrogenase-2 from *E. coli* to catalyze the reduction of NAD^+ in the presence of H_2—offering great promise for biocatalyzed redox reactions in the industry (Reeve et al., 2012). In cyanobacteria and purple sulfur photosynthetic bacteria, clade 3D hydrogenase is a heteropentameric enzyme comprising hydrogenase (HoxYH) and diaphorase (HoxFUE) moieties. There is no homologue of HoxI in these photosynthetic bacteria, and the role of HoxE remains yet to be elucidated. It shares high homology with NuoE of respiratory complex I, but it is bound loosely to the membrane and its link with respiration has not been demonstrated experimentally. Even though sequences of bidirectional hydrogenases from several microorganisms are now available and have been shown to be distributed ubiquitously in the environment (Barz et al., 2010), the physiological role of the enzyme is still a matter of debate. Extensive analyses have demonstrated that the bidirectional hydrogenase of *Synechocystis* sp. PCC 6803 and the purple sulfur photosynthetic bacteria *Thiocapsa roseopersicina* are mainly active under anaerobic conditions (Cournac et al., 2004; Rákhely et al., 2004). It has been proposed that the hydrogenase works as an electron valve, avoiding the formation of an excess of reducing equivalents (i.e., NADH) during dark fermentation or under elevated light densities by generating H_2 via the photosynthetic electron transport system (Appel et al., 2000). Bioengineering tools are used to increase biohydrogen production in cyanobacteria by directing electron flow toward the bidirectional hydrogenase and away from other electron-competing pathways (Tamagnini et al., 2007).

Group 4, Membrane-Associated, Energy-Converting [NiFe]-Hydrogenase

Hydrogenases belonging to group 4 are highly divergent from the other groups. In general, these multimeric enzymes reduce protons from water to dispose of excess reducing equivalents generated during fermentation through the anaerobic oxidation of C_1 (in)organic compounds, such as carbon monoxide and formate. These enzymes generally comprise four hydrophilic subunits and two integral membrane proteins. Identification of the genes encoding for the complex highlighted the similarity of subunits to the NADH:quinone oxidoreductase (complex I), suggesting that CO oxidation and H_2 evolution reactions are coupled to the formation of an energy-conservative proton motive force across the membrane. A prototype of these hydrogenases was first characterized in *E. coli*. During fermentation, enterobacteria convert an important proportion of the sugars into formate. Acidification of the cytoplasm and dispersion of reducing equivalents are counteracted by the so-called formate hydrogen-lyase reaction: $HCOO^- + H_2O \rightarrow HCO_3^- + H_2$. This reaction involves two enzymes: a formate dehydrogenase and hydrogenase-3, belonging to group 4 hydrogenases. Genetic modification of formate dehydrogenase and hydrogenases in *E. coli* led to significant enhancements of H_2 production, demonstrating the potential of metabolic engineering for biohydrogen fermentation applications (Maeda et al., 2008).

Group 4 hydrogenases are also widespread in facultative chemolithoautotrophic bacteria using CO as the sole energy and carbon source under anaerobic conditions. This metabolic flexibility has been observed in few nonsulfurous purple bacteria, such as *Rhodospirillum rubrum* and *Rhodocyclus gelatinosus*. In these photosynthetic bacteria, the presence of CO was shown to induce expression of a CO-dehydrogenase (CODH) and a hydrogenase catalyzing the net reaction $CO + H_2O \rightarrow CO_2 + H_2$. The CODH and the CO-induced hydrogenase are both inactivated by O_2 (Bonam et al., 1989). Reducing equivalents flow from the CODH to the hydrogenase through the iron–sulfur protein CooF bridging the CO oxidation/H_2 evolution system (Ensign and Ludden, 1991). The hydrogenase is thus decreasing the thermodynamic backpressure imposed by the electrochemical gradient generated across the membrane (Maness et al., 2005). This CODH–hydrogenase complex has been purified from the thermophilic Gram-positive bacterium *Carboxydothermus hydrogenoformans* (Soboh et al., 2002). Linkage between CODH and hydrogenase has inspired development of a biocatalyst catalyzing the water–gas shift reaction, resulting in the production of biohydrogen from CO and H_2O (Lazarus et al., 2009).

Two different group 4 hydrogenases are present in methanogenic archaea (Fig. 2). A first enzyme required for acetoclastic methanogenesis was characterized in *Methanosarcina barkeri*. The first type is the heterohexameric energy-converting hydrogenase (EchA-F), catalyzing the H_2-dependent reduction of ferredoxin coupled to protons translocation, leading to formation of a proton gradient across the cytoplasmic membrane for ATP generation through ATP synthase (Welte et al., 2010). Produced H_2 is then oxidized by the group 1 MBH coupled to the heterodisulfide reductase. The Ech hydrogenase is inactivated by O_2 and CO and was shown present in acetate, methanol, and H_2/CO_2-grown cells (Meuer et al., 1999). The second type of group 4 [NiFe]-hydrogenase is found in methanogens without cytochromes. These archaea possess ferredoxin-reducing, energy-converting hydrogenases (Eha and Ehb) catalyzing H_2 oxidation coupled to generation of a sodium ion motive force (Fig. 2). These hydrogenases contain 17–20 subunits and provide reduced ferredoxin for biosynthesis, as well as for the first step of methanogenesis—a function that is equivalent to Ech hydrogenase (Tersteegen and Hedderich, 1999). Eha is proposed to be coupled to formylmethanofuran dehydrogenase (FMD) via ferredoxin to catalyze the first reaction of methanogeneis from CO_2 and H_2 (Fig. 2). However, Ehb is proposed to be coupled to CODH-acetyl coenzyme A (a key enzyme in autotrophic CO_2 fixation) and a number of ferredoxin-dependent oxidoreductases that catalyze important anabolic reactions (Porat et al., 2006).

Finally, group 4 hydrogenases include the Mbh-type, energy-converting, H_2-producing enzyme first described in hyperthermophile *Pyrococcus furiosus* within the order Thermococcales (Sapra et al., 2000, 2003). These hyperthermophilic enzymes and their physiological role were the focus of an extensive review (Schut et al., 2013). In these hyperthermophile archeae, ferredoxin is the only electron acceptor utilized during the oxidation of sugar, and the role of Mbh is to replenish oxidized ferredoxin through H_2 production. This reaction is exergonic and coupled to the generation of an ion gradient across the membrane. Mutagenesis experiments have demonstrated that Mbh is essential for growth in the absence of elemental sulfur (Kanai et al., 2011).

Group 5, High-Affinity [NiFe]-Hydrogenase

Investigations dedicated to identify microorganisms responsible for the soil uptake of atmospheric H_2 led to the discovery of a novel hydrogenase (Constant et al., 2008, 2010). In contrast to knallgas bacteria demonstrating an (app)K_m of 1000 ppmv H_2, the H_2 oxidation activity of *Streptomyces* spp. is governed by an unusual kinetic characterized by an extremely high affinity, with an (app)K_m lower than 100 ppmv H_2. Genetic and biochemical investigations are in progress in the laboratories of the authors to unveil the molecular attributes conferring such unusual character to the enzyme. Even though high-affinity H_2-oxidizing actinobacteria are ubiquitous in soil, the exact role of the hydrogenase remains yet to be elucidated. Based on thermodynamic models, it was proposed that high-affinity, H_2-oxidizing bacteria harness the energy potential of atmospheric H_2 to supply long-term persistence (Constant et al., 2011). High-affinity hydrogenase might support formation of a microbial seed bank in the environment, ensuring the stability of microbial populations and the stability of ecosystem services.

[FE]-HYDROGENASES

[Fe]-hydrogenases (Hmd) are cytoplasmic enzymes composed of a homodimer of 38-kDa subunits. The enzyme is distributed unevenly in hydrogenotrophic methanogenic archaea (without cytochrome) utilizing $4H_2$ and CO_2 to generate methane. The enzyme differs from [NiFe]- and [FeFe]-hydrogenase by several aspects. Indeed, Hmd is devoid of any iron–sulfur cluster, does not mediate the reduction of one-electron acceptors (F_{420}, NAD, NADP, methylviologen), and does not catalyze the exchange between H_2 and protons of water in absence of an electron mediator or the conversion of *para* and *ortho* H_2 (Thauer et al., 1996; Zirngibl et al., 1992). The enzyme was first purified from the thermophilic methanogen *Methanothermobacter marburgensis*. The purified enzyme was shown to catalyze the reversible reduction of N_5,N_{10}-methenyl-5,6,7,8-tetrahydromethanopterin with H_2 to N_5,N_{10}-methylene-5,6,7,8-tetrahydromethanopterin and a proton (Schleucher et al., 1994; Schwörer et al., 1993; Zirngibl et al., 1990). In methanogens, H_4MPT serves as a C_1 compound carrier, leading to the reduction of CO_2 into CH_4 (Fig. 2). Over the course of methanogenesis, the C_1 compound originating from CO_2 is first added to H_4MPT in the oxidation state of formic acid and the carbon is subsequently stepwise reduced into methenyl and methyl. The methyl group is then transferred to coenzyme M and finally reduced into CH_4. The reducing equivalents required for these reduction steps are supplied by H_2. Hmd is not ubiquitous in hydrogenotrophic methanogens because the reduction of $CH \equiv H_4MPT^+$ to $CH_2 = MPT$ is also catalyzed by two other enzymes, namely, F_{420}-reducing [NiFe]-hydrogenase (Frh) and F_{420}-dependent methylene–H_4MPT dehydrogenase. Heterologous expression of the *hmd* gene in *E. coli* resulted in production of an inactive enzyme. However, high activity was restored by the addition of an ultrafiltrate from the active enzyme purified from methanogenic archaea denatured in urea, providing evidence for the occurrence of a cofactor bound to the enzyme (Buurman et al., 2000). Infrared spectroscopy demonstrated that two CO bind Fe into the cofactor (Lyon et al., 2004), and X-ray absorption spectroscopy revealed the presence of Fe with two CO, one sulfur, and one or two N/O ligands (Korbas et al., 2006). As observed in [NiFe]-hydrogenase and [FeFe]-hydrogenases, the Fe atom of the cofactor was low-spin redox inactive (Shima et al., 2005). The crystal structure of the Hmd confirmed previous spectroscopic analyzes, revealing that an iron atom is coordinated by the residue Cys_{176} of the holozyme, two CO, one guanylpyridol molecule, and a fifth ligand for which the chemical nature is unknown (Pilak et al., 2006; Shima et al., 2008). The genome of several methanogens contains genes specifying the Hmd holoenzyme, as well as one or two copies of paralog genes. Compelling evidence suggests that Hmd paralogs may bind the cofactor competitively, acting as a reservoir of cofactor when the H_2 level is low (Goldman et al., 2009; Shima and Thauer, 2007) or specifically bind archaeal aminoacyl-tRNA synthetases, being involved in protein synthesis (Lipman et al. 2003). Expression of the genes encoding two isozymes (HmdII, HmdIII) was not influenced by nickel, but H_2 partial pressure exerted a significant effect. The relative abundance of the HmdII transcript and protein was higher in cells grown under 5% H_2 than in 80% H_2-grown cells, while elevated H_2 partial pressure was necessary for HmdIII (Afting et al., 2000). The ecological advantage of Hmd may be related to a unique flexibility in H_2 metabolism, providing the archaea with a high capacity system to exploit elevated H_2 concentrations escaping from thermal vents or in close proximity of synthrophic H_2-producing bacteria. This is supported by two observations. First, Hmd gene expression and activity are maximal in the early growth stage, under conditions of excess H_2 (Morgan et al., 1997; Nölling et al., 1995; Vermeij et al., 1997). Second, the enzyme demonstrates low affinity for H_2, with a K_m in the range of 0.1–0.2 mM (Hartmann et al. 1996). In laboratory cultures, Hmd-specific activity was higher under nickel-limiting conditions than nonlimiting conditions (Afting et al., 1998), supporting the notion that the enzyme confers metabolic flexibility. In contrast to [NiFe]-hydrogenase, the maturation apparatus of Hmd has not been determined experimentally. Genes involved in the maturation of the enzyme have been identified *in silico* and are described in an extensive review on hydrogenases from methanogenic archaea (Thauer et al., 2010).

[FeFe]-HYDROGENASES

In stark contrast to [NiFe]-hydrogenases, which are distributed broadly in nature, the distribution of [FeFe]-hydrogenases is much more narrowly restricted. The gene encoding the catalytic core of the enzyme (*hydA*) is used as a functional biomarker to study distribution of the enzyme. An initial metagenomic survey of Sargasso Sea water showed that the natural occurrence and diversity of [FeFe]-hydrogenases in the environment are rather low (Meyer, 2004). Even a more recent survey using a BLAST search of over one billion base pairs of a nonredundant sequence found only 10 hits to a [FeFe]-hydrogenase bait sequence (Meyer, 2007). This can also be seen in studies of specific environments, although in some cases the *hydA* sequences obtained were not examined to verify that they are bona fide sequences containing the critical H cluster-binding residues. Another critique in the use of these approaches for studying [FeFe]-hydrogenase distribution is that none are capable of determining if the organism possessing the *hydA* sequence also possesses *hydE*, *hydF*, and *hydG*, thought to be required for [FeFe]-hydrogenase maturation (see later). Indeed, the observed *hydA* sequence may be an "orphan" sequence of unknown function, as a number of organisms with completely sequenced genomes have been shown to contain *hydA* without the accessory genes (Mulder et al., 2010). One environment of potential interest in obtaining *hydAs* of potential biotechnological interest is the upper (photic) region of a saline microbial mat. A large fraction of the putative *hydA* sequences obtained from this microenvironment was affiliated most closely with the Firmicutes (66.2% of the clones) and the Verrucomicrobia (24.6%), with a lower proportion related to the Bacteroidetes (9.2%) (Boyd et al., 2009). A degree of O_2 stability may be conferred by substitutions in the L1 motif, involved in the coordination of the oxygen-labile [4Fe-4S] subcluster of the H cluster, found in many of the putative HydA sequences recovered. Of course, because it is of interest to understand the genetic basis behind H_2 production during fermentations by mixed cultures, a number of studies have employed molecular techniques to examine this. Not too surprisingly, in one such study (Xing et al., 2008), genome sequences extracted from a system coproducing ethanol–H_2 gave 11 phylotypes closely related to *Ethanoligenens* and *Clostridium thermocellum*, in the Clostridiaceae family, well-known H_2 producers.

Thus, [FeFe]-hydrogenases are distributed much more narrowly than [NiFe]-hydrogenases. Nevertheless, they show a highly diverse modular structure (Calusinska et al., 2010; Meyer, 2007) (see later). An idea of the range of distribution of [FeFe]-hydrogenases can be obtained by examining complete genome sequences for what are likely to be functional [FeFe]-hydrogenases, that is, genomes that contain *hydE*, *hydF*, and *hydG* homologues as well as *hydA*. By this measure, [FeFe]-hydrogenases are widely found among the Gram-positive phylum Firmicutes, especially the order Clostridia, as well as in the Gram-negative γ-Proteobacteria (*Desulfovibrio*), with one example known in the δ-Proteobacteria (*Shewanella*) (Meyer, 2007), and are apparently absent from the α- and β-Proteobacteria. Aside from these *hydAs*, a surprisingly large number of "orphan" *hydAs*, that is, *hydAs* in genomes that appear to lack *hydE*, *hydF*, and *hydG* homologues, can be found; what their possible function might be and if they can be matured by an unrecognized maturation system are interesting, but presently unanswered questions.

[FeFe]-Hydrogenase Active Site Biosynthesis

As alluded to earlier, all [FeFe]-hydrogenases are thought to contain the same unique diiron active site. Relatively little was known until recently about the synthesis of this active H cluster. A great deal of research over the past few years has shed considerable light on synthesis of the active site H cluster and [FeFe]-hydrogenase maturation (Mulder et al., 2010; Peters and Broderick, 2012). The H cluster consists of a [4Fe-4S] subcluster linked by a cysteine thiolate to a modified 2Fe subcluster with unique nonprotein ligands, including CO and CN^- (Fig. 4). The hydrogenase maturation enzymes HydE, HydF, and HydG are together responsible for the synthesis of this 2Fe subcluster as well as the nonprotein ligands. In perhaps another example of convergent evolution, all three hydrogenases, [NiFe]-, [Fe]-, and [FeFe]-hydrogenases, require at least one inorganic, toxic ligand, CO, as an essential component of their active site. In addition, [NiFe]- and [FeFe]-hydrogenases require CN^-. How these toxic ligands are generated by cellular metabolism and inserted into the complex metal centers is of considerable interest. Two different routes, depending on whether the active site to

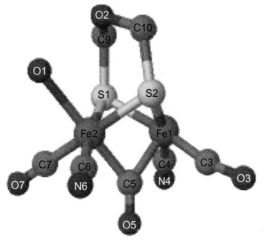

FIGURE 4 Active site structure of the [FeFe]-hydrogenase from *Clostridium pasteurianum* (pdb: 3C8Y) as determined by X-ray crystallography (Pandey et al., 2008). The bridging ligand was thought to be dithiomethylether when this structure was refined, but quantum refinement suggests that it is actually dithiomethylamine (Ryde et al., 2010). The figure was extracted from the pdb entry using the Jmol viewer. Shown are the Fe and S atoms, as well as CN^- and CO.

be made is for the [NiFe]-hydrogenase or the [FeFe]-hydrogenase, are used to generate the required cyanide. The CN^- component of the NiFe center comes from enzyme-bound thiocyanate generated from carbamoyl phosphate (Reissmann et al., 2003) and thus free cyanide as an intermediate is avoided. The CN^- of the active site of [FeFe]-hydrogenase (H cluster) is derived from a different route, the *S*-adenosylmethionine-dependent cleavage of tyrosine by HydG to give *p*-cresol and cyanide (Kuchenreuther et al., 2009). It is likely that free cyanide is not liberated and that the generated CN^- is passed immediately to HydF, which probably functions as a scaffold for the assembly of at least part of the H cluster (McGlynn et al., 2008). It has been shown that HydF contains a CO and CN^- ligated di-iron cluster (Czech et al., 2010) and that the 2Fe subcluster of the H cluster is synthesized on HydF from a [2Fe-2S] cluster framework in a process requiring HydE, HydG, and guanosine-5'-triphosphate (GTP) (Shepard et al., 2010). This suggests that the [4Fe-4S] subcluster of the active site can be formed by generalized host cell machinery, borne out by biochemical studies (Mulder et al., 2009), as well as recent determination of the structure of HydA$^{\Delta EFG}$ (HydA expressed in a host lacking HydE, F, and G), which revealed the existence of a preformed [4Fe-4S] subcluster (Mulder et al., 2010).

Thus, a variety of studies have demonstrated that the primary role of specialized hydrogenase maturation proteins, HydE, HydF, and HydG, is to synthesize the 2Fe subcluster with its CN^- and CO ligands and insert it into HydA containing a preformed [4Fe-4S] subcluster (Mulder et al., 2011). How the CO that forms an integral part of the H cluster is generated has been elucidated (Kuchenreuther et al., 2009; Shepard et al., 2010). In fact, the origin of the CO ligands had been unknown for all three hydrogenase active sites, although at least for the NiFe center it is thought to be derived from a different pathway than that used for synthesis of the CN^- ligand (Forzi and Sawers, 2007). Convincing evidence that both CN^- and CO ligands are synthesized from tyrosine has been presented in a study in which high-activity [FeFe]-hydrogenase was synthesized *in vitro* in a reaction requiring only *E. coli* cell lysates, *Shewanella oneidensis* maturases (HydE, F, and G), tyrosine, cysteine, *S*-adenosyl methionine, Fe^{+2}, S^{-2}, dithiothreitol, GTP, pyridoxal-5'-phosphate, and sodium dithionite

(Kuchenreuther et al., 2011). Recent evidence for dithiomethylamine as the previously unknown bridging ligand of the 2Fe subcluster has been presented (Ryde et al., 2010), and it is likely that HydG is involved in its synthesis (Pilet et al., 2009). Thus, the basic steps in [FeFe]-hydrogenase H cluster synthesis have been elucidated and the essential roles of HydE, F, and G in this process highlighted. Therefore, if, and how, orphan HydAs might be matured remains an open and intriguing question. Also, the mechanism behind the reported maturation of heterologously expressed HydAs in cyanobacteria (Asada et al., 2000; Berto et al., 2011) remains a deep mystery.

Diversity in [FeFe]-Hydrogenase Domain Structure and Interaction with Metabolism

Although [FeFe]-hydrogenases are of limited distribution, and although all contain a basic domain capable of H cluster binding, genomic sequence analysis has revealed a wide variety of different modular structures (Calusinska et al., 2010; Meyer, 2007) (Fig. 5). Many Clostridial genomes encode multiple [FeFe]-hydrogenases of different classes, many of which are characterized relatively poorly or not at all. Although they can be divided into monomeric, dimeric, trimeric, and even tetrameric forms, all, except group M2a, possess a basic underlying module consisting of an H domain, necessary for the catalytic site, and an F domain, consisting of the sequence necessary to ligand two [4Fe-4S] clusters, which conduct electrons from the redox partner to the active site (H cluster). [FeFe]-hydrogenases of group M2a are truncated and may coordinate only a single FeS cluster (Fig. 5A). Of course the main function of most members of this diverse protein family is probably H_2 production, and members of different classes may have evolved to interact differently with cellular metabolism. The organisms that are the best known fermenters, and H_2 producers, are the Clostridia, also the organisms with the richest content of hydrogenases. The Clostridia (strict anaerobes) are thought to principally use the pyruvate:ferredoxin oxidoreductase (PFOR) system to convert pyruvate to acetyl-CoA, and CO_2, producing reduced ferredoxin. The reduced ferredoxin transfers electrons to a [FeFe]-hydrogenase, driving H_2 evolution, giving 2 mol of H_2/mol of glucose (Fig. 6).

Monomeric Forms

Many different [FeFe]-hydrogenases are monomeric and include the best characterized hydrogenases, that of *Clostridium pasteurianum* and *Desulfovibrio desulfuricans* (from the M3b clade), well studied on the biochemical and structural levels. The types of hydrogenases exemplified by that of *C. pasteurianum* are widespread among the *Clostridium* and are likely to be the primary players in fermentative H_2 evolution, reacting with reduced ferredoxin. Thus, the [FeFe]-hydrogenases of clade M3b probably interact with ferredoxin. The [FeFe]-hydrogenases of clades M2a and M2e differ in their content of auxiliary FeS centers, suggesting that they interact with different partners. The other monomeric hydrogenases contain additional domains; either domains suggesting additional redox functions, such as previously unseen runs of Cys, suggesting coordination of an unknown metal center (M3a), and a complex rubredoxin domain (M2d), or domains suggestive of regulatory

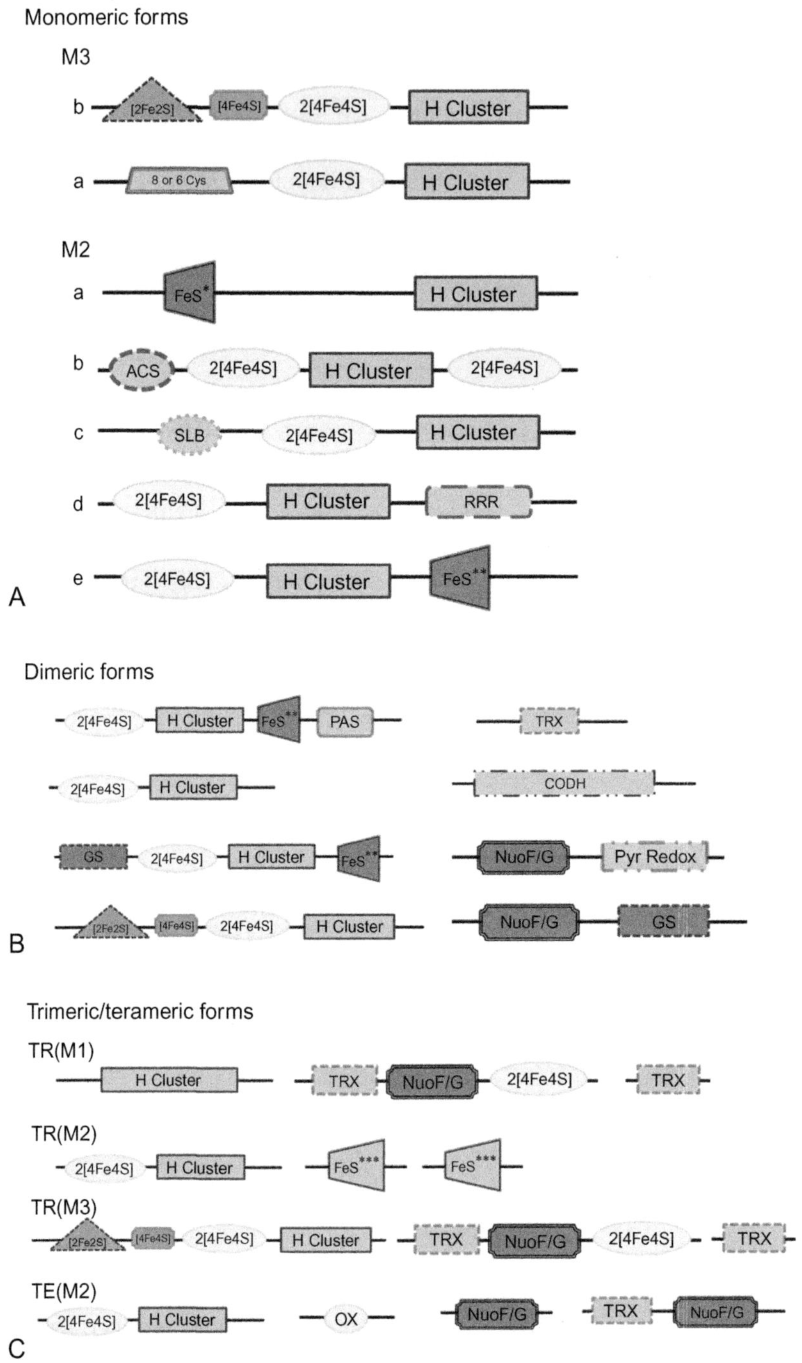

FIGURE 5 Modular domain structure of [FeFe]-hydrogenases. Shown are different hydrogenases found in genomes of various Clostridia (Calusinska et al. 2010), as well as in bacteria in general (Meyer, 2007). (A) Some of the different monomeric forms that can be discerned from sequence analysis. The top member of the M3 class is the ferredoxin-linked [FeFe]-hydrogenase from *Clostridium pasteurianum*, well studied at the protein level. A related hydrogenase in the M3 clade from *Clostridiun beijerinki* is also shown. In the M2 clade, many different monomeric forms can be found, some with additional domains that may either be involved in specialized redox reactions or in sensing and signaling. (B) Some of the dimeric forms that can be discerned in the genomes of some bacteria. (C) Several different trimeric and tetrameric forms that have been discovered. Many of the dimeric and trimeric forms contain Nuo subunits, suggesting a physiological role of NADPH in the function of these enzymes. Some Clostridial species can contain up to five different hydrogenase forms. Obviously much molecular and biochemical work remains to be done to elucidate the function of these various hydrogenases.

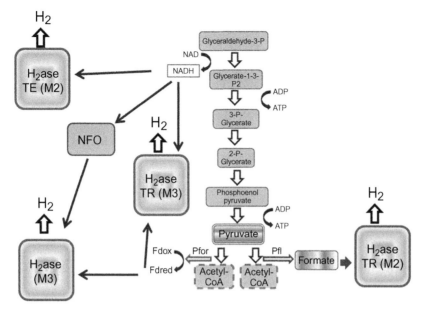

FIGURE 6 Various [FeFe]-hydrogenases and the pyruvate node of metabolism. Shown are the various possibilities for connection of different hydrogenases in Clostridia to glycolysis and pyruvate degradation suggested by the present evidence. Some aspects will require further verification. Hydrogenases are as noted in Figure 4 and discussed in the text. Some organisms can degrade at least part of the pyruvate to formate. If present, a trimeric hydrogenase may, perhaps anchored to the cytoplasmic membrane by an associated membrane protein, couple formate oxidation to proton reduction. The well-known monomeric *C. pasteurianum*-type hydrogenase (type M3) catalyzes H_2 production from reduced ferredoxin, either produced directly from pyruate or produced by NADH oxidation by NADH: ferredoxin oxidoreductase (NFO). NFO could be the soluble activity described several decades ago or the more recently described membrane complex of the Rnf family. Some multimeric hydrogenases [TE (M2)], similar to Hnd from the sulfate reducers, could possibly evolve H_2 directly from NAD(P)H at very low H_2 partial pressures. Finally, the newly described timeric bifurcating hydrogenase [TR(M3)] could simultaneously use NADH and reduced ferredoxin to evolve H_2 with the excess energy available from ferredoxin oxidation driving the unfavorable oxidation of NADH.

functions (PAS, SLBB). Nothing is known about these hydrogenases, except their presence in the genome, and additional characterization on the molecular level will be required to elucidate their function.

Multimeric [FeFe]-Hydrogenases

Many Clostridia species, as well as other bacteria, have been shown by genomic analysis to contain putative multimeric hydrogenases. These include dimeric, trimeric, and even one tetrameric hydrogenase. Dimeric hydrogenases appear to be of several different varieties (Fig. 5B). Many of these are novel, containing, for example, the [FeFe]-hydrogenase component of a CO-oxidizing system that may be a homologue to the CODH/group 4 [NiFe]-hydrogenase in nonsulfurous purple bacteria or [FeFe]-hydrogenase with a glutamate synthase interaction domain that may therefore be involved directly in supplying reducing equivalents from H_2 for nitrogen assimilation. None of these have been characterized on the protein or even mRNA level.

The newly described trimeric and tetrameric hydrogenases (Fig. 5C) are of considerable interest in fully understanding the metabolic basis of H_2 production by Clostridia and some other organisms, such as *Thermotoga maritima*, *Thermoanaerobacter tengcongensis*, and *Caldicellulosiruptor saccharolyticus* (Schut and Adams, 2009; Soboh et al., 2004; van de Werken et al., 2008). In terms of their modular structure, there appear to be two major types. One group [TR (M2)], carrying the basic [FeFe]-hydrogenase subunit and two additional subunits with [FeS] clusters, may be involved in H_2 production through formate oxidation, as suggested for *Eubacterium acidaminophilum* (Graentzdoerffer et al., 2003). Very little is known about formate metabolism in the Clostridia. *C. thermocellum* accumulates considerable amounts of formate during fermentative growth (Sparling et al., 2006), probably due to the fact that it appears to lack the TR(M2) enzyme system, as well as formate dehydrogenase.

Finally, a recent study shed light on a hydrogenase from a different class of trimeric enzymes [TR(M3)], encoded in the genomes of at least one-third of known H_2-producing bacteria, demonstrating unique properties that require a reexamination of the bioenergetics of the H_2-producing fermentations of many bacteria (Schut and Adams, 2009). This enzyme appears, at least in *Thermotogamaritima*, to use the simultaneous oxidation of NADH and reduced ferredoxin to reduce protons to H_2. In this way, the additional free energy available in reduced ferredoxin can drive the thermodynamically unfavorable flow of electrons from NADH to H_2. This type of bifurcating enzyme is proposed to be an example of a newly found form of energy conservation in bacteria, one that has already been found in the bifurcating system of *Clostridium kluyveri* that couples the energy gained during butyrate formation to drive ferredoxin reduction by NADH, leading to H_2 evolution (Herrmann et al., 2008). A similar trimeric hydrogenase is found in *Thermoanaerobacter tengcongensis*, but there it was proposed to be solely a NADH-dependent hydrogenase (Soboh et al., 2004). However, this enzyme may act as a bifurcating enzyme under conditions of low H_2 concentration. The possible importance of the trimeric bifurcating enzyme in H_2 evolution under normal fermentation conditions is highlighted by a study in which a knockout of the trimeric hydrogenase of the related *Thermoanaerobacterium saccharolyticum* was created. H_2 production was profoundly affected in this, but not one with a knockout of the more conventional type [FeFe]-hydrogenase, suggesting that this trimeric enzyme is, in fact, the primary one involved in H_2 fermentations (Shaw et al., 2009). Thus, H_2 production in many Clostridial-type fermentations may be more favorable than previously thought, where it was considered that 2 mol of H_2/mol of hexose would be derived from reduced ferredoxin and an additional maximum of 2 mol of H_2 by the oxidation of NADH, which could only occur at very low H_2 partial pressures. This would help explain, at least in part, what previously seemed like unusually high reported molar H_2 yields of some fermentations.

Multiple Pathways for Electron Flow to [FeFe]-Hydrogenase in Clostridia

Carbohydrates are the main feedstock for fermentations, leading to significant amounts of H_2. In general, the derived sugars are catabolized through the glycolytic pathway generating NADH and pyruvate, a key metabolite in many different fermentations. Pyruvate is then oxidized further by one of two different enzymes: pyruvate:formate lyase in enteric-type mixed acid fermentation and pyruvate:ferredoxin oxidoreductase in clostridial-type fermentations, generating acetyl-CoA in both cases. In Clostridia, reduced ferredoxin feeds electrons to the ferredoxin-dependent [FeFe]-hydrogenase, generating oxidized ferredoxin for further pyruvate utilization and 2 mol of H_2/mol of glucose.

Many organisms using the PFOR fermentation pathway are capable of producing additional H_2 by reoxidizing the NADH generated during glycolysis. Given that 2 mol of NADH are produced during the glycolysis of glucose, up to a maximum of two additional molecules of H_2 can be generated by this pathway. However, because the midpoint potential of the $H_2/2H^+$ (~450 mV at pH 7) is appreciably lower than that of the NADH/NAD couple (~320 mV), production of H_2 with electrons derived from NADH is only possible at a greatly reduced H_2 concentration or with the input of energy in some form. A variety of mechanisms can be proposed for driving H_2 production from NADH by a [FeFe]-hydrogenase.

Some organisms have been shown to contain heteromeric [FeFe]-hydrogenases, Hnd hydrogenases, whose subunits are homologues of some complex I proteins, suggesting that they might couple directly with NAD(P)(H). How they function is unclear at present and they may actually serve different functions in different organisms. In the Firmicutes, this enzyme may catalyze H_2 evolution from NADH directly, as suggested from a study with *Thermoanaerobacter tengcongensis* (Soboh et al., 2004). However, a study of a similar enzyme in *Desulfovibrio fructosovorans* suggested that the Hnd [FeFe]-hydrogenase functions physiologically in the reduction of NADP with H_2 (Malki et al. 1995).

In addition, Clostridia may possess pathways that allow them, under the proper conditions (low partial pressure of H_2), to couple NADH oxidation to proton reduction to H_2 through the intermediate reduction of ferredoxin (Fig. 6). In theory, these organisms are capable of producing an additional 2 mol of H_2/mol of glucose this way, giving a total of up to $4H_2$/glucose, albeit at very low H_2 partial pressures. Soluble NADH:ferredoxin oxidoreductase activity (Fig. 6) in crude extracts was reported a long time ago (Jungermann et al., 1973). However, the enzyme has yet to be purified, and a gene encoding this activity has never been identified. Two known pathways exist for increasing the thermodynamic driving force for the normally unfavorable reduction of ferredoxin by NADH. The Rnf complex, known to be present in some Clostridial species (Boiangiu et al., 2005), is a membrane-bound NADH:ferredoxin oxidoreductase.The Rnf complex, originally shown to be essential for photoheterotrophic nitrogen fixation, is encoded by seven genes in *Rhodobacter capsulatus* and is thought to drive ferredoxin reduction from NADH using an ion gradient (Schmehl et al., 1993). Thus, in some fermentative organisms, ferredoxin reduction by NADH could be driven by an Rnf complex using an existing ion gradient, making this a more favorable reaction. Recent work with the trimeric [FeFe]-hydrogenase from *Thermotoga maritime* shows another way in which NADH can be used to drive H_2 evolution. This enzyme has been reported to be a bifurcating enzyme, using NADH and reduced ferredoxin synergistically, coupling the excess energy available in reduced ferredoxin to "boost" the electron coming from NADH. Homologues of this enzyme are widespread (Schut and Adams, 2009). Because many Clostridia contain genes for this trimeric enzyme, this suggests that the energetics of H_2 production from NADH are more favorable, at a given H_2 partial pressure, than previously thought.

CONCLUSION

Thus nature seems to have independently evolved several different types of enzymes capable of H_2 oxidation or proton reduction, and within each type a surprising number of variants are found. These hydrogenases are of fundamental interest in understanding how different microorganisms have adapted to survive and thrive in their own ecological niches, as well as understanding how the basic process of H_2 oxidation or proton reduction on the enzyme level is coupled to various types of microbial metabolism. On an applied level, attempts to increase biological H_2 production will benefit from a greater understanding of these key enzymes in a number of ways. Knowledge about reactions at the active site may help in designing more effective artificial catalysts or in developing hydrogenases active in H_2 production in the presence of O_2. Furthermore, given the large numbers of different hydrogenases found in Clostridia, it would appear that much work remains to be done in understanding the molecular details of H_2 production in Clostridia, some of which may be useful in designing more efficient and more effective strains for biohydrogen production. In the future, synthetic biology may be used to create highly effective strains containing hydrogenases and pathways that link them efficiently to cellular metabolism.

Acknowledgments

PC acknowledges support from the NSERC-Discovery grant program supporting research on hydrogen metabolism and biogeochemistry in his laboratory. Research on hydrogen metabolism and advanced biofuels in the laboratory of PCH is funded by grants from NSERC and FQRNT.

References

Afting, C., Hochheimer, A., Thauer, R.K., 1998. Function of H_2-forming methylenetetrahydromethanopterin dehydrogenase from *Methanobacterium thermoautotrophicum* in coenzyme F_{420} reduction with H_2. Arch. Microbiol. 169 (3), 206–210.

Afting, C., Kremmer, E., Brucker, C., Hochheimer, A., Thauer, R.K., 2000. Regulation of the synthesis of H_2-forming methylenetetrahydromethanopterin dehydrogenase (Hmd) and of HmdII and HmdIII in *Methanothermobacter marburgensis*. Arch. Microbiol. 174 (4), 225–232.

Appel, J., Phunpruch, S., Steinmüller, K., Schulz, R., 2000. The bidirectional hydrogenase of *Synechocystis* sp. PCC 6803 works as an electron valve during photosynthesis. Arch. Microbiol 173 (5), 333–338.

Asada, Y., Koike, Y., Schnackenberg, J., Miyake, M., Uemura, I., Miyake, J., 2000. Heterologous expression of clostridial hydrogenase in the cyanobacterium *Synechococcus* PCC7942. Biochim. Biophys. Acta 1490 (3), 269–278.

Barz, M., Beimgraben, C., Staller, T., Germer, F., Opitz, F., Marquardt, C., et al., 2010. Distribution analysis of hydrogenases in surface waters of marine and freshwater environments. PLoS ONE 5 (11), e13846.

Berto, P., D'Adamo, S., Bergantino, E., Vallese, F., Giacometti, G.M., Costantini, P., 2011. The cyanobacterium *Synechocystis* sp. PCC 6803 is able to express an active [FeFe]-hydrogenase without additional maturation proteins. Biochem. Biophys. Res. Commun 405 (4), 678–683.

Black, L.K., Fu, C., Maier, R.J., 1994. Sequences and characterization of *hupU* and *hupV* genes of *Bradyrhizobium japonicum* encoding a possible nickel-sensing complex involved in hydrogenase expression. J. Bacteriol. 176 (22), 7102–7106.

Bleijlevens, B., Buhrke, T., van der Linden, E., Friedrich, B., Albracht, S.P.J., 2004. The auxiliary protein HypX provides oxygen tolerance to the soluble [NiFe]-hydrogenase of *Ralstonia eutropha* H16 by way of a cyanide ligand to nickel. J. Biol. Chem. 279 (45), 46686–46691.

Böck, A., King, P.W., Blokesch, M., Posewitz, M.C., 2006. Maturation of hydrogenases. In: Robert, K.P. (Ed.), Advances in Microbial Physiology, vol. 51. Academic Press, New York, pp. 1–225.

Boiangiu, C.D., Jayamani, E., Brügel, D., Herrmann, G., Kim, J., Forzi, L., et al., 2005. Sodium ion pumps and hydrogen production in glutamate fermenting anaerobic bacteria. J. Mol. Microbiol. Biotechnol. 10 (2–4), 105–119.

Bonam, D., Lehman, L., Roberts, G.P., Ludden, P.W., 1989. Regulation of carbon monoxide dehydrogenase and hydrogenase in *Rhodospirillum rubrum*: Effects of CO and oxygen on synthesis and activity. J. Bacteriol. 171 (6), 3102–3107.

Bothe, H., Schmitz, O., Yates, M.G., Newton, W.E., 2010. Nitrogen fixation and hydrogen metabolism in cyanobacteria. Microbiol. Mol. Biol. Rev. 74 (4), 529–551.

Boyd, E.S., Spear, J.R., Peters, J.W., 2009. [FeFe] Hydrogenase genetic diversity provides insight into molecular adaptation in a saline microbial mat community. Appl. Environ. Microbiol 75 (13), 4620–4623.

Brugna-Guiral, M., Tron, P., Nitschke, W., Stetter, K.-O., Burlat, B., Guigliarelli, B., et al., 2003. [NiFe] hydrogenases from the hyperthermophilic bacterium *Aquifex aeolicus*: Properties, function, and phylogenetics. Extremophiles 7 (2), 145–157.

Buhrke, T., Lenz, O., Krauss, N., Friedrich, B., 2005. Oxygen tolerance of the H_2-sensing [NiFe] hydrogenase from *Ralstonia eutropha* H16 is based on limited access of oxygen to the active site. J. Biol. Chem. 280 (25), 23791–23796.

Burgdorf, T., van der Linden, E., Bernhard, M., Yin, Q.Y., Back, J.W., Hartog, A.F., et al., 2005. The soluble NAD^+-reducing [NiFe]-hydrogenase from *Ralstonia eutropha* H16 consists of six subunits and can be specifically activated by NADPH. J. Bacteriol. 187 (9), 3122–3132.

Buurman, G., Shima, S., Thauer, R.K., 2000. The metal-free hydrogenase from methanogenic archaea: Evidence for a bound cofactor. FEBS Lett. 485 (2–3), 200–204.

Calusinska, M., Happe, T., Joris, B., Wilmotte, A., 2010. The surprising diversity of clostridial hydrogenases: A comparative genomic perspective. Microbiology 156 (6), 1575–1588.

Carrasco, C.D., Holliday, S.D., Hansel, A., Lindblad, P., Golden, J.W., 2005. Heterocyst-specific excision of the *Anabaena* sp. strain PCC 7120 *hupL* element requires *xisC*. J. Bacteriol 187 (17), 6031–6038.

Cavendish, H., 1766. Three papers, containing experiments on factitious air. Philos. Trans. R. Soc. Lond. 56, 141–184.

Constant, P., Chowdhury, S.P., Hesse, L., Pratscher, J., Conrad, R., 2011. Genome data mining and soil survey for the novel group 5 [NiFe]-hydrogenase to explore the diversity and ecological importance of presumptive high affinity H_2-oxidizing bacteria. Appl. Environ. Microbiol. 77 (17), 6027–6035.

Constant, P., Poissant, L., Villemur, R., 2008. Isolation of *Streptomyces* sp. PCB7, the first microorganism demonstrating high-affinity uptake of tropospheric H_2. ISME J 2 (10), 1066–1076.

Constant, P., Poissant, L., Villemur, R., 2009. Tropospheric H_2 budget and the response of its soil uptake under the changing environment. Sci. Total Environ 407 (6), 1809–1823.

Constant, P., Chowdhury, S.P., Pratscher, J., Conrad, R., 2010. Streptomycetes contributing to atmospheric molecular hydrogen soil uptake are widespread and encode a putative high-affinity [NiFe]-hydrogenase. Environ. Microbiol. 12 (3), 821–829.

Cournac, L., Guedeney, G., Peltier, G., Vignais, P.M., 2004. Sustained photoevolution of molecular hydrogen in a mutant of *Synechocystis* sp. strain PCC 6803 deficient in the type I NADPH-dehydrogenase complex. J. Bacteriol 186 (6), 1737–1746.

Czech, I., Silakov, A., Lubitz, W., Happe, T., 2010. The [FeFe]-hydrogenase maturase HydF from *Clostridium acetobutylicum* contains a CO and CN^- ligated iron cofactor. FEBS Lett. 584 (3), 638–642.

Dementin, S., Leroux, F., Cournac, L., Lacey, A.L., Volbeda, A., Léger, C., et al., 2009. Introduction of methionines in the gas channel makes [NiFe] hydrogenase aero-tolerant. J. Am. Chem. Soc. 131 (29), 10156–10164.

Dubini, A., Pye, R.L., Jack, R.L., Palmer, T., Sargent, F., 2002. How bacteria get energy from hydrogen: A genetic analysis of periplasmic hydrogen oxidation in *Escherichia coli*. Int. J. Hydrogen Energy 27 (11–12), 1413–1420.

Elsen, S., Colbeau, A., Chabert, J., Vignais, P.M., 1996. The *hupTUV* operon is involved in negative control of hydrogenase synthesis in *Rhodobacter capsulatus*. J. Bacteriol. 178 (17), 5174–5181.

Ensign, S.A., Ludden, P.W., 1991. Characterization of the CO oxidation/H_2 evolution system of *Rhodospirillum rubrum*: Role of a 22-kDa iron-sulfur protein in mediating electron transfer between carbon monoxide dehydrogenase and hydrogenase. J. Biol. Chem. 266 (27), 18395–18403.

Fontecilla-Camps, J.C., Volbeda, A., Cavazza, C., Nicolet, Y., 2007. Structure/Function relationships of [NiFe]- and [FeFe]-hydrogenases. Chem. Rev. 107 (10), 4273–4303.

Forzi, L., Sawers, R., 2007. Maturation of [NiFe]-hydrogenases in *Escherichia coli*. Biometals 20 (3), 565–578.

Fox, J.A., Livingston, D.J., Orme-Johnson, W.H., Walsh, C.T., 1987. 8-Hydroxy-5-deazaflavin-reducing hydrogenase from *Methanobacterium thermoautotrophicum*. 1. Purification and characterization. Biochemistry (Mosc) 26 (14), 4219–4227.

Friedrich, B., Fritsch, J., Lenz, O., 2011. Oxygen-tolerant hydrogenases in hydrogen-based technologies. Curr. Opin. Biotechnol. 22 (3), 358–364.

Fritsch, J., Scheerer, P., Frielingsdorf, S., Kroschinsky, S., Friedrich, B., Lenz, O., et al., 2011. The crystal structure of an oxygen-tolerant hydrogenase uncovers a novel iron-sulphur centre. Nature 479 (7372), 249–252.

Golding, A.L., Dong, Z., 2010. Hydrogen production by nitrogenase as a potential crop rotation benefit. Environ. Chem. Lett. 8 (2), 101–121.

Goldman, A.D., Leigh, J.A., Samudrala, R., 2009. Comprehensive computational analysis of Hmd enzymes and paralogs in methanogenic Archaea. BMC Evol. Biol. 9, 199.

Graentzdoerffer, A., Rauh, D., Pich, A., Andreesen, J.R., 2003. Molecular and biochemical characterization of two tungsten- and selenium-containing formate dehydrogenases from *Eubacterium acidaminophilum* that are associated with components of an iron-only hydrogenase. Arch. Microbiol. 179 (2), 116–130.

Gross, R., Simon, J., Theis, F., Kröger, A., 1998. Two membrane anchors of *Wolinella succinogenes* hydrogenase and their function in fumarate and polysulfide respiration. Arch. Microbiol. 170 (1), 50–58.

Gross, R., Simon, J., Kröger, A., 1999. The role of the twin-arginine motif in the signal peptide encoded by the *hydA* gene of the hydrogenase from *Wolinella succinogenes*. Arch. Microbiol. 172 (4), 227–232.

Guiral, M., Tron, P., Aubert, C., Gloter, A., Iobbi-Nivol, C., Giudici-Orticoni, M.T., 2005. A membrane-bound multi-enzyme, hydrogen-oxidizing, and sulfur-reducing complex from the hyperthermophilic bacterium *Aquifex aeolicus*. J. Biol. Chem. 280 (51), 42004–42015.

Hartmann, G.C., Klein, A.R., Linder, M., Thauer, R.K., 1996. Purification, properties and primary structure of H2-forming N5, N10-methylenetetrahydromethanopterin dehydrogenase from *Methanococcus thermolithotrophicus*. Arch. Microbiol. 165 (3), 187–193.

Herrmann, G., Jayamani, E., Mai, G., Buckel, W., 2008. Energy conservation via electron-transferring flavoprotein in anaerobic bacteria. J. Bacteriol. 190 (3), 784–791.

Horch, M., Lauterbach, L., Saggu, M., Hildebrandt, P., Lendzian, F., Bittl, R., et al., 2010. Probing the active site of an O_2-tolerant NAD^+-reducing [NiFe]-hydrogenase from *Ralstonia eutropha* H16 by *in situ* EPR and FTIR spectroscopy. Angew Chem. Int. Ed. Engl. 49 (43), 8026–8029.

Ide, T., Bäumer, S., Deppenmeier, U., 1999. Energy conservation by the H_2:heterodisulfide oxidoreductase from *Methanosarcina mazei* Gö1: Identification of two proton-translocating segments. J. Bacteriol. 181 (13), 4076–4080.

Ihara, M., Nishihara, H., Yoon, K.-S., Lenz, O., Friedrich, B., Nakamoto, H., et al., 2006. Light-driven hydrogen production by a hybrid complex of a [NiFe]-hydrogenase and the cyanobacterial photosystem I. Photochem. Photobiol. 82 (3), 676–682.

Ishii, M., Takishita, S., Iwasaki, T., Peerapornpisal, Y., Yoshino, J.-I, Kodama, T., et al., 2000. Purification and characterization of membrane-bound hydrogenase from *Hydrogenobacter thermophilus* strain TK-6, an obligately autotrophic, thermophilic, hydrogen-oxidizing bacterium. Biosci. Biotechnol. Biochem. 64 (3), 492–502.

Jones, A.K., Sillery, E., Albracht, S.P.J., Armstrong, F.A., 2002. Direct comparison of the electrocatalytic oxidation of hydrogen by an enzyme and a platinum catalyst. Chem. Commun. 8, 866–867.

Jungermann, K., Thauer, R.K., Leimenstoll, G., Decker, K., 1973. Function of reduced pyridine nucleotide-ferredoxin oxidoreductases in saccharolytic *Clostridia*. Biochim. Biophys. Acta 305 (2), 268–280.

Kanai, T., Matsuoka, R., Beppu, H., Nakajima, A., Okada, Y., Atomi, H., et al., 2011. Distinct physiological roles of the three [NiFe]-hydrogenase orthologs in the hyperthermophilic Archaeon *Thermococcus kodakarensis*. J. Bacteriol. 193 (12), 3109–3116.

Kleihues, L., Lenz, O., Bernhard, M., Buhrke, T., Friedrich, B., 2000. The H_2 sensor of *Ralstonia eutropha* is a member of the subclass of regulatory [NiFe] hydrogenases. J. Bacteriol. 182 (10), 2716–2724.

Korbas, M., Vogt, S., Meyer-Klaucke, W., Bill, E., Lyon, E.J., Thauer, R.K., et al., 2006. The iron-sulfur cluster-free hydrogenase (Hmd) is a metalloenzyme with a novel iron binding motif. J. Biol. Chem. 281 (41), 30804–30813.

Kröger, A., Biel, S., Simon, J., Gross, R., Unden, G., Lancaster, C.R.D., 2002. Fumarate respiration of *Wolinella succinogenes*: Enzymology, energetics and coupling mechanism. Biochim. Biophys. Acta 1553 (1–2), 23–38.

Kuchenreuther, J.M., Stapleton, J.A., Swartz, J.R., 2009. Tyrosine, cysteine, and *S*-adenosyl methionine stimulate *in vitro* [FeFe] hydrogenase activation. PLoS ONE 4 (10), e7565.

Kuchenreuther, J.M., George, S.J., Grady-Smith, C.S., Cramer, S.P., Swartz, J.R., 2011. Cell-free H-cluster synthesis and [FeFe] hydrogenase activation: All five CO and CN^- ligands derive from tyrosine. PLoS ONE 6 (5), e20346.

Lane, N., Allen, J.F., Martin, W., 2010. How did LUCA make a living? Chemiosmosis in the origin of life. Bioessays 32 (4), 271–280.

Laska, S., Lottspeich, F., Kletzin, A., 2003. Membrane-bound hydrogenase and sulfur reductase of the hyperthermophilic and acidophilic archaeon *Acidianus ambivalens*. Microbiology 149 (9), 2357–2371.

Lauterbach, L., Liu, J., Horch, M., Hummel, P., Schwarze, A., Haumann, M., et al., 2011. The hydrogenase subcomplex of the NAD⁺-reducing [NiFe] hydrogenase from *Ralstonia eutropha*: Insights into catalysis and redox interconversions. Eur. J. Inorg. Chem. 2011 (7), 1067–1079.

Lazarus, O., Woolerton, T.W., Parkin, A., Lukey, M.J., Reisner, E., Seravalli, J., et al., 2009. Water – gas shift reaction catalyzed by redox enzymes on conducting graphite platelets. J. Am. Chem. Soc. 131 (40), 14154–14155.

Lenz, O., Friedrich, B., 1998. A novel multicomponent regulatory system mediates H_2 sensing in *Alcaligenes eutrophus*. Proc. Natl. Acad. Sci. USA 95 (21), 12474–12479.

Lenz, O., Strack, A., Tran-Betcke, A., Friedrich, B., 1997. A hydrogen-sensing system in transcriptional regulation of hydrogenase gene expression in *Alcaligenes* species. J. Bacteriol. 179 (5), 1655–1663.

Lenz, O., Bernhard, M., Buhrke, T., Schwartz, E., Friedrich, B., 2002. The hydrogen-sensing apparatus in *Ralstonia eutropha*. J. Mol. Microbiol. Biotechnol. 4 (3), 255–262.

Lenz, O., Ludwig, M., Schubert, T., Bürstel, I., Ganskow, S., Goris, T., et al., 2010. H_2 conversion in the presence of O_2 as performed by the membrane-bound [NiFe]-hydrogenase of *Ralstonia eutropha*. Chem. Phys. Chem. 11 (6), 1107–1119.

Lindblad, P., Christensson, K., Lindberg, P., Fedorov, A., Pinto, F., Tsygankov, A., 2002. Photoproduction of H_2 by wildtype *Anabaena* PCC 7120 and a hydrogen uptake deficient mutant: From laboratory experiments to outdoor culture. Int. J. Hydrogen Energy 27 (11–12), 1271–1281.

Lipman, R.S.A., Chen, J., Evilia, C., Vitseva, O., Hou, Y.-M., 2003. Association of an aminoacyl-tRNA synthetase with a putative metabolic protein in Archaea. Biochemistry (Mosc) 42 (24), 7487–7496.

Livingston, D.J., Fox, J.A., Orme-Johnson, W., Walsh, C.T., 1987. 8-Hydroxy-5-deazaflavin-reducing hydrogenase from *Methanobacterium thermoautotrophicum*. 2. Kinetic and hydrogen-transfer studies. Derivation of a steady-state rate equation for deazaflavin-reducing hydrogenase. Biochemistry (Mosc) 26 (14), 4228–4237.

Lopes Pinto, F.A., Troshina, O., Lindblad, P., 2002. A brief look at three decades of research on cyanobacterial hydrogen evolution. Int. J. Hydrogen Energy 27 (11–12), 1209–1215.

Lyon, E.J., Shima, S., Buurman, G., Chowdhuri, S., Batschauer, A., Steinbach, K., et al., 2004. UV-A/blue-light inactivation of the 'metal-free' hydrogenase (Hmd) from methanogenic archaea. Eur. J. Biochem. 271 (1), 195–204.

Ma, K., Schicho, R.N., Kelly, R.M., Adams, M W., 1993. Hydrogenase of the hyperthermophile *Pyrococcus furiosus* is an elemental sulfur reductase or sulfhydrogenase: Evidence for a sulfur-reducing hydrogenase ancestor. Proc. Natl. Acad. Sci. USA 90 (11), 5341–5344.

Ma, K., Hao, Z., Adams, M.W.W., 1994. Hydrogen production from pyruvate by enzymes purified from the hyperthermophilic archaeon, *Pyrococcus furiosus*: A key role for NADPH. FEMS Microbiol. Lett. 122 (3), 245–250.

Ma, K., Weiss, R., Adams, M.W.W., 2000. Characterization of hydrogenase II from the hyperthermophilic archaeon *Pyrococcus furiosus* and assessment of its role in sulfur reduction. J. Bacteriol. 182 (7), 1864–1871.

Maeda, T., Sanchez-Torres, V., Wood, T.K., 2008. Metabolic engineering to enhance bacterial hydrogen production. Microb. Biotechnol. 1 (1), 30–39.

Malki, S., Saimmaime, I., De Luca, G., Rousset, M., Dermoun, Z., Belaich, J.P., 1995. Characterization of an operon encoding an NADP-reducing hydrogenase in *Desulfovibrio fructosovorans*. J. Bacteriol. 177 (10), 2628–2636.

Maness, P.-C., Huang, J., Smolinski, S., Tek, V., Vanzin, G., 2005. Energy generation from the CO oxidation-hydrogen production pathway in *Rubrivivax gelatinosus*. Appl. Environ. Microbiol. 71 (6), 2870–2874.

Martin, W., Muller, M., 1998. The hydrogen hypothesis for the first eukaryote. Nature 392 (6671), 37–41.

Matias, P.M., Pereira, I.A.C., Soares, C.M., Carrondo, M.A., 2005. Sulphate respiration from hydrogen in *Desulfovibrio* bacteria: A structural biology overview. Prog. Biophys. Mol. Biol. 89 (3), 292–329.

McGlynn, S.E., Shepard, E.M., Winslow, M.A., Naumov, A.V., Duschene, K.S., Posewitz, M.C., et al., 2008. HydF as a scaffold protein in [FeFe] hydrogenase H-cluster biosynthesis. FEBS Lett. 582 (15), 2183–2187.

Meuer, J., Bartoschek, S., Koch, J., Künkel, A., Hedderich, R., 1999. Purification and catalytic properties of Ech hydrogenase from *Methanosarcina barkeri*. Eur. J. Biochem. 265 (1), 325–335.

Meyer, J., 2004. Miraculous catch of iron–sulfur protein sequences in the Sargasso Sea. FEBS Lett. 570 (1–3), 1–6.

Meyer, J., 2007. [FeFe] hydrogenases and their evolution: A genomic perspective. Cell. Mol. Life Sci. 64 (9), 1063–1084.

Moreira, D., López-García, P., 1998. Symbiosis between methanogenic archaea and δ-Proteobacteria as the origin of eukaryotes: The syntrophic hypothesis. J. Mol. Evol. 47 (5), 517–530.

Morgan, R.M., Pihl, T.D., Nölling, J., Reeve, J.N., 1997. Hydrogen regulation of growth, growth yields, and methane gene transcription in *Methanobacterium thermoautotrophicum* deltaH. J. Bacteriol. 179 (3), 889–898.

Mulder, D.W., Ortillo, D.O., Gardenghi, D.J., Naumov, A.V., Ruebush, S.S., Szilagyi, R.K., et al., 2009. Activation of HydA$^{\Delta EFG}$ requires a preformed [4Fe-4S] cluster. Biochemistry (Mosc) 48 (26), 6240–6248.

Mulder, D.W., Boyd, E.S., Sarma, R., Lange, R.K., Endrizzi, J.A., Broderick, J.B., et al., 2010. Stepwise [FeFe]-hydrogenase H-cluster assembly revealed in the structure of HydA$^{\Delta EFG}$. Nature 465 (7295), 248–251.

Mulder, D.W., Shepard, E.M., Meuser, J.E., Joshi, N., King, P.W., Posewitz, M.C., et al., 2011. Insights into [FeFe]-hydrogenase structure, mechanism, and maturation. Structure 19 (8), 1038–1052.

Nölling, J., Pihl, T.D., Vriesema, A., Reeve, J.N., 1995. Organization and growth phase-dependent transcription of methane genes in two regions of the *Methanobacterium thermoautotrophicum* genome. J. Bacteriol. 177 (9), 2460–2468.

Novelli, P.C., Lang, P.M., Masarie, K.A., Hurst, D.F., Myers, R., Elkins, J.W., 1999. Molecular hydrogen in the troposphere: Global distribution and budget. J. Geophys. Res. 104 (D23), 30427–30444.

Ogata, H., Kellers, P., Lubitz, W., 2010. The crystal structure of the [NiFe] hydrogenase from the photosynthetic bacterium *Allochromatium vinosum*: Characterization of the oxidized enzyme (Ni-A state). J. Mol. Biol. 402 (2), 428–444.

Palágyi-Mészáros, L.S., Maróti, J., Latinovics, D., Balogh, T., Klement, É., Medzihradszky, K.F., et al., 2009. Electron-transfer subunits of the NiFe hydrogenases in *Thiocapsa roseopersicina* BBS. FEBS J. 276 (1), 164–174.

Pandelia, M.-E., Lubitz, W., Nitschke, W., 2012. Evolution and diversification of Group 1 [NiFe] hydrogenases: Is there a phylogenetic marker for O_2-tolerance? Biochim. Biophys. Acta 1817 (9), 1565–1575.

Pandey, A.S., Harris, T.V., Giles, L.J., Peters, J.W., Szilagyi, R.K., 2008. Dithiomethylether as a ligand in the hydrogenase H-cluster. J. Am. Chem. Soc. 130 (13), 4533–4540.

Peters, J.W., Broderick, J.B., 2012. Emerging paradigms for complex iron-sulfur cofactor assembly and insertion. Annu. Rev. Biochem. 81 (1), 429–450.

Peterson, R.B., Burris, R.H., 1978. Hydrogen metabolism in isolated heterocysts of *Anabaena* 7120. Arch. Microbiol. 116 (2), 125–132.

Pilak, O., Mamat, B., Vogt, S., Hagemeier, C.H., Thauer, R.K., Shima, S., et al., 2006. The crystal structure of the apoenzyme of the iron–sulphur cluster-free hydrogenase. J. Mol. Biol. 358 (3), 798–809.

Pilet, E., Nicolet, Y., Mathevon, C., Douki, T., Fontecilla-Camps, J.C., Fontecave, M., 2009. The role of the maturase HydG in [FeFe]-hydrogenase active site synthesis and assembly. FEBS Lett. 583 (3), 506–511.

Porat, I., Kim, W., Hendrickson, E.L., Xia, Q., Zhang, Y., Wang, T., et al., 2006. Disruption of the operon encoding Ehb hydrogenase limits anabolic CO_2 assimilation in the archaeon *Methanococcus maripaludis*. J. Bacteriol. 188 (4), 1373–1380.

Rakhely, G., Colbeau, A., Garin, J., Vignais, P.M., Kovacs, K.L., 1998. Unusual organization of the genes coding for HydSL, the stable [NiFe]hydrogenase in the photosynthetic bacterium *Thiocapsa roseopersicina* BBS. J. Bacteriol. 180 (6), 1460–1465.

Rákhely, G., Kovács, Á.T., Maróti, G., Fodor, B.D., Csanádi, G., Latinovics, D., et al., 2004. Cyanobacterial-type, heteropentameric, NAD$^+$-reducing NiFe hydrogenase in the purple sulfur photosynthetic bacterium *Thiocapsa roseopersicina*. Appl. Environ. Microbiol. 70 (2), 722–728.

Reeve, H.A., Lauterbach, L., Ash, P.A., Lenz, O., Vincent, K.A., 2012. A modular system for regeneration of NAD cofactors using graphite particles modified with hydrogenase and diaphorase moieties. Chem. Commun. 48 (10), 1589–1591.

Reissmann, S., Hochleitner, E., Wang, H., Paschos, A., Lottspeich, F., Glass, R.S., et al., 2003. Taming of a poison: Biosynthesis of the NiFe-hydrogenase cyanide ligands. Science 299 (5609), 1067–1070.

Römermann, D., Warrelmann, J., Bender, R.A., Friedrich, B., 1989. An *rpoN*-like gene of *Alcaligenes eutrophus* and *Pseudomonas facilis* controls expression of diverse metabolic pathways, including hydrogen oxidation. J. Bacteriol. 171 (2), 1093–1099.

Ryde, U., Greco, C., De Gioia, L., 2010. Quantum refinement of [FeFe] hydrogenase indicates a dithiomethylamine ligand. J. Am. Chem. Soc. 132 (13), 4512–4513.

Sapra, R., Verhagen, M.F.J.M., Adams, M.W.W., 2000. Purification and characterization of a membrane-bound hydrogenase from the hyperthermophilic archaeon *Pyrococcus furiosus*. J. Bacteriol. 182 (12), 3423–3428.

Sapra, R., Bagramyan, K., Adams, M.W.W., 2003. A simple energy-conserving system: Proton reduction coupled to proton translocation. Proc. Natl. Acad. Sci. USA 100 (13), 7545–7550.

Schleucher, J., Griesinger, C., Schwoerer, B., Thauer, R.K., 1994. H$_2$-forming N5, N10-methylenetetrahydromethanopterin dehydrogenase from Methanobacterium thermoautotrophicumcatalyzes a stereoselective hydride transfer as determined by two-dimensional NMR spectroscopy. Biochemistry (Mosc) 33 (13), 3986–3993.

Schmehl, M., Jahn, A., Meyer zu Vilsendorf, A., Hennecke, S., Masepohl, B., Schuppler, M., et al., 1993. Identification of a new class of nitrogen fixation genes in Rhodobacter capsalatus: A putative membrane complex involved in electron transport to nitrogenase. Mol. Gen. Genet. 241 (5), 602–615.

Schut, G.J., Adams, M.W.W., 2009. The iron-hydrogenase of Thermotoga maritimautilizes ferredoxin and NADH synergistically: A new perspective on anaerobic hydrogen production. J. Bacteriol. 191 (13), 4451–4457.

Schut, G.J., Nixon, W.J., Lipscomb, G.L., Scott, R.A., Adams, M., 2012. Mutational analyses of the enzymes involved in the metabolism of hydrogen by the hyperthermophilic archaeon Pyrococcus furiosus. Front. Microbiol. 3, 163.

Schut, G.J., Boyd, E.S., Peters, J.W., Adams, M.W., 2013. The modular respiratory complexes involved in hydrogen and sulfur metabolism by heterotrophic hyperthermophilic archaea and their evolutionary implications. FEMS Microbiol. Rev. 37 (2), 182–203.

Schwartz, E., Gerischer, U., Friedrich, B., 1998. Transcriptional regulation of Alcaligenes eutrophus hydrogenase genes. J. Bacteriol. 180 (12), 3197–3204.

Schwörer, B., Fernandez, V.M., Zirngibl, C., Thauer, R.K., 1993. H$_2$-forming N5, N10-methylenetetrahydromethanopterin dehydrogenase from Methanobacterium thermoautotrophicum. Eur. J. Biochem. 212 (1), 255–261.

Shaw, A.J., Hogsett, D.A., Lynd, L.R., 2009. Identification of the [FeFe]-hydrogenase responsible for hydrogen generation in Thermoanaerobacterium saccharolyticum and demonstration of increased ethanol yield via hydrogenase knockout. J. Bacteriol. 191 (20), 6457–6464.

Shepard, E.M., McGlynn, S.E., Bueling, A.L., Grady-Smith, C.S., George, S.J., Winslow, M.A., et al., 2010. Synthesis of the 2Fe subcluster of the [FeFe]-hydrogenase H cluster on the HydF scaffold. Proc. Natl. Acad. Sci. USA 107 (23), 10448–10453.

Shima, S., Thauer, R.K., 2007. A third type of hydrogenase catalyzing H$_2$ activation. Chem. Rec. 7 (1), 37–46.

Shima, S., Lyon, E.J., Thauer, R.K., Mienert, B., Bill, E., 2005. Mössbauer studies of the iron – sulfur cluster-free hydrogenase: The electronic state of the mononuclear Fe active site. J. Am. Chem. Soc. 127 (29), 10430–10435.

Shima, S., Pilak, O., Vogt, S., Schick, M., Stagni, M.S., Meyer-Klaucke, W., et al., 2008. The crystal structure of [Fe]-hydrogenase reveals the geometry of the active site. Science 321 (5888), 572–575.

Shomura, Y., Yoon, K.-S., Nishihara, H., Higuchi, Y., 2011. Structural basis for a [4Fe-3S] cluster in the oxygen-tolerant membrane-bound [NiFe]-hydrogenase. Nature 479 (7372), 253–256.

Silva, P.J., van den Ban, E.C.D., Wassink, H., Haaker, H., de Castro, B., Robb, F.T., et al., 2000. Enzymes of hydrogen metabolism in Pyrococcus furiosus. Eur. J. Biochem. 267 (22), 6541–6551.

Soboh, B., Linder, D., Hedderich, R., 2002. Purification and catalytic properties of a CO-oxidizing:H$_2$-evolving enzyme complex from Carboxydothermus hydrogenoformans. Eur. J. Biochem. 269 (22), 5712–5721.

Soboh, B., Linder, D., Hedderich, R., 2004. A multisubunit membrane-bound [NiFe] hydrogenase and an NADH-dependent Fe-only hydrogenase in the fermenting bacterium Thermoanaerobacter tengcongensis. Microbiology 150 (7), 2451–2463.

Sparling, R., Islam, R., Cicek, N., Carere, C., Chow, H., Levin, D.B., 2006. Formate synthesis by Clostridium thermocellum during anaerobic fermentation. Can. J. Microbiol. 52 (7), 681–688.

Stephenson, M., Stickland, L.H., 1931. Hydrogenase: A bacterial enzyme activating molecular hydrogen. Biochem. J. 25, 205–214.

Stojanowic, A., Mander, G.J., Duin, E.C., Hedderich, R., 2003. Physiological role of the F$_{420}$-non-reducing hydrogenase (Mvh) from Methanothermobacter marburgensis. Arch. Microbiol. 180 (3), 194–203.

Sun, J., Hopkins, R.C., Jenney Jr., F.E., McTernan, P.M., Adams, M.W.W., 2010. Heterologous expression and maturation of an NADP-dependent [NiFe]-hydrogenase: a key enzyme in biofuel production. PLoS ONE 5 (5), e10526.

Swanson, K.D., Duffus, B.R., Beard, T.E., Peters, J.W., Broderick, J.B., 2011. Cyanide and carbon monoxide ligand formation in hydrogenase biosynthesis. Eur. J. Inorg. Chem. 2011 (7), 935–947.

Tamagnini, P., Costa, J.-L., Almeida, L., Oliveira, M.-J., Salema, R., Lindblad, P., 2000. Diversity of cyanobacterial hydrogenases, a molecular approach. Curr. Microbiol. 40 (6), 356–361.

Tamagnini, P., Leitão, E., Oliveira, P., Ferreira, D., Pinto, F., Harris, D.J., et al., 2007. Cyanobacterial hydrogenases: Diversity, regulation and applications. FEMS Microbiol. Rev. 31 (6), 692–720.

Tersteegen, A., Hedderich, R., 1999. Methanobacterium thermoautotrophicum encodes two multisubunit membrane-bound [NiFe] hydrogenases. Eur. J. Biochem. 264 (3), 930–943.

Thauer, R.K., Klein, A.R., Hartmann, G.C., 1996. Reactions with molecular hydrogen in microorganisms: Evidence for a purely organic hydrogenation catalyst. Chem. Rev. 96 (7), 3031–3042.

Thauer, R.K., Kaster, A.-K., Goenrich, M., Schick, M., Hiromoto, T., Shima, S., 2010. Hydrogenases from methanogenic archaea, nickel, a novel cofactor, and H_2 storage. Annu. Rev. Biochem. 79 (1), 507–536.

van de Werken, H.J.G., Verhaart, M.R.A., VanFossen, A.L., Willquist, K., Lewis, D.L., Nichols, J.D., et al., 2008. Hydrogenomics of the extremely thermophilic bacterium *Caldicellulosiruptor saccharolyticus*. Appl. Environ. Microbiol. 74 (21), 6720–6729.

van der Linden, E., Burgdorf, T., Bernhard, M., Bleijlevens, B., Friedrich, B., Albracht, S.P.J., 2004. The soluble [NiFe]-hydrogenase from *Ralstonia eutropha* contains four cyanides in its active site, one of which is responsible for the insensitivity towards oxygen. J. Biol. Inorg. Chem. 9 (5), 616–626.

van Haaster, D.J., Silva, P.J., Hagedoorn, P.-L., Jongejan, J.A., Hagen, W.R., 2008. Reinvestigation of the steady-state kinetics and physiological function of the soluble NiFe-hydrogenase I of *Pyrococcus furiosus*. J. Bacteriol. 190 (5), 1584–1587.

Vermeij, P., Pennings, J.L., Maassen, S.M., Keltjens, J.T., Vogels, G.D., 1997. Cellular levels of factor 390 and methanogenic enzymes during growth of *Methanobacterium thermoautotrophicum* deltaH. J. Bacteriol. 179 (21), 6640–6648.

Vignais, P.M., Billoud, B., 2007. Occurrence, classification, and biological function of hydrogenases: An overview. Chem. Rev. 107 (10), 4206–4272.

Vignais, P.M., Colbeau, A., 2004. Molecular biology of microbial hydrogenases. Curr. Issues Mol. Biol. 6, 159–188.

Vincent, K.A., Cracknell, J.A., Lenz, O., Zebger, I., Friedrich, B., Armstrong, F.A., 2005. Electrocatalytic hydrogen oxidation by an enzyme at high carbon monoxide or oxygen levels. Proc. Natl. Acad. Sci. USA 102 (47), 16951–16954.

Volbeda, A., Charon, M.-H., Piras, C., Hatchikian, E.C., Frey, M., Fontecilla-Camps, J.C., 1995. Crystal structure of the nickel-iron hydrogenase from *Desulfovibrio gigas*. Nature 373 (6515), 580–587.

Volbeda, A., Amara, P., Darnault, C., Mouesca, J.-M., Parkin, A., Roessler, M.M., et al., 2012. X-ray crystallographic and computational studies of the O_2-tolerant [NiFe]-hydrogenase 1 from *Escherichia coli*. Proc. Natl. Acad. Sci. USA 109 (14), 5305–5310.

Welte, C., Krätzer, C., Deppenmeier, U., 2010. Involvement of Ech hydrogenase in energy conservation of *Methanosarcina mazei*. FEBS J. 277 (16), 3396–3403.

Weyman, P.D., Pratte, B., Thiel, T., 2008. Transcription of *hupSL* in *Anabaena variabilis* ATCC 29413 is regulated by NtcA and not by hydrogen. Appl. Environ. Microbiol. 74 (7), 2103–2110.

Wu, L.-F., Chanal, A., Rodrigue, A., 2000. Membrane targeting and translocation of bacterial hydrogenases. Arch. Microbiol. 173 (5), 319–324.

Xing, D., Ren, N., Rittmann, B.E., 2008. Genetic diversity of hydrogen-producing bacteria in an acidophilic ethanol-H_2-coproducing system, analyzed using the [Fe]-hydrogenase gene. Appl. Environ. Microbiol. 74 (4), 1232–1239.

Zhang, Y.H.P., Evans, B.R., Mielenz, J.R., Hopkins, R.C., Adams, M.W.W., 2007. High-yield hydrogen production from starch and water by a synthetic enzymatic pathway. PLoS ONE 2 (5), e456.

Zirngibl, C., Hedderich, R., Thauer, R.K., 1990. N5, N10-Methylenetetrahydromethanopterin dehydrogenase from *Methanobacterium thermoautotrophicum* has hydrogenase activity. FEBS Lett. 261 (1), 112–116.

Zirngibl, C., Van Dongen, W., Schwörer, B., Von Bünau, R., Richter, M., Klein, A., et al., 1992. H_2-forming methylene-tetrahydromethanopterin dehydrogenase, a novel type of hydrogenase without iron-sulfur clusters in methanogenic archaea. Eur. J. Biochem. 208 (2), 511–520.

Biohydrogen Production from Organic Wastes by Dark Fermentation

G. Balachandar, Namita Khanna, Debabrata Das

Department of Biotechnology,
Indian Institute of Technology, Kharagpur, India

INTRODUCTION

Present Energy Scenario

Fossil fuels consist of oil, coal, and natural gas. Presently, most countries rely on fossil fuels to feed their growing energy demands. However, the supply of these fossil fuel reserves is inherently finite. It is generally agreed that we will run out of petroleum within 50 years, natural gas within 65 years, and coal in about 200 years at the current rate of consumption (Soetaert and Vandamme, 2009). Currently, fossil fuels provide almost 80% of the world's energy supply. Out of this, oil provides about 40% of the total global energy needs and fuels 90% of the transport sector. However, noted petroleum experts (Campbell and Laherrere, 1998) predicted that oil production would peak around 2010 and thereafter decline. At present, as the underground oil reservoirs run dry, more oil producers are turning to unconventional sources such as tar, sands, and oil shale. Heavier crude oils, extracted from tar sands and shale, require the use of more energy-intensive methods as compared to conventional oil, resulting in increased costs of extraction and thus higher oil prices.

Moreover, fossil fuels are composed primarily of carbon and hydrogen atoms. On combustion they produce carbon dioxide, sulfur dioxide, and carbon monoxide as the main by-products. Over the years, these gaseous emissions have impacted our environment adversely and have contributed greatly to the effect of global warming. Changing weather patterns, droughts, and floods are a few manifestations of the adverse effect of global warming. The tipping point of atmospheric carbon dioxide has been estimated at 450 ppm with current levels reaching 370 ppm (Jacquet et al., 2007). It has been estimated that global carbon dioxide

(CO_2) emissions must be reduced by 50–85% by 2050 if global warming is to be confined between 2 and 2.4°C (Remme et al., 2011).

Today, wind power, hydel power, nuclear power, and energy from biomass are being developed as alternatives to the rapidly depleting and CO_2-intensive fossil fuels. Wind energy is well established and is expected to contribute significantly in the short term. With advancements in the field of photovolatics, solar energy is also expected to contribute in the long term. In addition, bioenergy, the renewable energy released from biomass, is expected to contribute significantly in the mid to long term. In fact, according to the International Energy Agency, bioenergy offers the possibility to meet 50% of the world energy needs in the 21st century.

Benefits of Renewable Economy

In contrast to fossil fuels, a renewable energy economy appears attractive, as it is safe, easily accessible, and sustainable. The benefits of a renewable energy economy are multifold as discussed here.

(i) *Pollution:* Renewable energy is known to emit far lower concentrations of pollutants as compared to fossil fuels. Coal mining, petroleum exploration, and refinement produce solid toxic wastes, such as mercury and other heavy metals (Allen et al., 2011).
The burning of coal to produce electricity uses large quantities of water and often discharges arsenic and lead into the surface waters. In addition, it releases carbon dioxide, sulfur dioxide, nitrogen oxides, and mercury into the air (Mastorakis et al., 2011). Gasoline and other petroleum products cause similar pollution. The release of these harmful pollutants into the atmosphere causes health hazards, such as respiratory disorders, and environmental issues, such as acid rain and ozone layer depletion.

(ii) *Climate change:* Indiscriminate burning of fossil fuels releases excessive greenhouse gases into the atmosphere. This has caused widespread climate change and global warming. Climate change may damage agriculture, cause extinction of exotic flora and fauna, imperil clean water supplies, and aid the spread of tropical diseases (Dalby, 2009). However, the use of biofuels may help curb the climate change.

(iii) *Accessibility:* Coal, natural gas, and oil reserves are limited. An unknown and restricted amount of each resource is buried deep underground or under the ocean. As more is harvested, finding new sources becomes more difficult and expensive. Moreover, drilling under the ocean floor can lead to catastrophic accidents as, for example, the British Petroleum oil spill of 2010 (Adams and Gabbatt, 2010). Renewable energy, by contrast, is as easy to find as wind or sunlight.

(iv) *Reliability, stability, and safety:* The daily price of oil depends on many factors, including political stability in historically volatile regions. Political strife has caused energy crises, including those that occurred in 1973 and 1979 (Hakes, 1998). Renewable energy can be produced locally and therefore is not vulnerable to distant political upheavals. Many of the safety concerns surrounding fossil fuels, such as explosions on oil platforms and collapsing coal mines, do not exist with renewable energy.

Toward a Carbon-Neutral Fuel

CO_2 emissions from fuels depend primarily on their carbon content and their hydrogen–carbon ratio. Over the years, the trend of fossil fuel usage tends toward a higher hydrogen by carbon (H/C) ratio. The higher the H/C ratio, the higher the energy efficiency of the fuel (Fig. 1) and the lower the CO_2 emissions from its combustion (Fig. 2). Primitive fuel, such as wood, had twice the carbon content as compared to its successor, coal. Wood had twice the carbon content as compared to coal. However, coal, with a lower H/C ratio, was twice as energy efficient as compared to wood. Later, coal was succeeded by oil, which had a still higher H/C ratio and thus benefited over wood and coal in having higher energy efficiency and lower CO_2 emissions. Natural gas has still lower carbon content as compared to oil (Hoffert et al., 1998). However, the ratio of carbon to hydrogen is still lower in biofuels. In fact, biofuels such as hydrogen have zero carbon–hydrogen ratios. Therefore, the use of hydrogen as fuel may achieve the goal of complete decarbonization. Thus, today the focus

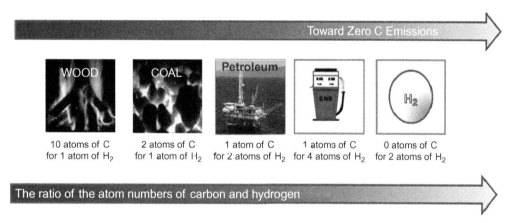

FIGURE 1 Toward zero carbon emissions. *From Hoffert et al. (1998)*

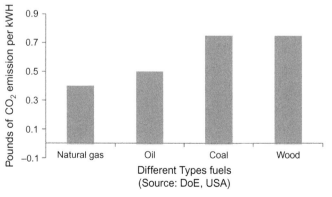

FIGURE 2 Comparison of CO_2 emissions from various fuels.

is more on the development of efficient hydrogen production technologies as alternatives to depleting and carbon-intensive fossil fuels.

Conventional Hydrogen Production Technologies and Limitations

Molecular hydrogen does not occur naturally in the Earth's crust. For practical applications, molecular hydrogen needs to be produced. At present, the total annual world hydrogen production is around 368 trillion cubic meters (Pandu and Joseph, 2012). Out of the total global production of hydrogen, 48% is produced from steam methane reforming, about 30% from oil/naphta reforming from refinery/chemical industrial off-gases, 18% from coal gasification, 3.9% from water electrolysis, and 0.1% from other sources (Baghchehsaree et al., 2010). These figures imply that globally 96% of the hydrogen production comes from fossils. However, these conventional processes are energy intensive and not always environmentally friendly. For example, methane steam reforming, coal gasification, and methane pyrolysis produce 0.25, 0.83, and 0.05 mol CO_2 mol^{-1} H_2, respectively (Abanades, 2012). Comparatively, hydrogen production from renewable resources appears to be more promising mainly due to three reasons: (i) they can utilize renewable energy resources, (ii) they are usually operated at ambient temperature and atmospheric pressure, and (iii) their production is not carbon intensive.

Biohydrogen Production Technology

Biological hydrogen production processes are not only environmentally friendly, but also inexhaustible (Benemann, 1997; Greenbaum, 1990). In addition, they can also be produced using waste as substrate. Thus, they may solve the dual problem of waste disposal and energy generation. Studies on hydrogen production have focused on direct and indirect biophotolysis mediated by cyanobacteria and green algae, photofermentative hydrogen production by photofermentative bacteria, and dark hydrogen production by fermentative bacteria. However, each of the processes has their own advantages and disadvantages. Major problems with biophotolysis of water and photofermentation are that they are light dependent and their hydrogen yield and rate of hydrogen production are too low for practical applications. Among the various other processes, dark fermentation appears to be more promising. It does not require light energy, requires moderate process conditions, and has lower energy demands. In addition, it can also use waste such as sewage sludge and distillery waste as substrates for hydrogen production. In addition, the hydrogen yield and the rate of hydrogen production from this process are more attractive as compared to the other processes (Das et al., 2008).

MICROBIOLOGY OF DARK FERMENTATIVE BACTERIA

In nature, hydrogen is produced by a wide variety of organisms. These organisms produce hydrogen mostly under anaerobic conditions. Among all the processes, dark fermentation is primarily responsible for most of the hydrogen produced. Hydrogen producers associated with this process are popularly known as dark fermentative microorganisms. These microorganisms may be classified based on their sensitivity to oxygen and temperature requirements. Those microorganisms that strictly require anaerobic conditions are called obligate anaerobes. Microorganisms that can sustain both anaerobic and aerobic environments are called facultative anaerobes. Facultative bacteria are always more advantageous for carrying out experimental work over obligate anaerobes as they are easier to cultivate and maintain in a laboratory. Further, based on their temperature needs, they may be classified further as mesophiles, which require ambient temperature for growth, or thermophiles that are adapted to higher temperatures. The major problem with thermophiles is that they are highly energy intensive (Das, 2009). However, they can produce hydrogen closer to the theoretical yield by overcoming the thermodynamic barrier. In nature, hydrogen can be produced either by pure microbial species or by a mixed community. Some members of the community may be hydrogen producing, while others may efficiently consume the hydrogen for energy purposes. In the laboratory, hydrogen studies may be carried out by either pure microbial species or mixed consortia. The choice of organisms depends mostly on the choice of fermentable substrate used.

Facultative Anaerobic Bacteria

In the presence of oxygen, facultative anaerobes can produce ATP by aerobic respiration. In the absence of oxygen, they can produce ATP by anaerobic fermentation. *Enterobacter* sp. are common facultative anaerobes that can produce hydrogen under anaerobic conditions. Members belonging to the family Enterobacteriaceae have several properties that favor H_2 production. Primarily, because they are facultative organisms, they are easier to handle in a reactor during anaerobic hydrogen production. Second, during the biohydrogen production process, a high partial pressure of hydrogen tends to build up in the reactor. Nakashimada et al. (2002) reported that these organisms can sustain higher concentrations of hydrogen. Moreover, they mostly possess either [FeFe]-hydrogenase or formate hydrogen lyase (FHL), which is responsible for a higher yield and higher rate of hydrogen production (Rachman et al., 1998).

Facultative anaerobes belonging to the Enterobacterieaceae family are known to produce a higher hydrogen yield and rate of hydrogen production. Two strains of Enterobacterieaceae have been studied extensively—*Enterobacter aerogenes* E.82005 isolated by Tanisho et al. (1987) from leaves of *Mirabilis jalapa* and *Enterobacter cloacae* IIT-BT 08 strain isolated by Kumar and Das (2000a) from leaf extracts. Under anaerobic batch cultivation, *E. aerogenes* E.82005 produced 1.0 mol H_2/mol glucose at a rate of 21 mmol/liter/h (Jo et al., 2008a,b). Studies were also carried out in a continuous mode of operation using molasses as the substrate. The average H_2 yield and the rate of hydrogen production were 1.5 mol H_2 mol^{-1} sucrose and 17 mmol H_2/liter/h, respectively (Tanisho and Ishiwata, 1994). However, extensive studies on hydrogen production were also reported using *E. cloacae* IIT-BT 08. It can grow and produce hydrogen using different carbon sources. In batch operations, the highest reported H_2 yield using glucose, sucrose, and cellobiose as substrates were 2.2, 6.0, and 5.4 mol H_2 per mole of substrate consumed, respectively. The maximum H_2 production rate was 35 mmol/liter/h using sucrose as the substrate (Kumar and Das, 2000b).

Obligate Anaerobic Bacteria

Until now, most of the research on hydrogen production has been carried out using obligate anaerobic bacteria because of their ability to utilize a wide range of carbohydrates, including different kinds of wastewater. Moreover, compared to facultative anaerobes, they also produce a higher rate of hydrogen production. Mostly *Clostridia* species have been widely used. They produce hydrogen mainly during the exponential growth phase. During the stationary phase, the metabolism of this organism shifts from hydrogen/acid production to solvent production during the stationary phase (Han and Shin, 2004). *Clostridia saccharoperbutylacetonicum, C. tyrobutyricum, C. butyricum, C. acetobutyricum, C. beijerinckii, C. thermolacticum, C. thermocellum*, and *C. paraputrificum* are promising examples of spore-forming hydrogen producers under anaerobic conditions. In their experiments, Lin et al. (2007) showed that different species of *Clostridium* can produce hydrogen in a range between 1.47 and 2.81 mol/mol glucose.

Thermophiles

Thermophiles are mostly obligate anaerobes found in various geothermally heated regions of the earth, such as hot springs and deep-sea hydrothermal vents. Their culture requirements differ depending on the source of isolation. Isolates from deep-sea volcanoes require a higher sodium chloride concentration, whereas isolates from volcanic vents require a higher sulfur concentration in growth and hydrogen production media (Schroder et al., 1994; van Niel et al., 2002). Further, because they are obligate anaerobes, reducing agents such as L-cystine HCl are required in media to remove even trace quantities of oxygen from the medium. Thermophiles can utilize a broad range of substrates such as cellulose, hemicelluloses, and pectin-containing biomass (van de Werken et al., 2008). Typical examples of this group include genera *Caldicellulosiuptor, Thermoanaerobacter*, and *Thermotoga*. Until now, the genus *Caldicellulosiuptor* contains five well-studied species, all of which are extremely thermophilic, Gram positive, and all but one cellulolytic (Onyenwoke et al., 2006). They have an unusual ability to degrade cellulose at high temperatures (up to 78°C) (Bredholt et al., 1999). Their major end metabolites include ethanol, lactate, and acetate. However, ethanol and lactate are rather low, although concomitantly, while hydrogen production is rather high (van Niel et al., 2003). The genus *Thermoanaerobacter* comprises 16 well-studied species that differ from *Caldicellulosiuptor* in that they cannot utilize cellulose, may form endospores, and may oxidize hydrogen using thiosulfate or Fe(III) as electron acceptors (Slobodkin et al., 1999). In contrast, members of the genus *Thermotogales* are Gram negative, anaerobic, thermophilic, and heterotrophic bacteria characterized by the presence of a toga, an outer sheath-like structure surrounding the cells. All members of this genus are able to utilize complex carbohydrate and proteins for growth and fermentative hydrogen production (van Ooteghem et al., 2002). Several members of this genus, such as *Thermotogamaritima, Thermotoganeapolitana*, and *Thermotogaelfii* (Nguyen et al., 2008; van Niel et al., 2002), have been isolated from waste material to carry out studies on hydrogen production (Schroder et al., 1994).

Mixed Cultures

Cocultures or mixed consortia are used mainly when a complex material, such as sewage sludge or cane molasses, is used as the substrate for hydrogen production. Use of consortium provides two essential functions. First, members of the consortium communicate with one another by trading metabolites or by exchanging dedicated molecular signals, which enables the second important feature—the division of labor for degrading the numerous complex substances. Thus, use of consortia may enhance substrate utilization. Numerous studies have reported the benefits of mixed consortia over pure culture while using complex substrates. Guwy et al. (1997) are of the opinion that mixed cultures for hydrogen production from organic waste may be more advantageous because pure cultures can become contaminated easily with hydrogen-consuming bacteria. In fact, in industries, due to economical reasons, hydrogen production using readily available complex feedstocks is usually carried out under nonsterile conditions. Mixed microbial consortia may address this problem, as they have been selected for growth and dominance under nonsterile conditions. They are potentially more robust to changes in environmental conditions such as pH and temperature.

Mixed cultures as inocula for hydrogen production can be isolated from various sources, such as fermented soybean meal or sludge from anaerobic digesters of municipal sewage or organic waste and sludge from kitchen wastewater. However, the major bottleneck in selecting these consortia is the presence of methanogens and hydrogen-consuming bacteria, within their community. In most cases, selected consortia are pretreated to suppress the activity of methanogens and remove hydrogen-consuming bacteria (Cheng et al., 2002; Shaw et al., 2008). Mostly mixed consortia contain *Clostridium* species. Production of hydrogen using mixed culture at higher temperature could be favorable to reaction kinetics, avoiding contamination by hydrogen-consuming bacteria (Zhang et al., 2003).

It is known that the culture of obligate anaerobes requires the presence of expensive reducing agents such as L-cysteine in growth and production media to maintain culture anaerobicity. Use of these chemicals increases costs and makes the process less economically favorable. In an interesting study, Yokoi et al. (2001) suggested a coculture of *C. butyricum* and *E. aerogenes* for hydrogen production to eliminate the use of such an expensive reducing agent to maintain the anaerobic environment. They suggested that any trace amounts of oxygen present in media may be consumed by facultative *E. aerogenes*. This would help create an anaerobic condition that may be favorable to propagate obligate *Clostridium butyricum*. They showed successfully hydrogen production using this coculture without the supplementation of reducing agents in production media. This approach may decrease the overall production cost considerably in case of large-scale hydrogen production.

BIOCHEMISTRY OF DARK FERMENTATIVE HYDROGEN PRODUCTION

Fermentation is a metabolic process that occurs under anaerobic conditions to regenerate the energy currency (ATP) of the cell. Moreover, under anaerobic conditions the tricarboxylic acid cycle is blocked. Therefore, fermentation disposes off the excess cellular reductant by the formation of reduced metabolic end products such as alcohol and acids. Similarly, hydrogen is also a reduced metabolic end product produced to maintain the cellular redox potential. Carbohydrates, mainly glucose, are the preferred carbon sources for the fermentation process that predominantly gives rise to acetic and butyric acids concomitantly with hydrogen gas. Complex organic polymers converted to glucose by hydrolysis. Glucose produces pyruvate via the glycolytic pathway to regenerate ATP. Subsequently, pyruvate may be involved in two different biochemical reactions leading to the formation of hydrogen shown in Figure 3.

In obligate anaerobes such as Clostridia (McCord et al., 1971) and thermophilic bacteria (Zeikus, 1977), pyruvate is oxidized to acetyl coenzyme A (acetyl-CoA) by pyruvate-ferredoxinoxidoreductase (Uyeda and Rabinowitz, 1971). Acetyl-CoA subsequently converts to acetyl phosphate with a concomitant generation of ATP and acetate. Oxidation of pyruvate to acetyl-CoA requires a reduction of ferredoxin (Fd). Reduced Fd is oxidized by

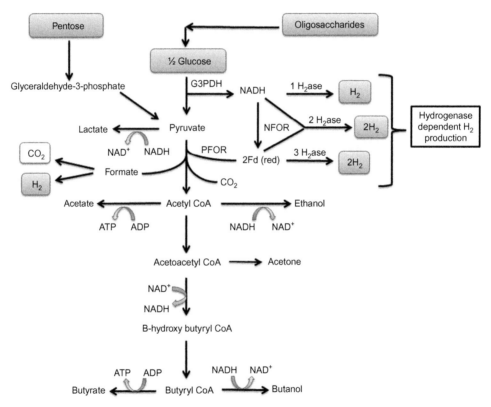

FIGURE 3 Metabolic insight on biohydrogen production. NFOR, NADH ferredoxin oxidoreductase; PFOR, pyruvate ferredoxin oxidoreductase. For hydrogen production, fermentable sugars need to get channelized to a glycolytic pathway so as to produce reducing equivalents such as NADH and $FADH_2$. These electron donors donate electrons to the hydrogenase enzyme to produce molecular hydrogen. Complex sugars such as polysaccharides are hydrolyzed to simpler sugars such as hexose. Then hexoses get utilized in the glycolytic pathway. In case of pentose sugars, these sugars were converted to glyceraldehyde-3-phosphate. This glyceraldehyde-3-phosphate is then channelized to the glycolytic pathway.

[FeFe]-hydrogenase and catalyzes the formation of H_2. The overall reaction is shown in Equations (1) and (2).

$$\text{Pyruvate} + \text{CoA} + 2\text{Fd (ox)} \rightarrow \text{Acetyl-CoA} + 2\text{Fd (red)} + CO_2 \tag{1}$$

$$2H^+ + \text{Fd (red)} \rightarrow H_2 + \text{Fd (ox)} \tag{2}$$

Four moles of hydrogen per mole of glucose is formed when pyruvate is oxidized to acetate as the sole metabolic end product (Benemann, 1996). However, only 2 mol of hydrogen per mole of glucose is produced when pyruvate is oxidized to butyrate. Therefore, for the organisms that follow a mixed acid pathway, a higher A/B ratio is critical for higher hydrogen production (Khanna et al., 2011). The overall biochemical reaction with acetic acid and butyric acid as metabolic end products is shown in Equations (3) and (4), respectively.

$$C_6H_{12}O_6 + 2H_2O \rightarrow 2CH_3COOH + 2CO_2 + 4H_2 \tag{3}$$

$$C_6H_{12}O_6 \rightarrow CH_3CH_2CH_2COOH + 2CO_2 + 2H_2 \tag{4}$$

The second type of biochemical reaction involved in hydrogen production occurs in a few facultative anaerobic bacteria such as *Escherichia coli*. In this pathway, pyruvate is oxidized to acetyl-CoA and formate. The reaction is catalyzed by the enzyme pyruvate formatelyase (Knappe and Sawers, 1990) [Eq. (5)]:

$$\text{Pyruvate} + \text{CoA} \rightarrow \text{acetyl-CoA} + \text{formate} \tag{5}$$

Subsequently, formate is cleaved by FHL to produce carbon dioxide and hydrogen [Eq. (6)]. Stephenson and Stickland (1932) first described this pathway in the 1930s.

$$HCOOH \rightarrow CO_2 + H_2 \tag{6}$$

In lactic acid fermentation, pyruvate is oxidized to lactate directly. No hydrogen is produced when ethanol, lactic, or propionic acid is the sole metabolic end product. Moreover, some facultative anaerobes such as Enteric bacteria may carry out anaerobic respiration instead of fermentation using nitrate or fumarate as terminal electron acceptors. Anaerobic respiration may hamper hydrogen production. Therefore, for carrying out studies on hydrogen production, media should be devoid of these electron acceptors.

Organic acids produced during dark fermentation can be used as a substrate for photofermentation. Photofermentative bacteria can oxidize organic acids to produce CO_2 and H_2. Therefore, higher hydrogen production can be obtained by following a two-stage process of dark fermentation followed by photofermentation (Das, 2009). Theoretically, 12 mol of hydrogen per mole of glucose can be produced from this hybrid process (Nath and Das, 2009).

$$2CH_3COOH + 4H_2O \rightarrow 8H_2 + 4CO_2 \tag{7}$$

THERMODYNAMICS OF BIOHYDROGEN PRODUCTION PROCESSES

Hydrogen yield is determined by thermodynamics of the process. The maximum hydrogen yield that can be obtained from glucose is 4 mol H_2/mol glucose. However, studies show that for most of the mesophiles, hydrogen yields are limited to 2 mol H_2/mol glucose. This is attributed to the fact that under standard conditions (1 M concentrations of the reactants, 25°C, pH 7) the complete oxidation of glucose to CO_2 and H_2 has a positive Gibbs energy change and, thus, will not proceed without an input of extra energy.

$$\text{Glucose} + 12H_2O \rightarrow 6HCO_3^- + 6H^+ + 12H_2 \quad \Delta G° = 32 kJ/mol \qquad (8)$$

Numerous experiments have shown that the production of acetate from glucose favors hydrogen production. The theoretical maximum of 4 mol of hydrogen can be produced only when glucose is oxidized into acetate with a concomitant production of hydrogen and CO_2. However, acetate production is not a redox reaction. Therefore, theoretically, this reaction is possible only if all the reducing equivalents are transferred to hydrogen. However, in practice, the reaction will be favored only when the hydrogen pressure is kept low either by hydrogen-consuming microorganisms or by stripping the reactor of hydrogen gas manually (Thauer et al., 1977).

During the fermentation of glucose, NADH and Fd are involved as the principal reducing equivalents. The midpoint redox potentials of $NAD^+/NADH$ and oxidized ferredoxin/reduced ferredoxin are −320 and −398 mV, respectively (Amend and Shock, 2001). The reducing equivalents are recycled to maintain the process continuity. Recycling of these electron carriers can be accomplished by a variety of reactions, such as the production of ethanol by the reduction of pyruvate by NADH. However, the likelihood of occurrence of these reactions is determined by the standard Gibbs free energy change ($\Delta G°$) of the individual conversions (Table 1). From Table 1 it can be observed that under standard conditions, the reduction of protons with NADH, resulting in molecular hydrogen, has a redox potential of around −414 mV. This potential is higher than that of the $NAD^+/NADH$ couple. Therefore, formation of molecular hydrogen by oxidation of NADH is thermodynamically constrained. Comparatively, oxidation of ferredoxin to produce molecular hydrogen is more favorable. However, formation of other reduced products such as ethanol, lactate is more feasible compared to hydrogen. However, in thermophiles, yields up to 4 mol H_2/mol glucose have been reported. The higher yields of thermophiles may be attributed to the higher temperature at which hydrogen production occurs. This is based on the following equation relating Gibb's free energy to enthalpy (H), temperature (T), and entropy (S) of the process.

$$\Delta G_o = \Delta H - T\Delta S_o \qquad (9)$$

TABLE 1   Gibbs Free Energy Values for Different Fermentative Reactions

Fermentative reactions involved in pathway	$\Delta G°$ (KJ/mol)	Comments
$NADH + H^+ + pyruvate^- \rightarrow NAD^+ + lactate^-$	−25.0	Highest positive Gibbs free energy; hence this reaction is most unfavorable
$2NADH + 2H^+ + acetyl\text{-}CoA \rightarrow 2NAD^+ + ethanol + CoA$	−27.5	
$NADH + H^+ + pyruvate^- + NH_4^+ \rightarrow NAD^+ + alanine + H_2O$	−36.7	
$NADH + H^+ \rightarrow NAD^+ + H_2$	+81.1	Highest negative Gibbs free energy; hence this reaction is most favorable
$2\ Ferredoxin\ (red) + 2H^+ \rightarrow 2\ ferredoxin(ox) + H_2$	+3.1	

At higher temperatures, the Gibbs free energy for the overall reaction of oxidation of glucose into acetate becomes more favorable. Second, it may also be reasoned that $\Delta G^{\circ\prime}$ values are calculated for standard conditions, which can differ considerably based on the physiological conditions. Moreover, several researchers also suggest that if the H_2 partial pressure is kept low [e.g., $P(H_2)$ of $10 - 2kPa$], even proton reduction with NADH may become an exergonic reaction (-4.7 kJ/mol). However, an intriguing question that surfaces is why these thermophiles have evolved for optimized H_2 production. Resolving this phenomenon would aid in designing highly efficient H_2 cell factories (Soetaert and Vandamme, 2009).

MOLECULAR BIOLOGY OF KEY ENZYMES INVOLVED IN DARK FERMENTATION PROCESS

Two key enzymes are involved in dark fermentative hydrogen production. Some organisms produce hydrogen through the formate hydrogen lyase pathway, whereas hydrogen production is catalyzed by [FeFe]-hydrogenase in others. The structure and characteristics of the two different enzymes are detailed here.

Formate Hydrogen Lyase

Formate is a key metabolite in the energy metabolism of several bacteria. It has a low redox potential of -420 mV, which permits it to serve as an electron donor for the anaerobic reduction of fumarate (Macy et al., 1976). However, depending on the physiological condition, formate may be secreted, oxidized aerobically, used as a reductant in anaerobic respiration with a variety of terminal reductases, or converted into CO_2 and H_2 (Ordal and Halvorson, 1939; Stephenson and Stickland, 1932). In the latter physiological situation, acidification is counteracted by the conversion of the formic acid into a neutral molecule, that is, hydrogen, which possesses a redox potential comparable to that of formic acid. A complex enzyme, formate hydrogen lyase, catalyzes this reaction.

The FHL complex consists of formate dehydrogenase, hydrogenase, and electron transfer carriers responsible for formic acid oxidation to CO_2 and H_2 (Bohm et al., 1990; Sauter et al., 1992). In this reaction, formic acid acts as the electron donor, whereas protons are the only terminal acceptor. Electron transfer from formate dehydrogenase to hydrogenase takes place via an electron carrier. Their nature is still unknown; however, it is suggested that hydrogenase subunits may act as these carriers (Leonhartsberger et al., 2002).

Formate hydrogen lyase contains five membrane-bound proteins and a unique selenocysteine amino acid within its active site, which together make heterologous expression difficult (Boyington et al., 1997; Sawers, 2005). Formate dehydrogenase is a selenocysteine and molybdenum-containing peripheral membrane protein located at the cytoplasmic side. It consists of several subunits. Among these, the large subunit, *FdhF*, exhibits catalytic properties and has a molecular mass of 79 kDa. Selenocysteine is located at position 140 (Zinoni et al., 1987) and is involved directly in active site formation (Boyington et al., 1997).

In addition, evidence shows that the large subunit contains one [4Fe-4S] cluster. It catalyzes oxidation of formic acid to CO_2 and a couple of redox equivalents $2 (H^+ + e^-)$, which involve NAD^+ formed in the fermentation process:

$$HCOOH \rightarrow CO_2 + 2(H^+ + e^-) \tag{10}$$

Hydrogenase of the formate hydrogen lyase complex exists in two forms—Hyd-3 (Sauter et al., 1992; Sawers et al., 1985) and Hyd-4—each of which consists of many subunits. Some of these subunits represent polytopic membrane proteins. Hyd-3 is a protein complex that consists of several large and small subunits, denominated as Hyc. This complex is rather stable. One of the most interesting large subunits is the nickel-containing iron–sulfur HycE protein with a molecular mass of 65 kDa (Balantine and Boxer, 1985). In some cases, nickel may be substituted by zinc or other divalent metals. This peripheral membrane protein, located at the cytoplasmic side, is responsible for equilibrium of the reaction of proton reduction:

$$NADH + H^+ \leftrightarrow NAD^+ + H_2 \tag{11}$$

The nature of natural electron donors and acceptors involved in the hydrogenase reaction remains to be clarified. It is suggested that small Hyd-3 subunits with molecular masses of 21–33 kDa may participate in such electron transfer (Sauter et al., 1992). Some of these subunits possessing α-helical sites are bound to the membrane tightly. It is also suggested that these subunits contain iron–sulfur clusters [4Fe-4S], which are involved in intramolecular electron transfer between redox centers. Interestingly, amino acid sequences of some of these small subunits share homology with components of the electron transport chains of bacteria, mitochondria, and plastids (e.g., NADH-ubiquinone-oxidoreductase, complex I) (Bohm et al., 1990). This suggests involvement of hydrogenase subunits in the formation of various electron transport chains.

At the genetic level, all the proteins of the active complex, except formate dehydrogenase (FDH-H), are encoded by the hyc operon (Hyc B C D E F G), while formate dehydrogenase-H is encoded by *fdhF* and is regulated by hydrogenase-3 (Hyd-3) and four other polypeptides all encoded by the *hyc* operon. Hyd-3 is composed of a cytoplasmically oriented large subunit, encoded by *hycE* and a small subunit encoded by *hycG*. The remaining four polypeptides, along with the product of hycG, are the component of the membrane integral electron transfer component. Expression of the FHL system is activated by a transcriptional activator FHL activator protein (FHL A) encoded by the *fhlA* gene.

Several researchers have shown enhancement of hydrogen production by engineering the formate hydrogen lyase pathway metabolically. Yoshida et al. (2006) developed the formate hydrogen lyase-overexpressing strain SR13 by combining FHL repressor (*hycA*) inactivation with FHL activator (*fhlA*) overexpression. They obtained 6.5- and 7.0-fold increases in the transcription of a large subunit of formate dehydrogenase (*fdhF*) and a large subunit of hydrogenase (*hycE*), respectively, compared to the wild-type strain. The genetic modification effectively resulted in a 2.8-fold increase in hydrogen productivity of SR13 compared to the wild-type strain.

Higher hydrogen production is related with a higher accumulation of formate within the cell. The intracellular level of formate is determined by the rates of biosynthesis and metabolism of formate, and also by formate transporter FocA, which is a membrane protein involved in formate transport into and out of the cell (Sawers, 2005). Knockout of the *focA* gene results in intracellular accumulation of formate (Suppmann and Sawers, 1994). FHL catalyzes the accumulated formate to release hydrogen and carbon dioxide. Fan et al. (2006) engineered *E. coli* strain ZF1 by disrupting the formate transporter gene. They obtained 14.9 μmol of hydrogen/mg of dry cell weight, respectively, compared to 9.8 μmol of hydrogen/mg of dry cell weight generated by wild-type *E. coli* strain W3110.

Hydrogenase

Hydrogenases, the enzymes responsible for the reversible conversion of molecular hydrogen into two protons and two electrons ($H_2 \leftrightarrow 2H^+ + 2e^-$), are complex metal coenzymes that can be classified into three groups based on the number and identity of the metals in their active sites: [NiFe]-, [FeFe]-, and [Fe]-hydrogenases (Vignais et al., 2001). They are not related phylogenetically but still share common properties at their active sites; for example, they all contain Fe and CO as a ligand to the Fe atom. Among these, [FeFe]-hydrogenase is known to be most potent in terms of hydrogen production. The enzymes are either monomeric, as in *Clostridium*, or multimeric, as in *Thermotogamaritima* and *Thermoanaerobactertengcongensis*, that consist of three and four subunits, respectively. The [FeFe]-hydrogenases are organized into modular domains. The accessory clusters, known as the F cluster, function as inter- and intramolecular electron transfer centers. The accessory cluster is linked electronically to the catalytic cluster known as the H cluster. Structural reports reveal that the H cluster is composed of a binuclear Fe–Fe cluster with two diatomic ligands bound to each of the metal centers and three bridging ligands between them consisting of two sulfur atoms and a small molecule. However, the accessory F cluster is composed of [FeS] clusters. The simplest [FeFe]-hydrogenases are the ones isolated from algae, as they contain only the H cluster and no extra FeS centers. [FeFe]-hydrogenase is known to require a minimum of three auxiliary proteins for maturation. The maturation operon comprises mainly HydE, HydF, and HydG, two of which belong to the class of radical *S*-adenosylmethionine enzymes, while the third belongs to the family of GTPases (Nicolet et al., 2010). They are considered to be sufficient to generate active enzyme in a heterologous host when their genes are coexpressed with the structural genes.

However, these enzymes are known to be oxygen sensitive and can be expressed only under anaerobic conditions. Several researchers have attempted to improve the yield of the dark fermentative hydrogen production process by modifying and/or overexpressing the hydrogenase enzyme. Karube et al. (1983) cloned and expressed the hydrogenase gene from *Clostridium butyricum* in *E. coli* strain H K16 (Hyd-lacking native hydrogenase activity). They obtained a 3- to 3.5-fold enhancement in hydrogenase activity as compared to wild-type *C. butyricum*.

In another study, the hydrogen yield of *E. coli* BL21 was enhanced by heterologous overexpression of *hydA* from *Enterobacter cloacae* IIT-BT 08. The recombinant *E. coli* BL-21 produced a yield of 3.12 mol of hydrogen per mole of glucose, which was higher than the wild-type strain *E. cloacae* IIT-BT-08. Moreover, the maximum hydrogen production rate was reported to be 5.6 mmol/mg protein/h from glucose with the recombinant strain using a continuous immobilized whole cell bioreactor in MYG medium. However, Morimoto et al. (2005) enhanced hydrogen production from *C. paraputrificum* by overexpressing the native

[FeFe]-hydrogenase gene homologously. As a result, they found that the recombinant clone produced 1.7 times as much hydrogen gas as the parental clone, along with a drastic reduction in lactic acid production.

FACTORS AFFECTING THE BIOHYDROGEN PROCESS

Effect of pH

pH is known to be one of the major environmental factors affecting metabolic pathways and the hydrogen yield. Most facultative anaerobes produce hydrogen by breakdown of glucose to pyruvate during glycolysis. Metabolites produced by the breakdown of pyruvate in turn determine the hydrogen yield (Antonopoulou et al., 2008). Metabolites such as ethanol and other alcohols contain additional hydrogen atoms not present in their corresponding acids. All the enzymes are active only in a certain specific range of pH and have maximum activity at the optimal pH (Craven, 1988; Lay et al., 1997). Thus, pH also affects the activity of the key enzyme hydrogenase (Dabrock et al., 1992).

Studies have shown that at higher concentrations of dissociated acids, due to increased ionic strength of the medium, the hydrogen production shifts from an acetogenic to a solventogenic phase (Khanal et al., 2004). However, the same inhibition effect is known to occur at low pH when nonpolar undissociated acids, present in the medium, penetrate the cell walls. In addition, in operations where a mixed consortia or sludge is used as inoculum, regulation of pH becomes critical in suppressing the activity of the hydrogen-consuming methanogens (Ginkel and Sung, 2001). The decrease in hydrogen production at low pH (lower than 5.0) is due to the increased formation of acidic metabolites, which destroys the ability of the cell to maintain internal pH (Bowles and Ellefson, 1985). It results in lowering the intracellular level of ATP, therefore inhibiting glucose uptake. Nath et al. (2006) observed that the most suitable pH for maximizing the rate of hydrogen production was dependent on both the type of microorganisms and the substrates used.

van Ginkel and Logan (2005) conducted an interesting study to evaluate the effect of undissociated acetic and butyric acid on hydrogen production by adding the acids externally. They found that cumulative hydrogen production reduced up to 93% and the yield of the process was also decreased significantly. Various studies have shown that a pH of media between 5 and 6 has been found suitable to obtain a higher hydrogen yield (Cao and Zhao, 2009; Ginkel and Sung, 2001; Ma et al., 2008). pH other than the optimum has been shown to suppress the hydrogen yields (Kumar and Das, 2000a). Thus, it is important to control the pH in order to produce higher hydrogen production.

Temperature

Dark hydrogen fermentation can be performed at different temperatures: mesophilic (25–40°C), thermophilic (40–65°C), extreme thermophilic (65–80°C), or hyperthermophilic (>80°C) (Levin et al., 2004). Analysis of the literature reveals that most of the studies (nearly 73%) at the laboratory scale have been conducted using mesophiles (Li and Fang, 2007). These studies show that temperature has a significant effect on hydrogen production. It affects the maximum specific growth rate of the microorganism, as well as the rate of substrate conversion during hydrogen production. Moreover, a higher temperature may lead to thermal inactivation of the enzymes responsible for the fermentative hydrogen production process. Numerous batch studies using mixed cultures have been carried out to show the dependency of hydrogen production on the operational temperature (Gavala et al., 2006; Lin et al., 2008).

Lee et al. (2006a) studied the potential of mixed culture on hydrogen production under varying temperatures ranging between 15 and 35°C. They found that the hydrogen yield and specific rate of hydrogen production increase with respect to temperature. At 35 ± 1°C, the maximum rate of hydrogen production and highest hydrogen yield of 574 mmol/liter/day and 1.70 mol H_2/mol glucose, respectively, were obtained. Similarly, Lin et al. (2008) operated a chemostat-type reactor to determine the suitable temperature for hydrogen production using mixed consortia. These studies were carried out in a range of 30 to 55°C. The highest H_2 production was obtained at 45°C. Hydrogen production at different temperatures was correlated with a shift in the metabolic pathways and changes in the microbial community. Thus, studies showed that understanding the temperature dependency of the microbial community is critical for optimizing the hydrogen production systems. However, limited information is available on the effect of temperature on the dynamics of microbial diversity in bioreactor systems (Singh and Pandey, 2011). Conversely, thermophiles operate at higher temperatures. Studies show that thermophiles produce higher hydrogen as compared to mesophiles (van Groenestijn et al., 2002). A thermodynamically higher temperature is favorable for hydrogen production due to an increase in the entropy of the system, which makes the process more energetic. Concomitantly, the hydrogen utilization process is affected negatively with an increase in temperature (Amend and Shock, 2001; Conrad and Wetter, 1990). However, from an economic perspective, extreme thermophilic conditions may not be viable due to intensive energy requirements needed to maintain the high temperatures (Hallenbeck, 2005).

Hydrogen Partial Pressure

Hydrogen production pathways are very sensitive to hydrogen partial pressure. Accumulation of hydrogen in the reactor headspace may increase the partial pressure of hydrogen in the reactor system. According to Le Chatlier's principle, due to hydrogen built up, the forward reaction will be inhibited. Thus a higher partial pressure of hydrogen in the reactor system will decrease the production of hydrogen. Concomitantly, metabolic pathways may also shift toward the production of more reduced end products, such as lactate, ethanol, acetone, and butanol. Studies have shown that the partial pressure of hydrogen is an important factor in case of continuous hydrogen production (Hawkes et al., 2007). Different studies have shown a correlation between the optimal partial pressure of hydrogen depending on the operational temperature used. Lee and Zinder (1988) obtained 50 kPa as the optimal operational pressure at 60°C. Similarly, different studies found 20 kPa at 70°C (van Niel et al., 2002) and 2 kPa at 98°C (Adams, 1990; Levin et al., 2004) as the optimal pressure. Thus, it is very important to remove excess hydrogen from the system to maintain hydrogen production. Different strategies of removing or separating hydrogen gas have been developed to avoid the negative effect of hydrogen accumulation in the gas and liquid phase. Several studies have shown that decreasing the hydrogen partial pressure can enhance the hydrogen production. Mizuno et al. (2000) and Tanisho et al. (1998) increased the hydrogen yield by 68% by sparging nitrogen gas. Lay (2000) increased the hydrogen yield by lowering the hydrogen partial pressure in the system.

Volatile Fatty Acid (VFA)

In fermentative hydrogen production, metabolic end products are known to affect the hydrogen yield. Mostly, ethanol, acetic acid, butyric acid, and propionic acid are produced as the dominant end metabolites (Lee et al., 2002). However, toward the stationary phase, as the increase in concentration of the soluble end metabolites increase the ionic strength of the medium, which causes cellular lysis. Some researchers have also shown that at higher acid concentrations, protons may permeate the cell membrane of hydrogen-producing bacteria and disrupt the physiological balance in the cell. Under such conditions, to restore the physiological balance in the cell, maintenance energy is utilized. Redirection of maintenance energy, however, compromises bacterial growth and hydrogen production significantly (Jones and Woods, 1986). To determine the inhibitory effect of VFAs on hydrogen production, Lee et al. (2002) added soluble metabolites in the medium. They found that on addition of ethanol, acetic acid, propionic acid, and butyric acid in the range of 0 to 300 mmol/liter hydrogen production potential, the substrate degradation efficiency, rate of hydrogen production, and hydrogen yield decreased significantly with increasing concentration of the VFAs.

Nutrients

Supplementation of nitrogen, phosphate, and other inorganic trace minerals is essential to maximize the hydrogen yield using carbohydrate as a substrate for hydrogen production. Nitrogen is an essential component of amino acids and is required for optimal growth of the microorganism. Studies have shown that with respect to hydrogen production, organic nitrogen appears more suitable as compared to inorganic nitrogen. The hydrogen yield of 2.4 mol H_2/mol glucose was obtained from starch by supplementing it with 0.1% (w/v) polypeptone. However, no significant improvement in hydrogen production was obtained when urea or other inorganic salts were used as a nitrogen source (Yokoi et al., 2001).

However, although supplementation of the nitrogen source increases the overall hydrogen yield, it also adds to the overall cost of production. Thus, research is presently underway to find less expensive alternatives that can substitute as the nitrogen source in hydrogen production media. In this regard, corn-steep liquor, which is a waste of the corn starch manufacturing process, appears promising and best alternative supplementation for peptone could be used as the nitrogen source (Yokoi et al., 2002). Additionally, the C/N ratio also plays a significant role in stabilizing the dark fermentation process and affecting the hydrogen productivity and specific hydrogen production rate (Lin and Lay, 2004). Optimal phosphate concentrations are also desirable to enhance the overall performance of the process. Phosphate acts as an important inorganic nutrient for optimal hydrogen production (Lin and Lay, 2004). Excess amounts of phosphate may lead to the production of more volatile fatty acids. However, higher VFA production is not desirable as it diverts the cellular reductants away from hydrogen production.

Metal Ions

For any fermentative process, supplementation of suitable metal ions in media is essential. These metal ions act as enzyme cofactors and are also involved in the cellular transport processes. Hydrogenase, the key enzyme involved in hydrogen production, contains a bimetallic Fe–Fe center. It is also surrounded by FeS protein clusters (Nicolet et al., 2010). Therefore, several researchers have studied the effect of supplementation of iron on biohydrogen production. For example, Lee et al. (2001) studied the effect of iron concentration on hydrogen fermentation and found that a higher Fe ion concentration influences the system positively. The study obtained the maximum rate of hydrogen production, 24 ml/g VSS/h, when the medium was supplemented with 4000 mg/liter $FeCl_2$.

In addition, during the glycolysis process, the magnesium ion acts as an important cofactor for 10 different enzymes, including hexokinase, phosphofructokinase, and phosphoglycerate kinase (Voet et al., 1999). In another study, Lin and Lay (2004) reported the effect of various trace elements such as Mg, Na, Zn, Fe, K, I, Co, Mn, Ni, Cu, Mo, and Ca for hydrogen production using *C. pasteurianum*. The study showed that suitable concentrations of Mg, Na, Zn, and Fe were necessary for a higher hydrogen yield. Based on these results, they proposed an optimal nutrient formulation containing (mg/liter) $MgCl_26H_2O$, 120; NaCl, 1000; $ZnCl_2$, 0.5; and $FeSO_47H_2O$, 3 (Wang and Wan, 2009).

Hydraulic Retention Time (HRT)

Fermentation time is considered the key factor in selecting microbial populations whose growth rates are able to sustain the mechanical dilutions created by the continuous volumetric flow. The specific growth rate of methane-producing bacteria is very low (about $0.0167–0.02\ h^{-1}$) when compared with hydrogen-producing bacteria ($0.172\ h^{-1}$) (Lo et al., 2009). Therefore, in studies where a mixed consortium such as sewage sludge is used, a short HRT is capable of inhibiting or terminating slow-growing methanogens from the reactor. This may eventually lead to a mixed population rich in hydrogen-producing bacteria. In an interesting study, Zhang et al. (2006) effectively increased the hydrogen production by lowering the HRT from 8 to 6 h. This decreased the production of propionate, resulting in higher hydrogen yields.

BIOHYDROGEN PRODUCTION FROM WASTES

Environmental Challenges in Waste Disposal

Disposing of waste has huge environmental impacts and can cause serious problems. Waste generation increases with population expansion and economic development. Improperly managed waste poses a risk to human health and the environment. Uncontrolled dumping and improper waste handling cause a variety of problems, including contaminating water, attracting insects and rodents, and increasing flooding due to blocked drainage canals or gullies. In addition, it may result in safety hazards from fires or explosions. Improper waste management also increases greenhouse gas emissions, which contribute to climate change.

Planning for and implementing a comprehensive program for waste collection, transport, and disposal, along with activities to prevent or recycle waste, can eliminate these problems. Wastes such as food and starch-based, cellulosic materials, dairy wastes, palm oil mill efflu-ent, and glycerol waste consist of a very high chemical oxygen demand (COD) and biological oxygen demand (BOD) (Chong et al., 2009). If these wastes are released in the environment without undergoing wastewater treatment, they could cause a potential threat to the aquatic fauna. Moreover, storage and transportation of these wastes are cumbersome and demand separate infrastructure. These problems in waste disposal suggest that effective disposal of the waste is required at or near the source. In view of this, fermentation of the waste to gen-erate clean energy, hydrogen, may be a plausible solution. As discussed previously, many organisms can utilize complex waste as a substrate to produce hydrogen. This may solve the dual purpose of waste disposal and clean energy generation. Hydrogen reactors may be set up at or near the site of waste generation to economize the process further (Kapdan and Kargi, 2006).

Biohydrogen Production from Organic Waste as Feedstock

One of the bottlenecks of producing hydrogen via the biological route is the economics of the process. To address this issue, production of biological hydrogen using waste as substrate is an inexpensive and promising approach. This would make the process cost-effective when compared with the well-established chemical and electrolytic hydrogen production process (Nath and Das, 2003). A wide range of wastes from different industries could be used as substrate as described next.

Sources of Wastewater

Important criteria required for the selection of waste materials as substrate for hydrogen production are sustainability, availability, organic content, cost, and biodegradability. Simple sugars such as glucose, sucrose, maltose, and lactose are readily biodegradable and most preferred substrates for hydrogen production, as shown in Tables 2 and 3. However, pure carbohydrate sources are expensive raw materials and as such are not economically viable as substrates for the hydrogen production process because they mostly may be consumed in other well-established processes. Complex waste resources offer greater economic poten-tial. Major complex waste resources that can be used as substrate for hydrogen production by dark fermentation process include domestic wastewater, paper mill wastewater, starch efflu-ent, food processing wastewater, rice winery wastewater, molasses-based wastewater, palm oil mill wastewater, glycerol-based wastewater, chemical wastewater, cattle wastewater, dairy process wastewater, and designed synthetic wastewater, as shown in Tables 4 and 5. There are still several different kinds of wastewaters that have not been explored for their potential to produce hydrogen, for example, oil industry wastewaters (Inan et al., 2004). How-ever, these wastewaters are not ideal substrates as they are mostly rich in only one type of substrate. In view of this, a combination of different wastewaters may provide an effective new substrate for hydrogen production. For example, a combination of wastewaters rich in carbohydrates with those rich in nitrogen content may achieve a higher hydrogen yield

TABLE 2 Biohydrogen Production from Pure Carbohydrates by Batch Dark Fermentation

Microorganism	Substrate	Substrate concentration	Hydrogen yield	Reference
Bacillus coagulans IIT-BT S1	Glucose	20 g/liter	2.3 mol/mol glucose	Kotay and Das, 2007
C. pasteurium	Sucrose	20 g COD/liter	4.8 mol/mol sucrose	Lin and Chang, 2004
Clostridium sp.	cellulose	25 g/liter	2.2 mmol/g cellulose	Lay, 2001
C. pasteurium	Starch	24 g/liter	106 ml/g starch	Lin and Shen, 2004
Clostridium sp.	Glucose	—	2.8 mol/mol hexose	Steven et al., 2005
C. cellobilparum	Glucose	—	2.7 mol/mol hexose	Chung, 1976
C. bejerickit	Glucose	—	2.8 mol/mol hexose	Lin et al., 2007
Clostridium sp.	Sucrose	—	1.6 mol/mol hexose	Zhao et al., 2008
Clostridium CGS2	Starch	—	2.0 mol/mol hexose	Chen et al., 2008
Coculture (*Clostridium* sp. + *Thermoana* sp.)	Cellulose	—	1.8 mol/mol hexose	Liu et al., 2008
E. cloacae IIT-BT-08	Glucose	10 g/liter	2.2 mol/mol glucose	Kumar and Das, 2000b
E. cloacae IIT-BT-08	Sucrose	10 g/liter	6 mol/mol sucrose	Kumar and Das, 2000b
E. cloacae IIT-BT-08	Cellobiose	10 g/liter	5.4 mol/mol cellobiose	Kumar and Das, 2000b
E. cloacae DM11	Glucose	10 g/liter	3.3 mol/mol glucose	Nath et al., 2006
E. cloacae BL-21	Glucose	10 g/liter	3.1 mol/mol glucose	Chittibabu et al., 2006
E. coli	Glucose	20 g/liter	4.73×10^{-8} mol/mol glucose	Podestá et al., 1997
E. aerogenes	Starch	20 g glucose/liter	1.1 mol/mol glucose	Fabiano and Perego, 2002
E. coli	Glucose	—	2.0 mol/mol hexose	Turcot et al., 2008
Klebsielle oxytoca HP1	Sucrose	50 mM	1.5 mol/mol sucrose	Minnan et al., 2005
K. oxytoca HP1	Glucose	50 mM	1 mol/mol glucose	Minnan et al., 2005
Sludge compost	Glucose	10 g/liter	2.1 mol/mol glucose	Morimoto et al., 2004
Slostridium sp. strain	Xylose	—	2.6 mol/mol hexose	Taguchi et al., 2005
Thermoanaerobacterium	Cellulose	5 g/liter	102 ml/g cellulose	Liu et al., 2003
Thermoanaerobacterium	Glucose	—	2.7 mol/mol hexose	Thong et al., 2008
Thermoanaerobacterium	Starch	4.6 g/liter	92 ml/g starch	Zhang et al., 2003
Mixed culture	Starch	1 g COD/liter	0.6 mol/mol starch	Logan et al., 2002

Continued

TABLE 2   Biohydrogen Production from Pure Carbohydrates by Batch Dark Fermentation—Cont'd

Microorganism	Substrate	Substrate concentration	Hydrogen yield	Reference
Mixed culture	Sugar beet juice	—	1.7 mol/mol hexose	Hussy et al., 2005
Mixed culture	Glucose	1 g COD/liter	0.9 mol/mol glucose	Logan et al., 2002
Mixed culture	Sucrose	6 g/liter	300 ml/g COD	Khanal et al., 2004
Mixed culture	Sucrose	1 g COD/liter	1.8 mol/mol sucrose	Logan et al., 2002
Mixed microflora	Sucrose	—	1.8 mol/mol hexose	Lin and Cheng, 2006
Mixed microflora	Starch	—	1.5 mol/mol hexose	Lin et al., 2008
Mixed microflora	Cellulose	—	2.4 mol/mol hexose	Ueno et al., 1995
Mixed culture	Glucose	10 g/liter	2.7 mol/mol glucose	Roy et al., 2012

TABLE 3   Biohydrogen Production from Pure Carbohydrates by Continuous Dark Fermentation

Microorganism	Substrate	Substrate concentration	Hydrogen yield	Reference
E. cloacae IIT-BT-08	Glucose	10 g/liter	75.6 mmol/liter/h	Kumar and Das, 2001
C. acetobutyricum	Glucose	—	2 mol/mol glucose	Chin et al., 2003
Clostridia sp.	Glucose	20 g COD/liter	1.7 mol/mol glucose	Liu et al., 2003
C. butyricum + E. aerogenes	Starch	20 g/liter	2.5 mol/mol glucose	Yokoi et al., 1998
C. butyricum + E. aerogenes	Starch	20 g/liter	2.6 mol/mol glucose	Yokoi et al., 1998
C. termolacticum	Lactose	29 mmol/liter	3 mol/mol lactose	Collet et al., 2004
Klebsiella oxytoca HP1	Sucrose	50 mM	3.6 mol/mol sucrose	Minnan et al., 2005
Thermococcus kodakaraensis KOD1	Starch	5 g/liter	3.3 mol/mol starch	Kanai et al., 2005
Mixed culture	Glucose	13.7 g/liter	1.2 mol/mol glucose	Oh et al., 2004a
Mixed culture	Sucrose	20 g COD/liter	3.5 mol/mol sucrose	Chen et al., 2001
Mixed culture	Sucrose		2.1 mol/mol sucrose	Lee et al., 2004
Mixed culture	Sucrose	20 g COD/liter	1.5 mol/mol sucrose	Chang and Lin, 2004
Mixed culture	Sucrose	20 g COD/liter	2.6 mol/mol glucose	Lin and Lay, 2004
Mixed culture	Sucrose	20 g COD/liter	1.5 mol/mol sucrose	Chen and Lin, 2003
Mixed culture	Starch	10 g/liter	0.8 mol/mol starch day	Hussy et al., 2000
Mixed culture	Starch	6 kg starch/m^3	1.3 liter/g starch COD	Lay, 2000

TABLE 4 Biohydrogen Production from Organic Wastes by Batch Dark Fermentation

Microorganism	Substrate	Hydrogen yield	Reference
Mixed culture	Apple processing wastewater	4.1 mmol H_2/g COD	van Ginkel et al., 2005
Mixed culture	Potato processing wastewater	5.7 mmol H_2/g COD	van Ginkel et al., 2005
Mixed culture	Food waste	57 ml H_2/g VS	Pan et al., 2008
Mixed culture	Preserved fruit soaking solution	3.7 mol H_2/mol hexose	Lay et al., 2010a
Coculture of *Clostirium sporosphaeroides* F52, and *C. pasteurianum* F40	Condensed molasses fermentation soluble	1.8 mol H_2/mol hexose	Hsiao et al., 2009
Mixed culture	Condensed molasses fermentation solubles	1.5 mol H_2/mol hexose	Wu and Lin, 2004
Mixed culture	Confectionery processing wastewater	6.9 mmol H_2/g COD	van Ginkel et al., 2005
Mixed culture	Rice slurry	346 ml H_2/g carbohydrate	Fang et al., 2006
Thermoanaerobacterium sp. mixed culture	Starch wastewater	92 ml H_2/g starch	Zhang et al., 2003
C. acetobutylicum X9 + *Ethanoligenens harbinense* B49	Microcrystalline cellulose	1.8 liter H_2/liter-palm oil mill effluent (POME)	Wang et al., 2008
Thermoanaerobacterium-rich sludge	POME	6.3 liter H_2/liter-POME	O-Thong et al., 2007
Mixed culture	POME	4.7 liter H_2/liter-POME	Atif et al., 2005
Mixed culture	POME	2.3 liter H_2/liter-POME	Atif et al., 2005
Mixed culture	Olive mill wastewater	0.5 mmol H_2/g COD	Eroglu et al., 2006
C. saccharoperbutylacetonicum ATCC27021	Cheese whey	2.7 mol H_2/mol lactose	Ferchichi et al., 2005
Mixed culture	Vinasse wastewater	24.9 mmol H_2/g COD	Fernandes et al., 2010
Mixed culture	Brewery wastewater	6.1 mmol H_2/g COD	Shi et al., 2010
Mixed culture	Distillery wastewater	—	Venkata Mohan et al., 2008
Coculture of *C. freundii* 01, *E. aerogens* E10, and *R. palustric* P2	Distillery effluent	2.7 mol H_2/mol hexose	Vatsala et al., 2008
Mixed culture	Glycerin wastewater	6.0 mmol H_2/g COD	Fernandes et al., 2010

Continued

TABLE 4 Biohydrogen Production from Organic Wastes by Batch Dark Fermentation—Cont'd

Microorganism	Substrate	Hydrogen yield	Reference
Mixed culture	Domestic sewage	6.0 mmol H_2/g COD	Fernandes et al., 2010
Mixed culture	Cattle wastewater	12.4 mmol H_2/g COD	Tang et al., 2008
Mixed culture	Lagoon wastewater	0.5 mol H_2/mol hexose	Oh et al., 2003
Mixed culture	Chemical wastewater and domestic sewage wastewater	1.2 mmol H_2/g COD	Venkata Mohan et al., 2007
Mixed culture	Probiotic wastewater	1.8 mol H_2/mol hexose	Sivaramakrishna et al., 2009
Mixed culture	Rice spent wash	464 ml H_2/g reducing sugar	Roy et al., 2012

TABLE 5 Biohydrogen Production from Organic Wastes by Continuous Dark Fermentation

Microorganism	Substrate	Hydrogen yield	Reference
Mixed culture	Food waste	0.4 liter H_2/g COD	Han and Shin, 2004
Mixed culture	Citric acid wastewater	0.8 mol H_2/mol hexose	Yang et al., 2006
Mixed culture	Molasses	5.6 m^3 H_2/m3 reactor/day	Ren et al., 2006
Mixed culture	Condensed molasses fermentation solubles	0.9 mol H_2/mol hexose	Lay et al., 2010b
Mixed culture	Sugar beet wastewater	1.7 mol H_2/mol hexose	Hussy et al., 2005
Sludge compost	Sugary wastewater	2.5 mol H_2/mol hexose	Ueno et al., 1996
Mixed culture	Rice winery wastewater	2.1 mol H_2/mol hexose	Yu et al., 2002
Mixed culture	Dairy waste	1.1 mmol H_2/m^3/min	Venkata Mohan et al., 2007
Mixed culture	Cheese processing wastewater	3.2 mmol H_2/g COD	Yang et al., 2007
Mixed culture	Cheese whey wastewater	22.0 mmol H_2/g COD	Azbar et al., 2009
Mixed culture	Olive pulp water	2.8 mol H_2/mol hexose	Koutrouli et al., 2006
Mixed culture	POME	0.4 liter H_2/g COD reduced	Vijayaraghavan and Ahmad, 2006
Mixed culture	Coffee drink manufacturing wastewater	0.20 mol H_2/mol hexose	Jung et al., 2010
Mixed culture	Purified terephthalic acid	19.3 mmol H_2/g COD	Zhu et al., 2010

(Huang et al., 2010). Apart from this, a combination of solid organic wastes and wastewater could also be considered a novel approach for biohydrogen production.

Agricultural-Based and Food Industry Wastes

Agricultural and food industries are the major sources that generate waste rich in starch, protein, and/or cellulose. However, the biodegradability of these wastes may be affected adversely due to their complex nature. As such, agricultural wastes are rich in cellulose and thus may require some pretreatment process (heat, acid or base treatment, etc.) prior to use as a substrate for hydrogen production. Food waste is also considered a potential substrate for hydrogen production as it contains about 90% volatile suspended solids favoring microbial degradation. Among food wastes, rice slurry is a starch-based waste from industry. It is rich in carbohydrate, protein, lipid, and water. Molasses is a sugar industry waste. It is a viscous by product of the sugar industry and is produced during the processing of sugarcane and sugar beets into sugar. It is good source of sucrose and can be degraded easily anaerobically (Chong et al., 2009). At present, generally the disposal of food waste is carried out by the process of land filling. However, the process imposes several environmental problems, such as fouling and pollution of ground water. In view of this, utilization of waste to produce clean energy appears highly promising and economically viable.

Lignocellulosic Wastes

The current production of biohydrogen relies on hydrogen from starch and sugars, but there has been considerable debate about its sustainability. In this context, biohydrogen produced from lignocellulosic biomass is an interesting alternative, as lignocellulosic raw materials do not compete with food crops and as such are also less expensive than conventional agricultural feedstocks. Lignocellulosic materials are composed mostly of cellulose, hemicelluloses, and lignin. These products of photosynthesis form the structural component of the plant cell wall (Soloman et al., 2007). Cellulose is the major constitute of the plant biomass and is readily available in agricultural residues and wastes (Table 6). However, bacteria cannot utilize cellulose directly. Thus, pretreatment of cellulose is required to convert from the complex form to simpler forms. However, cellulose-degrading bacteria are also present, which can be used to directly degrade cellulose into simpler sugars. These sugars may then be used as substrate for hydrogen production (De Vrije et al., 2002).

Sewage Sludge from Wastewater Treatment Plants

With an increasing global population, the annual production of sewage sludge is rising and this increase is deplorable. Sewage sludge can be considered and used as substrate for fermentative hydrogen production. It provides several advantages over other wastes based on its abundant availability and low cost. In addition, it contains several macronutrients (carbohydrates, proteins, and lipids) and micronutrients (vitamins and minerals). Therefore, it may be considered as suitable substrate for hydrogen fermentation. Despite these advantages, it has received little attention as a source of seed or inoculum (Kotay and Das, 2008). However, some of the organic nutrients present in sewage sludge make it unsuitable to be used directly by microbes during fermentation (Lay et al., 1999) Moreover, sludge as seed will contain methanogenic bacteria that will interfere with the hydrogen production process. Therefore, studies suggest that pretreatment processes prior to use are required to

TABLE 6   Biohydrogen Production from Different Bioreactors by Dark Fermentation

Reactor type[a]	Volume (liter)	Microorganism	Substrate	Hydrogen yield (mol/mol)	Hydrogen rate (liter/liter/h)	Reference
IBR	2.5	*C. tyrobutyricum*	Food waste	223 ml/g hexose	0.3	Jo et al., 2008b
UASB		*Clostridium* sp.	Sucrose	1.6	0.1	Zhao et al., 2008
UASB	—	Mixed culture	—	—	0.3	Chang and Lin, 2004
USAB	—	Anaerobic fermentative bacteria	—	—	0.05	Kotsopoulos et al., 2006
CIGSB	0.45–1.3	*Clostridium* sp.	Sucrose	2.0	9.3	Lee et al., 2006b
Granule-based CSTR	6.0	Fermentative bacteria	Glucose	1.8	3.26	Show et al., 2007
AFBR Biofilm reactor	1.4	Anaerobic fermentative bacteria	Glucose	0.4–1.7	7.6	Zhang et al., 2008a
AFBR	—	Anaerobic fermentative bacteria	—	—	1.8	Lin et al., 2009
AFBR	—	Mixed bacteria	—	—	2.4	Zhang et al., 2007c
AFBR Granule reactor	1.4	Anaerobic fermentative bacteria	Glucose	0.4–1.7	6.6	Zhang et al., 2008a
CIGSB	0.88	Mixed bacteria	Sucrose	1.9	7.7	Lee et al., 2006b
HAIB	0.850	Fermentative microflora	Glucose	2.5		Leite et al., 2008
MBR	1	Fermentative culture	Glucose Sucrose Fructose	1.3 2.1 2.7	1.5 1.4 1.4	Lee et al., 2007
PBR	—	Anaerobic fermentative bacteria	—	—	7.4	Lee et al., 2006a
PBR	—	Anaerobic fermentative bacteria	—	—	0.25	Palazzi et al., 2000
PBR	—	*E. cloacae* IIT-BT-08	Glucose	—	1.69	Kumar and Das, 2001
PBR	0.48	*E. cloacae* BL-21	Cane molasses	—	2.17	Chittibabu et al., 2006

Continued

TABLE 6 Biohydrogen Production from Different Bioreactors by Dark Fermentation—Cont'd

Reactor type[a]	Volume (liter)	Microorganism	Substrate	Hydrogen yield (mol/mol)	Hydrogen rate (liter/liter/h)	Reference
GAC-AFBR		Clostridium rich mixed culture	Glucose	1.2	2.36	Zhang et al., 2007c
DTFBR	8	Clostridium sp.	Sucrose	2.5	2.27	Lin et al., 2006
TBFR	—	Mixed culture	—	—	1.07	Oh et al., 2004b
CSTR	—	Mixed culture	Glucose	—	0.54	Fang and Liu, 2002
CSTR	—	Mixed culture	—	—	15.09	Wu et al., 2006
CSTR	—	Mixed culture	—	—	3.20	Zhang et al., 2007b
Column reactor	—	Anaerobic fermentative bacteria	—	—	7.49	Zhang et al., 2008b

[a] IBR, immobilized bioreactor; CIGSB, carrier-induced granular sludge bed reactor; AFBR, anaerobic-fluidized bed reactor; HAIB, horizontal flow anaerobic-immobilized biomass reactor; MBR, membrane bioreactor; GAC-AFBR, granular-activated carbon anaerobic-fluidized bed reactor; DTFBR, draft tube fluidized bed reactor; PBR, packed bed reactor; CSTR, continuous stirred tank reactor; UASB, upflow anaerobic sludge blanket reactor.

render it suitable as a substrate for dark hydrogen fermentation. As such, use of less expensive renewable biomass such as sewage sludge can lower the cost of hydrogen production considerably. In addition, it may also prove beneficial in effective waste management.

Glycerol Waste

Glycerol is the major by-product of the biodiesel industry. With the increase in biodiesel production, copious amounts of glycerol are being produced. As such, disposal of waste glycerol has become a major problem. Researchers have attempted to use glycerol as a substrate for hydrogen production and have obtained promising results. For example, Yang et al. (2008) used glycerol as a substrate in an immobilized bioreactor packed with polyurethane at thermophilic and mesophilic temperatures. Results showed that removal of dissolved organic carbon at the thermophilic condition was 86.7%. The study shows that hydrogen can be produced from glycerol waste using hydrogentrophic bacteria. In another study, Ito et al. (2005) used glycerol waste under mesophilic conditions to produce 80 mmol H_2 /liter/h by Enterobacter aerogenes HU-101.

Palm Oil Mill Effluent

In many tropical developing countries, palm oil is a major cash crop. During extraction of crude palm oil, mills generate a liquid waste popularly known as palm oil mill effluent (POME). On average, 0.9–1.5 m^3 POME is generated every 1 ton of palm oil produced. Researchers have shown that POME could be a potential source for biohydrogen production because it contains significant amounts of carbohydrates, ammonium nitrogen, phosphate, and phosphorous (Chong et al., 2009). In addition, it has a high organic content (BOD more than 20 g/liter and nitrogen more than 0.5 g/liter). O-Thong et al. (2007) carried out studies using POME as a substrate for hydrogen production with thermophilic microorganisms and the study produced 6.5 liter H_2/liter-POME under anaerobic, thermophilic conditions.

Pretreatment of Wastes

Complex substrates as such are inaccessible to the organism. They require the substrate to be available in simpler forms. Thus, complex materials require being pretreated to break them into smaller substrates that can be utilized effectively by the organism. As such, pretreatment technologies are diverse and numerous based on the types of feedstock available (Hendriks and Zeeman, 2009; Taherzadeh and Karimi, 2008). However, they can be classified into four main groups as shown in Figure 4. The different pretreatment methods have their own specific advantages and disadvantages. Among the different methods, thermochemical and chemical are currently the most effective methods and are used more popularly in the various industries. Significant considerations are given to the choice of the pretreatment

FIGURE 4   Pretreatment of cellulosic feedstock.

methods used in the study, as the choice of the pretreatment method may heavily influence cost and performance of the fermentation process. A combination of different pretreatment methods has been considered to achieve higher yields. Although considerable literature is available on the effects of the different pretreatments processes on biomass composition and hydrogen yields, very few references are available on the cost comparison of the pretreatment process.

Characterization of Wastes

The selection of waste as a feedstock for biohydrogen production is based on characterization of the waste by which the composition of the different waste streams is analyzed. In addition, precise characterization of wastes may help improve the understanding of the dynamics of the waste constituents in order to treat the waste. The characterization of wastes has varied based on sources of waste given in Table 7.

TABLE 7 Contents of Cellulose, Hemicellulose, and Lignin in Common Agricultural Residues and Wastes[a]

Lignocellulosic materials	Cellulose (%)	Lignin (%)	Hemicellulose (%)
Barley bran	23	21.4	32
Barley straw	31–45	14–19	27–38
Coastal Bermuda grass	25	6.4	35.7
Cotton seed hairs	80–95	0	5–20
Coconut fiber	36 43	41–45	1.5–2.5
Corn cobs	45	15	35
Grasses	25–40	10–30	35–50
Hardwoods stems	40–55	18–25	24–40
Leaves	15–20	0	80–85
Newspaper	40–55	18–30	25–40
Nut shells	25–30	30–40	25–30
Paper	85–99	0–15	0
Rice straw	32–47	5–24	19–27
Rice bran	35	17	25
Solid cattle manure	1.6–4.7	2.7–5.7	1.4–3.3
Softwood stems	45–50	25–35	25–35
Switch grass	45	12	31.4
Sorghum stalk	27	11	25
Wheat bran	30	15	50

[a] *From Reshamwala et al. (1995), Cheung and Anderson (1997), Boopathy (1998), Dewes and Hunsche (1998), and Sun and Cheng, (2002).*

REACTOR CONFIGURATIONS AND OPERATION

Several studies have been conducted with batch, semicontinuous, and continuous bioreactors on biohydrogen production. Mostly biohydrogen fermentation can be carried out in either batch or continuous mode. For initial optimization studies, batch mode fermentation is more suitable, but mostly continuous mode fermentation is preferred by industries. Many studies have been conducted with continuous stirred tank reactors (CSTR) because of simple construction, ease of operation, and effective homogeneous mixing and can keep the system at certain hydraulic retention time controls the microbial growth rate. In addition to the extensively studied CSTR, various types of bioreactors, such as anaerobic sequencing batch reactor (ASBR), membrane bioreactor (MBR), carrier-induced granular sludge bed reactor (CIGSB), agitated granular sludge bed reactor (AGSBR), fixed bed bioreactor, fluidized bed bioreactor (FBR), and upflow anaerobic sludge blanket bioreactor (UASB), have been developed with high production yields (Table 8). Several studies have found that higher hydrogen can be produced from these reactors due to the physical retention of microbial biomass, including the use of naturally forming flocs or granules of self-immobilized microbes, microbial immobilization on inert materials, microbial-based biofilms, or retentive membranes. Very few studies have been conducted on comparative analysis of these reactors whether difference in reactor configurations or operational parameters of the system. The choice of the bioreactors depends on the nature of the feedstocks, which could be converted with help of microorganisms into organic acids, alcohols, and biogas; they are used mostly in streams containing soluble organic wastes (Murnleitner et al., 2001). Some specialized bioreactors need to be developed with resistant to short-term fluctuations in operational parameters, reliable performance that is stable over long periods of time, and more robust for the improvement of biohydrogen production.

Upflow Anaerobic Sludge Blanket Bioreactor

The UASB bioreactor has been used extensively in laboratory or pilot scale studies for biohydrogen production from different waste materials (Chang and Lin, 2004). However, it has been effective in treating organic wastes and biogas production. The UASB reactor is used mainly for treating waste from sugar refining, beverage industry, food industry, distilleries industry, and paper and pulp industry. Applications of this technology are expanding to include treating waste from the chemical and petrochemical industry, textile industry, landfill leachates, and also domestic wastewater. Anaerobic granular sludge bed technology refers to a special kind of reactor concept for the anaerobic treatment of wastewater at a high rate. This concept was first initiated with the USAB reactor. Wastewaters pass upward through an anaerobic sludge bed where the microorganisms in the sludge come into contact with wastewater substrates. Microorganisms present in the sludge bed that can naturally form granules of 0.2 to 2 mm diameter have a high sedimentation velocity and thus resist wash out from the system even at a high hydraulic load. The resulting biogas production takes place under an anaerobic degradation process. The released gas bubbles upward motion causes hydraulic turbulence that provides reactor mixing without any mechanical parts. When comparing with other anaerobic bioreactors, it contains granule sludge and an internal

TABLE 8 Typical Industrial Wastewater Contaminants[a]

Industry	Characteristics of wastewaters
Food processing	High in dissolved organics; mainly protein, fat, and lactose
Meat and poultry processing	High in dissolved and suspended organics, including protein, blood, greases, fats, and manure
Fruit and vegetable canneries	High in dissolved and suspended organics from natural products
Breweries and distilleries	High in dissolved and suspended organics
Pharmaceuticals	High in dissolved and suspended organics, including some surfactants and biological agent
Organic chemicals	Dissolved organics, including acids, aldehydes, phenolics, and free and emulsified oils
Petroleum refining	Phenolics, free and emulsified oils, and other dissolved organics
Pulp and paper	Dissolved and suspended organics and inorganics
Plastics and resins	Dissolved organics, including acids, aldehydes, phenolics, cellulose, alcohols, surfactants, and oils
Explosives	Organic acids and alcohols, soaps, and oils
Rubber	Dissolved and suspended organics and oils
Textiles	Dissolved and suspended organics, fats, and oil
Leather tanning and finishing	Dissolved and suspended organics, fats and oils, organic nitrogen, hair, and fleshings
Coke and gas	High in phenolics, ammonia, and dissolved organics
Sugar	Dissolved sugar and protein
Yeast	Solid organics
Rice processing	Suspended and dissolved carbohydrates
Soft drinks	Suspended and dissolved carbohydrates
Fish processing	Organic solids
Pickles	Suspended solids and dissolved organics
Acids	Low pH
Detergents	Surfactants
Insecticides	Organics, benzene, acid, highly toxic
Steel	Acid, cyanogen, phenol, coke, oil
Metal plating	Metals, acid
Photographic products	Organic and inorganic reducing agents, alkaline
Oil	Sodium chloride, sulfur, phenol, oil
Glass	Suspended solids

[a] *From Metcalf and Eddy (1991), Droste (1997), Eckenfelder (1989), and Corbitt (1990).*

three-phase gas/sludge/liquid separator system device. The operating conditions of the UASB acidogenic reactor, such as concentration of solids in the feed, retention time, organic loading density, pH, and flow recirculation, were studied to maximize hydrogen production (Yu and Mu, 2006; Zhao et al., 2008). In order to predict the steady-state performance of granule-based hydrogen production, the UASB reactor can be simulated using a neural network and genetic algorithm, and a model was designed, trained, and validated (Mu and Yu, 2007).

Continuous Stirred Tank Reactor

A CSTR is used commonly for continuous hydrogen production. Due to a constant mixing pattern, the hydrogen-producing microbes are completely mixed and suspended in the reactor liquor. Under this condition, there is good substrate–microbes contact and mass transfer can be accomplished. However, due to problems of biomass washout, suspended bacterial populations are generally not able to operate at a high dilution rate in the reactor. Therefore, HRTs must be lesser than the maximum specific growth rate of the organism(s). When compared with granular sludge or immobilized cell-based bioreactors, the CSTR has lower hydrogen production rates. The volatile suspended solids retained in the bioreactor normally range between 1 and 4 g VSS/liter, depending on the HRT (Show et al., 2007, 2010). In order to evaluate the performance of continuous hydrogen-producing processes, the rate of hydrogen production has been considered. However, the CSTR is unable to maintain high levels of hydrogen-producing biomass due to its intrinsic structure. Thereby it always shows poor performance with regard to rate of hydrogen production at a short hydraulic retention time (Zhang et al., 2007a). Continuous biohydrogen production by immobilized cell systems has become more popular than suspended cell systems due to maintenance of higher biomass concentrations to achieve satisfactory rates of hydrogen production even at lower HRTs (Wu et al., 2002). Acidogenic anaerobic digestion of organic waste reaches a maximum rate of hydrogen production at lower hydraulic retention times (Kapdan and Kargi, 2006; Kotsopoulos et al., 2006).

Fluidized Bed Bioreactor

A fluidized bed bioreactor is a combination of the two most common, packed bed and stirred tank, continuous flow reactors. It has excellent heat and mass transfer characteristics. In FBR, cells are attached to solids in the form of biofilm or granules. Due to the dragging force of the upward wastewater flow, they can be maintained in suspension. This increases their catalytic activity, and they are known to cause higher degradation rates of organic wastes. FBR can work at both higher hydraulic loading rates and also at low HRTs (Zhang et al., 2008a). Anaerobic FBR is considered a more efficient reactor because it has less chance of particle washout when compared with a UASB reactor. Nowadays, FBR has received more attention in case of wastewater treatment and also hydrogen production. The main disadvantage of the anaerobic FBR is that it requires more energy in order to achieve fluidization in the bioreactor (Zhang et al., 2007c).

Fixed Bed Reactor

A fixed bed bioreactor is packed completely with carrier materials within the reactor. These carrier materials play a vital role in biomass retention and hydrogen production. However, the hydraulic mixing regime is less turbulent. This causes a lower rate of substrate conversion and lower hydrogen production due to higher mass transfer resistance (Chang et al., 2002). At the same time, pH gradient distribution along the reactor column causes a heterogeneous distribution of microbial activity. To overcome these bottlenecks, recirculation flow is recommended to increase both hydrogen production and substrate conversion rates. Kumar and Das (2001) reviewed hydrogen production using different reactor configurations. They found that at a HRT of 1.08 h, a rhomboid bioreactor with a convergent–divergent configuration produced a maximum rate of hydrogen production (1.60 liter/h/liter) as compared with a tapered reactor (1.46 liter/h/liter) and tubular reactor (1.40 liter/h/liter). The increased production rate could be attributed to higher turbulent mixing favoring mass transfer and lower gas hold-up.

Anaerobic Sequencing Batch Reactor

The ASBR was developed to complement the disadvantage of the sequencing batch reactor system. It is currently used for treating wastewater, such as swine manure, leachate, and dairy industry. The process consists of repetition of a cycle including five separate steps: feed, react, settle, draw, and idle. The ASBR has shown promising results in hydrogen production by changing the time of each cycle. The pH and cyclic duration of the operations mostly impacted fermentative hydrogen production (Chen et al., 2009). It is a newly developed technology and has been studied extensively due to its advantages, such as high efficiency for both COD removal and hydrogen production. In addition, the reactor is extremely easy to control and handle (Xiangwen et al., 2007). However, if the reactor is overloaded, the performance of the reactor turns down. This is the reason why the ASBR has not been used as widely in the wastewater treatment industry as other anaerobic processes (Cheong and Conly, 2007). The influence of the substrate loading rate in the ASBR using chemical wastewater as substrate was studied. Bhaskar et al. (2008) found that a change in the organic loading rate from 6.3 to 7.9 kg COD/m^3 day varied the hydrogen yield from 6.064 to 13.44 mol H$_2$/kg COD.

Membrane Bioreactor

The MBR is a most promising process combination of membrane filtration and activated sludge treatment for biomass retention. The main purpose of using a MBR is to retain the biomass concentration in the reactor. Additionally, it also offers the option of independent selection of hydraulic retention time and sludge retention time, which helps in greater control of operational parameters (Oh et al., 2004a). Increasing the sludge retention time further would enhance the retention time of the biomass. This improves the substrate utilization rate, but decreases the rate of hydrogen production. The biomass concentration has increased from 2.2 g/liter in a control reactor without membrane to 5.8 g/liter in an anaerobic MBR at a HRT of 3.3 h. Under such operating conditions, the rate of hydrogen production also increased from 0.50 to 0.64 liter/h/liter. A similar study showed the rate of hydrogen production in a range of 0.25–0.69 liter/h/liter using MBR (Li and Fang, 2007). The major disadvantage of MBR is membrane fouling and high operating costs.

Carrier-Induced Granular Sludge Bed Reactor

There are some reactors very efficient for biohydrogen production accomplished with the formation of self-floculating granular sludge or matrix, entrapped immobilized cells, such as a CIGSB (Lee et al., 2006), AGSBR (Lin and Lay, 2010), and CSABR with silicone-immobilized cells (Wu et al., 2006). These reactors are known to produce a high rate of hydrogen due to enhanced cell retention at high dilution rates. However, its major bottlenecks include inefficient mixing and stability of functional granules. Among these reactors, the CIGSB reactor has been shown very effective in case of hydrogen production (Lee et al., 2006b). However, an absence of mechanical agitation may cause a problem with poor mass transfer efficiency during operations. Lee et al., (2006b) addressed the issue by adjusting the height-to-diameter ratio and introducing an appropriate agitation device in the CIGSB reactor. By incorporating these changes they obtained a higher rate of hydrogen production (6.87 liter/h/liter) and a higher hydrogen yield (3.88 mol H_2/mol sucrose). Results indicated that the physical configuration of the reactor and proper upflow velocity can enhance the hydrogen production in a CIGSB reactor. A 400-liter pilot CIGSB reactor was examined with sucrose-based synthetic wastewater and fermentation wastewater (condensed molasses soluble) for biohydrogen production efficiency (Chang et al., 2009).

CHALLENGES

During the last decade, several efforts have been made to make the process more feasible. However, major hurdles remain to be overcome. Currently, low substrate conversion efficiency and low hydrogen yield are the biggest bottlenecks of the process. These challenges may be overcome by using efficient bioreactor design, process modification, suitable feedstock, and suitable microbial strain. Despite a large amount of research in the past and at present, these specific areas need to be concentrated for further enhancement of hydrogen production (Hallenbeck and Ghosh, 2009; Das and Verziroglu, 2001). Eco-friendly green biohydrogen production can be obtained using nonfood feedstock and waste organic materials by dark fermentation. Additionally, integrating the biohydrogen process to the conventional wastewater treatment process has many advantages, such as wastewater treatment and generation of clean energy. In the future, green wastes may be targeted as suitable feedstock for hydrogen production because of their natural abundance. Lignocellulose is the most abundant renewable natural resource. However, there has been a growing concern about biofuel generation using energy crops as feedstocks. As a result, efforts are being made to use nonedible crop residues and other organic wastes as substrates for hydrogen production (Juang et al., 2011; Kim and Lee, 2010).

Considering the metabolic pathways, production of higher VFA concentrations appears to be another bottleneck. The end metabolites produced by the organism compete for the same reductants as hydrogen. Therefore, redirection of the reductants toward soluble end metabolites decreases hydrogen production (Kumar et al., 2001). In view of this, several researchers are trying to redirect the metabolic pathway toward the production of lower end metabolites. Concomitantly, efforts are in progress to increase the acetate-to-butyrate ratio, as higher acetate production is associated with higher hydrogen yields.

Moreover, biohydrogen production is highly reliant on process conditions such as pH, hydraulic retention time, and temperature. Improving reactor design and determining the suitable process parameters could enhance the hydrogen yield. Stoichiometrically, fermentative bacteria can produce only relatively small amounts of hydrogen, typically less than 30%. Lower conversion rates affect the system economics and stability. Therefore, research is presently underway to enhance the hydrogen yield beyond 4 mol H_2/mol glucose. Studies include development of suitable hybrid processes such as the two-stage process of dark fermentation followed by the photofermentative production of hydrogen. In this process, the volatile fatty acids produced in the dark fermentative process can be used as the substrate in the photofermentative process. This may effectively increase the theoretical limit of hydrogen to 12 mol H_2/mol glucose (Nath et al., 2006). Similarly, hybrid processes including subsequent methane production or electrohydrogenesis are also being considered to maximize the gaseous energy recovery of the process (de Vrije and Claasen, 2003). With technology advancement, biohydrogen production may offer a sustainable alternative energy resource in the future.

CONCLUSION

Dark hydrogen production holds considerable promise. It is carried out by anaerobic mesophiles or thermophiles. Thermophiles appear more promising as they can overcome the thermodynamic limitation to produce a higher hydrogen yield. Both mesophiles and thermophiles are known to utilize a wide range of substrates, including waste resources for the generation of hydrogen. Formate hydrogen lyase and [FeFe]-hydrogenase are the principal enzymes that catalyze the process. However, to make the process more feasible, economical, and sustainable, hydrogen yields need to be increased beyond the theoretical maximum of 4 mol H_2/mol glucose. To obtain this, future research must be targeted toward genetically and metabolically improved strains and redirection of the metabolic pathways toward a lower production of VFAs and higher acetate-to-butyrate ratios.

Acknowledgments

The authors gratefully acknowledge the Department of Biotechnology (DBT), Ministry of New and Renewable Energy (MNRE), Government of India for the financial support to carry out research work on dark fermentation processes.

References

Abánades, A., 2012. The challenge of hydrogen production for the transition to a CO_2-free economy. Agronomy Res. Biosystem. Eng. 1, 11–16.
Adams, M.W., 1990. The structure and mechanism of iron-hydrogenase. Biochem. Biophys. Acta. 1020, 115–145.

Adams, R., Gabbatt, A., 2010. British petroleum oil spill report–as it happened'. Accessed from http://www.guard ian.co.uk/environment/blog/2010/sep/08/bp-oil-spill-report-live.

Allen, L., Cohen, M.J., David, A., Bart, M., 2011. Fossil fuel and water quality. In: Gleick, P.H. (Ed.), The World's Water. Island Press, pp. 73–96.

Amend, J.P., Shock, E.L., 2001. Energetics of overall metabolic reactions of thermophilic and hyperthermophilic Archea and Bacteria. Microbial Rev. 25, 175–243.

Antonopoulou, G., Stamatelatou, K., Venetsaneas, N., Kornaros, M., Lyberatos, G., 2008. Biohydrogen and methane production from cheese whey in a two-stage anaerobic process. Ind. Eng. Chem. Res. 47, 5227–5233.

Atif, A.A.Y., Fakhru',l-Razi, A., Ngan, M.A., Morimoto, M., Iyuke, S.E., Veziroglu, N.T., 2005. Fed batch production of hydrogen from palm oil mill effluent using anaerobic microflora. Int. J. Hydrogen Energy. 30, 1393–1397.

Azbar, N., Cetinkaya Dokgoz, F.T., Keskin, T., Korkmaz, K.S., Syed, H.M., 2009. Continuous fermentative hydrogen production from cheese whey wastewater under thermophilic anaerobic conditions. Int. J. Hydrogen Energy. 34, 7441–7447.

Baghchehsaree, B., Nakhla, G., Karamanev, D., Argyrios, M., 2010. Fermentative hydrogen production by diverse Microflora. Int. J. Hydrogen Energy. 35, 5021–5027.

Balantine, S.P., Boxer, D.H., 1985. Nickel-containing hydrogenase isoenzymes from anaerobically grown *Escherichia coli* K-12. J. Bacteriol. 163, 454–459.

Benemann, J.R., 1996. Hydrogen biotechnology: Progress and prospects. Nat. Biotechnol. 14, 1101–1103.

Benemann, J.R., 1997. Feasibility analysis of photobiological hydrogen production. Int. J. Hydrogen Energy. 22, 979–987.

Bohm, R., Sauter, M., Bock, A., 1990. Nucleotide sequence and expression of an operon in Escherichia coli coding for formate hydrogenlyase components. Mol. Microbiol. 4, 231–243.

Boopathy, R., 1998. Biological treatment of swine waste using anaerobic baffled reactors. Bioresour. Technol. 64, 1–6.

Bowles, L.K., Ellefson, W.L., 1985. Effects of butanol on *Clostridium acetobutylicum*. Appl. Environ. Microbiol. 50, 1165–1170.

Boyington, J.C., Gladyshev, V.N., Khangulov, S.V., Stadtman, T.C., Sun, P.D., 1997. Crystal structure of formate dehydrogenase H: Catalysis involving Mo, molybdopterin, selenocysteine, and Fe4SO4 cluster. Science 275, 1305–1308.

Bredholt, S., Sonne-Hansen, J., Nielsen, P., Mathrani, I.M., Ahring, B.K., 1999. Caldicellulosiruptor kristjanssonii sp. nov., a cellulolytic, extremely thermophilic, anaerobic bacterium. Int. J. Syst Bacteriol. 49 (3), 991–996.

Campbell, C.J., Laherrere, J.H., 1998. The end of cheap oil. Sci. Am. 278, 78–83.

Cao, X., Zhao, Y., 2009. The influence of sodium on biohydrogen production from food waste by anaerobic fermentation. J. Mater. Cycles Waste Manag. 11, 244–250.

Chang, F.Y., Lin, C.Y., 2004. Biohydrogen production using an up-flow anaerobic sludge blanket reactor. Int. J. Hydrogen Energy. 29 (1), 33–39.

Chang, J.S., Lee, K.S., Lin, P.J., 2002. Biohydrogen production with fixed-bed bioreactors. Int. J. Hydrogen Energy. 27, 1167–1174.

Chang, F.Y., Lay, C.H., Wu, J.H., Chen, C.H., Chang, J.S., Lin, C.Y., 2009. Performance of lab- and pilot-scale bioreactors for fermentative hydrogen production from concentrated molasses fermentation soluble. In: The 10th Asian Hydrogen Energy Conference. Daegu 55.

Chen, C.C., Lin, C.Y., 2003. Using sucrose as a substrate in an anaerobic hydrogen producing reactor. Adv. Environ. Res. 7, 695–699.

Chen, C.C., Lin, C.Y., Chang, J.S., 2001. Kinetics of hydrogen production with continuous anaerobic cultures utilizing sucrose as limiting substrate. Appl. Microbiol. Biotechnol. 57, 56–64.

Chen, S.D., Lee, K.S., Lo, Y.C., Chen, W.M., We, J.F., Lin, C.Y., et al., 2008. Batch and continuous biohydrogen production from starch hydrolysate by *Clostridium* species. Int. J. Hydrogen Energy. 33, 1803–1812.

Chen, S.Y., Chu, C.Y., Cheng, M.J., Lin, C.Y., 2009. The autonomous house: A bio-hydrogen based energy self-sufficient approach. Int. J. Environ. Res. Public Health. 6, 1515–1529.

Cheng, C.C., Lin, C.Y., Lin, M.C., 2002. Acid-base enrichment enhances anaerobic hydrogen production process. Appl. Microbiol. Biotechnol. 58, 224–228.

Cheong, D.Y., Conly, L.H., 2007. Effect of feeding strategy on the stability of anaerobic sequencing batch reactor responses to organic loading conditions. Bioresour. Technol. 42, 223–232.

Cheung, S.W., Anderson, B.C., 1997. Laboratory investigation of ethanol production from municipal primary wastewater. Bioresour. Technol. 59, 81–96.

Chin, H.L., Chen, Z.S., Chou, C.P., 2003. Fed-batch operation using *Clostridium acetobutylicum* suspension cultures as biocatalyst for enhancing hydrogen production. Biotechnol. Prog. 9, 383–388.

Chittibabu, G., Nath, K., Das, D., 2006. Feasibility studies on the fermentative hydrogen production by recombinant *Escherichia coli* BL-21. Process Biochem. 41, 682–688.

Chong, M.L., Rahim, R.A., Shirai, Y., Hassan, M.A., 2009. Biohydrogen production by *Clostridium butyricum* EB6 from palm oil mill effluent. Int. J. Hydrogen Energy. 34, 764–771.

Chung, K.T., 1976. Inhibitory effects of H2 on growth of *Clostridium cellobioparum*. Appl. Environ. Microbiol. 31, 342–348.

Collet, C., Adler, N., Schwitzguebel, J.P., Peringer, P., 2004. Hydrogen production by *Clostridium thermolacticum* during continuous fermentation of lactose. Int. J. Hydrogen Energy. 29, 1479–1485.

Conrad, R., Wetter, B., 1990. Influence of temperature on energetics of hydrogen metabolism in homoacetogenic, methanogenic, and other anaerobic-bacteria. Arch. Microbiol. 155 (1), 94–98.

Corbitt, R.A., 1990. The Standard Handbook of Environmental Engineering. McGraw-Hill, New York, pp. 6.1–6.49.

Craven, S.E., 1988. Increased sporulation of *Clostridium-perfringens* in a medium prepared with the prereduced anaerobically sterilized technique or with carbon-dioxide or carbonate. J. Food Prot. 51 (9), 700–706.

Dabrock, B., Bahl, H., Gottschalk, G., 1992. Parameters affecting solvent production by *Clostridium pasteurium*. Appl. Environ. Microbiol. 58, 1233–1239.

Dalby, S., 2009. Security and Environmental Change. Polity Press, Cambridge, UK.

Das, D., 2009. Advancements in biohydrogen production processes: A step towards commercialization. Int. J. Hydrogen Energy. 34, 7349–7357.

Das, D., Veziroglu, T.N., 2001. Hydrogen production by biological processes: A survey of literature. Int. J. Hydrogen Energy. 6, 13–28.

Das, D., Khanna, N., Veziroglu, T.N., 2008. Recent developments in biological hydrogen production processes. Chem. Indust. Chem. Eng. Q. 14, 57–67.

de Vrije, T., Claasen, P.A.M., 2003. Dark hydrogen fermentation. In: Reith, J.H., Wijffels, R.H., Barten, H. (Eds.), Bio-methane and Bio-hydrogen. Dutch Biological Hydrogen Foundation, Hague, pp. 103–123.

De Vrije, T., De Haas, G.G., Tan, G.B., Keijsers, E.R.P., Claassen, P.A.M., 2002. Pretreatment of Miscanthus for hydrogen production by Thermotoga elfii. Int. J. Hydrogen Energy. 27, 1381–1390.

Dewes, T., Hunsche, E., 1998. Composition and microbial degradability in the soil of farmyard manure from ecologically-managed farms. Biol. Agric. Hortic. 16, 251–268.

Droste, R.L., 1997. Theory and Practice of Water and Wastewater Treatment. John Wiley & Sons, New York, pp. 94–132, 157–180, 181–193.

Eckenfelder Jr., W.W., 1989. Industrial Water Pollution Control. McGraw-Hill, New York, pp. 300–311.

Eroglu, E., Eroglu, I., Gunduz, U., Turkerc, L., Yucel, M., 2006. Biological hydrogen production from olive mill waste-water with two stage process. Int. J. Hydrogen Energy. 31, 1527–1535.

Fabiano, B., Perego, P., 2002. Thermodynamic study and optimization of hydrogen production by *Enterobacter aerogenes*. Int. J. Hydrogen Energy. 27, 149–156.

Fan, Y.T., Zhang, Y.H., Zhang, S.F., Hou, H.W., Ren, B.Z., 2006. Efficient conversion of wheat straw wastes into biohydrogen gas by cow dung compost. Bioresour. Technol. 97, 500–505.

Fang, H.H.P., Liu, H., 2002. Effect of pH on hydrogen production from glucose by mixed culture. Bioresour. Technol. 82, 87–93.

Fang, H.H.P., Li, C.L., Zhang, T., 2006. Acidophilic biohydrogen production from rice slurry. Int. J. Hydrogen Energy. 31, 683–692.

Ferchichi, M., Crabbe, E., Gil, G.H., Hintz, W., Almadidy, A., 2005. Influence of initial pH on hydrogen production from cheese whey. J. Biotechnol. 120, 402–409.

Fernandes, B.S., Peixoto, G., Albrecht, F.U., del Aguila, N.K.S., Zaiat, M., 2010. Potential to produce biohydrogen from various wastewaters. Energy Sustain. Dev. 14, 143–148.

Gavala, H.N., Skiadas, I.V., Ahring, B.K., 2006. Biological hydrogen production in suspended and attached growth anaerobic reactor systems. Int. J. Hydrogen Energy. 31, 1164–1175.

Ginkel, S.V., Sung, S., 2001. Biohydrogen as a function of pH and substrate concentration. Environ. Sci. Technol. 35, 4726–4730.

Greenbaum, E., 1990. Hydrogen production by photosynthetic water splitting. In: Veziroglu, T.N., Takashashi, P.K. (Eds.), Hydrogen Energy Progress VIII. Pergamon Press, New York, pp. 743–754.

Guwy, A.J., Hawkes, F.R., Hawkes, D.L., Rozzi, A.G., 1997. Hydrogen production in a high rate fluidized bed anaerobic digester. Water Res. 21, 1291–1298.

Hakes, J.E., 1998. 25th Anniversary of the 1973 Oil Embargo. Accessed from The Energy Information administration. http://www.eia.gov/emeu/25opec/anniversary.html.

Hallenbeck, P.C., 2005. Fundamentals of the fermentative production of hydrogen. Water Sci. Technol. 52, 21–29.

Hallenbeck, P.C., Ghosh, D., 2009. Advances in fermentative biohydrogen production: The way forward? Trends Biotechnol. 27 (5), 287–297.

Han, S.K., Shin, H.S., 2004. Biohydrogen production by anaerobic fermentation of food waste. Int. J. Hydrogen Energy. 29, 569–577.

Hawkes, F.R., Hussy, I., Kyazze, G., Dinsdale, R., Hawkes, D.L., 2007. Continuous dark fermentative hydrogen production by mesophilic microflora: Principles and progress. Int. J. Hydrogen Energy. 32, 172–184.

Hendriks, A.T.W.M., Zeeman, G., 2009. Pretreatments to enhance the digestibility of lignocellulosic biomass. Bioresour. Technol. 100, 10–18.

Hoffert, M.I., Caldeira, K., Jain, A.K., Haites, E.F., Danny, L.D., Seth, H., et al., 1998. Energy implications of future stabilization of atmospheric CO_2 content. Nature 39 (5), 881–884.

Hsiao, C.L., Chang, J.J., Wu, J.H., Chin, W.C., Wen, F.S., Huang, C.C., et al., 2009. *Clostridium* strain co-cultures for biohydrogen production enhancement from condensed molasses fermentation solubles. Int. J. Hydrogen Energy. 34, 7173–7181.

Huang, C.Y., Hsieh, H., Lay, C.H., Chuang, Y.S., Kuo, A.Y., Chen, C.C., 2010. Biohydrogen production by anaerobic co-digestion of textile and food wastewaters. The 2010 Asian bio-hydrogen symposium and APEC advanced bio-hydrogen technology conference Taiwan.

Hussy, I., Hawkes, F.R., Dinsdale, R., Hawkes, D.L., 2000. Continuous fermentative hydrogen production from wheat starch co-product by mixed microflora. Biotechnol. Bioeng. 84, 619–626.

Hussy, I., Hawkes, F.R., Dinsdale, R., Hawkes, D.L., 2005. Continuous fermentative hydrogen production from sucrose and sugar beet. Int. J. Hydrogen Energy. 30, 471–483.

Inan, H., Dimoglo, A., Karpuzcu, M., 2004. Olive oil mill wastewater treatment by means of electro-coagulation. Separation Purific. Technol. 36, 23–31.

Ito, T., Nakashimada, Y., Senba, K., Matsui, T., Nishio, N., 2005. Hydrogen and ethanol production from glycerol containing wastes discharged after biodiesel manufacturing process. J. Biosci. Bioeng. 100, 260–265.

Jacquet, F., Bamiere, L., Bureau, J.C., Guindé, L., Guyomard, H., Treguer, D., 2007. Recent developments and prospects for the production of biofuels in the EU: Can they really be a part of the solution. In: Biofuels, Food and Feed Tradeoffs Conference. St. Louis, MO.

Jo, J., Lee, H., Park, D.S., Choe, D., Park, W., 2008a. Optimization of key process variables for enhanced hydrogen production by *Enterobacter aerogenes* using statistical methods. Bioresour. Technol. 99, 2061–2066.

Jo, J.H., Lee, D.S., Park, D., Park, J.M., 2008b. Biological hydrogen production by immobilized cells of *Clostridium tyrobutyricum* JM1 isolated from a food waste treatment process. Bioresour. Technol. 99, 6666–6672.

Jones, D.T., Woods, D.R., 1986. Acetone–butanol fermentation revisited. Microbiol. Rev. 50 (4), 484–524.

Juang, C.P., Whang, L.M., Cheng, H.H., 2011. Evaluation of bioenergy recovery processes treating organic residues from ethanol fermentation process. Bioresour. Technol. 102 (9), 5394–5399.

Jung, K.W., Kim, D.H., Shin, H.S., 2010. Continuous fermentative hydrogen production from coffee drink manufacturing wastewater by applying UASB reactor. Int. J. Hydrogen Energy. 35, 13370–13378.

Kanai, T., Imanaka, H., Nakajimam, A., Uwamori, K., Omori, Y., Fukui, T., et al., 2005. Continuous hydrogen production by the hyperthermophilic archaeon, *Thermococcus kodakaraensis* KOD1. J. Biotechnol. 116, 271–282.

Kapdan, I.K., Kargi, F., 2006. Biohydrogen production from waste materials. Enzyme Microb. Technol. 38, 569–582.

Karube, I., Urano, N., Yamada, T., Hirochika, H., Sakaguchi, K., 1983. Cloning and expression of the hydrogenase gene from *C. butyricum* in *E. coli*. Microbiol. Lett. 58, 119–122.

Khanal, S.K., Chen, W.H., Sung, L.L.S., 2004. Biological hydrogen production: Effects of pH and intermediate products. Int. J. Hydrogen Energy. 29, 1123–1131.

Khanna, N., Kotay, S.M., Gilbert, J.J., Das, D., 2011. Improvement of biohydrogen production by *Enterobacter cloacae* IIT-BT 08 under regulated pH. J. Biotechnol. 152, 9–15.

Kim, M.S., Lee, D.Y., 2010. Fermentative hydrogen production from tofu-processing waste and anaerobic digester sludge using microbial consortium. Bioresour. Technol. 101 (1), 48–52.

Knappe, J., Sawers, G., 1990. A radical-chemical route to acetyl-CoA: The anaerobically induced pyruvate formate-lyase system of *Escherichia coli*. Microbiol. Lett. 75, 383–398.

Kotay, S.M., Das, D., 2007. Microbial hydrogen production with *Bacillus coagulans* IIT-BT S1 isolated from anaerobic sewage sludge. Bioresour. Technol. 98, 1183–1190.

Kotay, S.M., Das, D., 2008. Optimization of nutrient and seed formulation for maximization of biohydrogen production from sewage sludge. In: International Workshop on Biohydrogen Production Technology 2008. IIT Kharagpur, India Paper: O33.

Kotsopoulos, A., Zeng, J., Angelidaki, I., 2006. Biohydrogen production in granular up-flow anaerobic sludge blanket (UASB) reactors with mixed cultures under hyper-thermophilic temperature (70°C). Biotechnol. Bioeng. 94 (2), 296–302.

Koutrouli, E.C., Gavala, H.N., Skiadas, I.V., Lyberatos, G., 2006. Mesophilic biohydrogen production from olive pulp. Process Safety Environ. Protect. 84, 285–289.

Kumar, N., Das, D., 2000a. Production and purification of alpha-amylase from hydrogen producing *Enterobacter cloacae* IIT-BT 08. Bioprocess Eng. 23, 205–208.

Kumar, N., Das, D., 2000b. Enhancement of hydrogen production by *Enterobacter cloacae* IIT-BT 08. Process Biochem. 35, 589–593.

Kumar, N., Das, D., 2001. Continuous hydrogen production by immobilized *Enterobacter cloacae* IIT-BT 08 using lignocellulosic materials as solid matrices. Enzyme Microbiol. Technol. 29, 280–287.

Kumar, N., Ghosh, A., Das, D., 2001. Redirection of biochemical pathways for the enhancement of H_2 production by *Enterobacter cloacae*. Biotechnol. Lett. 23, 537–541.

Lay, J.J., 2000. Modeling and optimization of anaerobic digested sludge concerting starch to hydrogen. Biotechnol. Bioeng. 68, 269–278.

Lay, J.J., 2001. Biohydrogen generation by mesophilic anaerobic fermentation of microcrystalline cellulose. Biotechnol. Bioeng. 74, 281–287.

Lay, J.J., Li, Y.Y., Noike, T., 1997. Influences of pH and moisture content on the methane production in high-solids sludge digestion. Water Res. 31 (6), 1518–1524.

Lay, J.J., Lee, Y.J., Noike, T., 1999. Feasibility of biological hydrogen production from organic fraction of municipal solid waste. Water Res. 33, 2579–2586.

Lay, C.H., Chen, C.C., Lin, H.C., Lin, C.Y., Lee, C.W., Lin, C.Y., 2010a. Optimizing pH and substrate concentration for fermentative hydrogen production from preserved fruits soaking solution. J. Environ. Eng. Manag. 20, 35–41.

Lay, C.H., Wu, J.H., Hsiao, C.L., Chang, J.J., Chen, C.C., Lin, C.Y., 2010b. Biohydrogen production from soluble condensed molasses fermentation using anaerobic fermentation. Int. J. Hydrogen Energy. 35, 13445–13451.

Lee, M.J., Zinder, S.H., 1988. Hydrogen partial pressures in a thermophilic acetate-oxidizing methanogenic coculture. Appl. Environ. Microbiol. 54 (6), 1457–1461.

Lee, Y.J., Miyahara, T., Noike, T., 2001. Effect of iron concentration on hydrogen fermentation. Bioresour. Technol. 80, 227–231.

Lee, Y.J., Miyahara, K., Noike, T., 2002. Effect of pH on microbial hydrogen fermentation. J. Chem. Technol. Biotechnol. 77, 694–698.

Lee, K.S., Lo, Y.S., Lo, Y.C., Lin, P.J., Chang, J.S., 2004. Operating strategies for biohydrogen production with high-rate anaerobic granular sludge bed bioreactor. Enzyme Microbial Technol. 35, 605–612.

Lee, K.S., Lin, P.J., Chang, J.S., 2006a. Temperature effects on biohydrogen production in a granular sludge bed induced by activated carbon carriers. Int. J. Hydrogen Energy. 31 (4), 465–472.

Lee, K.S., Lo, Y.C., Lin, P.J., Chang, J.S., 2006b. Improving biohydrogen production in a carrier-inducedgranular sludge bed by altering physical configuration and agitation pattern of the bioreactor. Int. J. Hydrogen Energy. 31, 1648–1657.

Lee, K.S., Lin, P.J., Fangchiang, K., Chang, J.S., 2007. Continuous hydrogen production by anaerobic mixed microflora using a hollow-fiber microfiltration membrane bioreactor. Int. J. Hydrogen Energy. 32, 950–957.

Leite, J.A.C., Fernandes, B.S., Pozzi, E., Barboza, M., Zaiat, M., 2008. Application of an anaerobic packed-bed bioreactor for the production of hydrogen and organic acids. Int. J. Hydrogen Energy. 33, 579–586.

Leonhartsberger, S., Korsa, I., Bock, A., 2002. The molecular biology of formate metabolism in *Enterobacteria*. J. Mol. Microbiol. Biotechnol. 4 (3), 269–276.

Levin, D.B., Pitt, L., Love, M., 2004. Biohydrogen production: Prospects and limitations to practical application. Int. J. Hydrogen Energy. 29 (2), 173–185.

Li, C.L., Fang, H.H.P., 2007. Fermentative hydrogen production from wastewater and solid wastes by mixed cultures. Crit. Rev. Environ. Sci. Technol. 37 (1), 1–39.

Lin, C.N., Wu, S.Y., Chang, J.S., 2006. Fermentative hydrogen production with a draft tube fluidized bed reactor containing silicone-gel-immobilized anaerobic sludge. Int. J. Hydrogen Energy. 31, 2200–2210.

Lin, C.N., Wu, S.Y., Chang, J.S., Chang, J.S., 2009. Biohydrogen production in a three phase fluidized bed bioreactor using sewage sludge immobilized by ethylene vinyl acetate copolymer. Bioresour. Technol. 100 (13), 3298–3301.

Lin, C.Y., Chang, C.C., Hung, C.H., 2008. Fermentative hydrogen production from starch using natural mixed culture. Int. J. Hydrogen Energy. 33, 2445–2453.

Lin, C.Y., Chang, R.C., 2004. Fermentative hydrogen production at ambient temperature. Int. J. Hydrogen Energy. 29, 715–720.

Lin, C.Y., Cheng, C.H., 2006. Fermentative hydrogen production from xylose using anaerobic mixed microflora. Int. J Hydrogen Energy. 31, 832–840.

Lin, C.Y., Lay, C.H., 2004. Carbon/nitrogen-ratio effect on fermentative hydrogen production by mixed microfora. Int. J. Hydrogen Energy. 29, 41–45.

Lin, C.Y., Lay, C.H., 2010. Research and development of biohydrogen production in Taiwan. In: Fang, H.H.P. (Ed.), Environmental Anaerobic Technology. Imperial College Press, London, pp. 331–344.

Lin, P.Y., Whang, L.M., Wu, Y.R., Ren, W.J., Hsiao, C.J., Li, S.L., 2007. Biological hydrogen production of the genus *Clostridium*: Metabolic study and mathematical model simulation. Int. J. Hydrogen Energy. 32, 1728–1735.

Liu, G., Shen, J., 2004. Effects of culture medium and medium conditions on hydrogen production from starch using anaerobic bacteria. J. Biosci. Bioeng. 98, 251–256.

Liu, H., Zhang, T., Fang, H.P.P., 2003. Thermophilic H_2 production from cellulose containing wastewater. Biotechnol. Lett. 25, 365–369.

Liu, Y., Yu, P., Song, X., Qu, Y., 2008. Hydrogen production from cellulose by co-culture of *Clostridium thermocellum* JN4 and *Thermoanaerobacterium thermosaccharolyticum* GD17. Int. J. Hydrogen Energy. 33, 2927–2933.

Lo, Y.C., Su, Y.C., Chen, C.Y., Chen, W.M., Lee, K.S., Chang, J.S., 2009. Biohydrogen production from cellulosic hydrolysate produced via temperature-shift-enhanced bacterial cellulose hydrolysis. Bioresour. Technol. 100 (23), 5802–5807.

Logan, B.E., Oh, S.E., Ginkel, S.V., 2002. Biological hydrogen production measured in batch anaerobic respirometer. Environ. Sci. Technol. 36, 2530–2535.

Ma, J., Ke, S., Chen, Y., 2008. Mesophilic biohydrogen production from food waste. Bioinfomat. Biomed. Eng 2841–2844.

Macy, J., Kulla, H., Gottschalk, G., 1976. Hydrogen-dependent anaerobic growth of *Escherichia coli* on L-malate: succinate formation. J. Bacteriol. 125, 423–428.

Mastorakis, N.E., Jeles, A., Bulucea, C.A., Bulucea, C.A., Brindusa, C., 2011. Evaluating the environmental impact of coal-fired power plants through wastewater pollutant vector. GEMESED'11. In: Proceedings of the 4th WSEAS International conference on energy and development–environmentpp. 181–187.

McCord, J.M., Keele, B.B., Fridovich, I., 1971. An enzyme-based theory of obligate anaerobiosis: The physiological function of superoxide dismutase. Natural Acad. Sci. 68, 1024–1027.

Metcalf and Eddy Inc., 1991. Wastewater Engineering: Treatment, Disposal, and Reuse. McGraw-Hill, New York, pp. 739–755.

Minnan, L., Jinli, H., Xiaobin, W., Huijuan, X., Jinzao, C., Chuannan, L., et al., 2005. Isolation and characterization of a high H2-producing strain *Klebsialle oxytoca* HP1 from a hot spring. Res. Microbiol. 156, 76–81.

Mizuno, O., Dinsdale, R., Hawkes, F.R., Hawkes, D.L., Noike, T., 2000. Enhancement of hydrogen production from glucose by nitrogen gas sparging. Bioresour. Technol. 73, 59–65.

Morimoto, M., Atsuko, M., Atif, A.A.Y., Ngan, M.A., Fakhru'l-Razi, A., Iyuke, S.E., et al., 2004. Biological production of hydrogen from glucose by natural anaerobic microflora. Int. J. Hydrogen Energy. 29, 709–713.

Morimoto, K., Kimura, T., Sakka, K., Ohmiya, K., 2005. Overexpression of a hydrogenase gene in *Clostridium paraputrificum* to enhance hydrogen gas production. Microbiol. Lett. 246, 229–234.

Mu, Y., Yu, H.Q., 2007. Simulation of biological hydrogen production in a UASB reactor using neural network and genetic algorithm. Int. J. Hydrogen Energy. 32, 3308–3314.

Murnleitner, E., Becker, T.M., Delgado, A., 2001. State detection and control of overloads in the anaerobic wastewater treatment using fuzzy logic. Water Res. 36, 201–211.

Nakashimada, Y., Rachman, M.A., Kakizono, T., Nishio, N., 2002. Hydrogen production of *Enterobacter aerogenes* altered by extracellular and intracellular redox states. Int. J. Hydrogen Energy. 27, 1399–1405.

Nath, K., Das, D., 2003. Hydrogen from biomass. Curr. Sci. 85, 265–271.

Nath, K., Das, D., 2009. Effect of light intensity and initial pH during hydrogen production by an integrated dark and photofermentation process. Int. J. Hydrogen Energy. 34, 7497–7501.

Nath, K., Kumar, A., Das, D., 2006. Effect of some environmental parameters on fermentative hydrogen production by Enterobacter cloacae DM11. Can. J. Microbiol. 52, 525–532.

Nguyen, T.A.D., Kim, J.P., Kim, M.S., Oh, Y.K., Sim, S.J., 2008. Optimization of hydrogen production by hyperthermophilic eubacteria, *Thermotoga maritime* and *Thermotoga neapolitana* in batch fermentation. Int. J. Hydrogen Energy. 33, 1483–1488.

Nicolet, Y., Cavazza, C., Fontecilla-Camps, J.C., 2010. Fe-only hydrogenases: Structure, future and evolution. J. Inorg. Biochem. 91, 1–8.

Oh, S.E., van Ginkel, S.W., Logan, B.E., 2003. The relative effectiveness of pH control and heat treatment for enhancing biohydrogen gas production. Environ. Sci. Technol. 37 (22), 5186–5190.

Oh, S.E., Lyer, P., Bruns, M.A., Logan, B.E., 2004a. Biological hydrogen production using a membrane bioreactor. Biotechnol. Bioeng. 87 (1), 119–127.

Oh, Y.K., Kim, S.H., Kim, M.S., Park, S., 2004b. Thermophilic biohydrogen production from glucose with trickling biofilter. Biotechnol. Bioeng. 88 (6), 690–698.

Onyenwoke, U., Lee, J.Y., Dabroswki, B.K., Weigel, J., 2006. Reclassification of *thermoaneroaberobium aeigenum* and amendation of the genus decryption. Int. J. Syst. Evol. Microbiol. 56, 1391–1395.

Ordal, E.J., Halvorson, H.O., 1939. A comparison of hydrogen production from sugars and formic acid by normal and variant strains of *Escherichia coli*. J. Bacteriol. 38, 199–220.

O-Thong, S., Prasertsan, P., Intrasungkha, N., Dhamwichukorn, S., Birkeland, N.K., 2007. Improvement of biohydrogen production and treatment efficiency on palm oil mill effluent with nutrient supplementation at thermophilic condition using an anerobic sequencing batch reactor. Enzy. Microb. Technol. 41, 583–590.

Palazzi, E., Fabiano, B., Perego, P., 2000. Process development of continuous hydrogen production by *Enterobacter aerogenes* in a packed column reactor. Bioproc. Eng. 22 (3), 205–213.

Pan, J., Zhang, R., El-Mashad, H.M., Sun, H., Ying, Y., 2008. Effect of food to microorganism ratio on biohydrogen production from food waste via anaerobic fermentation. Int. J. Hydrogen Energy. 33, 6968–6975.

Pandu, K., Joseph, S., 2012. Comparisons and limitations of biohydrogen production processes: A review. Int. J. Adv. Eng. Technol. 2, 342–356.

Podestá, J.J., Navarro, A.M.G., Estrella, C.N., Esteso, M.A., 1997. Electrochemical measurements of trace concentrations of biological hydrogen produced by *Enterobacteriaceae*. Inst. Pasteur. 148, 87–93.

Rachman, M.A., Nakashimada, Y., Kakizono, T., Nishio, N., 1998. Hydrogen production with high yield and high evolution rate by self-flocculated cells of *Enterobacter aerogenese* in a packed-bed reactor. Appl. Microbiol. Biotechnol. 49, 450–454.

Remme, U., Trudeau, N., Graczyk, D., Taylor, P., 2011. Technology development prospects for the Indian power sector. Accessed from http://www.iea.org/papers/2011/technology_development_india.pdf.

Ren, N.Q., Li, J.Z., Li, B.K., Wang, Y., Liu, S.R., 2006. Biohydrogen production from molasses by anaerobic fermentation with a pilot scale bioreactor system. Int. J. Hydrogen Energy. 31, 2147–2157.

Reshamwala, S., Shawky, B.T., Dale, B.E., 1995. Ethanol production from enzymatic hydrolysates of AFEX-treated coastal Bermuda grass and switchgrass. Appl. Biochem. Biotechnol. 51/52, 43–55.

Roy, S., Ghosh, S., Das, D., 2012. Improvement of hydrogen production with thermophilic mixed culture from rice spent wash of distillery industry. Int. J. Hydrogen Energy. 37 (21), 15867–15874.

Sauter, M., Böhm, R., Böck, A., 1992. Mutational analysis of the operon (hyc) determining hydrogenase 3 formation in *Escherichia coli*. Mol. Microbiol. 6, 1523–1532.

Sawers, R.G., 2005. Formate and its role in hydrogen production in *Escherichia coli*. Biochem. Soc. Trans. 33, 42–46.

Sawers, R.G., Ballantine, S.P., Boxer, D.H., 1985. Differential expression of hydrogenase isoenzymes in *Escherichia coli* K-12: Evidence for a third isoenzyme. J. Bacteriol. 164, 1324–1331.

Schroder, C., Selig, M., Schonheit, P., 1994. Glucose fermentation to acetate, CO_2 and H_2 in the anaerobic hyperthermophilic *eubacterium Thermotoga maritime*: Involvement of the Embden–Meyerhof pathway. Arch. Microbiol. 61, 460–470.

Shaw, A.J., Jenney Jr., F.E., Adams, M.W.W., Lyn, L.R., 2008. End product pathways in the xylose fermenting bacterium *Thermoanaerobacterium saccharolyticum*. Enzy. Microb. Technol. 42, 453–458.

Shi, X.Y., Jin, D.W., Sun, Q.Y., Li, W.W., 2010. Optimization of conditions for hydrogen production from brewery wastewater by anaerobic sludge using desirability function approach. Renew. Energy. 35, 1493–1498.

Show, K.Y., Zhang, Z.P., Tay, J.H., Liang, T.D., Lee, D.J., Jiang, W.J., 2007. Production of hydrogen in a granular sludge-based anaerobic continuous stirred tank reactor. Int. J. Hydrogen Energy. 32 (18), 4744–4753.

Show, K.Y., Zhang, Z.P., Tay, J.H., Liang, T.D., Lee, D.J., Ren, N., et al., 2010. Critical assessment of anaerobic processes for continuous biohydrogen production from organic wastewater. Int. J. Hydrogen Energy. 35 (24), 13350–13355.

Singh, P., Pandey, A., 2011. An evaluative report and challenges for fermentative biohydrogen production. Int. J Hydrogen Energy. 34, 13435–13444.

Sivaramakrishna, D., Sreekanth, D., Himabindu, V., Anjaneyulu, Y., 2009. Biological hydrogen production from probiotic wastewater as substrate by selectively enriched anaerobic mixed microflora. Renew. Energy. 34, 937–940.

Slobodkin, A.I., Tourova, T.P., Kuznetsov, B.B., Kostrikina, N.A., Bonch-Osmolovskaya, E.A., 1999. *Thermoanaerobacter siderophilus* sp. nov., a novel dissimilatory Fe(III)-reducing, anaerobic, thermophilic bacterium. Int. J. Syst. Evol. Microbiol. 49 (4), 1471–1478.

Soetaert, W., Vandamme, E.J., 2009. Biofuels. John Wiley and Sons Ltd., Great Britain, pp. 1–8.

Soloman, B.D., Barnes, J.R., Halvorsen, K.E., 2007. Grain and cellulosic ethanol: History, economics and energy policy. Biomass Bioenergy. 31, 416–425.

Stephenson, M., Stickland, L.H., 1932. Hydrogenlyases: Bacterial enzymes liberating molecular hydrogen. Biochem. J. 26, 712–724.

Steven, W., Ginkel, V., Logan, B., 2005. Increased biological hydrogen production with reduced organic loading. Water Res. 39, 3819–3826.

Sun, Y., Cheng, J., 2002. Hydrolysis of lignocellulosic materials for ethanol production: A review. Bioresour. Technol. 83, 1–11.

Suppmann, B., Sawers, G., 1994. Isolation and characterization of hypophosphite-resistant mutants of *Escherichia coli*: Identification of the FocA protein, encoded by the pfl operon, as a putative formate transporter. Mol. Microbiol. 11, 965–982.

Taguchi, F., Mizukai, N., Saito-Taki, T., Hasegawa, K., 2005. Hydrogen production from continuous fermentation of xylose during growth of *Clostridium* sp. Strain No. 2. Can. J. Microbiol. 41, 536–540.

Taherzadeh, M.J., Karimi, K., 2008. Pretreatment of lignocellulosic wastes to improve ethanol and biogas production: A review. Int. J. Mol. Sci. 9, 1621–1651.

Tang, G.L., Huang, J., Sun, Z.J., Tang, Q.Q., Yan, C.H., Liu, G.Q., 2008. Biohydrogen production from cattle wastewater by enriched anaerobic mixed consortia: Influence of fermentation temperature and pH. J. Biosci. Bioeng. 106, 80–87.

Tanisho, S., Ishiwata, Y., 1994. Continuous hydrogen production from molasses by the bacterium *Enterobacter aerogenes*. Int. J. Hydrogen Energy. 19, 807–812.

Tanisho, S., Suzuki, Y., Wakao, N., 1987. Fermentative hydrogen evolution by *Enterobacter aerogenes* strain E.82005. Int. J Hydrogen Energy. 12, 623–627.

Tanisho, S., Kuromoto, M., Kadokura, N., 1998. Effect of CO_2 removal on hydrogen production by fermentation. Int. J Hydrogen Energy. 23, 559–563.

Thauer, R.K., Jungermann, K., Decker, K., 1977. Energy conservation in chemotrophic anaerobic bacteria. Bacteriol. Rev. 41, 100–180.

Thong, S.O., Prasertsan, P., Karakashev, D., Angelidaki, I., 2008. Thermophilic fermentative hydrogen production by the newly isolated *Thermoanerobacterium thermosaccharolyticum* PSU-2. Int. J. Hydrogen Energy. 33, 1204–1214.

Turcot, J., Bisaillon, A., Hallenbeck, P.C., 2008. Hydrogen production by continuous cultures of *Escherichia coli* under different nutrient regimes. Int. J. Hydrogen Energy. 33, 1465–1470.

Ueno, Y., Kawai, T., Sato, S., Otsuka, S., Morimoto, M., 1995. Biological production of hydrogen from cellulose by natural anaerobic microflora. J. Ferment. Bioeng. 79, 395–397.

Ueno, Y., Otsuka, S., Morimoto, M., 1996. Hydrogen production from industrial wastewater by anaerobic microflora in chemostat culture. J. Ferment. Bioeng. 82 (2), 194–197.

Uyeda, K., Rabinowitz, J.C., 1971. Pyruvate-ferredoxin oxidoreductase. J. Biol. Chem. 246, 3111–3119.

van de Werken, H.J.G., Verhaart, M.R., VanFossen, A.L., Willquist, K., Lewis, D.L., Nichols, J.D., et al., 2008. Hydrogenomics of the extremely hermophilic Bacterium *Caldicellulosiruptor saccharolyticus*. Appl. Environ. Microbiol. 74 (21), 6720–6729.

van Ginkel, S., Logan, B.E., 2005. Inhibition of biohydrogen production by undissociated acetic and butyric acids. Environ. Sci. Technol. 39 (23), 9351–9356.

van Ginkel, S.W., Oh, S.E., Logan, B.E., 2005. Biohydrogen gas production from food processing and domestic wastewaters. Int. J. Hydrogen Energy. 30, 1535–1542.

van Groenestijn, J.W., Hazewinkel, J.H.O., Nienoord, M., Bussmann, P.J.T., 2002. Energy aspects of biological hydrogen production in high rate bioreactors operated in the thermophilic temperature range. Int. J. Hydrogen Energy. 27 (11–12), 1141–1147.

van Niel, E.W.J., Budde, M.A.W., de Haas, G.G., van der Wal, F.J., Claassen, P.A.M., Stams, A.J.M., 2002. Distinctive properties of high hydrogen producing extreme thermophiles, *Caldicellulosiruptor saccharolyticus* and *Thermotoga elfii*. Int. J. Hydrogen Energy. 27, 1391–1398.

van Niel, E.W.J., Claassen, P.A.M., Stams, A.J.M., 2003. Substrate and product inhibition of hydrogen production by the extreme thermophile, *Caldicellulosiruptor saccharolyticus*. Biotechnol. Bioeng. 81 (3), 255–262.

van Ooteghem, S.A., Beer, S.K., Yue, P.C., 2002. Hydrogen production by the thermophilic bacterium. *Thermotoga neapolitan*. Appl. Biochem. Biotechnol. 98 (100), 177–189.

Vatsala, T.M., Mohan, S., Manimaran, A., 2008. A pilot-scale study of biohydrogen production from distillery effluent using defined bacterial co-culture. Int. J. Hydrogen Energy. 33, 5404–5415.

Venkata Mohan, S., Lalit, Babu, V., Sarma, P.N., 2007. Anaerobic biohydrogen production from diary wastewater treatment in sequencing batch reactor (AnSBR): Effect of organic loading rate. Enzy. Microb. Technol. 41, 506–515.

Venkata Mohan, S., Lalit, Babu, V., Sarma, P.N., 2008. Effect of various pretreatment methods on anaerobic mixed microflora to enhance biohydrogen production utilizing dairy wastewater as substrate. Bioresour. Technol. 99, 59–67.

Vignais, P.M., Billoud, B., Meyer, J., 2001. Classification and phylogeny of hydrogenase. Microbiol. Rev. 25, 455–501.

Vijayaraghavan, K., Ahmad, D., 2006. Biohydrogen generation from palm oil mill effluent using anaerobic contact filter. Int. J. Hydrogen Energy. 31, 1284–1291.

Voet, D., Voet, J.G., Pratt, C.W., 1999. Fundamentals of Biochemistry. John Wiley, New York, p. 382.

Wang, J., Wan, W., 2009. Factors influencing fermentative hydrogen production: A review. Int. J. Hydrogen Energy. 34, 799–811.

Wang, A., Ren, N., Shi, Y., Lee, D.J., 2008. Bioaugmented hydrogen production from microcrystalline cellulose using co-culture: *Clostridium acetobutyricum* X9 and *Ethanoligenens harbinense* B49. Int. J Hydrogen Energy. 33, 912–917.

Wu, J.H., Lin, C.Y., 2004. Biohydrogen production by mesophilic fermentation of food wastewater. Water Sci. Technol. 49, 223–228.

Wu, S.Y., Lin, C.N., Chang, J.S., Lee, K.S., Lin, P.J., 2002. Microbial hydrogen production with immobilized sewage sludge. Biotechnol. Prog. 18, 921–926.

Wu, S.Y., Hung, C.H., Lin, C.N., Chen, H.W., Lee, A.S., Chang, J.S., 2006. Fermentative hydrogen production and bacterial community structure in high-rate anaerobic bioreactors containing silicone-immobilized and self-flocculated sludge. Biotechnol. Bioeng. 93, 934–946.

Xiangwen, S., Dangcong, P., Zhaohua, T., Xinghua, J., 2007. Treatment of brewery wastewater using anaerobic sequencing batch reactor (ASBR). Bioresour. Technol. 24, 1125–1134.

Yang, H., Shao, P., Lu, T., Shen, J., Wang, D., Xu, Z., et al., 2006. Continuous bio-hydrogen production from citric acid wastewater via facultative anaerobic bacteria. Int. J. Hydrogen Energy. 31, 1306–1316.

Yang, P., Zhang, R., McGarvey, J.A., Benemann, J.R., 2007. Biohydrogen production from cheese processing wastewater by anaerobic fermentation using mixed microbial communities. Int. J. Hydrogen Energy. 32, 4761–4771.

Yang, Y., Tsukahara, K., Sawayama, S., 2008. Biodegradation and methane production from glycerol-containing synthetic wastes with fixed-bed bioreactor under mesophilic and thermophilic anaerobic conditions. Proc. Biochem. 43, 362–367.

Yokoi, H., Tokushige, T., Hirose, J., Hayashi, S., Takasaki, Y., 1998. H_2 production from starch by mixed culture of *Clostridium buytricum* and *Enterobacter aerogenes*. Biotechnol. Lett. 20, 143–147.

Yokoi, H., Saitsu, A.S., Uchida, H., Hirose, J., Hayashi, S., Takasaki, Y., 2001. Microbial hydrogen production from sweet potato starch residue. J. Biosci. Bioeng. 91, 58–63.

Yokoi, H., Maki, R., Hirose, J., Hayashi, S., 2002. Microbial production of hydrogen from starch manufacturing wastes. Biomass Bioenergy. 22, 89–395.

Yoshida, A., Nishimura, T., Kawaguchi, H., Inui, M., Yukawa, H., 2006. Enhanced hydrogen production from glucose using *ldh*- and *frd*-inactivated *Escherichia coli* strains. Appl. Microbiol. Biotechnol. 73, 67–72.

Yu, H.Q., Mu, Y., 2006. Biological hydrogen production in a UASB reactor with granules. II. Reactor performance in 3-year operation. Biotechnol. Bioeng. 94, 988–995.

Yu, H.Q., Zhu, Z.H., Hu, W.R., Zhang, H.S., 2002. Hydrogen production from rice winery wastewater in an upflow anaerobic reactor by using mixed anaerobic cultures. Int. J. Hydrogen Energy. 27, 1359–1365.

Zeikus, J.G., 1977. The biology of methanogenic bacteria. Bacteriol. Rev. 41, 514–541.

Zhang, T., Liu, H., Fang, H.H.P., 2003. Biohydrogen production from starch in wastewater under thermophilic conditions. J. Environ. Manag. 69, 149–156.

Zhang, H., Bruns, M.A., Logan, B.E., 2006. Biological hydrogen production by *Clostridium acetobutylicum* in an unsaturated flow reactor. Water Res. 40, 728–734.

Zhang, M.L., Fan, Y.T., Xing, Y., Pan, C.M., Zhang, G.S., Lay, J.J., 2007a. Enhanced biohydrogen production from cornstalk wastes with acidification pretreatment by mixed anaerobic cultures. Biomass Bioenergy. 31, 250–254.

Zhang, Z.P., Show, K.Y., Tay, J.H., Liang, D.T., Lee, D.J., Jiang, W.J., 2007b. Rapid formation of hydrogen-producing granules in an anaerobic continuous stirred tank reactor induced by acid incubation. Biotechnol. Bioeng. 96 (6), 1040–1050.

Zhang, Z.P., Tay, J.H., Show, K.Y., Yan, R., Liang, D.T., Lee, D.J., et al., 2007c. Biohydrogen production in a granular activated carbon anaerobic fluidized bed reactor. Int. J. Hydrogen Energy. 32 (2), 185–191.

Zhang, Z.P., Show, K.Y., Tay, J.H., Liang, D.T., Lee, D.J., 2008a. Biohydrogen production with anaerobic fluidized bed reactors: A comparison of biofilm-based and granule-based systems. Int. J. Hydrogen Energy. 33, 1559–1564.

Zhang, Z.P., Show, K.Y., Tay, J.H., Liang, T.D., Lee, D.J., Wang, J.Y., 2008b. The role of acid incubation in rapid immobilization of hydrogen-producing culture in anaerobic upflow column reactors. Int. J. Hydrogen Energy. 33 (19), 5151–5160.

Zhao, B.H., Yue, Z.B., Zhao, Q.B., Mu, Y., Yu, H.Q., Harada, H., et al., 2008. Optimization of hydrogen production in a granule-based UASB reactor. Int. J. Hydrogen Energy. 33, 2454–2461.

Zhu, G.F., Wu, P., Wei, Q.S., Lin, J., Gao, Y.L., Liu, H.N., 2010. Biohydrogen production from purified terephthalic acid (PTA) processing wastewater by anaerobic fermentation using mixed microbial communities. Int. J Hydrogen Energy. 35, 8350–8356.

Zinoni, F., Birkmann, A., Leinfelder, W., Bock, A., 1987. Cotranslational insertion of selenocysteine into formate dehydrogenase from *Escherichia coli* directed by a UGA codon. Proc. Natl. Acad. Sci. USA 84, 3156–3160.

Photofermentative Biohydrogen Production

Patrick C. Hallenbeck

University of Montreal, Montreal, Quebec, Canada

INTRODUCTION

Hydrogen is under very active investigation as a promising future energy carrier, with much research and development on hydrogen storage and conversion. The sustainable use of hydrogen requires a renewable means of production, and a variety of microbial routes are available and under active investigation (Hallenbeck, 2011; Hallenbeck and Benemann, 2002). Photofermentation is particularly well suited for hydrogen production from particular waste streams containing organic acids, as well as being an attractive option for extracting additional hydrogen from effluents of dark hydrogen-producing fermentations, which can only produce hydrogen at low yields (maximum 33%) and thus produce large quantities of byproducts (Hallenbeck, 2005, 2009, 2012; Hallenbeck and Ghosh, 2009).

Nonsulfur photosynthetic bacteria, carrying out photofermentation, are capable of deriving hydrogen from substrates for which this is not possible using dark fermentation, as they are able to use additional energy available in light. In addition, the captured light energy, along with irreversible hydrogen production by nitrogenase, allows them to drive substrate conversion to completion. Therefore, the efficiency of light utilization is one of the major critical factors in photofermentation, and therefore developing effective photobioreactors with efficient internal light distribution is key. This would be especially true at a large (industrial) scale where self-shading inside the fermenter would have to be prevented to ensure maximum hydrogen production (Akkerman et al., 2002). This area has been reviewed (Gadhamshetty et al., 2011) and will not be elaborated on further in detail here.

Here, the basic biology, the underlying enzymology and metabolism, of the photofermentation process is reviewed, and studies on the use of various techniques to increase the conversion of substrates to hydrogen are discussed. Relevant studies have used either pure compounds, allowing a more detailed accounting of substrate conversion, or real waste streams. Any future practical application of a photofermentative process will require the use of various waste streams, a number of which are abundantly available, as otherwise substrate costs would be prohibitive.

PHOTOFERMENTATIVE ORGANISMS AND METABOLIC PATHWAYS

Photofermentative Organisms

Purple nonsulfur photosynthetic bacteria, members of the α-proteobacteria, appear particularly suited to a photoheterotrophic lifestyle. In the light, they generate the energy necessary for growth and survival photosynthetically, producing ATP through cyclic photophosphorylation using their single photosystem. Electrons necessary for their metabolic activities are obtained from fixed carbon compounds, inorganic ions (Fe^{2+}), or hydrogen. Since all bacteria in this class can fix CO_2 through the Calvin–Bassham–Benson cycle, photoautolithotrophic growth is possible on hydrogen or iron. Although photoheterotrophic growth was thought to be primarily carried out by these organisms, this growth mode may be more widespread than previously thought. A marine cyanobacterium that possesses only a single photosystem (water-splitting PSII is absent) has been described that leads a photoheterotrophic life in the surface waters of the oceans (Tripp et al., 2010).

Purple nonsulfur photosynthetic bacteria are found in a variety of natural environments and, not surprisingly, have been found to be able to use a broad spectrum of substrates (van Niel, 1944). Although the substrate range depends on the species, as a group these bacteria are able to use a wide variety of organic carbon compounds—pyruvate, acetate and other organic acids, amino acids, alcohols, and carbohydrates. Some are also able to use one-carbon compounds such as methanol and formate, whereas others can grow using aromatic compounds such as benzoate, cinnamate, chlorobenzoate, phenylacetate, and phenol (Harwood, 2008).

Photofermentative hydrogen production by photosynthetic bacteria was first reported in the late 1940s (Gest and Kamen, 1949). Production of hydrogen has been reported for many purple nonsulfur photosynthetic bacteria: *Rhodobacter sphaeroides* (Kapdan and Kargi, 2006; Koku et al., 2002), *Rhodobacter capsulatus* (He et al., 2005), *Rhodovulum sulfidophilum* W-1S (Maeda et al., 2003), and *Rhodopseudomonas palustris* (Barbosa et al., 2001). *Rhodobacter* species are potent hydrogen producers and have been widely used for photofermentative hydrogen production. Numerous studies have shown that these bacteria, depending on the species, have the ability to produce hydrogen from organic acids, simple sugars (glucose, fructose, and sucrose), and industrial and agricultural effluents (Adessi and De Philippis, 2012; Han et al., 2012). However, in general, hydrogen yields are highest with organic acids such as acetic, butyric (Fang et al., 2005), propionic (Shi and Yu, 2004), malic (Eroglu et al., 1999), and lactic (He et al., 2005). Thus, nonsulfur photosynthetic bacteria can be useful in the conversion of organic acid by-products formed during the anaerobic fermentation of organic wastes to H_2 and CO_2.

Enzymes and Metabolic Pathways

A schematic diagram showing the photofermentation of organic substrates by photo-synthetic bacteria is given in Figure 1. Photosynthetic bacteria produce chemical energy from sunlight (i.e., a proton gradient) and this energy can be used to drive reverse electron to nitrogenase, which reduces protons released through metabolism to hydrogen with the hydrolysis of ATP, produced by photophosphorylation (Abo-Hashesh and Hallenbeck, 2012b; Hallenbeck, 2011). Nitrogenase functions normally to catalyze the biological reduction of dinitrogen to ammonia with the release of one H_2 per N_2 reduced [Eq. (1)], but in the absence of other reducible substrates nitrogenase continues to turnover reducing protons to hydrogen [Eq. (2)].

$$N_2 + 8H^+ + 8e^- + 16\,ATP \rightarrow 2NH_3 + H_2 + 16ADP + 16\,Pi \tag{1}$$

$$2H^+ + 2e^- + 4ATP \rightarrow H_2 + 4ADP + 4Pi \tag{2}$$

Thus, a significant amount of energy is required for hydrogen production by nitrogenase, both high-energy electrons and ATP, with $2ATP/e^-$, or $4ATP/H_2$. While this might appear to be a pointless loss of energy, hydrogen production under these conditions appears to be a necessary metabolic response to the need to maintain redox balance (reviewed in Masepohl and Hallenbeck, 2010). Thus, strategies that might be applied in the future to improving

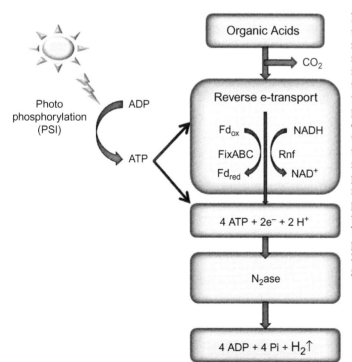

FIGURE 1 Schematic of photofermentation process. An outline of the metabolic processes involved in photofermentation is shown. Organic acids are imported by a TRAP transporter and metabolized through central metabolic pathways, giving off carbon dioxide and producing NADH. NADH is used to reduce ferredoxin, the electron donor to nitrogenase through a reverse electron transport pathway involving either membrane-bound RNF or FixABC complexes. These complexes use the proton motive force established by bacterial photosynthesis to drive this otherwise thermodynamically unfavorable reaction. The ATP necessary for the reduction of protons by nitrogenase, $2H^+ + 2e^- + 4ATP \rightarrow H_2 + 4ADP + 4\,Pi$, is also generated by bacterial photosynthesis.

TABLE 1   Advantages and Disadvantages of Photofermentative Hydrogen Production

Advantages	Disadvantages	Possible solutions
Uses light energy to produce hydrogen from otherwise unusable substrates	Poor light conversion efficiency	Create antenna mutants that harvest light more effectively
	Requires hydrogen-impermeable transparent photobioreactors	Develop high-tech, low-cost materials
Hydrogen production is catalyzed by nitrogenase and hence irreversible	Nitrogenase requires excess energy (ATP) and thus reduces efficiency	Replace nitrogenase with [FeFe]-hydrogenase
Can be used as second stage to derive more hydrogen from dark fermentation effluents	Extensive treatment of effluent to remove inhibition necessary	Direct photofermentation of sugars

photofermentative hydrogen production could include replacing the highly energy-demanding nitrogenase with a hydrogenase, which would not require ATP (Table 1, see later).

Because nitrogenase is the key enzyme in photofermentative hydrogen production, setting specific conditions must be met for hydrogen production, including particular requirements for its regulation, biosynthesis, and enzyme activity (Masepohl and Hallenbeck, 2010). Nitrogenase is an oxygen complex iron–sulfur molybdenum enzyme. The active enzyme is composed of two proteins—a homodimer bridged by a single 4Fe-4S cluster, called Fe protein, and a larger heterotetramer, called Mo-Fe protein, which contains two large, complex Fe-S centers. Mo-Fe protein is the site of N_2 fixation or proton reduction, which occurs on the Fe-Mo-co, a complex metallic center containing iron and molybdenum. Thus, there is a nutritional requirement for large quantities of iron as well as molybdenum, which is also a regulatory factor (Masepohl and Hallenbeck, 2010), as the Mo-nitrogenase is not made in its absence. In addition, nitrogenase synthesis is repressed by high concentrations of fixed nitrogen, especially ammonium, and oxygen. Substrate reduction requires high-energy electrons, furnished by cellular metabolism and specifically transferred to the Mo-Fe protein by the Fe protein in a process that requires chemical energy input in the form of ATP with $2ATP/e^-$, or $4ATP/H_2$.

Thus, effective photofermentation only takes place under anoxic conditions and when ammonium is present in only very limited quantities. Therefore, nitrogen for cell growth supplied in the form of amino acids, such as glutamate, or yeast extract is much better at supporting hydrogen production than ammonium. However, nitrogen limitation is also useful in another way, as excess cell synthesis, which would result in a reduction in light diffusion, is avoided. Small amounts of ammonium added to the culture medium cause inhibition of nitrogenase activity, as this causes ADP-ribosylation of the Fe protein, but the inhibitory action of ammonium is reversible, and hydrogen production activity can be recovered after ammonia is consumed (Masepohl and Hallenbeck, 2010).

As mentioned earlier, photoheterotrophic growth under nitrogen-limiting conditions leads to the nearly stoichiometric conversion of various organic acids to hydrogen through the action of nitrogenase. Although seemingly wasteful energetically, hydrogen production appears to be a response to the metabolic need to maintain redox balance (Hallenbeck, 2011). Depending on the oxidation state of the substrate, its assimilation can generate excess electrons as NAD(P)H, which must be disposed of in order for there to be continued growth and metabolic activity. Under nitrogen-deficient conditions, hydrogen production by nitrogenase allows the reoxidation of NADH. Under nitrogen-replete conditions, this need can be met by CO_2 fixation by RuBisCO, an otherwise seemingly paradoxical fixation of carbon dioxide, as it occurs simultaneously with the assimilation of an organic carbon source. The carbon flux through this metabolic shunt is significant, with as much as 68% of the CO_2 given off during growth on acetate being refixed through the Calvin cycle (McKinlay and Harwood, 2010). The metabolic necessity for this "electron valve" is so important that strains lacking RuBisCO readily develop mutations that allow nitrogenase expression even in the presence of fixed nitrogen (Hallenbeck, 2011; Laguna et al., 2010).

The energy required for nitrogenase activity, highly reducing electrons and ATP, must be met by cellular metabolism. Both forms of energy are produced by the photosynthetic activity of the organism, which generates a proton gradient via electron flow through the photosynthetic apparatus driven by captured light energy. This gradient is then used, by ATP synthase to generate ATP and by some form of reverse electron flow, to generate high-energy electrons (Hallenbeck, 2011) (Fig. 2).

However, hydrogen produced by nitrogenase can also be used as an electron donor when it is oxidized by a membrane-bound [NiFe]-hydrogenase, thus possibly reducing the net hydrogen evolved through nitrogenase activity (Kapdan and Kargi, 2006). Of course, this is only a major factor if electron acceptors are available to couple with hydrogen oxidation—not obviously the case under these conditions. Nevertheless, hydrogenase-negative mutants of photofermentative bacteria have been created in attempts to enhance hydrogen production and have been reported, under some conditions, to produce two to three times more hydrogen than wild-type cells (Kim et al., 2006). Other factors as well may affect photofermentative hydrogen production, including light intensity, carbon source, and microbial strain. Increasing light intensity, up to a certain point, can stimulate hydrogen yields as well as production rates. However, it has the opposite effect on light conversion efficiencies (Barbosa et al., 2001; Shi and Yu, 2005). Cultures are usually maintained under constant illumination, as hydrogen production under dark conditions is negligible (Koku et al., 2003; Oh et al., 2004). Several studies suggest that higher production rates might be obtained using light/dark cycles (Adessi et al., 2010), either alternating 14-h light/10-h dark cycles (Koku et al., 2003) or a 30-min light/dark cycle (Wakayama et al., 2000).

DIFFERENT PHOTOFERMENTATION BIOPROCESSES

Single-Stage Photofermentation Bioprocesses

Different substrates, including various waste streams, can be used for photofermentation by purple nonsulfur photosynthetic bacteria (Keskin et al., 2011). A number of different bioprocesses employing photofermentation can potentially be used. Of course, given the proven metabolic capacity of nonsulfur photosynthetic bacteria, organic acids and waste streams containing them can be directly converted in a single stage to hydrogen (and carbon dioxide). Photofermentative hydrogen production has been scaled up to large photobioreactors operated under natural, outdoor conditions (Adessi et al., 2012a,b; Androga et al., 2011; Avcioglu et al., 2011; Boran et al., 2010, 2012).

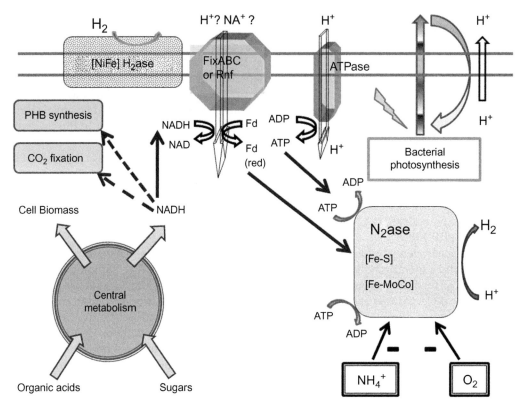

FIGURE 2   Metabolic interactions during photofermentation. The metabolic pathways leading to hydrogen production by nonsulfur photosynthetic bacteria are shown. Organic acids or sugars are imported and transformed through the pathways of central metabolism, with some of the carbon flux being used for biomass biosynthesis. Excess reducing equivalents (NADH) must be disposed of to allow for further metabolism. Nitrogenase is expressed under nitrogen-limiting conditions and reduces protons to hydrogen actively. Nitrogenase activity is supported by bacterial photosynthesis, which absorbs light and converts the captured energy into a proton gradient. During cyclic anoxygenic photosynthesis, a photon stimulates the excitation of bacteriochlorophyll in the reaction center; this energy causes the release of an electron that subsequently reduces the membrane quinone pool. The quinone cycle releases protons into the periplasmic space and reduces the cytochrome bc_1 complex, which reduces cytochrome c_2. Cytochrome c_2 in turn reduces the oxidized primary electron donors in the reaction center, thus closing the cycle. Protons that have accumulated in the periplasm form an electrochemical gradient, which is used by ATP-synthase to generate ATP. The electrochemical gradient is also used to drive the thermodynamically unfavorable reduction of ferredoxin by NADH. The mechanism(s) involved in this reverse electron flow is little understood, but apparently involves either Rnf or FixABC complexes (Hallenbeck, 2011). Electron flow to the hydrogen production pathway may be reduced by competing pathways such as carbon dioxide fixation or polyhydroxybutyrate (PHB) synthesis. *Figure taken from Keskin et al. (2011).*

Another substrate amenable to direct, single-stage conversion to hydrogen is glycerol, and such conversions are of interest given the very large quantities of waste glycerol being produced by the biodiesel industry. Moreover, the maximal yield of hydrogen that can be expected from dark fermentation of glycerol is 1 mol of H_2 per mole of glycerol and, due to the capacity of photofermentation to potentially drive substrate conversion to completion, photofermentative hydrogen yields from glycerol could possibly be much higher. Indeed, this is the case as shown by a series of studies carried out on the photofermentation of glycerol. In an initial study, 5 to 6 mol of H_2 per mole of glycerol was obtained with pure glycerol; yields were somewhat lower with a crude glycerol fraction, 4 mol of H_2 per mole of glycerol (Sabourin-Provost and Hallenbeck, 2009). Two additional studies were able to improve on these yields, giving mole of H_2 per mole of glycerol. When the effect of nitrogen source and glycerol concentration was studied more systematically, 6.1 mol of H_2 per mole of crude glycerol was obtained under optimal conditions (20 mM glycerol, 4 mM glutamate), a yield of 87% of the theoretical (7 mol of H_2 per mole of glycerol) (Ghosh et al., 2012c). More complete optimization using response surface methodology (RSM) was able to define conditions under which essentially complete (96%) conversion to hydrogen was achievable (Ghosh et al., 2012a).

In addition, although relatively little investigated, simple sugars can also serve directly as substrates for these organisms. Several studies have examined this possibility and have shown that relatively high yields can be obtained. In an initial study, 3.3 mol of H_2 per mole of glucose was obtained (Abo-Hashesh et al., 2011). Optimization of the batch culture conditions by RSM brought these yields to 5.5 mol of H_2 per mole of glucose (Ghosh et al. 2012b). These studies have been extended to show that it is possible to obtain 4 to 5 mol H_2/ mol hexose from sugar industry wastes in a batch process (Keskin and Hallenbeck, 2012a). In fact, it has been shown that continuous cultures give very high yields, with 9 mol of H_2 per mole of glucose being obtained at a hydraulic retention time of 48 h (Abo-Hashesh et al., 2013). These results thus compare very favorably, or even surpass, those obtained with two-stage or coculture approaches as discussed in what follows.

Two-Stage Photofermentation Bioprocesses

Sugars are obvious substrates for hydrogen production by dark fermentations, which normally also produce large quantities of organic acids. Complex carbohydrates, either starch or cellulose, found in waste streams must first be converted to simple sugars, either in a separate pretreatment or simultaneously in a consolidated bioprocess, before being converted by dark fermentation to hydrogen and organic acids. However, dark fermentations can only function at low molar yields, with a maximum of 4 mol H_2/mol hexose, or only 33% of the hydrogen that is chemically available (Abo-Hashesh and Hallenbeck, 2012b). Complete degradation of 1 mol of glucose can potentially yield 12 mol of hydrogen. However, Gibbs free energy of the complete oxidation of glucose into hydrogen and carbon dioxide shows that this reaction is not feasible thermodynamically [Eq.(3)].

$$C_6H_{12}O_6 + 6H_2O \rightarrow 12H_2 + 6CO_2 \qquad \Delta G^\circ = +3.2 \text{ kJ/mol} \qquad (3)$$

At 33% yield, dark fermentations therefore are impractical.

Thus, to be practical, it is essential that further energy, preferably hydrogen, be derived from the organic acids produced as side products. However, these are ideal substrates for photofermentation, as noted previously, and therefore several processes for coupling photofermentation with dark fermentation have been proposed and are under study. Thus, conversion of these organic acids to hydrogen by photofermentation can proceed either

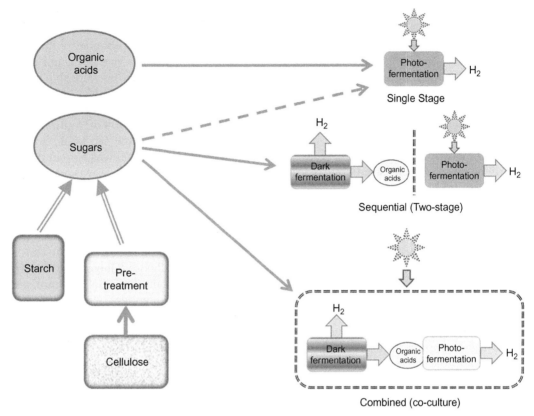

FIGURE 3 Different photofermentation bioprocesses. Different substrates, including various waste streams, can be used for photofermentation by purple nonsulfur photosynthetic bacteria. Organic acids and waste streams containing them can be converted directly in a single stage to hydrogen (and carbon dioxide). In addition, although relatively little investigated, simple sugars can also serve directly as substrates for these organisms (shown as dotted arrow). Also, sugars can serve as a substrate for hydrogen and organic acid production by dark fermentation. Complex carbohydrates, either starch or cellulose, found in waste streams must first be degraded to simple sugars before being converted by dark fermentation to hydrogen and organic acids. The effluent of this dark hydrogen fermentation can be suitably modified and converted in a second photofermentation stage. Alternatively, further conversion of these organic acids to hydrogen via photofermentation can occur simultaneously in a combined coculture process. *Figure taken from Keskin et al. (2011).*

simultaneously, in a combined, coculture process, or be suitably modified and converted in a second photofermentation stage (Fig. 3) (Adessi et al., 2012b; Keskin and Hallenbeck, 2012b; Tekucheva and Tsygankov, 2012).

Several studies have examined two-stage systems where a dark fermentation step with pure cultures is combined with a subsequent photofermentation step (Cheng et al., 2011). For example, dark fermentation with pure cultures such as *Caldicellulosiruptor saccharolyticus*, followed by photofermentation with either *Rhodobacter capsulatus*, a *R. capsulatus* hup⁻mutant, or *Rhodopseudomonas palustris*, have been used to produce hydrogen from beet molasses, glucose, potato steam peels, and molasses (Afsar et al., 2011; Laurinavichene et al., 2010; Ozgur et al., 2010b). Various sequential combinations have been shown to be effective in producing hydrogen from different sugars—*Clostridium butyricum* and *R. palustris* to produce hydrogen from glucose (Su et al., 2009), *Clostridium pasterianum* and *R. palustris* to produce hydrogen from sucrose (Alalayah et al., 2009), and *Clostridium saccharoperbutylacetonicum* and *Rhodobacter sphaeroides* to produce hydrogen from glucose (Alalayah et al., 2009). These combinations effectively resulted in higher hydrogen yields in comparison to single systems.

One of the main advantages of two-stage systems employing mixed cultures in the first stage is the ability to use a wider variety of organic wastes and wastewaters. Carbohydrate-rich raw materials, especially starch and cellulose containing renewable biomass resources, can and have been used in many studies of two-stage systems (see, e.g., Argun et al., 2009). Of course, when using complex substrates, such as agricultural wastes, it would be important to choose the best pretreatment option from the point of view of both system efficiency and cost. With pretreatment, many substrates become available for hydrogen with a combined two-stage approach. For example, acid pretreatment allowed the degradation of corn cob, a waste rich in cellulose, permitting hydrogen production in a combined process (Yang et al., 2010).

In fact, a large 5-year European project that ended recently, Hyvolution, was devoted to developing a two-stage process (Claassen and de Vrije 2006; Claassen et al., 2010; Ozgur et al., 2010b). These studies clearly demonstrated that additional hydrogen could be obtained with two-stage systems, with combined yields ranging from 5 to 7 mol H_2 per mole of hexose. However, these yields are lower than what would be needed for a practical system. Moreover, a number of problems were noted in directly using effluents from the first stage. These effluents are inhibitory to photofermentation in a number of respects: acidic pH, excess fixed nitrogen, and excessively high substrate concentrations. Therefore, effluents must be pretreated extensively before being used for photofermentation, with dilution, nutrient addition, pH adjustment, and, depending on the substrate used, even sterilization and centrifugation being necessary (Ozgur et al., 2010a). Thus, even though sequential dark photofermentation appears relatively straightforward in principle, it will require further extensive bioprocess development.

Another related process that can potentially overcome some of these problems is the use of cocultures where dark and photofermentations occur in the same reactor simultaneously (Fig. 3). This approach has been the subject of a number of studies (see, e.g., Asada et al., 2006; Fang et al., 2006; Lee et al., 2012; Liu et al., 2010). This approach would appear to have the advantage that the substrates for photofermentation are generated *in situ* by the dark fermenting organism and therefore would not build up to inhibitory levels if used immediately for photofermentation. Additionally, acidification caused by dark fermentation would possibly be neutralized by alkalinization due to photofermentation. However, as was seen to be the case for two-stage systems, overall yields of only 7 mol of H_2 per mole of hexose were obtained in the best cases. Also, this approach also poses a number of questions from the operational point of view. It is unclear how to provide a growth medium that would be appropriate for the two different metabolic types. In addition, it is not obvious how to develop a scenario for obtaining a good match in the growth rates between the two different organisms.

Finally, in an effort to circumvent the inherent problems of light-dependent processes, light saturation, low light conversion efficiencies, and technically and economically challenging photobioreactors, studies on microaerobic dark fermentation with nonsulfur photosynthetic bacteria have been initiated (Abo-Hashesh and Hallenbeck, 2012a). The idea is to replace the bacterial photosynthesis process with oxidative respiration as the source of energy, ATP and high-energy electrons, needed to drive nitrogenase activity. This initial study, really a proof of principle, showed that 0.6 mol of H_2 per mole of malate could be obtained at 4 to 8% hydrogen. While the yield is admittedly low, it is, however, remarkable as no dark process has been shown to produce hydrogen from organic acids such as malate. The technical challenge is to find conditions that maximize nitrogenase activity while at the same time minimizing oxygen inhibition of nitrogenase or wasteful excess oxidation of the substrate.

STRATEGIES FOR IMPROVING PHOTOFERMENTATION

A number of strategies may be applied in attempts to increase the practicality of photofermentation systems. A number of considerations apply to the efficiency of light utilization. Due to the light requirement, the type of reactor to be used is one aspect that deserves special attention as it can influence light distribution within the culture greatly. The physical characteristics of reactors can impact productivities significantly, and various designs are being investigated (Chen et al., 2010; Show et al., 2011). Beyond such purely physical considerations, several possible improvements can be made to the biology of the system. For example, most photosynthetic cultures saturate at relatively low light intensities due to self-shading. Thus, higher efficiencies at high-light intensities might be obtained using strains that have truncated photosynthetic antennas. This was borne out by one study, where a 50% decrease in LHII content gave a 50% increase in the hydrogen production rate at 300 W/m^2 (Kondo et al., 2002). This question has also been examined in several other studies, sometimes with contradictory results. A reduced pigment mutant of *R. sphaeroides* was reported to give higher hydrogen production, but this was only at low-light intensities (10 W/m^2) (Kim et al., 2006), opposite to what one would expect. Even more at odds with expectations, as mutants with less antenna pigment would be expected to greatly outperform the wild type under high-light conditions, the difference at a higher light intensity (100 W/m^2) was quite small. Similarly, a more recent study where reduced pigment was obtained through transposon insertion found only a very small improvement (Ma et al., 2012). Thus, this question is far from resolved and deserves further investigation.

Of course, as changes in light absorption efficiencies brought about by manipulating antenna complexes might be expected to affect overall rates of hydrogen production, other metabolic changes can possibly be made that could affect substrate conversion yields directly. In general, these involve increasing hydrogen yields by suppressing metabolic reactions that divert electrons from nitrogenase. Several different pathways are available for targeting (Fig. 2). Photofermentation is carried out under excess carbon conditions, necessary for nitrogenase expression. However, at high C/N, synthesis of the reduced carbon storage compound, polyhydroxybutyrate, is active. Thus, in theory, creating Phb- mutants should increase yields; however, in practice this has been difficult to demonstrate clearly. An initial study found no difference between a Phb- mutant of *R. sphaeroides* and the wild type when hydrogen production from lactate was examined (Franchi et al., 2004). A Phb- mutant of a *R. sphaeroides* strain when grown on malate gave a 34% increase in specific hydrogen production, but with only a 21% increase in volumetric hydrogen production (Kim et al., 2006). Additionally, when grown on a substrate that might be expected to show the largest difference, acetate, the mutant strain grew poorly for reasons that are not clear. A more extensive recent study where the effects of different carbon sources on hydrogen production by Phb- and wild-type strains found that in fact only some substrates, notably acetate and propionate, show an effect (Wu et al., 2012a,b). In some cases, the effect of an additional mutation in the uptake hydrogenase appeared additive to the Phb- mutation (Franchi et al., 2004; Kim et al., 2005). In fact, mutating the uptake hydrogenase is another strategy that has been widely applied in efforts to increase photofermentative hydrogen production. In general, Hup- strains show a small but significant increase in total hydrogen accumulation (11–40%) (Liu et al., 2010).

Another modification that might be carried out would be with the central catalyst, nitrogenase. As noted earlier, photofermentation carried out by nitrogenase inflicts a significant energy drain due to the requirement for ATP for hydrogen evolution by this enzyme ($4ATP/H_2$). Therefore, replacing nitrogenase with a [FeFe]-hydrogenase would, in theory, replace an enzyme with a slow turnover (6.4 s^{-1}) with one with a much higher turnover rate ($2000–6000 \text{ s}^{-1}$), possibly leading to increased volumetric production rates. In addition, and more importantly, elimination of the requirement to use a large fraction of the captured photons in ATP production would potentially lead to higher light conversion efficiencies. Strains with [FeFe]-hydrogenase would thus be postulated to show higher photosynthetic efficiencies for hydrogen production. Although some results have been presented that purport to show this (Kim et al., 2008a,b), the basis for this effect is unclear, as FeFe protein maturation is thought to require the accessory factors, HydE, F, and G (Posewitz et al., 2004), absent in photosynthetic bacteria. Thus, whether or not replacing nitrogenase with an active [FeFe]-hydrogenase would lead to greatly increased efficiencies remains to be fully investigated.

From what was discussed earlier, it is obvious that a great deal of improvement in photofermentative hydrogen production may possibly be made through extensive metabolic remodeling. Successful application of the required strategies, which necessitate single multiple mutations and the introduction of novel pathways and enzymes, requires careful consideration of the metabolic fluxes in the different pathways and their possible interaction. This can be guided by metabolic models, several of which are now available for photosynthetic bacteria (Golomysova et al., 2010; Hadicke et al., 2011; Imam et al., 2011).

Finally, a number of other modifications might be beneficial for some applications: strains that are derepressed for nitrogenase would produce hydrogen from nitrogen-rich substrates, strains with increased heat tolerance would perform better in some locations, and strains that are more tolerant to acid would thus be better for use with dark fermentation effluents (Khusnutdinova et al., 2012).

CONCLUSION

Purple nonsulfur photosynthetic bacteria have a versatile metabolism and have proved to be prodigious hydrogen producers from organic acids in a unique reaction called photofermentation. Many studies have shown effectively that this substrate range can be extended and that they can usefully supplement dark hydrogen fermentations in a two-stage process, increasing overall hydrogen yields. However, a number of significant challenges remain to be solved. Various strategies, in particular employing metabolic engineering, promise to lead to increased yields and efficiencies in the future.

Acknowledgments

I want to thank many students who have worked in my laboratory over the years on biofuel projects and the NSERC and FQRNT for research support.

References

Abo-Hashesh, M., Hallenbeck, P.C., 2012a. Fermentative hydrogen production. In: Hallenbeck, P.C. (Ed.), Microbial Technologies in Advanced Biofuels Production. Springer, New York, pp. 77–92.

Abo-Hashesh, M., Hallenbeck, P.C., 2012b. Microaerobic dark fermentative hydrogen production by the photosynthetic bacterium, *Rhodobacter capsulatus* JP91. Int. J. Low-Carbon Tech. 7, 97–103.

Abo-Hashesh, M., Ghosh, D., Tourigny, A., Taous, A., Hallenbeck, P.C., 2011. Single stage photofermentative hydrogen production from glucose: An attractive alternative to two stage photofermentation or co-culture approaches. Int. J. Hydrogen Energy 36, 13889–13895.

Abo-Hashesh, M., Desaunay, N., Hallenbeck, P.C., 2013. High yield single stage conversion of glucose to hydrogen by photofermentation with continuous cultures of Rhodobacter capsulatus JP91. Bioresour. Technol. 128, 513–517.

Adessi, A., De Philippis, R., 2012. Hydrogen production: Photofermentation. In: Hallenbeck, P.C. (Ed.), Microbial Technologies in Advanced Biofuels Production. Springer, New York, pp. 53–75.

Adessi, A., Fedini, A., De Philippis, R., 2010. Photobiological hydrogen production with *Rhodopseudomonas palustris* under light/dark cycles in lab and outdoor cultures. J. Biotechnol. 150, S15–S16.

Adessi, A., De Philippis, R., Hallenbeck, P.C., 2012a. Combined systems for maximum substrate conversion. In: Hallenbeck, P.C. (Ed.), Microbial Technologies in Advanced Biofuels Production. Springer, New York, pp. 107–126.

Adessi, A., Torzillo, G., Baccetti, E., De Philippis, R., 2012b. Sustained outdoor H-2 production with *Rhodopseudomonas palustris* cultures in a 50 L tubular photobioreactor. Int. J. Hydrogen Energy 37 (10), 8840–8849.

Afsar, N., Ozgur, E., Gurgan, M., Akkose, S., Yücel, M., Gündüz, P., et al., 2011. Hydrogen productivity of photosynthetic bacteria on dark fermenter effluent of potato steam peels hydrolysate. Int. J. Hydrogen Energy 36, 432–438.

Akkerman, I., Janssen, M., Rocha, J., Wijffels, R.H., 2002. Photobiological hydrogen production: Photochemical efficiency and bioreactor design. Int. J. Hydrogen Energy 27, 1195–1208.

Alalayah, W.M., Kalil, M.S., Kadhum, A., Jahim, J.M., Jaapar, S.Z.S., Alauj, N.M., 2009. Bio-hydrogen production using a two-stage fermentation process. Pakistan J. Bio. Sci. 12 (22), 1462–1467.

Androga, D.D., Ozgur, E., Guncluz, U., Yucel, M., Eroglu, I., 2011. Factors affecting the long term stability of biomass and hydrogen productivity in outdoor photofermentation. Int. J. Hydrogen Energy 36 (17), 11369–11378.

Argun, H., Kargi, F., Kapdan, I.K., 2009. Hydrogen production by combined dark and light fermentation of ground wheat solution. Int. J. Hydrogen Energy 34, 4304–4311.

Asada, Y., Tokumoto, M., Aihara, Y., Oku, M., Ishimi, K., Wakayama, T., et al., 2006. Hydrogen production by co-cultures of Lactobacillus and a photosynthetic bacterium, *Rhodobacter sphaeroides* RV. Int. J. Hydrogen Energy 31, 1509–1513.

Avcioglu, S.G., Ozgur, E., Eroglu, I., Yucel, M., Gunduz, U., 2011. Biohydrogen production in an outdoor panel photobioreactor on dark fermentation effluent of molasses. Int. J. Hydrogen Energy 36 (17), 11360–11368.

Barbosa, M.J., Rocha, J.M.S., Tramper, J., Wijffels, R.H., 2001. Acetate as a carbon source for hydrogen production by photosynthetic bacteria. J. Biotechnol. 85, 25–33.

Boran, E., Ozgur, E., van der Burg, J., Yucel, M., Gunduz, U., Eroglu, I., 2010. Biological hydrogen production by *Rhodobacter capsulatus* in solar tubular photo bioreactor. J. Cleaner Prod. 18, S29–S35.

Boran, E., Ozgur, E., Yucel, M., Gunduz, U., Eroglu, I., 2012. Biohydrogen production by *Rhodobacter capsulatus* in solar tubular photobioreactor on thick juice dark fermenter effluent. J. Cleaner Prod. 31, 150–157.

Chen, C.Y., Yeh, K.L., Lo, Y.C., Wang, H.M., Chang, J.S., 2010. Engineering strategies for the enhanced photo-H_2 production using effluents of dark fermentation process as substrate. Int. J. Hydrogen Energy 35, 13356–13364.

Cheng, J., Su, H., Zhou, J., Song, W., Cen, K., 2011. Hydrogen production by mixed bacteria through dark and photo fermentation. Int. J. Hydrogen Energy 35, 450–457.

Claassen, P.A.M., de Vrije, T., 2006. Non-thermal production of pure hydrogen from biomass: HYVOLUTION. Int. J. Hydrogen Energy 31, 1416–1423.

Claassen, P.A.M., de Vrije, T., Urbaniec, K., Grabarczyk, R., 2010. Development of a fermentation-based process for biomass conversion to hydrogen gas. Zuckerindustrie 135 (4), 218–221.

Eroglu, I., Aslan, K., Gündüz, U., Yücel, M., Türker, L., 1999. Substrate consumption rate for hydrogen production by *Rhodobacter sphaeroides* in a column photobioreactor. J. Biotechnol. 70, 103–113.

Fang, H.H.P., Liu, H., Zhang, T., 2005. Phototrophic hydrogen production from acetate and butyrate in wastewater. Int. J. Hydrogen Energy 30, 785–793.

Fang, H.H.P., Zhu, H.G., Zhang, T., 2006. Phototrophic hydrogen production from glucose by pure and co-cultures of *Clostridium butyricum* and *Rhodobacter sphaeroides*. Int. J. Hydrogen Energy 31, 2223–2230.

Franchi, E., Tosi, C., Scolla, G., Penna, G.D., Rodriguez, F., Pedroni, P.M., 2004. Metabolically engineered *Rhodobacter sphaeroides* RV strains for improved biohydrogen photoproduction combined with disposal of food wastes. Mar. Biotechnol. 6, 552–565.

Gadhamshetty, V., Sukumaran, A., Nirmalakhandan, N., 2011. Photoparameters in photofermentative biohydrogen production. Crit. Rev. Environ. Sci. Technol. 41 (1), 1–51.

Gest, H., Kamen, M.D., 1949. Photoproduction of molecular hydrogen by *Rhodospirillum rubrum*. Science 109 (2840), 558–559.

Ghosh, D., Sobro, I.F., Hallenbeck, P.C., 2012a. Stoichiometric conversion of biodiesel derived crude glycerol to hydrogen: Response surface methodology study of the effects of light intensity and crude glycerol and glutamate concentration. Bioresour. Technol. 106, 154–160.

Ghosh, D., Sobro, I.F., Hallenbeck, P.C., 2012b. Optimization of the hydrogen yield from single-stage photofermentation of glucose by *Rhodobacter capsulatus* JP91 using response surface methodology. Bioresour. Technol. 123, 199–206.

Ghosh, D., Tourigny, A., Hallenbeck, P.C., 2012c. Near stoichiometric reforming of biodiesel derived crude glycerol to hydrogen by photofermentation. Int. J. Hydrogen Energy 37 (3), 2273–2277.

Golomysova, A., Gomelsky, M., Ivanov, P.S., 2010. Flux balance analysis of photoheterotrophic growth of purple nonsulfur bacteria relevant to biohydrogen production. Int. J. Hydrogen Energy 35, 12751–12760.

Hadicke, O., Grammel, H., Klamt, S., 2011. Metabolic network modeling of redox balancing and biohydrogen production in purple nonsulfur bacteria. BMC Syst. Biol. 5.

Hallenbeck, P.C., 2005. Fundamentals of the fermentative production of hydrogen. Water Sci. Technol. 52 (1–2), 21–29.

Hallenbeck, P.C., 2009. Fermentative hydrogen production: Principles, progress, and prognosis. Int. J. Hydrogen Energy 34 (17), 7379–7389.

Hallenbeck, P.C., 2011. Microbial paths to renewable hydrogen production. Biofuels 2, 285–302.

Hallenbeck, P.C., 2012. Fundamentals of dark hydrogen fermentations: Multiple pathways and enzymes. In: Azbar, N., Levin, D.B. (Eds.), State of the Art and Progress in Production of Biohydrogen. Bentham Science Publishers, pp. 94–111.

Hallenbeck, P.C., Benemann, J.R., 2002. Biological hydrogen production; fundamentals and limiting processes. Int. J. Hydrogen Energy 27 (11–12), 1185–1193.

Hallenbeck, P.C., Ghosh, D., 2009. Advances in fermentative biohydrogen production: The way forward? Trends Biotechnol 27 (5), 287–297.

Han, H.L., Liu, B.Q., Yang, H.J., Shen, J.Q., 2012. Effect of carbon sources on the photobiological production of hydrogen using *Rhodobacter sphaeroides* RV. Int. J. Hydrogen Energy 37 (17), 12167–12174.

Harwood, C.S., 2008. Degradation of aromatic compounds by purple nonsulfur bacteria. In: Hunter, C.D., Daldal, F., Thurnauer, M.C., Beatty, J.T. (Eds.), Advances in Photosynthesis and Respiration, vol. 28. Springer, The Netherlands, pp. 577–594.

He, D., Bultel, Y., Magnin, J.P., Roux, C., Willison, J.C., 2005. Hydrogen photosynthesis by *Rhodobacter capsulatus* and its coupling to PEM fuel cell. J. Power Sources 141, 19–23.

Imam, S., Yilmaz, S., Sohmen, U., Gorzalski, A.S., Reed, J.L., Noguera, D.R., et al., 2011. iRsp1095: A genome-scale reconstruction of the *Rhodobacter sphaeroides* metabolic network. BMC Syst. Biol. 21 (5), 116.

Kapdan, I.K., Kargi, F., 2006. Bio-hydrogen production from waste materials. Enzyme Microb. Technol. 38, 569–582.

Keskin, T., Hallenbeck, P.C., 2012a. Hydrogen production from sugar industry wastes using single-stage photofermentation. Bioresour. Technol. 112, 131–136.

Keskin, T., Hallenbeck, P.C., 2012b. Enhancement of biohydrogen production by two-stage systems: Dark and photofermentation. In: Baskar, C., Baskar, S., Dhillon, R.S. (Eds.), Biomass Conversion. Springer, New York, pp. 313–340.

Keskin, T., Abo-Hashesh, M., Hallenbeck, P.C., 2011. Photofermentative hydrogen production from wastes. Bioresour. Technol. 102 (18), 8557–8568.

Khusnutdinova, A.N., Ovchenkova, E.P., Khristova, A.P., Laurinavichene, T.V., Shastik, E.S., Liu, J.G., et al., 2012. New tolerant strains of purple nonsulfur bacteria for hydrogen production in a two-stage integrated system. Int. J. Hydrogen Energy 37 (10), 8820–8827.

Kim, E.J., Yoo, S.B., Kim, M.S., Lee, J.K., 2005. Improvement of photoheterotrophic hydrogen production of *Rhodobacter sphaeroides* by removal of B800-850 light-harvesting complex. J. Microbiol. Biotechnol. 15 (5), 1115–1119.

Kim, M.S., Baek, J.S., Lee, J.K., 2006. Comparison of H_2 accumulation by *Rhodobacter sphaeroides* KD131 and its uptake hydrogenase and PHB synthase deficient mutant. Int. J. Hydrogen Energy 31, 12–17.

Kim, E.J., Kim, M.S., Lee, J.K., 2008a. Hydrogen evolution under photoheterotrophic and dark fermentative conditions by recombinant *Rhodobacter sphaeroides* containing the genes for fermentative pyruvate metabolism of *Rhodospirillum rubrum*. Int. J. Hydrogen Energy 33 (19), 5131–5136.

Kim, E.J., Lee, M.K., Kim, M.S., Lee, J.K., 2008b. Molecular hydrogen production by nitrogenase of *Rhodobacter sphaeroides* and by Fe-only hydrogenase of *Rhodospirillum rubrum*. Int. J. Hydrogen Energy 33 (5), 1516–1521.

Koku, H., Eroglu, I., Gunduz, U., Yucel, M., Turker, L., 2002. Aspects of the metabolism of hydrogen production by *Rhodobacter spheriodes*. Int. J. Hydrogen Energy 27, 1315–1329.

Koku, H., Eroglu, I., Gündüz, U., Yücel, M., Türker, L., 2003. Kinetics of biohydrogen production by the photosynthetic bacterium *Rhodobacter sphaeroides* O.U. 001. Int. J. Hydrogen Energy 28, 381–3888.

Kondo, T., Arakawa, M., Hirai, T., Wakayama, T., Hara, M., Miyaye, J., 2002. Enhancement of hydrogen production by a photosynthetic bacterium mutant with reduced pigment. J. Biosci. Bioeng. 93 (2), 145–150.

Laguna, R., Joshi, G.S., Dangel, A.W., Amanda, K., Luther, A.K., Tabita, F.R., 2010. Integrative control of carbon, nitrogen, hydrogen, and sulfur metabolism: The central role of the Calvin–Benson–Bassham cycle. In: Hallenbeck, P.C. (Ed.), Recent Advances in Phototrophic Prokaryotes. Springer, New York, pp. 265–271.

Laurinavichene, T.V., Belokopytov, B.F., Laurinavichius, K.S., Tekucheva, D.N., Seibert, M., Tsygankov, A.A., 2010. Towards the integration of dark- and photo-fermentative waste treatment. 3. Potato as substrate for sequential dark fermentation and light-driven H-2 production. Int. J. Hydrogen Energy 35 (16), 8536–8543.

Lee, J.Y., Chen, X.J., Lee, E.J., Min, K.S., 2012. Effects of pH and carbon sources on biohydrogen production by co-culture of *Clostridium butyricum* and *Rhodobacter sphaeroides*. J. Microbiol. Biotechnol. 22 (3), 400–406.

Liu, B.F., Ren, N.Q., Xie, G.J., Ding, J., Guo, W.Q., Xing, D.F., 2010. Enhanced bio-hydrogen production by the combination of dark and photo-fermentation in batch culture. Bioresour. Technol. 101, 5325–5329.

Ma, C., Guo, L.J., Yang, H.H., 2012. Improved photo-hydrogen production by transposon mutant of *Rhodobacter capsulatus* with reduced pigment. Int. J. Hydrogen Energy 37 (17), 12229–12233.

Maeda, I., Miyasaka, H., Umeda, F., Kawase, M., Yagi, K., 2003. Maximization of hydrogen production ability in high-density suspension of Rhodovulum sulfidophilum cells using intracellular poly (3-hydroxcbutyrate) as sole substrate. Biotechnol. Bioeng. 81, 474–481.

Masepohl, B., Hallenbeck, P.C., 2010. Nitrogen and molybdenum control of nitrogen fixation in the phototrophic bacterium Rhodobacter capsulatus. In: Hallenbeck, P.C. (Ed.), Recent Advances Phototrophic Prokaryotes, vol. 675. Springer, pp. 49–70.

McKinlay, J.B., Harwood, C.S., 2010. Carbon dioxide fixation as a central redox cofactor recycling mechanism in bacteria. Proc. Natl. Acad. Sci. USA 107 (26), 11669–11675.

Oh, Y.K., Seol, E.H., Kim, M.S., Park, S., 2004. Photoproduction of hydrogen from acetate by a chemoheterotrophic bacterium Rhodopseudomonas palustris P4. Int. J. Hydrogen Energy 29, 1115–1121.

Ozgur, E., Afsar, N., Vrije, T., Yücel, M., Gündüz, U., Claassen, P., et al., 2010a. Potential use of thermophilic dark fermentation effluents in photofermentative hydrogen production by *Rhodobacter capsulatus*. J. Cleaner Prod. 18, S23–S28.

Ozgur, E., Mars, A., Peksel, B., Lowerse, A., Afsar, N., Vrije, T., et al., 2010b. Biohydrogen production from beet molasses by sequential dark and photofermentation. Int. J. Hydrogen Energy 35, 511–517.

Posewitz, M.C., King, P.W., Smolinski, S.L., Zhang, L.P., Seibert, M., Ghirardi, M.L., 2004. Discovery of two novel radical S-adenosylmethionine proteins required for the assembly of an active Fe hydrogenase. J. Biol. Chem. 279 (24), 25711–25720.

Sabourin-Provost, G., Hallenbeck, P.C., 2009. High yield conversion of a crude glycerol fraction from biodiesel production to hydrogen by photofermentation. Bioresour. Technol. 100 (14), 3513–3517.

Shi, X.Y., Yu, H.Q., 2004. Hydrogen production from propionate by *Rhodopseudomonas capsulatus*. Appl. Biochem. Biotechnol. 117, 143–154.

Shi, X.Y., Yu, H.Q., 2005. Response surface analysis on the effect of cell concentration and light intensity on hydrogen production by *Rhodopseudomonas capsulate*. Process Biochem 40, 2475–2481.

Show, K.Y., Lee, D.J., Chang, J.S., 2011. Bioreactor and process design for biohydrogen production. Bioresour. Technol. 102, 8524–8533.

Su, H., Cheng, J., Zhou, J., Song, W., Cen, K., 2009. Combination of dark and photo-fermentation to enhance hydrogen production and energy conversion efficiency. Int. J. Hydrogen Energy 34, 8846–8853.

Tekucheva, D.N., Tsygankov, A.A., 2012. Combined biological hydrogen-producing systems: A review. Appl. Biochem. Microbiol. 48 (4), 319–337.

Tripp, H.J., Bench, S.R., Turk, K.A., Foster, R.A., Desany, B.A., Niazi, F., et al., 2010. Metabolic streamlining in an open-ocean nitrogen-fixing cyanobacterium. Nature 464 (7285), 90–94.

van Niel, C.B., 1944. The culture, general physiology, morphology and classification of the non-sulfur purple and brown bacteria. Bacteriol. Rev. 8, 1–118.

Wakayama, T., Nakada, E., Asada, Y., Miyake, J., 2000. Effect of light/dark cycle on bacterial hydrogen production by *Rhodobacter sphaeroides* RV. Appl. Biochem. Biotechnol. 8486, 431–440.

Wu, S.C., Liou, S.Z., Lee, C.M., 2012a. Correlation between bio-hydrogen production and polyhydroxybutyrate (PHB) synthesis by *Rhodopseudomonas palustris* WP3-5. Bioresour. Technol. 113, 44–50.

Wu, T.Y., Hay, J.X.W., Kong, L.B., Juan, J.C., Jahim, J.M., 2012b. Recent advances in reuse of waste material as substrate to produce biohydrogen by purple non-sulfur (PNS) bacteria. Renew. Sustain. Energy Rev. 16 (5), 3117–3122.

Yang, H., Guo, L., Liu, F., 2010. Enhanced bio-hydrogen production from corncob by a two-step process: Dark- and photo-fermentation. Bioresour. Technol. 101 (6), 2049–2052.

8

Biohydrogen Production from Algae

Man Kee Lam, Keat Teong Lee

School of Chemical Engineering, Universiti Sains Malaysia,
Pulau Pinang, Malaysia

INTRODUCTION

Heavy dependence on fossil fuels as a primary energy source has damaged the Earth's ecosystem significantly. Fossil fuels are nonrenewable and will be exhausted under a century if new oil wells are not found (Kulkarni and Dalai, 2006). Global industrialization and population growth are the main drivers that caused rapid energy exhaustion, which has consequently led to continual increases in the world price of petroleum and escalating tensions involving oil-producing countries (Kessel, 2000; Mathews and Wang, 2009). In addition, the combustion of fossil fuels has created numerous environmental issues, such as greenhouse gas (GHG) effects, which are responsible for the melting of arctic ice, rising sea levels, extreme heat waves, and frequent occurrences of droughts and desertification. Hence, identifying alternative renewable energy sources has become one of the key challenges of this century in attaining a more sustainable and greener energy for the future. Renewable energy sources such as solar, wind, hydro, and energy from biomass and waste have been developed and used successfully by different nations to limit the use of fossil fuels (Lam et al., 2010). Nevertheless, only energy produced from the combustion of biomass and wastes has made a significant contribution, indicating immense opportunities for biomass in the energy sector. In other words, research and development on green technologies using biomass to generate energy should be promoted extensively to strengthen this renewable source for diversification of the energy supply in the global fuel market.

Hydrogen (H_2) is considered the cleanest renewable fuel because its combustion by-product is only water vapor with no carbon dioxide (CO_2) emitted to the atmosphere (Brentner et al., 2010). H_2 has a higher specific energy content (142 MJ/kg) than methane (56 MJ/kg), natural gas (54 MJ/kg), and gasoline (47 MJ/kg) (Lam and Lee, 2011). These characteristics make H_2 a good candidate to be an important fuel in the future. To date, approximately 96% of the global H_2 supply has been derived from nonrenewable and chemical-based processes such as steam

161

methane reforming (SMR), petroleum refining, and coal gasification (Balat, 2008; Srirangan et al., 2011). Attention has been given to SMR because the process is cost-effective and currently available on a commercial scale (Jin et al., 2008). However, the SMR reaction is an endothermic reaction in which external energy (e.g., combustion of fuels) must be supplied to initiate the reaction to occur (Fan et al., 2009). Furthermore, using methane for H_2 production is not sustainable in the long term because most methane is currently derived from fossil fuel, which is nonrenewable.

On the contrary, biological H_2 production from photosynthetic microorganisms has the potential to reduce production costs and environmental impact considerably because its production utilizes sunlight as the driving force to split water into H_2 and O_2 under controlled conditions (Brentner et al., 2010; Eroglu and Melis, 2011). Research on H_2 metabolism by green microalgae was first explored by Gaffron and Rubin (1942) and thereafter, for photosynthetic bacteria, by Gest and Kamen (1949). The capability of these microorganisms to grow under both aerobic and anaerobic conditions has made them easier to shift within different cultivation modes to initiate H_2 production (Melis et al., 2007). Unlike biodiesel and bioethanol production from edible crops, biological H_2 production does not compete with agricultural land for food production, which is an important consideration in the transition toward cleaner and greener energy development. There are several mechanisms for producing H_2 from microalgae, including direct biophotolysis, indirect biophotolysis, photofermentation, and dark fermentation (Hallenbeck and Benemann, 2002). However, these methods must be improved through intensive research to make biological H_2 production economically viable without requiring large energy inputs (Mathews and Wang, 2009). Thus, the scope of this chapter is on the metabolic pathway for H_2 production from microalgae and the steps taken to overcome production bottlenecks. The discussion also include factors to be considered when scaling up H_2 production from microalgae and a potential integrated system to enhance the overall life cycle energy balance of microalgae-based biohydrogen.

BIOHYDROGEN GENERATION BY MICROALGAE

Direct Biophotolysis

Direct biophotolysis by microalgae is the most studied method for splitting water into H_2 and O_2 using sunlight as the energy source (Mathews and Wang, 2009; Oh et al., 2011). This is the simplest reaction for producing H_2, involving only water and light with no emission of GHG (Kim and Kim, 2011). When light is absorbed by photosynthetic microalgae, it enhances the oxidation of H_2O molecules by photosystem II (PSII), and the released electrons and protons are transported to the chloroplast ferredoxin via photosystem I (Melis et al., 2007). Reduced ferredoxin (FD) acts as an electron donor to [FeFe]-hydrogenases (Ghirardi et al., 2009), which reversibly facilitates the reduction of protons (H^+) to H_2 molecules, as shown in Equation (1) (Eroglu and Melis, 2011; Melis, 2007; Melis et al., 2000). The [FeFe]-hydrogenases in microalgae are monomeric enzymes with molecular masses of approximately 48 kDa and have a unique prosthetic group (H-group) in the active center, which results in a 100-fold higher enzyme activity than [NiFe]-hydrogenases and metal-free

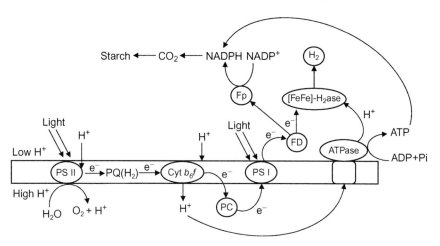

FIGURE 1 H_2 production pathway by photosynthetic microalgae. ATPase, ATP synthase; cyt *b6f*; cytochrome *b6f* complex; Fd, ferredoxin; Fp, ferredoxin-NADP reductase; H_2ase, hydrogenase; PC, plastocyanin; PQ, plastoquinones; $PQ(H_2)$, reduced plastoquinones; PS, photosystem. *Modified from Akkerman et al. (2002) Ghirardi et al. (2009), and Oh et al. (2011).*

hydrogenases (Ghirardi et al., 2009; Happe et al., 2002). However, the complicated reaction system for direct biophotolysis requires the input of a large amount of free energy ($\Delta G = + 237 \text{ kJ/mol}$), resulting in slowing the H_2 production rate by microalgae (Kim and Kim, 2011). A better understanding of H_2 production by microalgae via direct biophotolysis pathway is illustrated in Figure 1.

$$2H^+ + 2FD^- \leftrightarrow H_2 + 2FD \tag{1}$$

Limitations and Strategies to Improve H_2 Evolution via Direct Photolysis

Sparging with Inert Gas

One of the bottlenecks in direct biophotolysis is the very short period of H_2 evolution by microalgae, typically ranging from several seconds to a few minutes (Melis, 2007). This short period is caused by the simultaneous production of H_2 and O_2 molecules upon H_2O oxidation with a theoretical ratio of $H_2:O_2 = 2:1$ (Greenbaum et al., 1983; Melis, 2007; Spruit, 1958). O_2 is a powerful inhibitor of the [FeFe]-hydrogenase enzymatic reaction and is also a suppressor of [FeFe]-hydrogenase (Hy*d*A) gene expression (Ghirardi et al., 2009; Melis et al., 2007). Specifically, O_2 is speculated to bind to an iron atom with the double Fe subcluster, which leads to inhibition of the proton binding required for H_2 generation (Srirangan et al., 2011). This interferes with H_2 metabolism in microalgae cells and results in a low yield of H_2 production (Levin et al., 2004; Mathews and Wang, 2009; Melis, 2007). To overcome this limitation, removing O_2 from the system temporally or spatially is necessary to allow continuous activation of [FeFe]-hydrogenases in H_2 metabolism (Levin et al., 2004). This can be achieved by constantly sparging inert gas into the microalgae cultivation system to keep both H_2 and O_2 gases at low partial pressure (Mathews and Wang, 2009). The partial pressure of O_2 should be maintained at 0.1% or lower to effectively maintain the activity of [FeFe]-hydrogenases to produce H_2 for long period. In reality, the sparging approach is impractical because a large quantity of inert gas is required to dilute the gas stream, leading to a high power input for gas transfer and consequently extra costs for the process (Hallenbeck and Benemann, 2002; Mathews and Wang, 2009). In addition, the extra volume (photobioreactor) required to accommodate the additional quantity of inert gas will be large, causing the process to be economically infeasible for commercialization. The energy required to separate diluted H_2 from O_2 and inert gas is another issue that results in the inert gas sparging approach being unsustainable.

Sulfate Deprivation

Another possible method for optimizing H_2 production from microalgae via direct biophotolysis is to cultivate microalgae under sulfate-deprived conditions (Melis et al., 2000). This is a two-stage process in which O_2 and H_2 production are separated temporally. Initially, microalgae are allowed to grow through normal photosynthesis with O_2 generation while bio-fixing CO_2 from the atmosphere. This stage allows the microalgae to attain high bulk density. The microalgae are then transferred to a new medium with minimal or no sulfate content to obtain low oxygen partial pressures and catalyze H_2 evolution metabolism (Melis, 2007; Melis et al., 2000). When microalgae are cultivated under limited sulfate conditions, synthesis of the PSII D1 polypeptide chain (32 kDa) is impeded, interfering with the repair of PSII from photo-oxidative damage (Melis, 2007; Wykoff et al., 1998). Thus, the microalgae cells can maintain a high respiration rate exceeding the photosynthesis rate, creating an anaerobic condition that allows H_2 production for 4 to 5 days (Ghirardi et al., 2009; Melis, 2007). In this situation, [FeFe]-hydrogenases are not inhibited significantly by O_2 molecules because the O_2 is taken up quickly during respiration. For example, when *Chlamydomonas reinhardtii* were deprived of sulfate, the photosynthesis rate was decreased by 5–10% of the initial value after 15–20 h (Wykoff et al., 1998). However, respiration activity was sustained over the first 70 h of sulfate deprivation and declined only slightly thereafter (Melis et al., 2000). After 150 h, about 87% (v/v) H_2 was recorded in the headspace of the photobioreactor, indicating the feasibility of sulfate deprivation for improving the H_2 production rate by microalgae (Melis et al., 2000). Table 1 shows H_2 production rates under sulfate-deprived conditions for different microalgae strains.

However, continuous evolution of H_2 under sulfate-deprived conditions cannot be sustained for a long period because substrates are utilized for catabolism (Melis et al., 2000; Zhang et al., 2002). In a study performed by Zhang et al. (2002), the protein and starch content in microalgae cells were increased about 40 and 600%, respectively, from their initial values for the first 24 and 30 h of sulfate deprivation, but declined gradually thereafter. This provides clear evidence that substantial protein and starch catabolism occurs during H_2 evolution under sulfate-deprived conditions (Zhang et al., 2002). At this point, the microalgae should be recultivated in a nutrient-rich medium to replace the needed metabolites before again being subjected to a sulfate-free medium for a subsequent cycle of H_2 production (Ghirardi et al., 2009). The starch content in microalgae cells decreases because a substantial amount is consumed to support the mitochondrial respiration of the cell under anaerobic conditions in the cultivation system (Zhang et al., 2002). Mitochondrial respiration is responsible for absorbing the small amount of O_2 released from residue photosynthetic activity, which

TABLE 1 H_2 Producing Microalgae by Means of Sulfate-Deprived Method

Microalgae strain	H_2 production rate (ml H_2 liter^{-1} h^{-1})	Duration H_2 (h)	Reference
Chlamydomonas reinhardtii	2	150	(Melis et al., 2000)
C. reinhardtii	4.3	552	(Laurinavichene et al., 2006)
C. reinhardtii	0.58	4000	(Fedorov et al., 2005)
C. reinhardtii	0.17–0.87	65–107	(Tsygankov et al., 2006)
C. reinhardtii	2.24	96	(Jo et al., 2006)
C. reinhardtii CC124	1.29	117	(Faraloni et al., 2011)
C. reinhardtii L159I-N230Y	1.81	285	(Torzillo et al. (2009))
Chlorella autotrophica IOAC689S	0.11	<7	(He et al., 2012)
C. protothecoides IOAC038F	2.93	<6	(He et al., 2012)
C. salina Mt	0.5	120	(Chader et al., 2009)
C. sorokiniana Ce	1.35	222	(Chader et al., 2009)
Chlorella sp. Pt6	0.8	120	(Chader et al., 2009)
Chlorella sp. IOAC085F	1.33	<6	(He et al., 2012)
Nannochloropsis sp. IOAC676S	0.018	7–19	(He et al., 2012)
Platymonas subcordiformis	0.04	5	(Guan et al., 2004)
Scenedesmus obliquus IOAC039F	0.052	<6	(He et al., 2012)
Tetraselmis striata IOAC707S	0.34	<7	(He et al., 2012)

helps maintain the low partial pressure of O_2 in the culture medium (Melis and Happe, 2001). Microalgae are expected to decay under anaerobic conditions because of the absence of electron transport from H_2O oxidation, which indirectly inhibits oxidative phosphorylation in mitochondria (Melis, 2007). However, the expression of [FeFe]-hydrogenases in H_2 evolution permits an alternative but slow rate of electron transport in the thylakoid membrane of photosynthesis, which is coupled to electron transport in mitochondria, leading to ATP generation in both organelles (Melis, 2007; Zhang et al., 2002). The ATP is probably used by the microalgae cells for sustaining basic life activities, prolonging their survival under sulfate-deprived conditions (Melis, 2007). Figure 2 illustrates the electron transport pathways for photosynthesis and respiration by microalgae when deprived of sulfate. Hence, H_2 production by microalgae cultivated with a limited sulfate medium involves a four-way coordinated interaction of oxygenic photosynthesis, mitochondrial respiration, catabolism of endogenous substrate, and electron transport via the [FeFe]-hydrogenase pathway (Melis, 2007; Melis and Happe, 2001; Zhang et al., 2002). More fundamental research is urgently needed to address the relationship of these four processes and subsequently improve the stability of H_2 production in green microalgae.

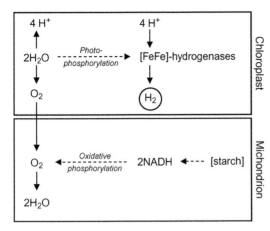

FIGURE 2 Coordinated photosynthetic and respiratory electron transport coupled with phosphorylation during H_2 production by microalgae. Electrons generated from oxidation of H_2O through photosynthesis are delivered to [FeFe]-hydrogenases, leading to photophosphorylation and H_2 evolution. The O_2 generated by this process is responsible for driving coordinate oxidative phosphorylation during mitochondrial respiration. At this point, electrons are derived upon endogenous substrate (starch) catabolism, generating reductant (NADH) and CO_2. Production of H_2 by the chloroplast permits the sustained operation of this coordinated photosynthesis–respiration function in green microalgae and allows the generation of ATP by the two bioenergetic organelles in the cell. *Modified from Melis and Happe (2001).*

Mutagenesis and O_2-Tolerant Hydrogenases

Protein and genetic engineering is an alternative approach to enhance H_2 production from microalgae (Brentner et al., 2010; Ghirardi et al., 2000). As discussed previously, [FeFe]-hydrogenases play a vital role in biologically catalyzing H_2 evolution from microalgae cells; however, this enzyme is extremely sensitive to the presence of O_2 (Melis, 2007). Therefore, identifying and isolating O_2-resistant [FeFe]-hydrogenases from nature could be the solution to this problem (Oh et al., 2011). In a study done by Ghirardi et al. (2000), a classical random mutagenesis approach was conducted to identify O_2-tolerant hydrogenases in the green microalgae *C. reinhardtii*. Initially, the mutagenized microalgae cells were subjected to two different selective pressures. The first pressure was selected for the cells to metabolize H_2, while the second pressure was selected for the cells to release H_2 in the presence of O_2. The surviving microalgae cells were subsequently subjected to additional screening using a chemochromic sensor that detects H_2 evolution activity in microalgae (Flynn et al., 2002; Ghirardi et al., 2000). Finally, mutated microalgae cells that recorded a high tolerance toward O_2 were isolated for H_2 production in a larger scale. Through this mutagenesis and isolation procedures, the selected mutants had two to nine times higher O_2 tolerance than the wild type, indicating the possibility of sustaining H_2 evolution by microalgae under aerobic conditions for longer periods (Flynn et al., 2002; Ghirardi et al., 2000). However, more significant breakthroughs and the identification of O_2-tolerant [FeFe]-hydrogenases from different microalgae strains for higher H_2 production rate are required to accelerate its commercial development.

In addition to isolating O_2-tolerant hydrogenases, genetic engineering can be used to modify the electron transport pathway in the H_2 evolution system so that the electrons are directed to the biophotolysis system, which enhances the H_2 production rate (Lee and Greenbaum, 2003; Oh et al., 2011). Lee and Greenbaum (2003) discovered a new competitive pathway for background O_2 sensitivity in *C. reinhardtii* that was 10 times more sensitive than the classic O_2 sensitivity of [FeFe]-hydrogenases. From the study, O_2 that apparently serves as a terminal electron acceptor was in competition with the H_2 production pathway for photosynthetically generated electrons from direct biophotolysis (water splitting).

Subsequently, this O_2-sensitive H_2 production electron transport pathway was inhibited by 3-(3,4-dichloropheneyl)-1,1-dimethylurea (DCMU). DCMU is a strong chemical inhibitor that blocks the transport of electrons acquired from PSII water splitting to PSI (Lee and Greenbaum, 2003). Therefore, genetic insertion of a programmed polypeptide protein channel with a hydrogenase promoter into the thylakoid membrane of microalgae cells could potentially dissipate the proton gradient across the membrane without ATP formation (ATP is produced during proton transfer) (Kim and Kim, 2011; Lee and Greenbaum, 2003). Through this strategy, the proton gradient that hinders electron transfer from water to [FeFe]-hydrogenases could be avoided and simultaneously eliminate the newly discovered O_2-sensitive pathway (Lee and Greenbaum, 2003).

In another study done by Kruse et al. (2005), mutated *C. reinhardtii* (with given name *Stm6*) was constructed by random gene insertion. This strain had modified respiratory metabolism and was unable to perform cyclic electron transport around PSI. This led to the accumulation of starch inside the cells that enhanced substrate availability for a prolonged H_2 production period (see earlier discussion). The starch accumulation was probably caused by the inhibition of energy consumption by mitochondrial respiration under anaerobic conditions during illumination (Kruse et al., 2005). In addition, this mutated strain showed low levels of dissolved O_2 concentration (30–40% of parental strain), which helped reduce the inhibitory effects of O_2 toward [FeFe]-hydrogenases activity (Kruse et al., 2005). Hence, H_2 production by *Stm6* was sustained for a longer period with a production rate 5–13 times higher than the parental strain.

Enhanced Light-Capturing Efficiency

Under an ideal environment, microalgae can absorb and use photon energy with nearly 100% efficiency to generate H_2 from sunlight and water (Greenbaum, 1988; Ley and Mauzerall, 1982; Melis et al., 2007). Unfortunately, under real conditions, because the light intensity is high but the absorption capacity of photosynthetic pigments (chlorophyll a) in microalgae cells is limited, the photon energy could not be converted efficiently to chemical energy. Instead, up to 80% of the energy is wasted by dissipation in the form of fluorescence and heat (Hallenbeck et al., 2012; Kosourov et al., 2011; Polle et al., 2002). Furthermore, the shading effect created by dense cultures and low-light penetration into the cultivation system might bring down the efficiency to less than 1% and often closer to 0.1% (Hallenbeck et al., 2012). As a consequence, upscaling a microalgae cultivation system for H_2 production would be problematic, as the cultivation would probably take place in an enclosed photobioreactor and cultivated outdoor with abundant sunlight (Melis et al., 2007). To overcome this problem, a possible strategy is to engineer a microalgae mutant with a limited amount of chlorophyll (smaller light-harvesting antenna size) to permit greater transmittance of irradiance through a high-density culture, as shown in Figure 3 (Kirst et al., 2012; Kosourov et al., 2011; Melis, 2009; Melis et al., 2007; Polle et al., 2002, 2003). Hence, less photon energy would be wasted at the surface of the culture and sunlight could penetrate deeper into the culture, leading to an overall increase in biomass productivity and H_2 production (Benemann, 2000).

In a study done by Polle et al. (2002), DNA insertional and chemical mutagenesis of *C. reinhardtii* was carried out to isolate *tla1*, a stable transformant that has a truncated light-harvesting chlorophyll antenna size. From biochemical analyses, this mutated strain

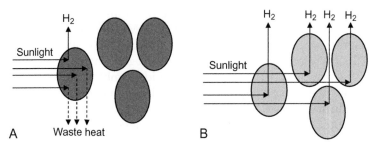

FIGURE 3 Sunlight absorption and dissipation by green microalgae. (a) Dark-green microalgae-only individual cells on the surface absorb the sunlight (more than can be utilized by photosynthesis) and dissipate in the form of waste heat, resulting in low productivity of H_2. (b) Truncated chlorophyll antenna size—individual cells have a limited light-absorbing capacity, thus permitting greater light penetration and H_2 production by cells deeper in the culture. *Modified from Melis (2009).*

was extremely deficient in chlorophyll content (but not photosystem), being 50 and 65% lower in functional chlorophyll antenna size of PSII and PSI, respectively (Melis et al., 2007; Polle et al., 2000, 2001, 2002). Therefore, the *tla1* strain exhibited higher saturating light intensity for photosynthesis and a greater solar conversion efficiency than the wild type. This unique characteristic permits mass cultivation of microalgae under direct sunlight and thus enhances energy sustainability in the cultivation system. However, a high photosynthetic conversion efficiency does not necessary translate to a high H_2 production rate because of the sensitivity of [FeFe]-hydrogenases toward O_2 (Hallenbeck et al., 2012). The microalgae must still be deprived of sulfate to initiate the anaerobic conditions for subsequent H_2 production.

In another study carried out by Kosourov et al. (2011), a *tla1* mutant (CC-4169) was immobilized within thin alginate films (≈ 300 μm) and cultivated under sulfate-deprived conditions. Under limited light conditions (light intensity of 19 μE m^{-2} s^{-1}), CC-4169 demonstrated an extremely low yield of H_2, being four times lower than the parental strain. This result was expected because CC-4169 had smaller light-harvesting antenna, thus lowering the photosynthetic efficiency. However, when the light intensity was increased to 285 and 350 μE m^{-2} s^{-1}, the mutant exhibited four and more than six times H_2 production than the parental strain, respectively. Therefore, the hypothesis was confirmed that truncating light-harvesting antenna in microalgae cells could enhance light conversion efficiency under high-light intensity and result in higher H_2 production per unit of illuminated culture surface area (Kosourov et al., 2011). A similar result was reported by Torzillo et al. (2009) in which the *C. reinhardtii* D1 mutant with a substitution of double amino acids exhibited a positive result in H_2 production because of a low chlorophyll content, higher photosynthetic capacity and conversion efficiency, and high respiration rate under anaerobic conditions. Therefore, the H_2 production rate from this mutant was higher compared to the wild type—1.81 ml H_2/liter/h versus 0.30 ml H_2/liter/h—and H_2 production was able to proceed for much longer—285 h versus 53 h (Torzillo et al., 2009). In short, microalgae with truncated light-harvesting chlorophyll antenna play a key role in improving H_2 production and have a greater chance of mass cultivation under direct sunlight (Melis et al., 2007; Torzillo et al., 2009).

Indirect Biophotolysis

For indirect biophotolysis, the problem of [FeFe]-hydrogenase inactivation by O_2 is potentially eliminated through the temporal or spatial separation of O_2 and H_2 evolution (Benemann, 2000; Hallenbeck and Benemann, 2002; Manish and Banerjee, 2008). In the first stage, microalgae are allowed to grow under normal cultivation conditions with light as the driving energy source to fix CO_2 into carbohydrate (Benemann, 2000). During this process, microalgae are allowed to reproduce as much as possible to increase the overall total carbohydrate in the cells while producing O_2 as a by-product (Mathews and Wang, 2009). In the second stage, commonly referred to as anaerobic dark fermentation, carbohydrate in microalgae cells is used as a substrate by the nitrogenase enzyme (nitrogen fixation) to produce H_2 (Benemann, 2000; Hallenbeck and Benemann, 2002; Pilon et al., 2011). The overall reactions for the both stages are shown in Equations (2) and (3), respectively. Compared to direct biophotolysis, indirect biophotolysis has the advantage of using water as the electron donor and inorganic carbon as the carbon source, but O_2 inhibition during H_2 production is not significant (Brentner et al., 2010; Oh et al., 2011). However, because of the multiple steps involved in converting solar energy to H_2 and the requirement for ATP by the nitrogenase enzyme, the maximum conversion efficiency of light to H_2 via indirect biophotolysis is only 16.3% (Pilon et al., 2011; Prince and Kheshgi, 2005). It should be noted that the two-stage process using sulfate deprivation to create anaerobic conditions for H_2 production (see earlier discussion) is considered to be direct biophotolysis and not indirect biophotolysis, as photosynthesis is still functioning inside the cells and delivering electrons to the hydrogenases during the anaerobic phase (Brentner et al., 2010; Hallenbeck and Benemann, 2002; Hallenbeck and Ghosh, 2009).

$$12\,H_2O + 6\,CO_2 \rightarrow C_6H_{12}O_6 + 6\,O_2 \tag{2}$$

$$C_6H_{12}O_6 + 12\,H_2O \rightarrow 12\,H_2 + 6\,CO_2 \tag{3}$$

Cyanobacteria (or blue-green algae) are well known for H_2 production via indirect biophotolysis (Das and Veziroğlu, 2001; Eroglu and Melis, 2011; Levin et al., 2004; Pilon et al., 2011). Cyanobacteria strains that can produce H_2 are mostly filamentous and contain specialized nitrogen-fixing cells known as heterocysts (Eroglu and Melis, 2011). In the heterocyst, nitrogenase is protected from O_2 by a thick cell wall that has a low diffusivity of O_2. With a high respiration rate, O_2 is normally utilized, creating an ideal anaerobic environment for higher H_2 production (Das and Veziroğlu, 2001). However, the lack of heterocysts in green microalgae causes it to be inefficient in fixing nitrogen under anaerobic dark conditions, which results in limited amounts of H_2 generated though indirect biophotolysis. In a study done by Ohta et al. (1987), green microalgae *Chlamydomonas* MGA 161 was initially grown using a light intensity of $25\,W/m^2$ with continuous sparging with air containing 5% CO_2. Once microalgae growth reached the stationary phase, the culture was transferred to a medium that was flushed with pure N_2 gas and cultivated under a dark environment to initiate anaerobic fermentation. After 12 h of incubation, 1.5 mmol H_2/g dry wt was detected with degradation of starch at a rate of 0.08 mmol glucose/g dry wt/h. Apart from H_2, other fermentative by-products were also produced, such as CO_2, acetate, ethanol, formate, and glycerol with mole ratios of 0.9, 0.73, 0.33, 0.13, and 0.05 for every 1 mol of H_2 produced, respectively (Ohta et al., 1987).

Integrated System

To date, neither direct nor indirect biophotolysis of microalgae has been performed at a commercial scale for sustainable H_2 production, even though significant breakthroughs and improvements have been achieved. This is mainly because several technical problems still remain unsolved, such as low yields of H_2, sensitivity of hydrogenases toward O_2, high cost of photobioreactor, and technological challenges in genetic and molecular engineering (Ghirardi et al., 2009). In contrast, dark fermentation is superior for H_2 production, but the process is limited to fermentative bacteria (e.g., *Escherichia coli*, *Clostridium*, and *Sporolactobacillus*) and rarely to green microalgae (Brentner et al., 2010). For instance, the H_2 production rate through the dark fermentation of *E. coli* SR 13 was 300 liter H_2/liter/h (Yoshida et al., 2005), whereas direct biophotolysis of *C. reinhardtii* only yielded 0.012 liter H_2/liter/h (Laurinavichene et al., 2006). Nevertheless, bacteria required substrates such as glucose and sucrose during dark fermentation as an energy source to grow, and the cost of these substrates may prevent the implementation of this technology at a commercial scale for H_2 production (Kapdan and Kargi, 2006). To overcome this problem, the use of low-cost substrates such as industrial food wastes or wastewater appears to be a potential solution; however, several other problems may surface, such as the need to pretreat the low-cost substrates to provide a suitable environment for the bacteria to grow, possibility of contamination and competition for substrate utilization by other bacteria that do not produce H_2, and variations and inconsistencies in substrate concentrations.

Integrating direct biophotolysis by microalgae and dark fermentation by fermentative bacteria could circumvent the disadvantages of each individual process and enhance the overall H_2 production process. As shown in Figure 4, microalgae are cultivated initially in the first stage by utilizing light energy to produce H_2 through direct biophotolysis. After the microalgae have stopped producing H_2, the biomass is utilized as a substrate by bacteria to continue H_2 production via the dark fermentation process. During dark fermentation, CO_2 is emitted along with H_2, as shown in Equation (4). If a gas separation unit is added into the system to separate CO_2 from H_2, CO_2 could be used as a carbon source to grow microalgae in the first stage. This strategy not only creates a symbiotic relationship between indirect biophotolysis and dark fermentation, but also improves the energy conversion efficiency and CO_2 utilization in this integrated system. Furthermore, including a photofermentation

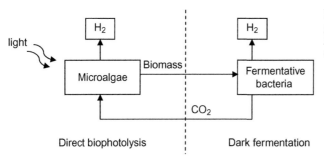

FIGURE 4 Schematic diagram of integrated biological H_2 production processes. *Modified from Eroglu and Melis (2011) and Ghirardi et al. (2009).*

process using purple nonsulfur bacteria such as *Rhodospirillum rubrum*, *Rhodopseudomonas palustris*, *Rhodobacter sphaeroides*, and *Rhodobacter capsulatus* in the integrated system appears to be another potential approach to enhance solar energy utilization and H_2 productivity (Eroglu and Melis, 2011; Ghirardi et al., 2009).

$$\text{Dark fermentation}: C_6H_{12}O_6 + 2H_2O \rightarrow 2CH_3COOH + 2CO_2 + 4H_2 \qquad (4)$$

Although the integrated system is still far from commercialization, several studies have revealed the potential of using microalgae starch as a substrate for fermentative bacteria to produce H_2 during dark fermentation (Efremenko et al. 2012; Nguyen et al., 2010; Sun et al., 2011). In a study done by Nguyen et al. (2010), starch accumulated in *C. reinhardtii* cells was utilized by the hyperthermophilic bacterium *Thermotoga neapolitana* as an alternative source of substrate for H_2 production. This bacteria is rod shaped, nonspore forming, and able to ferment at a high temperature, which means that it is less prone to interference by contaminants, as other bacteria may not grow at high temperatures (Nguyen et al., 2010). However, for the microalgae biomass to be used effectively as a substrate during dark fermentation, two important pretreatment steps must be taken: (1) the microalgae cell wall is rigid and disruption of the cell wall is required to release the starch and (2) starch is a macromolecule and thus hydrolyzing the starch into a simple reducing sugar is required to enhance its utilization. When the microalgae biomass was mixed directly with hyperthermophilic bacteria without any pretreatment, no H_2 was produced under dark fermentation conditions. Nevertheless, when microalgae cells were disrupted with methanol, H_2 evolution was detected at a production rate of 53 ml/h/liter. In addition, when the disrupted microalgae cells were subjected to enzymatic hydrolysis (α-amylase, Termamyl 120L) and subsequently proceeded to dark fermentation, the H_2 production rate was increased dramatically to 227 ml/h/liter. It is clear that pretreatment of the microalgae biomass played a crucial role in allowing fermentative bacteria to access and convert intracellular starch for H_2 production. However, it should be noted that the expensive pretreatment steps may increase the overall H_2 production costs significantly and make the process economically unsustainable.

Another example of an integrated microalgae H_2 system was demonstrated by Mussgnug et al. (2010). In their study, green microalgae *C. reinhardtii* were allowed to produce H_2 under sulfate-deprived conditions as established by Melis et al. (2000). After the microalgae had stopped producing H_2, the microalgae cells were harvested and used as a substrate in a subsequent anaerobic fermentation process. However, the targeted fermentative product was methane, which is an alternative biogas to H_2. Through this integrated system, the methane yield was increased to 123% compared to the control (Mussgnug et al., 2010). When *C. reinhardtii* was cultivated under sulfate-deprived conditions, fermentative compounds such as starch increased within the cells as a response to induction of the H_2 production cycle (Melis, 2007; Mussgnug et al., 2010; Zhang et al., 2002). Simultaneously, the risk of emitting H_2S during anaerobic fermentation will also be reduced, resulting in a higher purity of methane. Nevertheless, allowing microalgae to produce H_2 in the first stage for a prolonged period exposed the risk of decreasing starch content in the microalgae cells due to mitochondrial respiration under anaerobic conditions (Zhang et al., 2002). As a result, limited starch would be available for second-stage fermentation, leading to a possible low yield of biogas production.

Hence, more studies are needed to address the possibilities of this integrated system, typically in term of energy balance and cost-effectiveness.

LIFE CYCLE ASSESSMENT (LCA) OF H_2 PRODUCTION FROM MICROALGAE

Although H_2 production from microalgae has the potential to contribute significantly in diversifying global renewable energy resources, there is still the question of sustainability of the process for long-term operation. Up until now, hydrogen evolution from microalgae has been reported primarily on a laboratory scale (Kruse et al., 2005; Romagnoli et al., 2011), while pilot and commercial scales are rarely found in the literature. As a result, a critical understanding of the overall process operation is very limited, mainly in terms of energy balance and global warming potential. Microalgae H_2 production involves not only H_2 evolution during cultivation, but also other processing stages, such as nutrient sources and preparation, biomass harvesting and drying, water purification and reuse, H_2 purification, and compression, which are equally important in addressing its feasibility for scaling up (Campbell et al., 2011; Collet et al., 2011; Lardon et al., 2009; Pittman et al., 2011; Singh and Gu, 2010). In this regard, performing a LCA provides a platform to benchmark the environmental repercussions of H_2 production from microalgae through an in-depth inventory and impact assessments for each processing stage (Romagnoli et al., 2011). LCA is currently accepted as an effective tool for researchers and policy makers for understanding the real environmental potential of a particular product. It can also be used to indicate if the production of a particular product can lead to negative environmental phenomena, such as eutrophication, global warming, ozone depletion, human and marine toxicity, land competition, and photochemical oxidation, so that precautionary steps can be considered to reduce the negative impacts (Alvarenga et al., 2012; Collet et al., 2011; Lam et al., 2009; Renó et al., 2011). In addition, an energy balance can also be calculated to determine the energy hotspot for all stages within the system boundary of the LCA (Lam and Lee, 2012).

Romagnoli et al. (2011) carried out an LCA study of H_2 production from *C. reinhardtii* when scaled up to a commercial level. In that study, H_2 production through direct biophotolysis, coupled with sulfate deprivation, was used as a baseline to benchmark GHG emissions compared to natural gas, heavy fuel oil, and coal. It was estimated that the H_2 production yield from *C. reinhardtii* would be 109.5 tonne/year or equivalent to 15549 GJ/year, assuming the higher heating value of H_2 was 142 MJ/kg. CO_2 emissions from the combustion of natural gas, heavy fuel oil, and coal, which could be avoided, were 8.4, 10.3, and 25.5 tonne CO_2/year, respectively. This important finding has opened a new direction in the microalgae biofuel industry and provides an excellent platform for evaluation of a stronger and wider H_2 market strategy for policy makers and investors (Romagnoli et al., 2011). However, it is still unclear if H_2 production from microalgae can have a positive energy balance. Several recent LCA studies of microalgae biofuel production showed that intensive energy was consumed, especially in the cultivation system (e.g., nutrient production, air and water pumping) and in biomass processing (e.g., harvesting and drying) (Clarens et al., 2010; Jorquera et al., 2010; Razon and Tan, 2011; Stephenson et al., 2010).

Nutrient Sources

According to an LCA study done by Clarens et al. (2010), about 50% of overall energy use and GHG emissions were associated with the use of chemical (inorganic) fertilizers. Although chemical fertilizers have the advantage of reducing contamination in the cultivation medium and promote healthy growth of microalgae, their long-term usage is definitely unsustainable and not beneficial to the environment. On the contrary, using wastewater as an alternative source of nutrients to cultivate microalgae seems to be more practical and economically viable. Secondary and tertiary wastewaters normally contain significant amounts of nitrate and orthophosphate, which are not removed during primary treatment. Removing these compounds typically requires an additional 60 to 80% of the energy consumed in a typical wastewater treatment plant (Clarens et al., 2010; Maurer et al., 2003). Thus, cultivating microalgae in wastewater offers inexpensive wastewater treatment technology, as well as a means of reducing the need of chemical fertilizers and their associated life cycle burden (Clarens et al., 2010).

In view of the potential of H$_2$ evolution from *C. reinhardtii*, cultivating this microalgae strain using wastewater could synergize a better energy output and improve the sustainability of the microalgae cultivation system. Kong et al. (2010) reported the possibility of cultivating *C. reinhardtii* in wastewater discharged from industrial facilities and households. It was found that the microalgae could not grow exponentially when cultivated with treated wastewater (effluent), which contains low concentrations of nitrogen and phosphorus. On the contrary, when the microalgae were supplied with concentrated wastewater that contains a high level of nutrient source, their growth rate improved significantly, with a 2.5-fold increment in biomass quantity (Kong et al., 2010). It is necessary to enhance microalgae growth with a nutrient-rich medium to obtain a higher cell division rate before the microalgae are subjected to sulfate-deprived medium for H$_2$ production. Using wastewater as a nutrient source is ultimately a win–win approach in obtaining a high density of microalgae biomass, while also playing an important role in purifying the wastewater before discharge. Issues that need to be addressed prior to commercializing this strategy include contamination by fungus and bacteria in wastewater that may annihilate the entire population of microalgae, a possible inconsistency in the nutrient content of the wastewater that may affect the growth rate of microalgae, and the wastewater after microalgae cultivation needing to be monitored before being discharged to water sources to avoid microalgae bloom.

After being grown to a high density, the microalgae cells are ready to be transferred to a new medium with a limited sulfate concentration for H$_2$ production. In this regard, wastewater is no longer suitable to be used as the culture medium unless sulfate is removed from the wastewater. Depending on the discharge sources, wastewater normally contains a high concentration of sulfate, as it is true for pharmaceutical wastewater (Fox and Venkatasubbiah, 1996), textile wastewater (Selcuk, 2005), and molasses wastewater (Hilton and Archer, 1988). However, chemical processes for removing sulfate from wastewater, such as precipitation by iron (or other metals), is relatively expensive and generates large amounts of solid wastes (Silva et al., 2011; Tait et al., 2009). This makes the process impractical for integration with the microalgae H$_2$ production system. In fact, there is no supporting evidence to date that wastewater with sulfate removed is effective in enhancing H$_2$ evolution from microalgae. Apart from that, most works with microalgae have used chemical fertilizers and were conducted under controlled environments. Hence, more research on this subject is needed, particularly on the effects of fungal and bacterial contamination and inconsistent nutrient compositions in wastewater toward the H_2 evolution rate of microalgae.

Photobioreactor

The photobioreactor design is equally as important as supplemental nutrients in optimizing H_2 production from microalgae, especially through the direct biophotolysis method. A good photobioreactor should have the following characteristics: (1) effective illumination area, (2) optimal gas–liquid transfer rate, (3) easy to operate, (4) low potential for contamination, (5) low capital and production costs, and (6) minimal land area requirements (Xu et al., 2009). There are various type of photobioreactors designed mainly for microalgae biomass production, such as (1) an open system, which is normally known as raceway pond, and (2) a closed system, which includes tubular, bubble column, flat plat, and fermentor (Dasgupta et al., 2010). Table 2 summarizes the design characteristics of these photobioreactors.

To date, the raceway pond appears to be the most feasible method for mass cultivation of a microalgae biomass. The system consists of a closed-loop recirculation channel (oval shape) in which mixing is provided by paddle wheels. The depth of the pond is usually 0.2 to 0.5 m to ensure that microalgae receive sufficient exposure to sunlight (Brennan and Owende, 2010; Chisti, 2007; Greenwell et al., 2010). The system is relatively low cost compared to a closed system and requires a minimal energy input to operate. However, it is impossible to create an anaerobic environment in the raceway pond for H_2 evolution from microalgae (Dasgupta et al., 2010). Even if H_2 evolution is possible, H_2 collection would become problematic because of the large cultivation area involved. In this case, an extremely high-energy input to operate the air pump is expected, which could potentially result in a net negative energy balance in the life cycle of microalgae H_2 production.

A closed photobioreactor is an alternative for scaling up H_2 production from microalgae. Generally, a closed photobioreactor offers several advantages compared to raceway ponds, including better contamination control, higher microalgae biomass yield per unit reactor volume, and the possibility of using a single-strain culture for a prolonged duration (Chisti, 2007; Greenwell et al., 2010). A closed photobioreactor has a high surface area to volume (A/V) ratio that allows higher solar energy conversion efficiency by microalgae to H_2 through direct biophotolysis. Among all closed photobioreactors, the horizontal tubular/air-lift, helical tubular, α-shaped photobioreactor exhibited a relatively higher A/V ratio and is currently being tested under laboratory-scale conditions for H_2 production from cyanobacteria and microalgae (Dasgupta et al., 2010). However, LCA studies have shown that closed photobioreactors used substantial amounts of energy and resulted in an unsustainable practice for producing biofuel from microalgae (Jorquera et al., 2010; Stephenson et al., 2010). For example, it was estimated that the energy input to operate an air-lift photobioreactor was around 350% higher compared to a raceway pond (Lam and Lee, 2012). Furthermore, an energy assessment carried out by Stephenson et al. (2010) indicated that using an air-lift photobioreactor to cultivate microalgae for biodiesel production could lead to a net negative energy balance in the associated life cycle boundary. Most of

TABLE 2 Different Photobioreactor Designs for Microalgae Cultivation (Chisti, 2007; Mata et al., 2010; Sierra et al., 2008; Ugwu et al., 2008; Xu et al., 2009)

	Raceway pond	Flat plate	Tubular	Vertical-column
Conceptual design				
Design characteristics	Consists of close-loop recirculation channel (oval shape)	Bioreactor with rectangular shape	Consists of array of straight, coiled, or looped transparent tubes	Bioreactor with vertical-arranged cylindrical column
	Usually built using concrete or compacted earth-lined pond with white plastic	Usually made of transparent plastic or glass	Tubes are usually made of transparent plastic or glass	The columns are usually made of transparent plastic or glass
	Mixing and circulation are provided by paddle wheels	Usually coupled with gas sparger	Usually coupled with pump or air-lift technology	Usually coupled with pump or air-lift technology
	The depth of the pond is usually 0.2 to 0.5 m to ensure microalgae receive adequate exposure to sunlight	The light path (depth) is dependent on microalgae strain; range between 1.3 and 10 cm	Tube diameter is limited (0.1 m) to increase the surface/volume ratio	Optimum column diameter is 0.2 m with 4 m height
Advantages	Easy to construct and operate	Large illumination surface area gives maximum utilization of solar energy	Large illumination surface area	High mass transfer rate with good mixing

Continued

TABLE 2 Different Photobioreactor Designs for Microalgae Cultivation (Chisti, 2007; Mata et al., 2010; Sierra et al., 2008; Ugwu et al., 2008; Xu et al., 2009)—Cont'd

	Raceway pond	Flat plate	Tubular	Vertical-column
	Low energy input and low cost	Low concentration of dissolved oxygen	Relatively higher biomass productivity	Compact, easy to operate, and relatively low cost
		Can be position vertically or inclined at an optimum angle facing the sun	Potential of cell damage is minimized if airlift system is used	Lower power consumption
		Lower power consumption		
Disadvantages	Water loss due to high evaporation rate	Scale-up requires many compartments and support materials	Requires large land area because long tubes are used	Small illumination surface area
	Difficulty in controlling temperature and pH	Difficulty in controlling culture temperature	Potential in accumulating high concentration of O_2 (poison to microalgae) in culture medium if tubes are too long	Cell sedimentation may occur if air-lift system is not used
	Susceptible to contamination		Decreasing CO_2 concentration along the tube may cause microalgae to be deprived of carbon source	
			Mixing is problematic in extended tubes	

the energy is used for powering pumps so that sufficient mixing and optimum gas–liquid transfer are achieved. Based on current technology maturity, an air-lift photobioreactor is not up to a commercialization stage yet, unless proper modifications are made to reduce the overall operating energy consumption.

To overcome the limitations of open and close photobioreactors, integrating both types of photobioreactors into one system may be a possible solution for commercial H$_2$ production from microalgae. For the first stage, microalgae are cultivated in a nutrient-rich medium in a raceway pond to promote the growth of microalgae until a certain quantity of cells is achieved. In the subsequent stage, the microalgae cells are harvested and transferred immediately to a sulfate-deficient medium in a closed photobioreactor to permit H$_2$ production under anaerobic conditions. To achieve anaerobic conditions, the cultivation must be flushed initially with N$_2$ and incubated in a dark environment for several hours before exposing to light sources to initiate H$_2$ evolution. Mixing of a microalgae culture utilizing air bubbles is no longer possible in the second stage because of the need to have an anaerobic environment. Thus, a mechanical agitator in the closed photobioreactor is needed to provide homogeneous mixing and to allow microalgae to receive sufficient light (Giannelli et al., 2009; Rosello Sastre et al., 2007). In this regard, a vertical column-type photobioreactor is the most plausible option, as it allows easy installation of a mechanical agitator in the center of the column and facilitates H$_2$ collection at the headspace. Hence, choosing a suitable agitator and optimizing the agitation speed are important in minimizing the shear stress on microalgae cells that may subsequently affect the H$_2$ production rate (Sobczuk et al., 2006; Ugwu et al., 2008). Other types of closed photobioreactors, such as flat plat, helical tubular, and α shaped, are also suitable for cultivating microalgae, but are expected to consume more energy because of the difficulty of installing a mechanical agitator in these photobioreactors, resulting in the need for a heavy-duty water pump to circulate the culture to avoid settling.

Harvesting Technology

To promote H$_2$ evolution through direct biophotolysis, microalgae need to be harvested from the nutrient-rich medium before transferring to the sulfate-limited medium. Harvesting microalgae is challenging because they are small (generally, 3–30 μm) and suspended in liquid (Molina Grima et al., 2003). Centrifugation and filtration are among the most effective methods for harvesting microalgae cells and are applied commonly at the laboratory scale. However, it is not feasible to implement these two methods on a commercial scale because of their high-energy consumption and high capital and maintenance costs (Lam and Lee, 2012). A LCA study has highlighted that using filtration and centrifugation to harvest microalgae biomass contributed 88.6 and 92.7%, respectively, to the entire energy input of the system boundary (Sander and Murthy, 2010). This important finding should not be ignored, as it may have a significant impact on the overall energy balance in producing H$_2$ from microalgae.

Coagulation and flocculation offer an alternative and relatively low-energy method to harvest microalgae cells. Microalgae cells always carry negative charges, which cause them to repel each other and remain suspended in liquid even if mixing is not provided. The negative charges surrounding the microalgae cells can be neutralized by introducing a positively charged coagulant into the culture medium. At the same time, a flocculant can also be added to promote agglomeration by creating bridges between the neutralized cells, creating dense flocs that settle naturally due to gravity (de Godos et al., 2011). Nevertheless, high doses of conventional flocculants (e.g., multivalent salts) are required to achieve satisfactory results, which may cause contamination of microalgae cells (Chen et al., 2011; Renault et al., 2009). In addition, some of the microalgae cells may not survive when being transferred to a new medium with a limited sulfate content and, consequently, affect the H_2 production rate. Although no scientific research work or assessment has been carried out to justify this hypothesis, flocculant toxicity should not be ignored, especially when harvested microalgae cells are used for the second phase of cultivation. Nevertheless, a handful of researches are currently focused on the development of bioflocculants and immobilization technologies to reduce toxicity effects after harvesting and permit longer H_2 evolution from microalgae (Hahn et al., 2007; Kosourov and Seibert, 2009; Laurinavichene et al., 2006; Lee et al., 2009; Renault et al., 2009).

Immobilization of microalgae for H_2 production was described by Hahn et al. (2007). C. reinhardtii was immobilized onto fumed silica particles to produce H_2 in a two-step cycle. Immobilized microalgae were initially grown in a sulfate-rich environment to promote cell growth in the silica particles, and then the silica particles were filtered and moved to a sulfate-free environment for H_2 evolution. Because the silica particles are much denser than water and can be filtered easily, the microalgae cells were transferred readily to the sulfate-deficient medium without requiring much energy input. Results from the study indicated that H_2 evolution from microalgae was not affected by immobilization, and the H_2 production rate was nearly the same as that of free cells. In fact, this immobilization technology can be integrated easily into existing photobioreactor designs with little or no modification (Hahn et al., 2007). This technology could eliminate high-energy harvesting methods such as centrifugation, leading to a better energy balance for H_2 production from microalgae.

Another important finding regarding the immobilization of microalgae for H_2 production was reported by Kosourov and Seibert (2009). From their study, immobilization of C. reinhardtii within thin alginate films exhibited an extraordinary observation in which the microalgae were reported to produce H_2 even under atmospheric conditions (aerobic) instead of anaerobic conditions. This could be because the alginate films acted as shields that restrict O_2 diffusion into the microalgae cells, thus protecting the hydrogenases from deactivation (Kosourov and Seibert, 2009; Sabra et al., 2000). However, it was found that when a high density of cells was embedded within the alginate films, the H_2 production rate was reduced significantly. One possible explanation for this observation was that a high population of microalgae cells increased the porosity of alginate films (Kosourov and Seibert, 2009), resulting in more O_2 diffusing into the alginate films and inhibiting hydrogenases from producing H_2. On the contrary, higher H_2 production rates were observed in alginate films with a low cell density, suggesting a more efficient light utilization per unit matrix area by the microalgae and the elimination of any shading effects.

Process Integration with Biodiesel Production from Microalgae

To date, it has not been feasible to commercialize H_2 production from microalgae because of the low volume of H_2 produced, high costs, high energy input, and sensitivity of microalgae toward the surrounding environment for inducing or promoting H_2 production. Therefore, integration with a H_2 production system from bacteria is suggested to enhance H_2 productivity and to drive a positive energy balance for the system (see earlier discussion). It is also possible to improve the process energy balance for the microalgae culture by expanding and diversifying biofuel production from the microalgae biomass.

Apart from carbohydrates and proteins, certain microalgae strains are able to accumulate significant amount of lipids inside their cells, in which these lipids can be extracted and converted to biodiesel. Biodiesel is currently recognized as a green and alternative renewable diesel fuel that has attracted vast interest from researchers, governments, and local and international traders. Some of the advantages of using biodiesel instead of fossil diesel are that it is a nontoxic fuel, biodegradable, and has a lower emission of greenhouse gas when burned in diesel engines (Demirbas, 2009). H_2-producing microalgae such as *C. reinhardtii* could accumulate lipids up to a content of 25% (Kong et al., 2010; Morowvat et al., 2010), which makes it a potential microalgae strain for diversifying the types of biofuel that can be produced. In other words, after *C. reinhardtii* has stopped producing H_2, the microalgae cells can be harvested, dried, and followed by subsequent lipid extraction for biodiesel production. Various lipid extraction methods have been reported, including mechanical press, chemical solvent extraction, and supercritical fluid extraction. However, only chemical solvent extraction (hexane, methanol, chloroform, and mixed solvent) has demonstrated the highest efficiency due to high selectivity and solubility of the solvent toward lipids. This allows even interlipids to be extracted out through diffusion across microalgae cell walls (Ranjan et al., 2010). The extracted lipids are then subjected to a transesterification reaction in which biodiesel is produced as the main product while glycerol is the by-product (Lam et al., 2010). Through this strategy, the overall process chain to produce H_2 from microalgae would become more sustainable and feasible in term of energy balance and production cost. Even though research on this integrated system is limited, there are a number of potential advantages to this process.

CONCLUSIONS

Photosynthetic microalgae may prove to be an appealing alternative for biological H_2 production because of their natural ability to capture solar energy and store it as chemical energy for water splitting. Significant breakthroughs have been made to improve the H_2 productivity from microalgae, including genetic engineering and cultivation methods. *C. reinhardtii* appears to be the most promising strain to produce H_2 through direct biophotolysis when cultivated in a sulfate-limited medium. However, the H_2 production rate is still far lower compared to chemical-based process such as steam methane reforming, which impedes the commercialization potential of this technology. Several issues and factors need to be considered carefully before scaling up microalgae-based H_2 production, such as nutrient source, photobioreactor design, and harvesting methods. Integrated systems with diversified biofuel (e.g., biodiesel) production from microalgae may be a possible solution to improving the overall life cycle energy balance and economic feasibility of this industry instead of solely producing H_2.

Acknowledgment

The authors acknowledge the funding given by Universiti Sains Malaysia (Research University Grant No.814146, Postgraduate Research Grant Scheme No. 8044031 and USM Vice-Chancellor's Award) for facilitating the write-up of this chapter.

References

Akkerman, I., Janssen, M., Rocha, J., Wijffels, R.H., 2002. Photobiological hydrogen production: Photochemical efficiency and bioreactor design. Int. J. Hydrogen Energy 27, 1195–1208.

Alvarenga, R.A.F.D., da Silva Júnior, V.P., Soares, S.R., 2012. Comparison of the ecological footprint and a life cycle impact assessment method for a case study on Brazilian broiler feed production. J. Cleaner Prod. 28, 25–32.

Balat, M., 2008. Potential importance of hydrogen as a future solution to environmental and transportation problems. Int. J. Hydrogen Energy 33, 4013–4029.

Benemann, J.R., 2000. Hydrogen production by microalgae. J. Appl. Phycol. 12, 291–300.

Brennan, L., Owende, P., 2010. Biofuels from microalgae: A review of technologies for production, processing, and extractions of biofuels and co-products. Renew. Sustain. Energy Rev. 14, 557–577.

Brentner, L.B., Jordan, P.A., Zimmerman, J.B., 2010. Challenges in developing biohydrogen as a sustainable energy source: Implications for a research agenda. Environ. Sci. Technol. 44, 2243–2254.

Campbell, P.K., Beer, T., Batten, D., 2011. Life cycle assessment of biodiesel production from microalgae in ponds. Bioresour. Technol. 102, 50–56.

Chader, S., Hacene, H., Agathos, S.N., 2009. Study of hydrogen production by three strains of *Chlorella* isolated from the soil in the Algerian Sahara. Int. J. Hydrogen Energy 34, 4941–4946.

Chen, C.Y., Yeh, K.L., Aisyah, R., Lee, D.J., Chang, J.S., 2011. Cultivation, photobioreactor design and harvesting of microalgae for biodiesel production: A critical review. Bioresour. Technol. 102, 71–81.

Chisti, Y., 2007. Biodiesel from microalgae. Biotechnol. Adv. 25, 294–306.

Clarens, A.F., Resurreccion, E.P., White, M.A., Colosi, L.M., 2010. Environmental life cycle comparison of algae to other bioenergy feedstocks. Environ. Sci. Technol. 44, 1813–1819.

Collet, P., Hélias Arnaud, A., Lardon, L., Ras, M., Goy, R.A., Steyer, J.P., 2011. Life-cycle assessment of microalgae culture coupled to biogas production. Bioresour. Technol. 102, 207–214.

Das, D., Veziroğlu, T.N., 2001. Hydrogen production by biological processes: A survey of literature. Int. J. Hydrogen Energy 26, 13–28.

Dasgupta, C.N., Jose Gilbert, J., Lindblad, P., Heidorn, T., Borgvang, S.A., Skjanes, K., et al., 2010. Recent trends on the development of photobiological processes and photobioreactors for the improvement of hydrogen production. Int. J. Hydrogen Energy 35, 10218–10238.

de Godos, I., Guzman, H.O., Soto, R., García-Encina, P.A., Becares, E., Muñoz, R., et al., 2011. Coagulation/flocculation-based removal of algal–bacterial biomass from piggery wastewater treatment. Bioresour. Technol. 102, 923–927.

Demirbas, A., 2009. Progress and recent trends in biodiesel fuels. Energy Convers. Manage. 50, 14–34.

Efremenko, E.N., Nikolskaya, A.B., Lyagin, I.V., Senko, O.V., Makhlis, T.A., Stepanov, N.A., et al., 2012. Production of biofuels from pretreated microalgae biomass by anaerobic fermentation with immobilized *Clostridium acetobutylicum* cells. Bioresour. Technol. 114, 342–348.

Eroglu, E., Melis, A., 2011. Photobiological hydrogen production: Recent advances and state of the art. Bioresour. Technol. 102, 8403–8413.

Fan, M.S., Abdullah, A.Z., Bhatia, S., 2009. Catalytic technology for carbon dioxide reforming of methane to synthesis gas. ChemCatChem 1, 192–208.

Faraloni, C., Ena, A., Pintucci, C., Torzillo, G., 2011. Enhanced hydrogen production by means of sulfur-deprived *Chlamydomonas reinhardtii* cultures grown in pretreated olive mill wastewater. Int. J. Hydrogen Energy 36, 5920–5931.

Fedorov, A.S., Kosourov, S., Ghirardi, M.L., Seibert, M., 2005. Continuous hydrogen photoproduction by *Chlamydomonas reinhardtii*: Using a novel two-stage, sulfate-limited chemostat system. Appl. Biochem. Biotechnol. Enzyme Eng. Biotechnol. 121, 403–412.

Flynn, T., Ghirardi, M.L., Seibert, M., 2002. Accumulation of O_2-tolerant phenotypes in H_2-producing strains of *Chlamydomonas reinhardtii* by sequential applications of chemical mutagenesis and selection. Int. J. Hydrogen Energy 27, 1421–1430.

Fox, P., Venkatasubbiah, V., 1996. Coupled anaerobic/aerobic treatment of high-sulfate wastewater with sulfate reduction and biological sulfide oxidation. Water Sci. Technol. 34, 359–366.

Gaffron, H., Rubin, J., 1942. Fermentative and photochemical production of hydrogen in algae. J. Gen. Physiol. 26, 219–240.

Gest, H., Kamen, M.D., 1949. Studies on the metabolism of photosynthetic bacteria. IV. Photochemical production of molecular hydrogen by growing cultures of photosynthetic bacteria. J. Bacteriol. 59, 239–245.

Ghirardi, M.L., Zhang, L., Lee, J.W., Flynn, T., Seibert, M., Greenbaum, E., et al., 2000. Microalgae: A green source of renewable H_2. Trends Biotechnol. 18, 506–511.

Ghirardi, M.L., Dubini, A., Yu, J., Maness, P.C., 2009. Photobiological hydrogen-producing systems. Chem. Soc. Rev. 38, 52–61.

Giannelli, L., Scoma, A., Torzillo, G., 2009. Interplay between light intensity, chlorophyll concentration and culture mixing on the hydrogen production in sulfur-deprived *Chlamydomonas reinhardtii* cultures grown in laboratory photobioreactors. Biotechnol. Bioeng. 104, 76–90.

Greenbaum, E., 1988. Energetic efficiency of hydrogen photoevolution by algal water splitting. Biophys. J. 54, 365–368.

Greenbaum, E., Guillard, R.R.L., Sunda, W.G., 1983. Hydrogen and oxygen photoproduction by marine algae. Photochem. Photobiol. 37, 649–655.

Greenwell, H.C., Laurens, L.M.L., Shields, R.J., Lovitt, R.W., Flynn, K.J., 2010. Placing microalgae on the biofuels priority list: A review of the technological challenges. J. R. Soc. Interface 7, 703–726.

Guan, Y., Deng, M., Yu, X., Zhang, W., 2004. Two-stage photo-biological production of hydrogen by marine green alga *Platymonas subcordiformis*. Biochem. Eng. J. 19, 69–73.

Hahn, J.J., Ghirardi, M.L., Jacoby, W.A., 2007. Immobilized algal cells used for hydrogen production. Biochem. Eng. J. 37, 75–79.

Hallenbeck, P.C., Benemann, J.R., 2002. Biological hydrogen production; Fundamentals and limiting processes. Int. J. Hydrogen Energy 27, 1185–1193.

Hallenbeck, P.C., Ghosh, D., 2009. Advances in fermentative biohydrogen production: The way forward? Trends Biotechnol. 27, 287–297.

Hallenbeck, P.C., Abo-Hashesh, M., Ghosh, D., 2012. Strategies for improving biological hydrogen production. Bioresour. Technol. 110, 1–9.

Happe, T., Hemschemeier, A., Winkler, M., Kaminski, A., 2002. Hydrogenases in green algae: Do they save the algaes life and solve our energy problems? Trends Plant Sci. 7, 246–250.

He, M., Li, L., Liu, J., 2012. Isolation of wild microalgae from natural water bodies for high hydrogen producing strains. Int. J. Hydrogen Energy 37, 4046–4056.

Hilton, M.G., Archer, D.B., 1988. Anaerobic digestion of a sulfate-rich molasses wastewater: Inhibition of hydrogen sulfide production. Biotechnol. Bioeng. 31, 885–888.

Jin, H., Xu, Y., Lin, R., Han, W., 2008. A proposal for a novel multi-functional energy system for the production of hydrogen and power. Int. J. Hydrogen Energy 33, 9–19.

Jo, J.H., Lee, D.S., Park, J.M., 2006. Modeling and optimization of photosynthetic hydrogen gas production by green alga *Chlamydomonas reinhardtii* in sulfur-deprived circumstance. Biotechnol. Prog. 22, 431–437.

Jorquera, O., Kiperstok, A., Sales, E.A., Embiruçu, M., Ghirardi, M.L., 2010. Comparative energy life-cycle analyses of microalgal biomass production in open ponds and photobioreactors. Bioresour. Technol. 101, 1406–1413.

Kapdan, I.K., Kargi, F., 2006. Bio-hydrogen production from waste materials. Enzyme Microbial Technol. 38, 569–582.

Kessel, D.G., 2000. Global warming: Facts, assessment, countermeasures. J. Pet. Sci. Eng. 26, 157–168.

Kim, D.H., Kim, M.S., 2011. Hydrogenases for biological hydrogen production. Bioresour. Technol. 102, 8423–8431.

Kirst, H., García-Cerdán, J.G., Zurbriggen, A., Melis, A., 2012. Assembly of the light-harvesting chlorophyll antenna in the green alga *Chlamydomonas reinhardtii* requires expression of the TLA2-CpFTSY gene. Plant Physiol. 158, 930–945.

Kong, Q.X., Li, L., Martinez, B., Chen, P., Ruan, R., 2010. Culture of microalgae *Chlamydomonas reinhardtii* in wastewater for biomass feedstock production. Appl. Biochem. Biotechnol. 160, 9–18.

Kosourov, S.N., Seibert, M., 2009. Hydrogen photoproduction by nutrient-deprived *Chlamydomonas reinhardtii* cells immobilized within thin alginate films under aerobic and anaerobic conditions. Biotechnol. Bioeng. 102, 50–58.

Kosourov, S.N., Ghirardi, M.L., Seibert, M., 2011. A truncated antenna mutant of *Chlamydomonas reinhardtii* can produce more hydrogen than the parental strain. Int. J. Hydrogen Energy 36, 2044–2048.

Kruse, O., Rupprecht, J., Bader, K.P., Thomas-Hall, S., Schenk, P.M., Finazzi, G., et al., 2005. Improved photobiological H_2 production in engineered green algal cells. J. Biol. Chem. 280, 34170–34177.

Kulkarni, M.G., Dalai, A.K., 2006. Waste cooking oil: An economical source for biodiesel. Ind. Eng. Chem. Res. 45, 2901–2913.

Lam, M.K., Lee, K.T., 2011. Renewable and sustainable bioenergies production from palm oil mill effluent (POME): Win-win strategies toward better environmental protection. Biotechnol. Adv. 29, 124–141.

Lam, M.K., Lee, K.T., 2012. Microalgae biofuels: A critical review of issues, problems and the way forward. Biotechnol. Adv. 30, 673–690.

Lam, M.K., Lee, K.T., Rahmanmohamed, A., 2009. Life cycle assessment for the production of biodiesel: A case study in Malaysia for palm oil versus jatropha oil. Biofuels Bioprod. Biorefin. 3, 601–612.

Lam, M.K., Lee, K.T., Mohamed, A.R., 2010. Homogeneous, heterogeneous and enzymatic catalysis for transesterification of high free fatty acid oil (waste cooking oil) to biodiesel: A review. Biotechnol. Adv. 28, 500–518.

Lardon, L., Hélias, A., Sialve, B., Steyer, J.P., Bernard, O., 2009. Life-cycle assessment of biodiesel production from microalgae. Environ. Sci. Technol. 43, 6475–6481.

Laurinavichene, T.V., Fedorov, A.S., Ghirardi, M.L., Seibert, M., Tsygankov, A.A., 2006. Demonstration of sustained hydrogen photoproduction by immobilized, sulfur-deprived *Chlamydomonas reinhardtii* cells. Int. J. Hydrogen Energy 31, 659–667.

Lee, J.W., Greenbaum, E., 2003. A new oxygen sensitivity and its potential application in photosynthetic H_2 production. Appl. Biochem. Biotechnol. Enzyme Eng. Biotechnol. 106, 303–314.

Lee, A., Lewis, D., Ashman, P., 2009. Microbial flocculation, a potentially low-cost harvesting technique for marine microalgae for the production of biodiesel. J. Appl. Phycol. 21, 559–567.

Levin, D.B., Pitt, L., Love, M., 2004. Biohydrogen production: Prospects and limitations to practical application. Int. J. Hydrogen Energy 29, 173–185.

Ley, A.C., Mauzerall, D.C., 1982. Absolute absorption cross-sections for photosystem II and the minimum quantum requirement for photosynthesis in *Chlorella vulgaris*. BBA Bioenerg. 680, 95–106.

Manish, S., Banerjee, R., 2008. Comparison of biohydrogen production processes. Int. J. Hydrogen Energy 33, 279–286.

Mata, T.M., Martins, A.A., Caetano, N.S., 2010. Microalgae for biodiesel production and other applications: A review. Renew. Sustain. Energy Rev. 14, 217–232.

Mathews, J., Wang, G., 2009. Metabolic pathway engineering for enhanced biohydrogen production. Int. J. Hydrogen Energy 34, 7404–7416.

Maurer, M., Schwegler, P., Larsen, T.A., 2003. Nutrients in urine: Energetic aspects of removal and recovery. Water Sci. Technol. 48, 37–46.

Melis, A., 2007. Photosynthetic H_2 metabolism in *Chlamydomonas reinhardtii* (unicellular green algae). Planta 226, 1075–1086.

Melis, A., 2009. Solar energy conversion efficiencies in photosynthesis: Minimizing the chlorophyll antennae to maximize efficiency. Plant Sci. 177, 272–280.

Melis, A., Happe, T., 2001. Hydrogen production: Green algae as a source of energy. Plant Physiol. 127, 740–748.

Melis, A., Zhang, L., Forestier, M., Ghirardi, M.L., Seibert, M., 2000. Sustained photobiological hydrogen gas production upon reversible inactivation of oxygen evolution in the green alga *Chlamydomonas reinhardtii*. Plant Physiol. 122, 127–135.

Melis, A., Seibert, M., Ghirardi, M.L., 2007. Hydrogen fuel production by transgenic microalgae. In: León, R., Galván, A., Fernández, E. (Eds.), Transgenic Microalgae as Green Cell Factories. Springer, New York, pp. 110–121.

Molina Grima, E., Belarbi, E.H., Acién Fernández, F.G., Robles Medina, A., Chisti, Y., 2003. Recovery of microalgal biomass and metabolites: Process options and economics. Biotechnol. Adv. 20, 491–515.

Morowvat, M.H., Rasoul-Amini, S., Ghasemi, Y., 2010. Chlamydomonas as a "new" organism for biodiesel production. Bioresour. Technol. 101, 2059–2062.

Mussgnug, J.H., Klassen, V., Schlüter, A., Kruse, O., 2010. Microalgae as substrates for fermentative biogas production in a combined biorefinery concept. J. Biotechnol. 150, 51–56.

Nguyen, T.A.D., Kim, K.R., Nguyen, M.T., Kim, M.S., Kim, D., Sim, S.J., 2010. Enhancement of fermentative hydrogen production from green algal biomass of *Thermotoga neapolitana* by various pretreatment methods. Int. J. Hydrogen Energy 35, 13035–13040.

Oh, Y.K., Raj, S.M., Jung, G.Y., Park, S., 2011. Current status of the metabolic engineering of microorganisms for biohydrogen production. Bioresour. Technol. 102, 8357–8367.

Ohta, S., Miyamoto, K., Miura, Y., 1987. Hydrogen evolution as a consumption mode of reducing equivalents in green algal fermentation. Plant Physiol. 83, 1022–1026.

Pilon, L., Berberoğlu, H., Kandilian, R., 2011. Radiation transfer in photobiological carbon dioxide fixation and fuel production by microalgae. J. Quant. Spectrosc. Radiat. Transfer 112, 2639–2660.

Pittman, J.K., Dean, A.P., Osundeko, O., 2011. The potential of sustainable algal biofuel production using wastewater resources. Bioresour. Technol. 102, 17–25.

Polle, J.E.W., Benemann, J.R., Tanaka, A., Melis, A., 2000. Photosynthetic apparatus organization and function in the wild type and a chlorophyll b-less mutant of *Chlamydomonas reinhardtii*: Dependence on carbon source. Planta 211, 335–344.

Polle, J.E.W., Niyogi, K.K., Melis, A., 2001. Absence of lutein, violaxanthin and neoxanthin affects the functional chlorophyll antenna size of photosystem-II but not that of photosystem-I in the green alga *Chlamydomonas reinhardtii*. Plant Cell Physiol. 42, 482–491.

Polle, J.E.W., Kanakagiri, S., Jin, E., Masuda, T., Melis, A., 2002. Truncated chlorophyll antenna size of the photosystems: A practical method to improve microalgal productivity and hydrogen production in mass culture. Int. J. Hydrogen Energy 27, 1257–1264.

Polle, J.E.W., Kanakagiri, S.D., Melis, A., 2003. Tla1, a DNA insertional transformant of the green alga *Chlamydomonas reinhardtii* with a truncated light-harvesting chlorophyll antenna size. Planta 217, 49–59.

Prince, R.C., Kheshgi, H.S., 2005. The photobiological production of hydrogen: Potential efficiency and effectiveness as a renewable fuel. Crit. Rev. Microbiol. 31, 19–31.

Ranjan, A., Patil, C., Moholkar, V.S., 2010. Mechanistic assessment of microalgal lipid extraction. Ind. Eng. Chem. Res. 49, 2979–2985.

Razon, L.F., Tan, R.R., 2011. Net energy analysis of the production of biodiesel and biogas from the microalgae: *Haematococcus pluvialis* and *Nannochloropsis*. Appl. Energy 88, 3507–3514.

Renault, F., Sancey, B., Badot, P.M., Crini, G., 2009. Chitosan for coagulation/flocculation processes: An eco-friendly approach. Eur. Polym. J. 45, 1337–1348.

Renó, M.L.G., Lora, E.E.S., Palacio, J.C.E., Venturini, O.J., Buchgeister, J., Almazan, O., 2011. A LCA (life cycle assessment) of the methanol production from sugarcane bagasse. Energy 36, 3716 3726.

Romagnoli, F., Blumberga, D., Pilicka, I., 2011. Life cycle assessment of biohydrogen production in photosynthetic processes. Int. J. Hydrogen Energy 36, 7866–7871.

Rosello Sastre, R., Csögör, Z., Perner-Nochta, I., Fleck-Schneider, P., Posten, C., 2007. Scale-down of microalgae cultivations in tubular photo-bioreactors: A conceptual approach. J. Biotechnol. 132, 127–133.

Sabra, W., Zeng, A.P., Lunsdorf, H., Deckwer, W.D., 2000. Effect of oxygen on formation and structure of *Azotobacter vinelandii* alginate and its role in protecting nitrogenase. Appl. Environ. Microbiol. 66, 4037–4044.

Sander, K., Murthy, G.S., 2010. Life cycle analysis of algae biodiesel. Int. J. Life Cycle Assess. 15, 704–714.

Selcuk, H., 2005. Decolorization and detoxification of textile wastewater by ozonation and coagulation processes. Dyes Pigments 64, 217–222.

Sierra, E., Acién, F.G., Fernández, J.M., García, J.L., González, C., Molina, E., 2008. Characterization of a flat plate photobioreactor for the production of microalgae. Chem. Eng. J. 138, 136–147.

Silva, A.J., Domingues, M.R., Hirasawa, J.S., Varesche, M.B., Foresti, E., Zaiat, M., 2011. Kinetic modeling and microbial assessment by fluorescent in situ hybridization in anaerobic sequencing batch biofilm reactors treating sulfate-rich wastewater. Braz. J. Chem. Eng. 28, 209–219.

Singh, J., Gu, S., 2010. Commercialization potential of microalgae for biofuels production. Renew. Sustain. Energy Rev. 14, 2596–2610.

Sobczuk, T., Camacho, F., Grima, E., Chisti, Y., 2006. Effects of agitation on the microalgae *Phaeodactylum tricornutum* and *Porphyridium cruentum*. Bioprocess Biosyst. Eng. 28, 243–250.

Spruit, C.P., 1958. Simultaneous photoproduction of hydrogen and oxygen by Chlorella. Meded Landbouwhogesch Wageningen 58, 1–17.

Srirangan, K., Pyne, M.E., Perry Chou, C., 2011. Biochemical and genetic engineering strategies to enhance hydrogen production in photosynthetic algae and cyanobacteria. Bioresour. Technol. 102, 8589–8604.

Stephenson, A.L., Kazamia, E., Dennis, J.S., Howe, C.J., Scott, S.A., Smith, A.G., 2010. Life-cycle assessment of potential algal biodiesel production in the united kingdom: A comparison of raceways and air-lift tubular bioreactors. Energy Fuels 24, 4062–4077.

Sun, J., Yuan, X., Shi, X., Chu, C., Guo, R., Kong, H., 2011. Fermentation of *Chlorella* sp. for anaerobic bio-hydrogen production: Influences of inoculum–substrate ratio, volatile fatty acids and NADH. Bioresour. Technol. 102, 10480–10485.

Tait, S., Clarke, W.P., Keller, J., Batstone, D.J., 2009. Removal of sulfate from high-strength wastewater by crystallisation. Water Res. 43, 762–772.

Torzillo, G., Scoma, A., Faraloni, C., Ena, A., Johanningmeier, U., 2009. Increased hydrogen photoproduction by means of a sulfur-deprived *Chlamydomonas reinhardtii* D1 protein mutant. Int. J. Hydrogen Energy 34, 4529–4536.

Tsygankov, A.A., Kosourov, S.N., Tolstygina, I.V., Ghirardi, M.L., Seibert, M., 2006. Hydrogen production by sulfur-deprived *Chlamydomonas reinhardtii* under photoautotrophic conditions. Int. J. Hydrogen Energy 31, 1574–1584.

Ugwu, C.U., Aoyagi, H., Uchiyama, H., 2008. Photobioreactors for mass cultivation of algae. Bioresour. Technol. 99, 4021–4028.

Wykoff, D.D., Davies, J.P., Melis, A., Grossman, A.R., 1998. The regulation of photosynthetic electron transport during nutrient deprivation in *Chlamydomonas reinhardtii*. Plant Physiol. 117, 129–139.

Xu, L., Weathers, P.J., Xiong, X.R., Liu, C.Z., 2009. Microalgal bioreactors: Challenges and opportunities. Eng. Life Sci. 9, 178–189.

Yoshida, A., Nishimura, T., Kawaguchi, H., Inui, M., Yukawa, H., 2005. Enhanced hydrogen production from formic acid by formate hydrogen lyase-overexpressing *Escherichia coli* strains. Appl. Environ. Microbiol. 71, 6762–6768.

Zhang, L., Happe, T., Melis, A., 2002. Biochemical and morphological characterization of sulfur-deprived and H_2-producing *Chlamydomonas reinhardtii* (green alga). Planta 214, 552–561.

Biohydrogen from Renewable Resources

Ganesh D. Saratale,[*,†] *Rijuta G. Saratale,*[‡] *and Jo-Shu Chang*[§]

[*]Department of Biochemistry, Shivaji University, Kolhapur, India
[†]Department of Environmental Biotechnology, Shivaji University, Kolhapur, India
[‡]Department of Biotechnology, Shivaji University, Kolhapur, India
[§]Department of Chemical Engineering, National Cheng Kung University,
Tainan, Taiwan

INTRODUCTION

Definition and Background of Biohydrogen

Biofuels are the only alternative energy source for the foreseeable future and can still form the basis of sustainable development in terms of socioeconomic and environmental concerns (Demirbas, 2007). Biofuels produced from renewable resources represent the features of CO_2 recycling, eco-friendly, cost competitive with fossil fuels, biodegradable, and sustainable. Hence, they become important and promising alternative energy sources for fossil fuels to protect the biosphere and prevent more localized forms of pollution (Puppan, 2002). Renewable resources are distributed more evenly than fossil and nuclear resources. Energy flows from renewable resources are more than three orders of magnitude higher than current global energy use. Although the worldwide annual production of biofuels has increased from 4.4 to 50.1 billion liters, political and public support for biofuels has been countermined. It has been reported that the use of food crops or croplands for biofuel production results in food shortages and increased prices of staple food crops such as maize and rice (James et al., 2008; Keeney and Hertel, 2008). Nevertheless, biofuels (particularly biohydrogen) from renewable carbon sources are particularly attractive based on bioresource sustainability, economic feasibility, and minimized food security and land usage issues (Ragauskas et al., 2006; Schubert, 2006; Slade et al., 2009).

Production of hydrogen from renewable resources such as water, organic wastes, or biomass, either fermentatively or photobiologically, is termed "biohydrogen" (Asada and

Miyake, 1999; Das and Veziroğlu, 2001; Ghirardi et al., 2000; Koku et al., 2002). Methods adopted to produce H_2 from biological methods can be classified into the following groups: (1) biophotolysis of water using algae/cyanobacteria, (2) photodecomposition (photofermentation) of organic compounds using photosynthetic bacteria, (3) dark fermentative hydrogen production using anaerobic or facultative anaerobic bacteria, and (4) bioelectrohydrogenesis or microbial fuel cell (MFC) (Asada and Miyake, 1999; Chen et al., 2008b; Das and Veziroğlu, 2001; Ghirardi et al., 2000; Kim et al., 2011; Koku et al., 2002). Mainly, three kinds of microorganisms are capable of hydrogen production: cyanobacteria or green algae, photosynthetic bacteria, and anaerobic bacteria. Cyanobacteria/green algae directly decompose water to hydrogen and oxygen in the presence of light energy by photosynthesis. Algal hydrogen production could be considered an economical and sustainable method in terms of water utilization as a renewable resource and CO_2 consumption as one of the air pollutants. This reaction requires only water and sunlight and generates oxygen for the earth. This is very attractive from the viewpoint of environmental protection. However, natural-borne organisms of these species examined so far show rather low rates of hydrogen production due to the complicated reaction systems and inhibition of the hydrogenase enzyme by oxygen (Ghirardi et al., 2000; Guan et al., 2004; Laurinavichene et al., 2002; Melis et al., 2000). In contrast, dark and photofermentations are considered to be more advantageous due to simultaneous waste treatment and hydrogen gas production. Photosynthetic bacteria utilize organic substrates such as organic acids instead of water as the starting compound for hydrogen production. Compared to algal hydrolysis, photosynthetic bacteria require much less free energy (+8.5 kJ/mol hydrogen for lactate) to produce hydrogen and they can completely degrade organic substances toward mineralization. However, this process requires high activation energy to drive nitrogenase, which is the enzyme responsible for hydrogen production in photosynthetic bacteria, and the consequence is low solar conversion efficiencies, typically not much higher than that for algal biophotolysis systems (Fascetti and Todini, 1995; He et al., 2005). In addition, phototrophic hydrogen production with photosynthetic bacteria is extremely sensitive to ammonia and oxygen contents, making it difficult in practical applications (Chen et al., 2008a). Dark fermentation seems to be more economically feasible for its capability of using complicated feedstock, but it has poor chemical oxygen demand (COD) removal efficiency, as its effluent requires further treatment prior to being discharged.

Biohydrogen production presumes paramount importance as an alternative and renewable bioenergy resource of hydrogen with low energy and high efficiency, thereby being considered a promising way of producing hydrogen (Hallenbeck, 2005). In addition, biohydrogen production also includes energy security reasons, environmental concerns, foreign exchange savings, socioeconomic issues related to the rural sectors of all countries in the world (Hallenbeck, 2005; Saratale et al., 2008; Wang and Wan, 2009). The key advantages of biological hydrogen production are (1) process catalyzed by microorganisms in an aqueous environment at ambient temperature and pressure, (2) inexpensive, (3) low energy requirement, and (4) well suited for decentralized energy production in small-scale installations in locations where biomass or wastes are available, thus avoiding energy expenditure and costs for transport (Lo et al., 2009; Saratale et al., 2010).

Over the past quarter century, many hundreds of publications have appeared on microbial H_2 production, but advances toward practical applications have been minimal (Hallenbeck and Benemann, 2002; Show et al., 2008). From an economic point of view, it is hard to expect that biohydrogen production can compete with chemically synthesized hydrogen in the next few decades (Akkerman et al., 2002; Kotay and Das, 2008). However, hydrogen production from waste materials or renewable resources using specific microorganisms with dark fermentation represents an economically feasible process and becomes an important area in bioenergy production. In the past, microbial communities in the bioreactor were usually regarded as a "black box." Nowadays, many reviews focusing on the fundamentals of biohydrogen production, metabolism, and bioreactor design has already been published (Akkerman et al., 2002; Hallenbeck and Benemann, 2002; Koku et al., 2002; Kotay and Das, 2008; Saratale et al., 2008, 2011; Vijayaraghavan and Soom, 2007; Wang and Wan, 2009).

Hydrogen Gas Production by Dark Fermentation

In this chapter, the focus is on biohydrogen generation from renewable resources as a feedstock. The technology used for converting renewable resources into hydrogen is mainly via the dark fermentation process, which is more feasible when dealing with complex carbon sources such as lignocellulosic feedstock, organic wastes, or wastewater. After the mid-1990s, much attention has been paid to hydrogen production by the anaerobic dark fermentation system, which has the best potential for practical applications (Levin et al. 2004). Dark fermentative hydrogen production have some basic advantages, including process simplicity on technical grounds, low energy requirements, higher rates of H_2 production, economically feasible or better process economy, and ability to generate hydrogen from a large number of carbohydrates (or other organic materials) obtained frequently as refuse or waste products. It can produce H_2 all day long without light. It produces valuable metabolites such as butyric, lactic, and acetic acids as by-products. Because it is an anaerobic process, there is no O_2 limitation problem (Benemann, 1996; Levin et al., 2004; Nandi and Sengupta, 1998). A variety of microbes can be used for dark hydrogen fermentation. They are mainly anaerobic bacteria that include *Clostridium* sp. and facultative anaerobes, for example, *Enterobacter* and *Bacillus* sp., as well as bacterial consortium from organic wastes, for example, anaerobic digester sludge, soil, and animal feces (Kapdan and Kargi, 2006). The major soluble metabolites from dark fermentation include volatile fatty acids and alcohols, and their further decomposition is not possible under anaerobic conditions (Fascetti et al., 1995; Hallenbeck and Benemann, 2002; Lee et al, 2006). Anaerobic bacteria utilize organic substances as the sole source of electrons and energy, converting them into hydrogen. Reactions involved in hydrogen production [Eqs. (1) and (2)] are rapid, and these processes are useful for treating large quantities of organic waste by using an appropriate fermenter.

$$\text{Glucose} + 2H_2O \rightarrow 2\text{Acetate} + 2CO_2 + 4H_2 \quad \Delta G = -184.2 \text{ kJ} \tag{1}$$

$$\text{Glucose} \rightarrow \text{Butyrate} + 2CO_2 + 2H_2 \quad \Delta G = -257.1 \text{ kJ} \tag{2}$$

Thus, the theoretically maximal hydrogen yield from dark fermentation is 4 mol H_2/mol glucose. In addition, because dark fermentation is only an incomplete degradation of organic substrates, the production of hydrogen gas is accompanied by the formation of acetate and/or butyrate with a stoichiometrical ratio of 2 mol of H_2 per mole of acetate or butyrate. The production cost of biohydrogen production by dark fermentation is 340 times lower than the photosynthetic process and thus is considered to be more viable commercially (Atif et al., 2005). Most recent studies on hydrogen production used pure isolated anaerobic bacteria as the hydrogen producer (Levin et al., 2006; Liu et al., 2008). In some cases, the process uses mixed microflora or acclimated sewage sludge for hydrogen production (Chen et al., 2001; Ueno et al., 1995). Use of less expensive raw materials and efficient biological hydrogen production processes will surely make them more competitive with conventional H_2 generation processes in the near future (Das and Veziroglu, 2001). Moreover, intensive hydrogen research work has already been carried out on the advancement of these processes, such as the development of genetically modified microorganisms for the utilization of different waste materials, metabolic engineering, improvement of the bioreactor design, and development of two-stage processes (dark and photofermentation, etc.) for higher H_2 production performance.

Biohydrogen-Producing Microorganisms

Pure Culture Systems

A wide range of microorganisms are capable of producing hydrogen via dark fermentation. A literature survey suggests that dark fermentative hydrogen production is manifested by diverse group bacteria, including facultative anaerobes, facultative aerobes, and obligate anaerobes. This includes strict anaerobes (*Clostridia*, methylotrophs, rumen bacteria, methanogenic bacteria, archaea), facultative anaerobes (*Escherichia coli*, *Enterobacter*, *Citrobacter*), and even aerobes (*Alcaligenes*, *Bacillus*). Among hydrogen-producing bacteria, *Clostridium* sp. and *Enterobacter* are the most widely studied. Mainly the obligate anaerobes and spore-forming organisms such as *Clostridium buytricum* (sweet potato starch) (Yokoi et al., 2001), *C. thermolacticum* (lactose) (Collet et al., 2004), *C. pasteurianum* (starch) (Liu and Shen, 2004), *C. paraputrificum* M-21 (chitinous waste) (Evvyernie et al., 2001), *C. butyricum* CGS5 (rice husk hydrolysate) (Lo et al., 2009), and *C. bifermentants* (wastewater sludge) (Wang et al., 2003b) are generating hydrogen gas during the exponential growth phase. A study of the microbial community of mesophilic hydrogen producing sludge showed the presence of *Clostridia* species up to 64.6%, indicating that these species are dominant microbes for hydrogen production (Fang et al., 2002). Hydrogen production by *Thermotogales* sp. and *Bacillus* sp. were detected in mesophilic acidogenic cultures (Shin et al., 2004). In anaerobic granular sludge, some anaerobic cultures such as *Actinomyces* sp. and *Porphyromonos* sp. have been detected along with *Clostridium* sp. and these strains showed a hydrogen yield in between 1 and 1.2 mmol/mol glucose when cultivated under anaerobic conditions (Oh et al., 2003). Facultative anaerobes such as *E. coli* and species of genus *Enterobacter*, such as *E. aerogenes* (Tanisho and Ishiwata, 1994; Yokoi et al., 2001) and *E. cloacae* (Kumar and Das, 2001; Kumar et al, 2001), have also been used for hydrogen production. *Enterobacter* species have the ability to metabolize carbohydrates and produce some valuable products, such as gaseous H_2 and CO_2, mixture of acids, ethanol, and 2,3-butanediol. The capacity of hydrogen production of *E. aerogenes* using different substrates has been widely studied (Fabiano and Perego, 2002; Palazzi et al., 2000; Rachman et al., 1998; Tanisho and Ishiwata, 1994; Yokoi et al., 1997). Enhancement of hydrogen production (2.2 mol H_2/mol glucose) using *E. cloacae* ITT-BY 08 has been reported earlier (Kumar and Das, 2001). Isolated *Klebsiella* sp. HE1 showing an ability to produce 2,3-butanediol, ethanol, and hydrogen using sucrose as a substrate under the dark fermentation process was reported (Wu et al., 2008).

In recent years, extensive research has also been carried out in hydrogen production at high temperature, using thermophilic or hyperthermophilic bacteria, as the increase of temperature in principle improves the reaction kinetics and also enhances the assimilation of complicated substrates, such as lignocellulosic materials or other waste biomass. The H_2-producing thermophiles that have been studied include *Caldicellulosiruptor saccharolyticus* (van Niel et al., 2002), *Thermoanaerobacterium thermosaccharolyticum* (Thong et al., 2008), *Thermotoga* sp. (de Vrije et al., 2002; Schroder et al., 1994), and *Desulfotomaculum geothermicum* (Shin et al., 2004). The use of hyperthermophilic bacteria for biohydrogen production has attracted increasing interest due to their many outstanding properties. For example, they can grow successfully at very high temperatures, thus making hydrogen fermentation less sensitive to contaminations by undesirable intruders. Moreover, hyperthermophiles show a better resistance to high hydrogen partial pressures (Nguyen et al., 2010), which is one of most inhibitory factors in fermentative hydrogen production processes. Hence, higher hydrogen yields could be achieved by thermophiles as a result of the reduction of undesired by-products as well as the repeatability of the process. Although using a pure culture for biohydrogen production has some advantages, pure cultures can be quite sensitive to contaminations and thus their use demands the presence of aseptic conditions, which increases the overall cost of the process significantly.

Mixed Microbial Cultures

Considering an engineering standpoint, the selection of mixed cultures is considered to be favorable for a full-scale application. This is due to the fact that the control and operation of the process are facilitated when no medium sterilization is required, thus reducing the overall cost, whereas it also allows for a broader choice of feedstock selection (Valdez-Vazquez et al., 2005). It was observed that a single microorganism has less hydrolytic activities required to degrade complex wastes for biohydrogen production. The degradation of organic matter in anaerobic environments by microbial consortia involves the cooperation of different species existing in the system and synergistic metabolic activity, which generate a stable, self-regulating fermentation. For this purpose, worldwide efforts are being made to design a suitable mixed microbial consortium capable of decomposing various organic waste streams and to improve the hydrogen production yields and rates by using these cultures. Therefore, in general, a mixed culture is more suitable for biohydrogen production from renewable feedstocks, such as biomass and organic wastes, which are more complicated in their chemical compositions.

Mixed consortia can be derived from a variety of different natural sources, such as sewage sludge (Chen et al., 2002; Noike and Mizuno, 2000), anaerobically digested sludge (Koutrouli et al., 2006; Sparling et al., 1997), acclimated sludge (Lin and Lay, 2005), compost (Khanal et al., 2004; Ueno, et al., 1995), animal manure (Lay et al., 2003b), and soil (Logan et al., 2002; Van Ginkel et al., 2001). Even indigenous microorganisms found in certain wastes (Antonopoulou et al., 2008a,b; Lay et al., 1999; Noike and Mizuno, 2000; Wang et al., 2003b) may be useful for the direct utilization of starch and cellulose containing agricultural wastes, food industry wastes, carbohydrate-rich industrial wastewaters, and organic wastewater for biohydrogen production (Datar et al., 2007; Kapdan and Kargi, 2006).

Hydrogen production using microbial consortia has some major considerations. For instance, the developed microbial consortium is likely to be quite complex, and the species composition of the consortium might vary with substrate. Therefore, despite the advantages of mixed cultures in terms of economical viability of a process, their use always lurks the possible predominance of nonhydrogen-producing species such as methanogens, homoacetogens, and lactic acid bacteria; a case that could eventually lead to the dramatic failure of the viability of the process. Before preparation of the inoculum of various sludges, different pretreatment methods, such as incubation at high temperatures or acidic conditions or a combination of these, are necessary to remove the methanogens (Valdez-Vazquez and Poggi-Varaldo, 2009). In terms of seed pretreatment, heat treatment is generally the most common practice (Lay et al., 1999; Shizas and Bagley, 2005). By subjecting seed cultures to high temperatures, only the spore-forming acidogenic microorganisms survive the thermal shock, whereas the methanogenic nonspore-forming bacteria die. The use of temperatures reported are 75 °C (Wang et al., 2007), 100 °C (Lin and Cheng, 2006; Mu et al., 2007; Ntaikou et al., 2009; Wang et al., 2007), or even 121 °C (Wang et al., 2003), with different incubation periods between 15 min and 2 h (Gavala et al., 2006; Lin and Cheng, 2006; Mu et al., 2007; Wang et al., 2007). Alternatively to heat treatment, an acid/base treatment of the seed has also been suggested as a possible way for ensuring the dominance of hydrogen-producing bacteria (Chen et al., 2002; Lee et al., 2007). The principle of this method is to maintain the seed microorganisms for a prolonged period of time at very acidic or very basic conditions, which would eventually lead to the removal of methanogens that cannot survive such extreme pH values. These methods favor the survival of spore-forming *Clostridia* and are found to be effective in enriching for hydrogen producers.

RENEWABLE RESOURCES FOR BIOHYDROGEN PRODUCTION

Biomass Resources

Consideration of biomass resources, as well as their classification, is based on fuel quality and conversion technology. The eligible sources of biomass are crop residues, agricultural waste, forest residues, waste from processing of agricultural products, livestock residues, energy crops, algae biomass, wood waste, food and food processing waste, biomass-based components of municipal solid waste, and industrial waste. Agricultural residues include a wide range of plant materials produced along with the main product of the crop. Various raw materials, including corn stover (Cao et al., 2009; Datar et al., 2007), wheat bran (Noike et al., 2003), wheat straw (Kongjan and Angelidaki, 2010), rice straw (Lo et al., 2010), rice bran (Noike et al., 2003), sweet sorghum (Panagiotopoulos et al., 2010), potato steam peels (Mars et al., 2010), cassava stillage (Luo et al., 2010), sugarcane bagasse (Pattra et al., 2008; Saratale et al., 2010), and beer lees (Cui et al., 2010), can be utilized for hydrogen production. Livestock wastes include wet animal manure for biogas production and dry manure such as poultry litter that can be used in thermochemical conversion technologies. Several energy crops include perennial grasses (e.g., *Miscanthus*, switchgrass, giant reed, cardoon) for the production of lignocellulosic material, sugar, and starch crops. The average majority of biomass energy is produced from wood and wood wastes (64%), followed by municipal solid waste (24%), agricultural waste (5%), and landfill gases (5%) (Demirbas, 2010). Research in the United States suggests that average yields of 60 GJ/ha (net) could be achieved in some regions using biomass (Demirbas, 2007). Estimations suggest that average yields could reach nearly 37 dry t/ha in 2050, without the need for genetically modified crops, being around two and a half times today's average yield. To be acceptable, biofuel feedstocks must be produced sustainably in terms of agricultural practices, forest management, protection of biodiverse ecosystems, responsible and efficient use of water, and free of exploitation of landowners and importantly do not compete with food and fiber. However, care should be taken to ensure that biofuels should benefit the national economy of a developing country and also support the poorest people mainly in rural areas. Converting lignocellulosic materials into biohydrogen has been a hot topic recently as it addresses the way of producing biohydrogen in an economic and sustainable way (Cheng et al. 2011; Saratale et al., 2008, 2011). However, due to the complexity in the chemical structure of such lignocellulosic feedstock, efficient pretreatment and process design should be done to improve the yield and productivity of the cellulosic biohydrogen (Cheng et al., 2011).

Organic Wastes or Wastewaters

The biotransformation of wastes and wastewater toward hydrogen can be considered quite significant concerning both environmental (pollution control, renewable energy) and economical (resources recovery, low total cost waste management) parameters. Criteria according to which waste/wastewater would be characterized as an efficient feedstock for hydrogen generation include a high concentration of degradable organic compounds, a high proportion of readily fermentable compounds such as sugars, and a low concentration of inhibitor concentration to microbiological growth and activity. Many types of organic wastes and wastewaters can be used for biohydrogen production via anaerobic fermentation, such as biomass-based components of municipal solid wastes (MSW), food processing waste, dairy wastes, palm oil mill waste, and glycerol waste.

PRETREATMENT OF BIOMASS AND ORGANIC WASTES

Purposes of the pretreatment of biomass are to remove lignin and hemicellulose, reduce cellulose crystallinity, and increase the porosity of the materials (McMillan, 1994). Features of an effective pretreatment strategy include breaking the lignocellulosic complex; decreasing cellulose crystallinity; preserving hemicellulose sugars; limiting the formation of degradation products that are inhibitory to hydrolysis and fermentation; minimizing energy inputs and use of extraneous chemicals; requiring a simple setup; minimizing the production of toxic and hazardous wastes; and generating a minimum amount of wastewater and cost-effective treatment. Literature reports a number of pretreatment options that have been tried for various biomass types (Kumar et al., 2008; Saratale et al., 2008; Zhang et al., 2009). Several methods have been used to treat cellulosic feedstock (polysaccharides to corresponding

TABLE 1   Biomass Feedstock for Hydrogen Production Using Different Conversion Technologies and Its Merits and Demerits

Biomass feedstock used	Major conversion technology	Merits	Demerits
Almond shell Pine sawdust Crumb rubber Rice straw Danish wheat straw	Thermochemical gasification	Maximum conversion can be achieved	Significant gas conditioning is required Removal of tars is important
Microalgae Tea waste Peanut shell Maple sawdust slurry	Pyrolysis	Produces bio-oil, which is the basis of several processes for developing fuels, chemicals, and materials	Chances of catalyst deactivation
Starch biomass slurry Composted municipal refuse	Solar gasification	Good hydrogen yield	Requires effective collector plates
Kraft lignin MSW	Superficial conversion	Can process sewage sludge, which is difficult to gasify	Selection of superentical medium
Paper and pulp waste	Microbial conversion	Wastewater can also be treated simultaneously; also generates some useful secondary metabolites	Selection of suitable microorganism

monomers) and each generates a different pretreatment product stream (Table 1) (Chandrakant and Bisaria, 1998).

In addition to pretreatment for lignocellulosic materials, waste-activated sludge from municipal wastewater treatment plants contains high levels of organic matter (thus is a potential substrate for hydrogen production), which requires pretreatment to improve its assimilability. After appropriate pretreatments such as ultrasonication, acidification, freezing and thawing, sterilization, methanogenic inhibition, and microwaving, the ability of hydrogen-producing bacteria to produce hydrogen from waste-activated sludge can be improved (Ting et al., 2004; Wang et al., 2003a,b).

Physicochemical Methods

Physical pretreatment (mechanical comminution and pyrolysis) was found to be effective in breaking down the cellulose crystallinity but requires more cost for power and gives all three major compounds in one product stream (Cadoche and Lopez, 1989). Chemical methods such as ozonolysis, acid hydrolysis, alkaline hydrolysis, oxidative delignification, and solvent extraction are also effective pretreatment procedures, but require more energy and chemicals than biological processes and may cause secondary pollution problems (Sivers and Zacchi, 1995). Dilute acid hydrolysis has been developed successfully for the pretreatment of lignocellulosic materials. Dilute sulfuric acid pretreatment can achieve high reaction rates and improve cellulose hydrolysis significantly (Esteghlalian et al., 1997). At moderate temperature, direct saccharification suffered from low yields because of sugar decomposition, whereas at higher temperatures in dilute acid, treatment is favorable for cellulose hydrolysis (McMillan, 1994). Some bases can also be used for the pretreatment of lignocellulosic materials, and the effect of alkaline pretreatment depends on the lignin content of the materials (Fan et al., 1987; McMillan, 1994). The mechanism of alkaline hydrolysis is believed to be saponification of intermolecular ester bonds cross-linking xylan hemicellulosic and other components, for example, lignin and other hemicellulose. Dilute alkaline treatment caused swelling, leading to an increase in internal surface area, a decrease in the degree of polymerization, a decrease in crystallinity, separation of structural linkages between lignin and carbohydrates, and disruption of the lignin structure (Fan et al., 1987; Saratale et al., 2008). Steam explosion is recognized as one of the most cost-effective pretreatment processes for hardwoods and agricultural residues but, having limitation due to incomplete disruption of the lignin–carbohydrate matrix, generates compounds that may be inhibitory to microorganisms used in downstream processes (Mackie et al., 1985). To remove the inhibitory products, the pretreated biomass needs to be washed by water; however, the water wash decreases the overall saccharification yields. Ammonia fiber explosion pretreatment for various lignocellulosic materials shows better performance (Holtzapple et al., 1991), but ammonia makes the process expensive and also causes secondary pollution problems. Although all these methods, in general, have a potential for cellulose hydrolysis, they usually involve complicated procedures, generate unwanted by-products/pollutants, or are economically unfeasible (Mes-Hartree et al., 1988). In addition, some substrate pretreatment methods were reported for fermentative hydrogen production from wastewater sludge. Among which freezing and thawing (for *Clostridium bifermentants*) (Ting et al., 2004; Wang et al., 2003a,b) and sterilization (for *Pseudomonas* sp. GZ1) (Guo et al., 2008) are superior pretreatment methods of wastewater sludge for fermentative hydrogen production. In the case of *Eubacterium multiforme* and *Paenibacillus polymyxa,* alkaline pretreatment with an initial pH of 11 was the optimum method (Cai et al., 2004). This study demonstrates that the optimum pretreatment method for waste-activated sludge may be dependent on the inoculum used for fermentative hydrogen production

Biological Methods

Microbial Hydrolysis of Cellulosic Materials

Effective degradation of lignocellulosic biomass could be achieved by the use of cellulolytic microorganisms, which can produce cellulolytic enzymes (cellulases, xylanase, etc.) for the hydrolysis of the lignocellulosic materials. Filamentous fungi are able to adapt their metabolism rapidly to varying carbon and nitrogen sources achieved through the production of a large set of intra- and extracellular enzymes able to degrade various complex organic pollutants. Several studies have shown that white-rot fungi are the most effective microorganisms for the pretreatment of lignocelluloses, such as wood chips, wheat straw, Bermuda grass, and softwood (Akin et al. 1995; Hatakka, 1983; Saratale et al., 2011). Some reports suggest that application of a mixed bacterial culture is also useful for the efficient hydrolysis of lignocellulosic waste (Datar et al., 2007). Processes using fungal and bacterial systems for cellulose hydrolysis are inexpensive and easy to operate. However, they suffer the drawback of consumption of the hydrolyzed products (such as reducing sugars) due to cell growth.

To overcome these problems, some studies suggest a two-stage process using a mixed or pure microbial culture for hydrolysis and subsequent fermentative bioenergy production (Lo et al. 2008, 2009; Saratale et al., 2010). The efficiency of the two-stage process can be enhanced by a temperature shift strategy, in which after certain microbial growth, the temperature is increased to favor the cellulase enzyme system but bacterial growth is inhibited, resulting in higher biohydrogen production (Lo et al. 2009). In addition, the selected microbial communities used to harvest energy from lignocellulosic waste must be resilient to fluctuations in environmental conditions, variations in nutrient and energy inputs, and intrusion by microbial invaders that might consume the desired energy product. It was observed that different microorganisms can grow on lignocellulosic biomass under anaerobic conditions and convert this abundant organic matter into useful forms of energy such as methane, hydrogen, or even electricity.

Enzymatic Pretreatment of Cellulosic Materials

Conventional physicochemical methods for cellulose hydrolysis not only require large inputs of energy but also make secondary pollution. Enzymatic pretreatment is the overwhelming choice for lignocellulosic waste treatment due to ease of use, high efficiency, and low energy requirement. Biological pretreatment of cellulosic materials could be achieved using cellulolytic enzymes directly or using cellulolytic microorganisms for a combined microbial and enzymatic hydrolysis. Enzymatic hydrolysis of cellulose is carried out by cellulose-hydrolyzing enzymes, cellulases, which consist of a mixture of enzymes that hydrolyze crystalline/amorphous cellulose to fermentable sugars (Duff and Murray, 1996). The interaction between hydrolytic enzymes and cellulosic substrates is complex, in part due to the significant number of possible interactions in the system involving a multienzyme complex that adheres to a multicomponent-insoluble biomass substrate and acts catalytically upon it (Rabinovich et al., 2002; Zhang et al., 2006). There are several advantages of enzymatic hydrolysis, including little energy requirements and mild reaction conditions, high substrate specificity, high yield of sugars, and high hydrolysis efficiency. However, compared to chemical processes, enzymatic hydrolysis has certain disadvantages, including a low hydrolysis rate and high enzyme cost. Enzymatic hydrolysis of cellulose consists of three steps: adsorption of cellulase enzymes onto the surface of the cellulose, biodegradation of cellulose to fermentable sugars, and desorption of cellulases. Retardation of cellulase activity during hydrolysis may be because of the irreversible adsorption of cellulase on cellulose (Converse et al., 1988; Zhang et al., 2006). Conventional chemical pretreatment using acid or alkali could disrupt the crystalline structure of lignocelluloses and make cellulose more accessible to the enzymes for the conversion of polysaccharides into fermentable sugars (Mosier et al., 2005). Conducting such treatment at high temperature with thermostable enzymes has the advantages of enhancing enzyme penetration and disorganization of lignocellulosic cell walls (Turner et al., 2007). Supplementation of surfactants (e.g., Tween 20 and Tween 80) during hydrolysis is capable of modifying the cellulose surface property and minimizing the irreversible binding of cellulase on cellulose (Eriksson et al., 2002). The addition of polymers such as polyethylene glycol can also effectively increase the enzymatic hydrolysis of lignocelluloses due to a higher availability of enzymes for cellulose degradation (Borjesson et al., 2007). In lignin-containing substrates, addition of bovine serum albumin reduced adsorption of cellulase on lignin, resulting in an increase in activity (Ferreira et al., 2009). Saratale et al. (2010) reported that addition of certain metal additives, such as Mn^{2+}, could effectively enhance the multicomponent cellulase enzyme system of *Cellulomonas biazotea* NCIM-2550. Additional research efforts have been taken to improve the cellulase enzyme system by studying the cellulase structure and mechanism of action, the reconstitution of cellulase mixtures (cocktails), enzyme immobilization, and random mutagenesis, as well as genetic engineering approaches for cost-effective cellulase enzyme production (Cherry and Fidantsef, 2003; Himmel et al., 1993; Tao and Cornish, 2002; Zhang et al., 2006).

CONVERTING RENEWABLE RESOURCES INTO BIOHYDROGEN

Crop and Forest Residues

Waste from agro-processing industries accumulates at the mills where the crop is prepared for consumption. These includes bagasse residue from sugarcane, rice husks, cottonseed hulls, palm, coconut, ground nut, cashew, wheat, coffee processing waste, barley, and oats straw. The types of crop residue that play a significant role as biomass fuels are relatively few. The single largest category of crops is cereals, with global production of 1800 Tg in 1985 (Food and Agriculture Organization, 1986). Wheat, rice, maize, barley, and millet and sorghum account for 28, 25, 27, 10, and 6%, respectively, of these crops. The waste products, which are the main contributors to biomass burning, are wheat residue, rice straw and hulls, barley residue, maize stalks and leaves, and millet and sorghum stalks. Four minor crops provide residue from processing that is used frequently as fuel: palm empty fruit bunch and palm fiber, palm shells, coconut residue, groundnut shells, and coffee residue. Moisture content of the delivered feedstock is particularly important for thermochemical and biochemical system processes and also shows an impact on the delivered cost of energy to the processing plants. The moisture content of freshly harvested cereal straw residues (about 10–20%), maize stover (about 20–30%), bagasse and rice straw (about 40–50%) and woody biomass have over 50% moisture content. These feedstocks are mainly lignocellulose-based materials. Hence, pretreatment may be required for enhancing the biohydrogen conversion efficiency. There are some reports describing using crop residues as feedstock for biohydrogen production via dark fermentation (Table 2).

Direct hydrogen production processes, that is, one-stage H_2 production from lignocellulosic biomass, have been found to be more cost-effective and commercially feasible. In fact, some thermophilic H_2-producing bacteria, such as *Clostridium thermocellum*, can produce cellulolytic enzymes for effective hydrolysis, and subsequently produce hydrogen. Some studies reported that application of thermophilic microorganisms, such as *Thermococcus kodakaraensis* KOD1, *Clostridium thermolacticum*, and *Clostridium thermocellum* JN4, could be useful for cellulosic biohydrogen production (Kanai et al., 2005; Liu et al., 2008). Magnusson et al. (2008) reported that *C. thermocellum* ATCC 27405 could use hulls as the carbon source to produce hydrogen without any pretreatment, and the maximum hydrogen yield was 0.22 mol H_2/mol glucose. Rice straw, wheat straw, and corn stover are the most abundant and potential agricultural wastes for biofuel production (Kim and Dale, 2004). However, most biofuel-producing microorganisms cannot utilize cellulose or hemicellulose directly as a carbon source to grow and produce biofuel. Therefore, pretreatment and hydrolysis of the

TABLE 2 Production of Hydrogen by Utilizing Different Biomass-Based Waste Materials by Pure and Mixed Microbial Culture under Dark Fermentation

Nature of waste	Type of waste	Microorganism	Operation mode	Maximum H_2 yield	Reference
Crop and forest residues	Corn stover	*Thermoanaerobacterium thermosaccharolyticum* W16	Batch	2.24 mol H_2/mol hexose	Cao et al., 2009
	Corn stover	*Clostridium butyricum* AS1.209	Batch	68 ml H_2/g corn straw	Li and Chen, 2007
	Rice straw	*C. butyricum* CGS5	Batch	0.76 mol H_2/mol hexose	Lo et al., 2010
	Sugarcane bagasse	*C. butyricum*	Batch	1.73 mol H_2/mol hexose	Pattra et al., 2008
	Sweet sorghum	*Caldicellulosiruptor saccharolyticus*	Batch	2.6 mol H_2/mol hexose	Panagiotopoulos et al., 2010
	Potato steam peels	*C. saccharolyticus*	Batch	2.43.8 mol H_2/mol glucose	Mars et al., 2010
	Potato steam peels	*Thermotoga neapolitana*	Batch	2.43.8 mol H_2/mol glucose	Mars et al., 2010
	Sweet sorghum residues	*Rumicoccus albus*	Batch	2.59	Ntaikou et al., 2008
	Wheat straw	*Caldicellulosiruptor saccharolyticus*	Batch	3.8 (44.7 liter/kg dry biomass)	Ivanova et al., 2009
	Maize leaves	*C. saccharolyticus*	Batch	3.6 (81.5 liter/kg dry biomass)	Ivanova et al., 2009
	Barley straw	*C. saccharolyticus*	Batch	—	Panagiotopoulos et al., 2009
	Corn stalks	*C. saccharolyticus*	Batch	—	Panagiotopoulos et al., 2009
	Rice husk hydrolysate	*Clostridium butyricum* CGS5	Batch	17.24 mmol H_2/g rice husk hydrolysate	Lo et al., 2009
	Corn stover	Mixed mesophilic cultures	Continuous	3 mol H_2 mol glucose	Datar et al., 2007
	Fobber maize juice	Mixed mesophilic cultures	Continuous	69.4	Kyazze et al., 2008

Substrate	Culture	Mode	Yield	Reference
Bagasse	Mixed thermophilic cultures	Batch	13.39	Chairattanamanokorn et al., 2009
Corn stover	Activated sludge	Batch	2.84–3.0 mol H_2/mol glucose	Datar et al., 2007
Corn stover	Activated sludge	Batch	1.53 mol H_2/mol glucose	Liu and Cheng, 2010
Wheat straw	Mixed microflora	Batch	68.1 ml H_2/g TVS	Fan et al., 2006
Wheat straw	Activated sludge	UASB	212.0 ml H_2/g sugar	Kongjan and Angelidaki, 2010
Wheat bran	Mixed microflora	CSTR	1.30 mol H_2/mol hexose	Noike et al., 2003
Korean rice straw	*Thermotoga neapolitana*	Batch	0.41–0.49	Nguyen et al., 2010
Rice bran	Mixed microflora	CSTR	0.98 mol H_2/mol hexose	Noike et al., 2003
Sweet sorghum	Mixed microflora	Batch	15.13–127.26 ml H_2/g TVS	Shi et al., 2011
Barley hulls	*Clostridium thermocellum*	Batch	1.27 mol/mol hexose	Magnusson et al., 2008
Hemp stem	*Clostridium* AK14	Batch	0.60–0.70 mol H_2/mol hexose	Almarsdottir et al., 2010
Grass	*Clostridium* AK14	Batch	0.80–0.90 mol H_2/mol hexose	Almarsdottir et al., 2010
Paper	*Clostridium* AK14	Batch	0.10–0.40 mol H_2/mol hexose	Almarsdottir et al., 2010
Barley straw	*Clostridium* AK14	Batch	0.70–0.80 mol H_2/mol hexose	Almarsdottir et al., 2010
Wheat straw	Mixed culture	Continuous	1.70 mol H_2/mol hexose	Kongjan and Angelidaki, 2010
Wheat straw	Mixed culture	Continuous	1.51 mol H_2/mol hexose	Kongjan and Angelidaki, 2010
Wheat straw	Mixed culture	Continuous	1.00 mol H_2/mol hexose	Kongjan and Angelidaki, 2010
Wheat straw	Mixed culture	Batch	1.2–2.6 mol H_2/mol hexose	Kongjan and Angelidaki, 2010
Wheat straw	Mixed culture	Continuous	1.42	Kongjan and Angelidaki, 2010
Wheat straw	Mixed culture	Batch	2.54	Kongjan and Angelidaki, 2010

Continued

TABLE 2 Production of Hydrogen by Utilizing Different Biomass-Based Waste Materials by Pure and Mixed Microbial Culture under Dark Fermentation—Cont'd

Nature of waste	Type of waste	Microorganism	Operation mode	Maximum H_2 yield	Reference
	Poplar leaves	Mixed microflora	Batch	15.0044.92 ml/g poplar leaves	Cui et al., 2010
	Beer lees	Mixed microflora	Batch	53.03 ml/g dry beer lees	Cui et al., 2010
Waste and wastewater	Sugar factory wastewater	Mixed thermophilic culture	Continuous	2.6 mol/mol hexose	Ueno et al., 1996
	OFMSW	Mixed mesophilic culture	Batch	0.15 liter/g OFMSW	Lay et al., 1999
	Rice winery wastewater	Mixed culture	Continuous	2.14 mol/mol hexose	Yu et al., 2002
	Food waste–sewage sludge	Mixed mesophilic culture	Batch	122.9 ml/g COD carbohydrate	Kim et al., 2004
	Food waste	Mixed thermophilic culture	Batch	1.8 mol/mol hexose	Shin et al., 2004
	Food waste	Mixed culture	Continuous	2.5 to 2.8 mol/mol hexose	Chu et al., 2008
	Cheese whey	*Clostridium saccharoperbutylacetonicum*	Batch	7.89 mmol/g lactose	Ferchichi et al., 2005
	Potato processing wastewater	Mixed mesophilic culture	Batch	2.8 liter/liter wastewater	Van Ginkel et al., 2005
	Cheese whey	Mixed mesophilic culture	Batch	10 mM/g COD	Yang et al. 2007
	Dairy wastewater	Mixed mesophilic culture	Continuous	–	Venkata Mohan et al., 2007
	Molasses	Mixed mesophilic culture	Continuous	–	Ren et al., 2007
	Cheese whey	Mixed mesophilic culture	Batch	5.9 mol/mol lactose	Davila-Vazquez et al., 2008
	Cheese whey	Mixed mesophilic indigenous microbial culture	Continuous	0.9 mol/mol hexose	Antonopoulou et al., 2008b
	Olive pulp	Mixed mesophilic culture	Continuous	0.19 mol/kg TS	Koutrouli et al., 2009

Substrate	Culture	Mode	Yield	Reference
Olive oil mill wastewater	Mixed mesophilic culture	Continuous	196.2 ml/g hexose	Ntaikou et al., 2009
Wastepaper	*Ruminococcus albus*	Batch	2.29 mol/mol hexose (282.76 liter/kg dry biomass)	Van Ginkel et al., 2005
Apple processing wastewater	Mixed culture	Batch	0.9 liter H_2/liter medium (0.1 liter H_2/g COD)	Van Ginkel et al., 2005
Potato processing wastewater	Mixed culture	Batch	2.1 liter H_2/liter medium (0.1 liter H_2/g COD)	Han and Shin, 2004
Food waste	Mixed culture	Continuous	0.39 liter H_2/g COD	Pan et al., 2008
Food waste	Mixed culture	batch	57 ml H_2/g VS	Pan et al., 2008
Molasses	Mixed culture	Continuous	5.57 m^3 H_2/m^3 reactor/day	Ren et al., 2007
Rice slurry	Mixed culture	Batch	346 ml H_2/g carbohydrate	Fang et al., 2006
Starch wastewater	*Thermoanaerobacterium* sp. mixed culture	Batch	92 ml H_2/g starch	Zhang et al., 2003
Cheese whey (49.2 g lactose/liter)	*C. saccharoperbutylacetonicum* ATCC27021	Batch	2.7 mol H_2/mol lactose	Ferchichi et al., 2005
Dairy waste	Mixed culture	HRT 24 h	1.105 mmol H_2/m^3/min	Venkata Mohan et al., 2007
Cheese-processing wastewater	Mixed culture	HRT 24 h	2.4 mM H_2/gCOD	Yang et al. 2007
POME	*Thermoanaerobacterium*-rich sludge	Batch	1.8 liter H_2/liter-POME	Wang et al., 2009
POME	Mixed culture	Batch	6.33 liter H_2/liter-POME	Thong et al., 2007
POME	Mixed culture	Repeated batch	4.7 liter H_2/liter-POME	Atif et al., 2005
POME	Mixed culture	HRT 5d	2.3 liter H_2/liter-POME	Atif et al., 2005

Continued

TABLE 2 Production of Hydrogen by Utilizing Different Biomass-Based Waste Materials by Pure and Mixed Microbial Culture under Dark Fermentation—Cont'd

Nature of waste	Type of waste	Microorganism	Operation mode	Maximum H_2 yield	Reference
Energy crops	Miscanthus	Thermotoga elfii	Batch	1.1	De Vrije et al., 2002
	Wheat starch	Mixed mesophilic cultures	Continuous	1.26	Hussy et al., 2003
	Sugar beet juice	Mixed mesophilic cultures	Continuous	1.9	Hussy et al., 2005
	Corn starch	Mixed mesophilic cultures	Continuous	0.51	Arooj et al., 2008
	Sweet sorghum extract	Indigenous microbial mesophilic culture	Continuous	0.86	Antonopoulou et al., 2008a
	Sweet sorghum stalks	Rumicoccus albus	Batch	3.15 (59 liter/kg wet biomass)	Ntoikou et al., 2008
	Sweet sorghum extract	R. albus	Batch	2.61	Ntoikou et al., 2008
	Ryegrass	Mixed mesophilic cultures	Continuous	82	Kyazze et al., 2008
	Sweet sorghum	Caldicellulosiruptor saccharolyticus	Batch	1.75 (30.17 liter/kg dry biomass)	Ivanova et al., 2009
	Sugar beet extract	C. saccharolyticus	Batch	–	Panagiotopoulos et al., 2009
	Barley grains	C. saccharolyticus	Batch	–	Panagiotopoulos et al., 2009
	Corn grains	C. saccharolyticus	Batch	–	Panagiotopoulos et al., 2009
	Miscanthus	Thermotoga neapolitana	Batch	3.2	De vrije et al., 2009
	Miscanthus	Caldicellulosiruptor saccharolyticus	Batch	3.4	De vrije et al., 2009
Microalgae	Scenedesmus sp.	Activated sludge	Batch	45.54 ml/g volatile solid	Yang et al., 2010
	Laminaria japonica	Clostridium and Bacillus species	Batch	44 ml H_2/g dry algae	Lee et al., 2010
	Taihu blue algae	Activated sludge	Batch	105 ml H_2/g volatile solid	Yang et al., 2010

lignocellulosic feedstock are often required prior to fermentative biofuel production. The cellulosic biofuel can be produced by either physicochemical pretreatment (such as acid, alkaline) or biological pretreatment (such as enzyme) processes, which are discussed in detail further. Some investigators have used *Clostridium butyricum* to utilize sugarcane bagasse (pretreated by sulfuric acid) to produce hydrogen, getting a maximum hydrogen yield of approximately 1.73 mol H_2/mol hexose. A combination of physicochemical pretreatment and biological pretreatments (such as enzymes) can also achieve a high level of efficiency, particularly crop residues to hydrogen. Using 2% Vicozyme L (a mixed enzyme) to pretreat poplar leaves, the hydrogen yield was approximately three times greater than that from the raw substrate (Cui et al., 2010). Lo et al. (2010) also used the enzymes (produced from *Acinetobacter junii* F6-02) to hydrolyze rice straw, and then the hydrolysate was employed to conduct hydrogen fermentation by *C. butyricum* CGS5. Results showed that the hydrogen yield was 0.76 mol H_2/mol hexose. Moreover, some investigators proposed that hydrogen can be produced effectively from cellulose in a two-stage process, in which cellulose hydrolysis is carried out in the first stage and the hydrolysate rich in hexose equivalents is converted effectively into hydrogen in the second stage (Mosier et al., 2005; Saratale et al., 2008, 2010).

Forestry residues comprise materials left behind after trees are fallen and trimmed, sawdust, bark and waste generated at sawmills, and further wastes generated during paper production. It also includes undergrowth and fallen trees in forests that can be removed to help maintain a forest in good health and reduce the danger of fire. Paper and sawmill industries produce considerable quantities of waste, which can be turned economically into energy because they are generated at the processing plant during processing of the wood. Total estimated global generating capacity from all these forest residues is around 10,000 MW, a figure that probably underestimates the gross potential significantly (Berndes et al., 2003). Many wood and paper processing plants already utilize their waste to generate both heat and electricity but more could be exploited, while existing plants are often not the most efficient available. There are several examples using forest residues to produce biohydrogen as indicated in Table 2. Due to higher lignin contents in forestry residues in general when compared to agricultural wastes, more intensive pretreatments may be required. As a result, the yield and productivity of biohydrogen are in general lower using forest residues, while the hydrogen production cost could be higher.

Livestock Residues

Another potential source of energy is found in livestock residues. Estimates of the quantities of these available also vary widely. The major proportion of livestock residues exists in the form of dung. Animal dung is a poor quality fuel used by rural communities only when nothing better is available. It is unlikely that dung could ever become a viable source of electricity production. Where environmental legislation requires these effluents to be treated, it can be cost-effective to use some form of digester to turn the effluent into biogas, which can be burned to produce heat and power. The cost-effectiveness of energy generation from such livestock residues depends on the size of the operation. In addition, swine wastewater is also good source of energy, which is also utilized for methane production and hydrogen production. Because livestock waste is protein rich, the yield of hydrogen production is limited.

Treating swine waste produced a very low hydrogen yield of 3 ml/H_2/g/COD equivalent to only 0.2% energy recovery (Kotsopoulos et al., 2006). Thus livestock waste residues could be useful for methane production or for acid generation in a conventional two-stage process. However, the addition of livestock waste in carbohydrate-rich waste seems feasible for biohydrogen production as they are rich in nutrients. Generally, livestock waste has a higher pH (pH 7.5) so addition of these wastes in other waste material, for example, food waste, could be useful in maintaining the pH of the waste materials (Cheng et al., 2002; Valdez-Vazquez et al., 2005).

Energy Crops

The idea of exploiting whole crops for energy production had already emerged since the early 1980s (Helsel and Wedin, 1981; Lipinsky and Kresovich, 1982). Energy crops refer to certain plants that are cultivated solely for further exploitation of their biomass (either whole or part of it) as feedstock for energy production via combustion or biotransformed to biofuels. Perennial plants such as switchgrass, *Miscanthus*, and Napier grass have a high photosynthetic capacity, as well as water and nitrogen use efficiency (Somerville et al., 2010). In addition, some short-rotation forest species (e.g., *Eucalyptus*, poplars, *Robinia*) are also being considered specifically for the purpose of accumulating biomass. These crops can be high yielding when grown under good conditions and harvested over long seasons to provide a steady supply stream at the processing plant, thus avoiding the costly storage of large biomass volumes for several months between harvests. The importance or sustainability of such processes increases only if the crops are produced at low-cost means with minimum nutrient and water requirements, are resistant to environmental stresses, and have higher biomass yields. For hydrogen production via dark fermentation, such plants should have a high sugar content and low lignin content (Hawkes et al., 2002; Saratale et al., 2008). On the basis of chemical composition, energy crops used for fermentative hydrogen production can be divided into sugar-based crops (e.g., sweet sorghum, sugarcane, and sugar beet), starch-based crops (e.g., corn and wheat), and lignocellulose-based crops, including herbaceous (e.g., switch grass and fodder grass) and woody (e.g., *Miscanthus* and poplar). The results of hydrogen production via dark fermentation with different types of energy crops are depicted in Table 2. In Europe, fast-growing willow and poplars are showing promise as energy crops. In the United States, eucalyptus and cottonwood trees are being grown in southern states, while poplar, willow, and alfalfa have been investigated in more northern states. More significantly, perhaps, native switch grasses have shown promise in the Midwest prairie states. A literature survey suggests that energy crops can be quite sufficient for hydrogen generation; however; continuously rising food prices, sustainability doubts, and energy-equation challenges have led to a backlash against the use of energy crops as feedstocks for biofuel generation.

Algal Feedstocks

Algae are the fastest growers of the plant kingdom. When photosynthesizing, certain species can produce and store large amounts of carbohydrates and up to 50% by weight of oil as triglycerides inside the cell. Microalgae are autotrophic photosynthetic organisms—considered the fastest-growing plant species known (Wayman and Parekh, 1990). They can tolerate a wide range of pH and temperature conditions in diverse habitats, including freshwater and seawater (Harun et al., 2010). Microalgae are considered one of the most promising feedstocks for biofuel production (Wijffels and Barbosa, 2010). Microalgae display greater sustainability and have a number of commercial advantages over first- and second-generation feedstocks, including their rapid growth rate, cultivability with no need of soil, high capturing ability for CO_2 and other greenhouse gases, resulting in an overall reduction in the net gaseous emissions during the entire life cycle of the fuel, and a very short harvesting cycle (1–10 days) (Harun et al., 2010; Wayman and Parekh, 1990). Moreover, microalgae possess high carbohydrate and lipid contents for converting to biofuels. Consequently, microalgae have been considered a third-generation feedstock for biofuel production. Carbohydrates from green microalgae come from cellulose (on cell wall) and starch (in cytoplasm). Enzymatic saccharification of microalgal polysaccharides is necessary prior to fermentation, but the harsh pretreatment used for agricultural wastes may not be necessary in this case due to the absence of lignin and hemicellulose in the microalgae-based cellulosic materials. Therefore, algal biomass is very suitable as a feedstock to produce various biofuels, for example, biodiesel, hydrogen, methane, and bioethanol (Williams and Laurens, 2010).

For bioethanol or biohydrogen production, microalgae with a high carbohydrate content are required. Some microalgae contain over 30% of carbohydrate content, such as *Scenedesmus dimorphus*, *Spirogyra* sp., *Prymnesium parvum*, *Porphyridium cruentum*, and *Anabaena cylindrical* (Harun et al., 2010; Singh and Cu, 2010). Demirbas (2010) demonstrated the production of hydrogen from *Cladophora fracta* and *Chlorella protothecoid* through pyrolysis and steam gasification. In addition, the residual microalgae biomass (e.g., *Scenedesmus*) derived from oil extraction processes was also used to produce hydrogen (Yang et al., 2010). In contrast to lignocellulosic materials, which are very difficult to saccharify, algae-based carbohydrates are hydrolyzed much more easily, as the carbohydrates in algae are mainly starch and celluloses that are not associated with lignin. This addresses the advantage of using algal feedstock to produce biohydrogen through fermentation when compared with using lignocellulosic feedstock. However, still very few reports are available in the literature mentioning the conversion of algal biomass into biohydrogen.

Wastes and Wastewater

The conversion of organic wastes into hydrogen is attractive from both pollution control and energy recovery points of view. However, only a few studies have been conducted for hydrogen production from real wastewater due to challenges associated with inhibition and microbial shifts (Hafez et al., 2009; Kapdan and Kargi, 2006; Noike et al., 2003). Literature studies using wastewaters from rice winery (Yu et al., 2002), noodle (Noike, 2002), sugar (Kim, 2002; Ueno et al., 1996), sugar beet (Hussy et al., 2005), and molasses manufacturing (Logan et al., 2002); food processing (Van Ginkel et al., 2005); and filtered leachate of municipal solid wastes (Wang et al., 2003b) have been reported. The highest hydrogen yield of 321 ml H_2/g COD was demonstrated by Ueno et al. (1996) for the treatment of sugar factory wastewater in a continuous stirred tank reactor (CSTR) with 63% glucose conversion efficiency. With the exception of the two batch studies with soil microorganisms as seed

(Logan et al., 2002; Van Ginkel et al., 2005), most of the studies achieved around 200 ml H_2/g COD. Venkata Mohan et al. (2007) tested the biohydrogen production potential of a composite chemical wastewater (CW) in conjunction with cosubstrates in batch experiments at 29 °C using sequentially pretreated (heat shock at 100 °C for 2 h and acid at pH 3 for 24 h) anaerobic mixed consortia as the inoculum. Those authors found that a 40%/60% mixture of CW/DSW gave the highest yield (1.25 mmol/g COD) and highest relative H_2 production rate (0.08 mmol/g COD/h), followed by a 40%/60% mixture of CW/SW + 1 g/liter glucose. Venkata Mohan et al. (2007) operated a continuous H_2 producing system at 28 °C under acidophilic conditions (pH 6.0) utilizing dairy wastewater with heat shock (100 °C, 2 h) and acid (pH 3.0, 24 h) pretreatments on the mixed consortia. The authors achieved significant hydrogen production and high COD removal efficiencies (>60% at an organic loading rates ranging from 2.4 to 3.5 g COD/liter/day), proving that the dairy wastewater served as the main carbon source for the bacteria. Similar results were obtained elsewhere (Oh and Logan, 2005; Yang et al., 2006) using cereal processing and citric acid wastewaters. Moreover, some complex substrates (e.g., waste-activated sludge, primary sludge, hog manure) are not ideal for fermentative hydrogen production due to their complex structures; however, after pretreatment, they can be used by hydrogen-producing bacteria. For example, Zhang et al. (2007) reported that the hydrogen yield from cornstalk wastes after acidification pretreatment was 46-fold than that from raw cornstalk wastes.

Complex solid wastes, such as wastes from kitchen, food processing, mixed wastes, and municipal wastes, could act as good feedstocks for fermentative hydrogen production. Sewage generated from the municipal or industrial sewage cleaning process has a heating value of approximately 19,000 kJ/kg and acts as a good source of biofuels. In addition to carbohydrates, such wastes usually contain high concentrations of proteins and fats, and thus their conversion efficiencies to hydrogen are comparatively lower than those obtained from carbohydrate-based wastewaters. Among different types of wastes, the organic fraction of municipal solid waste (OFMSW) could be considered quite promising as a potential feedstock for hydrogen production, as it can represent up to 70% of the total MSW produced, consisting of paper (up to 40%), garden residues, food wastes, and wood (Fengel and Wegener; 1984; Ntaikou et al., 2009).

Food processing wastes are the most abundant common solid waste treated for fermentative hydrogen production (Dong et al., 2009; Lay et al., 2003; Noike et al., 2003). Different food processing industrial wastewaters, such as rice winery, noodle, sugar, and molasses manufacturing, olive mill wastewater, olive pulp, and cheese whey, have been tested successfully for hydrogen production at the laboratory scale (Table 2). Without any pretreatment, quite high yields of hydrogen have been achieved from different food industry wastes. In most cases dilution of the raw waste has to be performed to lower the organic loading which shows inhibitory effect on the process. (Antonopoulou et al., 2008a,b; Davila-Vazquez et al., 2008; Koutrouli et al., 2006; Ntaikou et al., 2009; Venetsaneas et al., 2009). Moreover, such wastes have a quite complex chemical composition, including different organic and inorganic substances with varying concentrations of carbohydrates, proteins, and lipids that may limit the reproducibility of a proposed process. Research suggests that carbohydrate-rich substrates produced more than 20 times more hydrogen than lipid-rich fats and skins or protein-rich substrates (Lay et al., 2003). Consequently, defining the exact conditions for efficient treatment of a certain waste is not always possible. Coming from the food industry, such wastewaters theoretically meet all three aforementioned criteria, which can make them ideal candidates for hydrogen production via microbial processes.

Food wastes from the industry and households contain high levels of carbohydrate and protein. Generally, food wastes from the industry are mostly treated anaerobically (Kim et al., 2008). The organic constituent, especially carbohydrate in food wastes, could be a potential substrate for anaerobic hydrogen production. Using seed sludge from an anaerobic digester from a wastewater treatment plant, Kim et al. (2004) reported the production of hydrogen gas using food wastes and sewage sludge,. The specific hydrogen production potential of food wastes was found to be higher than sewage sludge. However, the hydrogen production potential increased as sewage sludge composition was increased up to 13–19% of volatile solids. A maximum specific hydrogen production potential of 122.9 ml H_2/g carbohydrate-COD was found at 87:13 (food waste:sewage sludge). Moreover, a comparative study was carried out by Pan et al. (2008) on hydrogen production at mesophilic (35 ± 2 °C) and thermophilic (50 ± 2 °C) conditions in which the hydrogen yield from the thermophilic acidogenic culture (57 ml H_2/g VS) was higher than that from the mesophilic (39 ml H_2/g VS) culture at all tested food to microorganism ratios. The potential of hydrogen production from highly concentrated, carbohydrate-rich wastewaters was reviewed by Van Ginkel et al. (2001). The biogas produced using wastewater from apple processing and potato processing industries contained 60% hydrogen with no methane generation. Fang et al. (2006) conducted an experiment on hydrogen production using rice slurry containing 5.5 g carbohydrate/liter using *Clostridium* sp.-rich seed sludge. The hydrogen yield of 346 ml H_2/g carbohydrate was obtained at a low pH of 4.5, which was 62.6% of the theoretical yield at 553 ml hydrogen per gram of polysaccharides. The operation of hydrogen production at low pH was not common but could be attributed by the choice of seed sludge. Molasses is another rich-carbohydrate substrate and a good source of sucrose. It is degraded easily anaerobically. A maximum hydrogen production rate of 5.57 m^3 H_2/m^3 reactor/day, with a specific hydrogen production rate of 0.75 m^3 H_2/kg MLVSS/day, was obtained using molasses-based sugar refinery wastewater. The hydrogen yield reached 26.13 mol H_2/kg COD removed within an organic loading rate (OLR) range of 35–55 kg COD/m^3 reactor/day (Ren et al., 2007). Zhang et al. (2003) reported hydrogen production from starch wastewater at a thermophilic condition (55 °C); the maximum hydrogen yield of 92 ml H_2/g starch was achieved at pH 6.0, whereas the maximum specific hydrogen production rate of 365 ml/g VSS day was found at pH 7.0. An addition of polypeptone (nitrogen source) increased hydrogen production to 2.4 mol H_2/mol glucose from starch wastewater (Yokoi et al., 2001). Table 2 summarizes the production of hydrogen using food wastes and starch-based wastewater. The carbohydrate content of the food waste is influenced directly by the hydrogen production; however, carbohydrates could be converted into various acids during storage, which results in a reduction in the hydrogen production potential (Li et al., 2006). In addition, hydrogen production using food waste is also problematic due to the presence of indigenous microflora, mainly lactic acid producing bacteria (Kim et al., 2008). Pretreatment by heat, alkali, or acid useful for the removal of lactic acid producing indigenous microflora favors the environment for the growth of spore-forming bacteria such as *Clostridium* sp. Among which alkali treated food waste found to be suitable for stable hydrogen production. (Kim et al., 2008). Thus the feasibility of food waste for hydrogen production is not so meaningful.

Lactose-rich wastewater can be found in cheese and dairy industry wastewater. Cheese whey contains about 5% lactose, which can be a substrate for fermentative hydrogen production purposes. Several technologies have been applied to convert lactose-rich wastewater to other products, even though utilization and disposal of wastewater are still problems in the dairy industry (Feng et al., 2005; Nath et al., 2008). Some scientists conducted experiments to study the possibility of hydrogen production from crude cheese whey using *Clostridium saccharoperbutylacetonicum* (Ferchichi et al., 2005). It was also observed that this hydrogen production rate was affected by pH with the optimum at a mild acidic range. Davila-Vazquez et al. (2008) suggested that, in the case of cheese whey, rich in readily fermentable sugar wastewater, a pH between 6 and 7 is optimum for obtaining the highest hydrogen yield and production rate, respectively, whereas Yang et al. (2007) suggested that a pH in the range of 4–5 is the best in terms of both hydrogen yield and rate. In both cases, mixed mesophilic cultures were used. The highest hydrogen production potential and yield of hydrogen were achieved at 1432 ml and 2.7 mol H_2/mol lactose, respectively, at pH 6. An increase of yield of hydrogen production was achieved at 3 mol H_2/mol lactose *Clostridium thermolacticum* at continuous mode (Collet et al., 2004). Venkata Mohan et al. (2007) evaluated the potential of hydrogen production from dairy wastewater coupled with wastewater treatment in which the highest hydrogen gas production was found with an organic loading rate of 3.5 kg COD/m^3 day, with a yield of hydrogen production at 1.105 mmol H_2/m^3/min and 64.7% COD removal. Yang et al. (2007) showed the effect of food to microorganism (F/M) ratio (at 1.0–1.5) on hydrogen production. The maximum yield of hydrogen was between 1.8 and 2.3 mM/g COD fed for the different loading rates at a hydraulic retention time (HRT) of 24 h. It was also observed that during hydrogen fermentation, *Lactobacillus* sp. (50%) dominated the reactor with the presence of *Clostridium* sp. (5%).

Palm oil is a major cash crop in many tropical developing countries, including Malaysia, Sri Lanka, and some Asian countries. During the extraction of crude palm oil (CPO), mills generate liquid waste, namely palm oil mill effluent (POME). It was estimated that after processing of 1 ton of fresh fruit bunch, it generates about a ton of POME with COD in the range of 70–100 kg COD/liter (Thong et al., 2007). Some investigators studied the possibility of hydrogen production from raw POME (Atif et al., 2005). It was observed that the production of hydrogen was comparable with published data of hydrogen production from carbohydrate-rich wastewater. An optimized thermophilic condition was developed with the prediction of hydrogen gas at 6.5 liter H_2/liter-POME (Thong et al., 2007). Thong et al. (2008) had observed the diversity of microbes in anaerobic sludge for hydrogen production at thermophilic conditions. Thermophilic hydrogen-producing bacteria were identified as *Thermoanaerobacterium thermosaccharolyticum*.

Glycerol is generated in large amounts (about 10%) during the production of biodiesel. Estimations suggest that after transesterification of vegetable oils or animal fats along with biodiesel, crude glycerol is generated. The increase of biodiesel production from vegetable oils and fats has led to the generation of large quantities of glycerol, raising the problem of treating glycerol-containing waste efficiently. Hydrogen could be possibly produced from glycerol waste with indigenous hydrogentrophic bacteria. Ito et al. (2005) studied hydrogen and ethanol production from biodiesel manufacturing waste using *Enterobacter aerogenes* HU-101. The hydrogen production rate reaches a maximum at 63 mmol/liter/h using porous ceramics as support materials to immobilize cells to fix cells in the reactor. Experimental results showed that the addition of yeast extract and tryptone is needed to accelerate the production of hydrogen, as the waste is, by definition, poor in nitrogen, an element necessary for microbial growth, and biodiesel waste should be diluted due to the high salt content. It was also observed that yeast extract, NH_4Cl, KCl, and $CaCl_2$ were found to be the most important components influencing hydrogen production from glycerol (Liu and Fang, 2007). Akutsu et al. (2009) conducted a study using mixed mesophilic consortium as the inoculum, in which the maximum obtained yield was 38.1 ml H_2/g-$COD_{glycerol}$, with 1,3-propanediol being the main by-product, followed by acetate, whereas no butyrate was detected. Akutsu et al. (2009) based on microbial metabolism; assume that the maximum theoretical yield is 3 mol H_2/mol glycerol. Some studies reported that the maximum theoretical hydrogen yield from glycerol has a value of 1 mol H_2/mol glycerol, based on the reaction of glycerol conversion to H_2, CO_2, and ethanol (Ito et al., 2005). Sakai and Yagishita (2007) reported that the maximum observed hydrogen yield was about 0.77 mol H_2/mol glycerol, with ethanol being the dominant by-product, whereas no propanediol was detected. Using a pure culture of *Klebsiella pneumoniae* during fermentative hydrogen production gave a hydrogen yield of 0.53 mol H_2/mol glycerol, the maximum hydrogen production rate was 17.8 mmol/liter/h, and 1,3-propanediol was the main by-product reported (Liu and Fang, 2007). It was supposed that the maximum theoretical hydrogen yield that can be obtained from carbohydrates is 4 (Nandi and Sengupta, 1998). However, the maximum possible theoretical hydrogen yield from glycerol is still under question. Based on the findings from those studies, it can be assumed that the exploitation of crude glycerol toward hydrogen can be more efficient when based on enterobacteria fermentations. It has to be noted, however, that this is a rather new field that needs to be optimized further.

KEY ISSUES ON BIOHYDROGEN CONVERSION TECHNOLOGIES USING BIOMASS AND WASTES

Environmental and Operational Parameters

The organic loading rate of the wastewater also influences the H_2 production pattern, apart from other wastewater characteristics. Domestic sewage addition showed a positive effect on the acidogenic fermentation process due to the supplementation of additional micronutrients, organic matter, and microbial biomass in the direction of enhancing process efficiency. Fan et al. (2006) investigated the HRT (from 48 h to 8 h) effect on hydrogen production from brewery waste by cattle dung compost in a CSTR system at pH 5.5 in which a maximum H_2 production rate and H_2 yield were obtained at an HRT of 18 h. In these studies, they expressed that the lower yield at a high HRT was probably due to the interference of methanogens and inhibition via major intermediates of acidogenic and alcoholic products. The biomass activity was also HRT dependent, with each gram of biomass producing 65–145 mmol H_2/day. H_2 yields were inversely proportional to the glucose feeding rate, while the highest H_2 yields were observed at the lowest glucose loading rate (Fang and Liu, 2004). A marked reduction in the H_2 production rate was observed with an increase in OLR when chemical wastewater was used as the substrate (Wood and Jungermann, 1972). H_2 production was also found to decrease with an increase in OLR when dairy wastewater was used as the substrate (Han and Shin, 2004). Decreased H_2 production may also be due to end product inhibition by overaccumulated (supersaturated) soluble metabolites in the liquid phase at high OLRs (Pattra et al., 2008). However, each wastewater has its own threshold value, which relates to the system microenvironment and desired output (Chen et al., 2002; Han and Shin, 2004; Momirlan and Veziroglu, 2005; Wood and Jungermann, 1972).

Another major factor affecting dark fermentative hydrogen production is pH. Most investigators reported that the maximum hydrogen yield or specific hydrogen production rate was obtained at a pH between 5.0 and 6.0 (Chen et al., 2001; Lay et al., 1999), whereas some reported a pH range between 6.8 and 8.0 (Collet et al., 2004; Fabiano and Perego, 2002; Lay, 2001; Lin and Cheng, 2006; Liu and Shen, 2004). During dark fermentative hydrogen production, formation of organic acids depletes the buffering capacity of the medium and results in a low final pH. This lower pH inhibits hydrogen production, as pH affects the activity of the iron-containing hydrogenase enzyme and metabolic pathway (Dabrock et al., 1992). Medium pH also affects the hydrogen production yield, biogas content, type of organic acids produced, and specific hydrogen production rate (Khanal et al., 2004). It was also observed that pH affects the growth rate of microorganisms and causes drastic shifts in the relative number of different species in a heterogeneous population present in the hydrogen-producing reactors (Horiuchi et al., 1999). Thus the foregoing results suggest that, in an appropriate range, increasing pH could increase the ability of hydrogen-producing bacteria to produce hydrogen during fermentative hydrogen production. Fan et al. (2006) investigated the pH effect on hydrogen production from brewery waste by a cattle dung compost in a CSTR system at HRT $= 18$ h. Results indicated that maximum H_2 production (47% H_2 concentration, 43 ml H_2/g COD_{added}, and 3.1 liter H_2/liter reactor day) was achieved at pH 5.5.

Temperature is also an important environmental factor influencing the growth rate and metabolic activities of hydrogen-producing bacteria, as well as fermentative hydrogen production (Bailey and Ollis, 1986). Most biohydrogen-producing studies were performed under mesophilic conditions, with some under thermophilic conditions at a certain constant temperature. In comparable studies reported in the literature, the optimal temperature for H_2 production via dark fermentation varied widely, mainly depending on the type of hydrogen producers and carbon substrates. Nevertheless, optimal temperatures were in the range of 37 to 45 °C for pure cultures of *Clostridium* or *Enterobacter* species, whereas mixed bacterial flora gave distinct optimal temperatures and were more effective at thermophilic conditions. Semicontinuous H_2 production at mesophilic and thermophilic conditions was studied by Valdez-Vazquez et al. (2005), in which they found that the volumetric hydrogen production rate (VHPR) was 60% greater at thermophilic than at mesophilic conditions. The result suggests that this behavior might be due to the optimal temperature for the enzyme hydrogenase (50 and 70 °C) present in thermophilic *Clostridia* sp. Yu et al. (2002) studied the effect of temperature (20–55 °C) on hydrogen production from rice winery wastewater by mixed anaerobic cultures in an upflow anaerobic reactor. It has been shown that within an appropriate range, the increasing temperature could enhance the ability of hydrogen-producing bacteria to produce hydrogen during fermentative hydrogen production. However, at a much higher temperature, a decrease in hydrogen yield was observed (Mu et al. 2007).

Bioreactor Design

A literature survey suggests that most of the studies on fermentative hydrogen production were conducted in batch mode because of its simplicity and easy operation and control.

Enhancing H_2 production efficiency, stability, and sustainability is thus a major challenge to batch hydrogen systems. Batch mode operation, coupled with a biofilm configuration, combines the operational advantages of both systems and helps maintain stable and robust cultures suitable for treating highly variable wastewater (Fengel, 1984; Gottschalk, 1986; Kirkpatrick et al., 2001; Lynd et al., 2005; Valdez-Vazquez et al., 2005; van Wyk and Mohulatsi, 2003; Wang and Wan, 2009; Wood and Jungermann, 1972).

Reactor configuration is considered to be crucial for the overall performance of fermentative hydrogen production. It influences the reactor microenvironment, the prevailing microbial population, the established hydrodynamic behavior, the contact between substrate and consortia, and so on (Venkata Mohan et al., 2007). Various modes of reactor operation, such as batch, fed batch, semibatch/continuous, periodic discontinuous batch (sequencing batch operation), and continuous, have been used to produce hydrogen from renewable resources. When using an energy crop as feedstock, about a 25% improvement in H_2 production and substrate degradation efficiency was reported with the batch mode operation compared to the corresponding continuous mode operation (Ivanova et al., 2009). In addition, a fed-batch mode of operation with acidic pH showed highest H_2 production. This high efficiency observed in fed-batch mode operation may be attributed to the reduced accumulation of soluble metabolic intermediates formed during acidogenic fermentation due to the fill-draw mode operation (Chen et al., 2002; Han and Shin, 2004; Van Ginkel et al., 2001; Wang and Wan, 2009). Poor biomass retention/cell washout encountered during a continuous mode operation can be prevented to some extent with a batch mode operation (Harlander, 2008; Ivanova et al., 2009; Rajaram and Verma, 1990). Continuous flow stirred tank reactors are the most frequently used reactor for continuous production of hydrogen from organic matters (Chen et al., 2008b; Das and Veziroglu, 2001; Gavala et al., 2006; Hawkes et al., 2002; Lee et al., 2007; Levin et al., 2004; Wang and Wan, 2009; Wu et al., 2008; Zhang et al., 2007). In a CSTR operation, the biomass is well suspended in the mixed liquor and the substrate is fed at a relatively higher rate than that used in batch or fed-batch operations. Therefore, the productivity could be higher with CSTR due to a higher organic loading rate. However, renewable feedstocks usually contain a high solid content. This would raise the difficulty for CSTR operation. Also, the efficiency of assimilation of renewable resources is usually quite low. That means the cell growth rate and substrate utilization rate could be very low. As a consequence, the continuous culture could be washed out easily unless the hydraulic retention time is long enough. Hence, it might not be suitable to produce biohydrogen from renewable substrates on continuous mode.

There are also several variations of hybrid and attached growth reactors, such as the upflow anaerobic sludge blanket reactor (UASB) (Jeison and Chamy, 1999), the packed bed reactor (Li et al., 2006), the membrane bioreactor (Chang et al., 2002; Cheong et al., 2007), the anaerobic fluidized bed reactor (Zhang et al., 2007), and the carrier-induced granular sludge bed (Lee et al., 2006). It has also been reported that immobilization of biomass on artificial granules made of various support materials, such as cuprammonium rayon (Hallenbeck et al., 2005), polyvinyl alcohol, polyacrylamide, and anionic silica sol (Draude et al., 2001; Kim et al., 2005), could enhance the performance of biohydrogen production in some cases. A direct comparison of the performance of the different types of reactor configurations that have been studied in terms of hydrogen productivity is not possible, as the operational parameters (feedstocks and operating conditions), along with the reactor configuration, in all these studies are quite different. Again, the high solid content of renewable feedstocks, using attached growth, granular sludge, or immobilized cell systems, will face the challenge of poor mass transfer efficiency. However, if renewable feedstocks are pretreated or hydrolyzed properly, this problem could be resolved.

Upflow anaerobic sludge blanket reactor technology is being used extensively for effluents from different sources such as distilleries, food processing units, tanneries, and municipal wastewater. The active biomass in the form of sludge granules is retained in the reactor by direct settling for achieving high cell retention times, thereby achieving highly cost-effective designs. A major advantage of this technology is the comparatively less investment requirements when compared to a packed-bed or a fluidized-bed system. However, a long start-up period and the requirement for a sufficient amount of granular seed sludge for a faster start up are among notable disadvantages. Vijayaraghavan et al. (2006) demonstrated H_2 production from solid waste consisting of jackfruit peel with microflora isolated from cow dung in upflow anaerobic contact filter-packed rigid circular porous plastic balls of 40 mm diameter. Yu et al. (2002) investigated continuous hydrogen production from a high-strength rice winery wastewater by a mixed bacterial flora, using an upflow reactor. The hydrogen yield was in the range of 1.37–2.14 mol/mol of hexose. Jeison and Chamy, 1999 examined the hydrogen production from sucrose with an UASB reactor and demonstrated the feasibility of using the UASB system for hydrogen production. Yu and Mu (2006) evaluated the performance of a UASB for H_2 production from sucrose-rich synthetic wastewater at various substrate concentrations (5.33–28.07 g COD/liter) and HRTs (3–30 h) for over 3 years. Experimental results showed that the H_2 production rate increased with both increasing substrate concentration and decreasing HRT. The H_2 yield was in the range of 0.49–1.44 mol-H_2/mol glucose. In a traditional UASB reactor, channeling of wastewater through the bed is encountered frequently, resulting in a poor substrate–biomass contact (Jeison and Chamy, 1999). To overcome these problems, expanded granular sludge bed reactors are operated at high liquid upflow velocities by increasing the height/diameter ratio and recirculating the effluent liquid, thereby providing better mixing of the reactor contents and allowing more efficient contact between substrate and biomass (Hwu et al., 1998). Francese et al. (1998) also showed that upflow velocity affects physical characteristics and specific activity of granules.

Therefore, by studying and comparing the different configurations of reactors to draw a conclusion in regard to what configuration is better even under a specific set of conditions, many factors, particularly hydrogen yield and hydrogen production rate, depend significantly on experimental conditions, such as temperature, pH, substrate concentration, type of substrate, metal ion, and HRT, as well as long-term stability of the reactor and scale-up performance, which directly influence the economics of fermentative hydrogen production.

ECONOMICS OF BIOHYDROGEN PRODUCTION AND PERSPECTIVES

There are many reports in the literature about biohydrogen production, but only a few of them are related to the economic analyses of biohydrogen production. In general, the molar yield of hydrogen and the cost of the feedstock are the two main barriers for fermentative hydrogen production. The main challenges in fermentative hydrogen production are that only 15% of the energy from the organic source can be obtained in the form of hydrogen and a lower hydrogen yield (Logan, 2004). Hydrogen molar yields can be increased by applying metabolic engineering efforts. Moreover, the improvement of hydrogen production by gene manipulation is focused mainly on the disruption of endogenous genes and not introducing new activities in the microorganisms. New pathways must be discovered to directly take full advantage of the 12 mol of H_2 available in a mole of hexose. de Vrije et al. (2002) reported the cost of hydrogen production by utilizing locally produced lignocellulosic feedstock. The plant was set at a production capacity of 425 $Nm_3 H_2$/h and consisted of a thermobioreactor (95 m^3) for hydrogen fermentation followed by a photobioreactor (300 m^3) for the conversion of acetic acid to hydrogen and CO_2. Economic analysis resulted in an estimated overall cost of €2.74/kg H_2. This cost is based on acquisition of biomass at zero value, zero hydrolysis costs, and excludes personnel costs and costs for civil works, all of them with potential cost factors.

The costs of H_2 generated from biological processes and other available conventional processes are depicted in Table 3. Biological hydrogen was comparatively higher than that of hydrogen from pyrolysis. Among the alternative energies, at present, biomass and biofuel are the ones closer to the parity in conventional and distributed systems, respectively. Increased efforts in the development of advanced technologies to improve the technical feasibility and scalability of hydrogen production based on renewable energy, higher carbon emissions, and large investment growth in renewable energies could make cost parities to be reached in the future. Hydrogen production based on renewable technologies avoids fuel prices in uncertainties, mainly produced in the natural gas market, and massive investment in hydrogen production with renewable technologies could be produced before the parities are reached.

Regarding feedstock costs, commercially produced food products, such as corn and sugar, are not economical for hydrogen production (Benemann, 1996). However, by-products from agricultural crops or industrial processes with no or low value represent a valuable

TABLE 3 Comparison of Unit Cost of Hydrogen Production Processes with Different Conventional Processes

Materials used	Name of production process	Unit cost of energy content of fuel U.S. $/MBTU^{-1}
H_2O organic acids	Photobiological hydrogen	10
Molasses	Fermentative hydrogen	10
Coal, biomass	Hydrogen production by pyrolysis	4
H_2O	Hydrogen production by advanced electrolysis	11
Electrolysis and water splitting	Hydrogen production from nuclear energy	12–19
Biomass	Hydrogen production by biomass gasification	44–82
Wind mill	Hydrogen production from wind energy	34
Solar energy	Hydrogen production from photovoltaic power station	42
H_2O	Hydrogen production from thermal decomposition of steam	13
Organic acids	Hydrogen production from photochemical process	21

resource for energy production. Wastewater has a great potential for economic production of hydrogen; in the United States, the organic content in wastewater produced annually by humans and animals is equivalent to 0.41 quadrillion British thermal units (Logan, 2004). From Table 3, it was proved that hydrogen production by conventional methods is not affordable except in the case of pyrolysis. The cost of biological hydrogen production was still not cost-effective when compared to conventional hydrogen production methods so there is a need to develop some strategies, such as development of the two-step fermentation process (dark and photofermentation) or the use of modified microbial fuel cells (de Vrije et al., 2009; Logan, 2004). It was supposed that by applying these coupled processes, more hydrogen per mole of substrate can be achieved. There is no doubt that many technical and engineering challenges have to be solved before economic barriers can be meaningfully considered.

CONCLUDING REMARKS

Biohydrogen production has been established as a prospective alternative and integral component of green sustainable energy because of its eco-friendly nature. However, a number of obstacles must be overcome if this potential is to be realized on a practical scale. One attractive route that was discussed in detail in this chapter is the use of dark fermentative hydrogen production using biomass waste materials and other renewable resources as feedstock. Biohydrogen from renewable represents both energy sustainability and waste reduction, and could also reduce the production costs. However, several factors influencing the performance of dark fermentative hydrogen production from renewable feedstock should be identified and optimized. This chapter described various approaches to cope with this problem using either pure cultures or microbial consortia in a variety of reactor configurations with different substrates. Fermentative hydrogen production from waste materials could be competitive with fossil fuel-derived H_2, providing a plausible approach to practical hydrogen production. Processing of some biomass feedstock is too costly and therefore there is a need to develop low-cost methods for growing, harvesting, transporting, and pretreating energy crops and/or biomass waste products. There was no clear contender for a robust, industrially capable microorganism that can be engineered metabolically to produce more hydrogen using the dark fermentative approach. Several engineering issues need to be addressed, which include the appropriate bioreactor design, sustaining a steady continuous H_2 production rate in the long term, scale up of biohydrogen production processes, preventing interspecies H_2 transfer in nonsterile conditions, and separation/purification of H_2. There is also a need to conduct research on the sensitivity of the hydrogenase enzyme to O_2 and H_2 partial pressure, which are known to severely decrease the efficiency of the biohydrogen-producing processes. There is also a demand for studying the feasible strategies on the improvement of economics of the process by combination of H_2 production with other processes. To improve the yield and rates of biohydrogen production, numerous improvements in the development of bioreactor design, engineering of hydrogenase enzymes, and genetic modification of microorganism are required. Many studies are currently trying to achieve these technical and scientific advancements for better output as a futuristic goal. Independent of the source of hydrogen, many logistical and market challenges must also be overcome before a hydrogen economy can become a reality. It was supposed that biohydrogen is the key to accelerating the coming of the "hydrogen economy" and solving the problems of "global warming" and the "shortage of fossil fuel" simultaneously.

Acknowledgement

This chapter is dedicated to the memory of Advocate Mahesh D. Saratale. GDS acknowledges the Department of Science and Technology, India and University Grants Comission, India, for their support.

References

Akin, D.E., Rigsby, L.L., Sethuraman, A., Morrison III., W.H., Gamble, G.R., Eriksson, K.E.L., 1995. Alterations in structure, chemistry, and biodegradability of grass lignocellulose treated with the white rot fungi *Ceriporiopsis subvermispora* and *Cyathus stercoreus*. Appl. Environ. Microbiol. 61, 1591–1598.

Almarsdottir, A.R., Taraceviz, A., Gunnarsson, I., Orlygsson, J., 2010. Hydrogen production from sugars and complex biomass by *Clostridium* species, AK14, isolated from Icelandic hot spring. Iceland. Agric. Sci. 23, 61–71.

Akkerman, I., Janssen, M., Rocha, J., Wij1els, R.H., 2002. Photobiological hydrogen production: photochemical efficiency and bioreactor design. Int. J. Hydrogen Energy 27, 1195–1208.

Akutsu, Y., Lee, D.Y., Li, Y.Y., Noike, T., 2009. Hydrogen production potentials and fermentative characteristics of various substrates with different heat-pretreated natural microflora. Int. J. Hydrogen Energy 34, 5365–5372.

Antonopoulou, G., Gavala, H.N., Skiadas, I.V., Angelopoulos, K., Lyberatos, G., 2008a. Biofuels generation from sweet sorghum: Fermentative hydrogen production and anaerobic digestion of the remaining biomass. Bioresour. Technol. 99, 110–119.

Antonopoulou, G., Stamatelatou, K., Venetsaneas, N., Kornaros, M., Lyberatos, G., 2008b. Biohydrogen and methane production from cheese whey in a two-stage anaerobic process. Ind. Eng. Chem. Res. 47, 5227–5233.

Arooj, M.F., Han, S.K., Kim, S.H., Kim, D.H., Shin, H.S., 2008. Effect of HRT on ASBR converting starch into biological hydrogen. Int. J. Hydrogen Energy 33, 6509 6514.

Asada, Y., Miyake, J., 1999. Photobiological hydrogen production. J. Biosci. Biotechnol. 88, 1–6.

Atif, A.A.Y., Fakhru'l-Razi, A., Ngan, M.A., Morimoto, M., Iyukeand, S.E., Veziroglu, N.T., 2005. Fed batch production of hydrogen from palm oil mill effluent using anaerobic microflora. Int. J. Hydrogen Energy 30, 1393–1397.

Bailey, J.E., Ollis, D.F., 1986. Biochemical Engineering Fundamentals, second ed. McGraw-Hill, New York.

Benemann, J., 1996. Hydrogen biotechnology: Progress and prospects. Nat. Biotechnol. 14, 1101–1103.

Berndes, G., Hoogwijk, M., van den Broek, R., 2003. The contribution of biomass in the future global energy supply: A review of 17 studies. Biomass Bioenergy 25, 1–28.

Borjesson, J., Peterson, R., Tjerneld, F., 2007. Enhanced enzymatic conversion of softwood lignocellulose by poly (ethylene glycol) addition. Enzyme Microb. Technol. 40, 754–762.

Cadoche, L., Lopez, G.D., 1989. Assessment of size reduction as a preliminary step in the production of ethanol from lignocellulosic wastes. Biol. Wastes 30, 153–157.

Cai, M.L., Liu, J.X., Wei, Y.S., 2004. Enhanced biohydrogen production from sewage sludge with alkaline pretreatment. Environ. Sci. Technol. 38, 3195–3202.

Cao, G.L., Ren, N.Q., Wang, A.J., Lee, D.J., Guo, W.Q., Liu, B.F., et al., 2009. Acid hydrolysis of corn stover for biohydrogen production using *Thermoanaerobacterium thermosaccharolyticum* W16. Int. J. Hydrogen Energy 34, 7182–7188.

Chairattanamanokorn, P., Penthamkeerati, P., Reungsang, A., Lo, Y.C., Lu, W.B., Chang, J.S., 2009. Production of biohydrogen from hydrolyzed bagasse with thermally preheated sludge. Int. J. Hydrogen Energy 34, 7612–7617.

Chandrakant, P., Bisaria, V.S., 1998. Simultaneous bioconversion of cellulose and hemicellulose to ethanol. Crit. Rev. Biotechnol. 18, 295–331.

Chang, J.S., Lee, K.S., Lin, P.J., 2002. Biohydrogen production with fixed-bed bioreactors. Int. J. Hydrogen Energy 27, 1167–1174.

Chen, C.C., Lin, C.Y., Chang, J.S., 2001. Kinetics of hydrogen production with continuous anaerobic cultures utilizing sucrose as the limiting substrate. Appl. Microbiol. Biotechnol. 57, 56–64.

Chen, C.C., Lin, C.Y., Lin, M.C., 2002. Acid-base enrichment enhances anaerobic hydrogen production process. Appl. Microbiol. Biotechnol. 58, 224–228.

Chen, C.Y., Saratale, G.D., Lee, C.M., Chen, P.C., Chang, J.S., 2008a. Phototrophic hydrogen production in photobioreactors coupled with solar-energy-excited optical fiber. Int. J. Hydrogen Energy 33, 6886–6895.

Chen, S.D., Lee, K.S., Lo, Y.C., Chen, W.M., Wu, J.F., Lin, C.Y., et al., 2008b. Batch and continuous biohydrogen production from starch hydrolysate by *Clostridium* species. Int. J. Hydrogen Energy 33, 1803–1812.

Cheng, S.S., Chang, S.M., Chen, S.T., 2002. Effects of volatile fatty acids to a thermophilic anaerobic hydrogen fermentation process degrading peptone. Water Sci. Technol. 46, 209–214.

Cheng, C.L., Lo, Y.C.A., Lee, K.S., Lee, D.J., Lin, C.Y., Chang, J.S., 2011. Biohydrogen production from lignocellulosic feedstock. Bioresour. Technol. 102, 8514–8523.

Cheong, D.Y., Hansen, C.L., Stevens, D.K., 2007. Production of biohydrogen by mesophilic anaerobic fermentation in an acid-phase sequencing batch reactor. Biotechnol. Bioeng. 96, 421–432.

Cherry, J.R., Fidantsef, A.L., 2003. Directed evolution of industrial enzymes: An update. Curr. Opin. Biotechnol. 14, 438–443.

Chu, C.F., Li, Y.Y., Xu, K.Q., Ebie, Y., Inamori, Y., Kong, H.N., 2008. A pH- and temperature-phased two-stage process for hydrogen and methane production from food waste. Int. J. Hydrogen Energy 33, 4739–4746.

Collet, C., Adler, N., Schwitzguebel, J.P., Peringer, P., 2004. Hydrogen production by *Clostridium thermolacticum* during continuous fermentation of lactose. Int. J. Hydrogen Energy 29, 1479–1485.

Converse, A.O., Matsuno, R., Tanaka, M., Taniguchi, M., 1988. A model for enzyme adsorption and hydrolysis of microcrystalline cellulose with slow deactivation of the adsorbed enzyme. Biotechnol. Bioeng. 32, 38–45.

Cui, M.J.Z.L., Yuan, X.H., Zhi, L.L., Wei, J.Q., 2010. Shen biohydrogen production from poplar leaves pretreated by different methods using anaerobic mixed bacteria. Int. J. Hydrogen Energy 35, 4041–4047.

Dabrock, B., Bahl, H., Gottschalk, G., 1992. Parameters affecting solvent production by *Clostridium pasteurium*. Appl. Environ. Microbiol. 58, 1233–1239.

Das, D., Veziroglu, T.N., 2001. Hydrogen production by biological processes: A survey of literature. Int. J. Hydrogen Energy 26, 13–28.

Datar, R., Huang, J., Maness, P.C., Mohagheghi, A., Czernik, S., Chornet, E., 2007. Hydrogen production from the fermentation of corn stover biomass pretreated with a steam-explosion process. Int. J. Hydrogen Energy 32, 932–939.

Davila-Vazquez, G., Alatriste-Mondragon, F., de Leon-Rodriguez, A., Razo-Flores, E., 2008. Fermentative hydrogen production in batch experiments using lactose, cheese whey and glucose: Influence of initial substrate concentration and pH. Int. J. Hydrogen Energy 33, 4989–4997.

de Vrije, T., de Haas, G.G., Tan, G.B., Keijsers, E.R.P., Claassen, P.A.M., 2002. Pretreatment of *Miscanthus* for hydrogen production by *Thermotoga elfii*. Int. J. Hydrogen Energy 27, 1381–1390.

de Vrije, T., Bakker, R.R., Budde, M.A.W., Lai, M.H., Mars, A.E., Claassen, P.A.M., 2009. Efficient hydrogen production from the lignocellulosic energy crop Miscanthus by the extreme thermophilic bacteria *Caldicellulosiruptor saccharolyticus* and *Thermotoga neapolitana*. Biotechnol. Biofuels 2, 12–18.

Demirbas, A., 2007. Progress and recent trends in biofuels. Prog. Energy Combust. Sci. 33, 1–18.

Demirbas, A., 2010. Hydrogen from mosses and algae via pyrolysis and steam gasification. Energy Sources A 32, 172–179.

Dong, L., Zhenhong, Y., Yongming, S., Xiaoying, K., Yu, Z., 2009. Hydrogen production characteristics of the organic fraction of municipal solid wastes by anaerobic mixed culture fermentation. Int. J. Hydrogen Energy 34, 812–820.

Draude, K.M., Kurniawan, C.B., Duff, S.T.B., 2001. Effect of oxygen delignification on the rate and extent of enzymatic hydrolysis of lignocellulosic material. Bioresour. Technol. 79, 113–120.

Duff, S.J.B., Murray, W.D., 1996. Bioconversion of forest products industry waste cellulosics to fuel ethanol: A review. Bioresour. Technol. 55, 1–33.

Eriksson, T., Borjesson, J., Tjerneld, F., 2002. Mechanism of surfactant effect in enzymatic hydrolysis of lignocellulose. Enzyme Microb. Technol. 31, 353–364.

Esteghlalian, A., Hashimoto, A.G., Fenske, J.J., Penner, M.H., 1997. Modeling and optimization of the dilute sulphuric acid pretreatment of corn stover, poplar and switchgrass. Bioresour. Technol. 59, 129–137.

Evvyernie, D., Morimoto, K., Karita, S., Kimura, T., Sakka, K., Ohmiya, K., 2001. Conversion of chitinous wastes to hydrogen gas by *Clostridium paraputrificum* M-21. J. Biosci. Bioeng. 91, 339–343.

Fabiano, B., Perego, P., 2002. Thermodynamic study and optimization of hydrogen production by *Enterobacter aerogenes*. Int. J. Hydrogen Energy 27, 149–156.

Fan, L.T., Gharpuray, M.M., Lee, Y.H., 1987. Cellulose Hydrolysis, vol. 3. Springer-Verlag, Berlin.

Fan, Y., Zhang, G., Guo, X., Xing, Y., Fan, M., 2006. Biohydrogen-production from beer lees biomass by cow dung compost. Biomass Bioenergy 30, 493–496.

Fang, H.H.P., Liu, H., 2004. Biohydrogen production from wastewater by granular sludge. In: 1st International Symposium on Green Energy Revolution, Nagaoka, Japan. pp. 31–36.

Fang, H.H.P., Liu, H., Zhang, T., 2002. Characterization of a hydrogen producing granular sludge. Biotechnol. Bioeng. 78, 44–52.

Fang, H.H.P., Li, C., Zhang, T., 2006. Acidophilic biohydrogen production from rice slurry. Int. J. Hydrogen Energy 31, 683–692.

Fascetti, E., Todini, O., 1995. *Rhodobacter sphaeroids* RV cultivation and hydrogen production in a one and two stage chemostat. Appl. Microbial. Biotechnol. 44, 300–305.

Fascetti, E., D'Addario, E., Todini, O., Robertiello, A., 1998. Photosynthetic hydrogen evolution with volatile organic acids derived from the fermentation of source selected municipal solid wastes. Int. J. Hydrogen Energy 23, 753–760.

Feng, G.L., Letey, J., Chang, A.C., Campbell, M., 2005. Simulating dairy liquid waste management options as a nitrogen source for crops. Agri. Ecosys. Environ. 110, 219–229.

Fengel, D., Wegener, G., 1984. Wood: Chemistry, Ultrastructure, Reactions. De Gruyter, Berlin.

Ferchichi, M., Crabbe, V., Gil, G.H., Hintz, W., Almadidy, A., 2005. Influence of initial pH on hydrogen production from cheese whey. J. Biotechnol. 120, 402–409.

Ferreira, S., Duarte, A.P., Ribeiro, M.H.L., Queiroz, J.A., Domingues, F.C., 2009. Response surface optimization of enzymatic hydrolysis of *Cistus ladanifer* and *Cytisus striatus* for bioethanol production. Biochem. Eng. J. 45, 192–200.

Food and Agriculture Organization, 1986. Production Yearbook, vol. 39. Food and Agriculture Organization of the United Nations, Rome.

Francese, A., Cordoba, P., Duran, J., Sineriz, F., 1998. High upflow velocity and organic loading rate improves granulation in upflow anaerobic sludge blanket reactors. World J. Microbiol. Biotechnol. 14, 337–341.

Gavala, H.N., Skiadas, I.V., Ahring, B.K., 2006. Biological hydrogen production in suspended and attached growth anaerobic reactor systems. Int. J. Hydrogen Energy 31, 1164–1175.

Ghirardi, M.L., Zhang, L., Lee, J.W., Flynn, T., Seibert, M., Greenbaum, E., et al., 2000. Microalgae: A green source of renewable H_2. Tibtech 18, 506–511.

Gottschalk, G., 1986. Bacterial fermentations. In: Bacterial Metabolism. Springer, New York, pp. 237–239.

Guan, Y., Deng, M., Yu, X., Zang, W., 2004 Two stage photo-production of hydrogen by marine green algae *Platymonas subcordiformis*. Biochem. Eng. J. 19, 69–73.

Guo, L., Li, X.M., Bo, X., Yang, Q., Zeng, G.M., Liao, D.X., et al., 2008. Impacts of sterilization, microwave and ultrasonication pretreatment on hydrogen producing using waste sludge. Bioresour. Technol. 99, 3651–3658.

Hafez, H., Nakhla, G., El Naggar, H., 2009. Biological hydrogen production from corn-syrup waste using a novel system. Energies 2, 445–455.

Hallenbeck, P.C., 2005. Fundamentals of the fermentative production of hydrogen. Water Sci. Technol. 52, 21–29.

Hallenbeck, P.C., Benemann, J.R., 2002. Biological hydrogen production; fundamentals and limiting processes. Int. J. Hydrogen Energy 27, 1185–1193.

Han, S.K., Shin, H.S., 2004. Performance of an innovative two-stage process converting food waste to hydrogen and methane. J. Air Waste Manag. 54, 242–249.

Harlander, K., 2008. Food vs. fuel—A turning point for bioethanol? Acta Agron. Hung. 56, 429–433.

Harun, R., Danquah, M.K., Forde, G.M., 2010. Microalgal biomass as a fermentation feedstock for bioethanol production. J. Chem. Technol. Biotechnol. 85, 199–203.

Hatakka, A.I., 1983. Pretreatment of wheat straw by white-rot fungi for enzymatic saccharification of cellulose. Appl. Microbiol. Biotechnol. 18, 350–357.

Hawkes, F.R., Dinsdale, R., Hawkes, D.L., Hussy, I., 2002. Sustainable fermentative hydrogen production: Challenges for process optimization. Int. J. Hydrogen Energy 27, 1339–1347.

He, D., Bultel, Y., Magnin, J.P., Roux, C., Willison, J.C., 2005. Hydrogen photosynthesis by *Rhodobacter capsulatus* and its coupling to PEM fuel cell. J. Power Sources 141, 19–23.

Helsel, Z.R., Wedin, W.F., 1981. Direct combustion energy from crops and crop residues produced in Iowa. Energy Agric. 1, 317–329.

Himmel, M.E., Adney, W.S., Baker, J.O., Nieves, R.A., Thomas, S.R., 1993. Cellulases: structure, function and applications. In: Wyman, C.E. (Ed.), Handbook on Bioethanol. Taylor & Francis, Washington, DC, pp. 144–161.

Holtzapple, M.T., Jun, J.H., Ashok, G., Patibandla, S.L., Dale, B.E., 1991. The ammonia freeze explosion (AFEX) process: A practical lignocellulose pretreatment. Appl. Biochem. Biotechnol. 28–29, 59–74.

Horiuchi, J., Shimizu, T., Kanno, T., Kobayashi, M., 1999. Dynamic behavior in response to pH shift during anaerobic acidogenesis with a chemstat culture. Biotech. Tech. 13, 155–157.

Hussy, I., Hawkes, F.R., Dinsdale, R., Hawkes, D.L., 2003. Continuous fermentative hydrogen production from a wheat starch co-product by mixed microflora. Biotechnol. Bioeng. 84, 619–626.

Hussy, I., Hawkes, F.R., Dinsdale, R., Hawkes, D.L., 2005. Continuous fermentative hydrogen production from sucrose and sugarbeet. Int. J. Hydrogen Energy 30, 471–483.

Hwu, C.S., van Lier, J.B., Lettinga, G., 1998. Physicochemical and biological performance of expanded granular sludge bed reactors treating long-chain fatty acids. Process Biochem. 33, 75–81.

Ito, T., Nakashimada, Y., Senba, K., Matsui, T., Nishio, N., 2005. Hydrogen and ethanol production from glycerol-containing wastes discharged after biodiesel manufacturing process. J. Biosci. Bioeng. 100, 260–265.

Ivanova, G., Rakhely, G., Kovacs, K.L., 2009. Thermophilic biohydrogen production from energy plants by *Caldicellulosiruptor saccharolyticus* and comparison with related studies. Int. J. Hydrogen Energy 34, 3659–3670.

James, W.E., Jha, S., Sumulong, L., Son, H.H., Hasan, R., Khan, M.E., et al., 2008. Food Prices and Inflation in Developing Asia: Is Poverty Reduction Coming to an End?. Asian Development Bank, Manila, Philippines. http://www.adb.org//reports/food-prices-inflation/food-prices-inflation.pdf.

Jeison, D., Chamy, R., 1999. Comparison of the behaviour of expanded granular sludge bed (EGSB) and upflow anaerobic sludge blanket (UASB) reactors in dilute and concentrated wastewater treatment. Water Sci. Technol. 40, 91–97.

Kanai, T., Imanaka, H., Nakajima, A., Uwamori, K., Omori, Y., Fukui, T., et al., 2005. Continuous hydrogen production by the hyperthermophilic archaeon, *Thermococcus kodakaraensis* KOD1. J. Biotechnol. 116, 271–282.

Kapdan, I.K., Kargi, F., 2006. Biohydrogen production from waste materials. Enzyme Microb. Technol. 38, 569–582.

Keeney, R., Hertel, T.W., 2008. The indirect land use impacts of US biofuel policies: The importance of acreage, yield, and bilateral trade responses. Center for Global Trade Analysis, Purdue University, West Lafayette, Indiana. http://www.gtap.agecon.purdue.edu/resources/download/3904.pdf.

Khanal, S.K., Chen, W.H., Li, L., Sung, S., 2004. Biological hydrogen production: Effects of pH and intermediate products. Int. J. Hydrogen Energy 29, 1123–1131.

Kim, M.S., 2002. An integrated system for the biological hydrogen production from organic wastes and waste-waters. Paper read at International Symposium on Hydrogen and Methane Fermentation of Organic Waste, Tokyo, Japan.

Kim, S., Dale, B.E., 2004. Global potential bioethanol production from wasted crops and crop residules. Biomass Bioenergy 26, 361–375.

Kim, S.-H., Han, S.-K., Shin, H.-S., 2004. Feasibility of biohydrogen production by anaerobic co-digestion of food waste and sewage sludge. Int. J. Hydrogen Energy 29, 1607–1616.

Kim, O., Kim, Y.H., Ryu, J.Y., Song, B.K., Kim, I.H., Yeom, S.H., 2005. Immobilization methods for continuous hydrogen gas production biofilm formation versus granulation. Process Biochem. 40, 1331–1337.

Kim, J.K., Han, G.H., Oh, B.R., Chun, Y.N., Eom, C.Y., Kim, S.W., 2008. Volumetric scale-up of a three stage fermentation system for food waste treatment. Bioresour. Technol. 99, 4394–4399.

Kim, J.R., Kim, J.Y., Han, S.B., Park, K.W., Saratale, G.D., Oh, S.E., 2011. Application of Co-naphthalocyanine(CoNPc) as 5 alternative cathode catalyst and support structure for microbial fuel cells. Bioresour. Technol. 102, 342–347.

Kirkpatrick, C., Maurer, M.L., Oyelakin, N.E., Yoncheva, Y.N., Mauer, R., Slonczewski, J.L., 2001. Acetate and formate stress: opposite responses in the proteome of *Escherichia coli*. J. Bacteriol. 183, 6466–6477.

Koku, H., Eroğlu, İ., Gündüz, U., Yücel, M., Türker, L., 2002. Aspects of metabolism of hydrogen production by *Rhodobacter sphaeroides*. Int. J. Hydrogen Energy 27, 1315–1329.

Kongjan, P., Angelidaki, I., 2010. Extreme thermophilic biohydrogen production from wheat straw hydrolysate using mixed culture fermentation: Effect of reactor configuration. Bioresour. Technol. 101, 7789–7796.

Kotay, S.M., Das, D., 2008. Biohydrogen as a renewable energy resource prospects and potentials. Int. J. Hydrogen Energy 33, 258–263.

Kotsopoulos, T.A., Zeng, R.J., Angelidaki, I., 2006. Biohydrogen production in granular up-flow anaerobic sludge blanket (UASB) reactors with mixed cultures under hyper-thermophilic temperature (70 °C). Biotechnol. Bioeng. 94, 296–302.

Koutrouli, E.C., Gavala, H.N., Skiadas, I.V., Lyberatos, G., 2006. Mesophilic biohydrogen production from olive pulp. Proc. Saf. Environ. Prot. 84, 285–289.

Koutrouli, E.C., Kalfas, H., Gavala, H.N., Skiadas, I.V., Stamatelatou, K., Lyberatos, G., 2009. Hydrogen and methane production through two-stage mesophilic anaerobic digestion of olive pulp. Bioresour. Technol. 100, 3718–3723.

Kumar, N., Das, D., 2001. Continuous hydrogen production by immobilized *Enterobacter cloacae* IIT-BT 08 using lignocellulosic materials as solid matrices. Enzyme Microb. Technol. 29, 280–287.

Kumar, N., Ghosh, A., Das, D., 2001. Redirection of biochemical pathways for the enhancement of H_2 production by *Enterobacter cloacae*. Biotechnol. Lett. 23, 537–541.

Kumar, R., Singh, S., Singh, O.V., 2008. Bioconversion of lignocellulosic biomass: Biochemical and molecular perspectives. J. Ind. Microbiol. Biotechnol. 35, 377–391.

Kyazze, G., Dinsdale, R., Hawkes, F.R., Guwy, A.J., Premier, G.C., Donnison, I.S., 2008. Direct fermentation of fodder maize, chicory fructans and perennial ryegrass to hydrogen using mixed microflora. Bioresour. Technol. 99, 8833–8839.

Laurinavichene, T.V., Tolstygina, I.V., Galiulina, R.R., Ghirardi, M., Seibert, M., Tsygankov, A.A., 2002. Dilution methods to deprive *Chlamydomonas reinhardtii* cultures of sulfur for subsequent hydrogen photoproduction. Int. J. Hydrogen Energy 27, 1245–1249.

Lay, J.J., Lee, Y.J., Noike, T., 1999. Feasibility of biological hydrogen production from organic fraction of municipal solid waste. Water Res. 33, 2579–2586.

Lay, J.J., Fan, K.S., Chang, I.J., Ku, C.H., 2003. Influence of chemical nature of organic wastes on their conversion to hydrogen by heat-shock digested sludge. Int. J. Hydrogen Energy 28, 1361–1367.

Lee, K.S., Lo, Y.C., Lin, P.J., Chang, J.S., 2006. Improving biohydrogen production in a carrier-induced granular sludge bed by altering physical configuration and agitation pattern of the bioreactor. Int. J. Hydrogen Energy 31, 1648–1657.

Lee, K.S., Lin, P.J., Fang, K., Chang, J.S., 2007. Continuous hydrogen production by anaerobic mixed microflora using a hollow-fiber microfiltration membrane bioreactor. Int. J. Hydrogen Energy 32, 950–957.

Lee, H.S., Vermaas, W.F.J., Rittmann, B.E., 2010. Biological hydrogen production: prospects and challenges. Trends Biotechnol. 28, 262–271.

Levin, D.B., Pitt, L., Love, M., 2004. Biohydrogen production: Prospects and limitations to practical application. Int. J. Hydrogen Energy 29, 173–185.

Levin, D.B., Islam, R., Cicek, N., Sparling, R., 2006. Hydrogen production by *Clostridium thermocellum* 27405 from cellulosic biomass substrates. Int. J. Hydrogen Energy 31, 1496–1503.

Li, D., Chen, H., 2007. Biological hydrogen production from steam-exploded straw by simultaneous saccharification and fermentation. Int. J. Hydrogen Energy 32, 1742–1748.

Li, C., Zhang, T., Fang, H.H.P., 2006. Fermentative hydrogen production in packed-bed and packaging-free upflow reactors. Water Sci. Technol. 54, 95–103.

Lin, C.Y., Cheng, C.H., 2006. Fermentative hydrogen production from xylose using anaerobic mixed microflora. Int. J. Hydrogen Energy 31, 832–840.

Lin, C.Y., Lay, C.H., 2005. A nutrient formulation for fermentative hydrogen production using anaerobic sewage sludge microflora. Int. J. Hydrogen Energy 30, 285–292.

Lipinsky, E.S., Kresovich, S., 1982. Sugar crops as a solar energy converter. Experientia 38, 13–18.

Liu, C., Cheng, X., 2010. Improved hydrogen production via thermophilic fermentation of corn stover by microwave-assisted acid pretreatment. Int. J. Hydrogen Energy 35, 8945–8952.

Liu, F., Fang, B., 2007. Optimization of bio-hydrogen production from biodiesel wastes by *Klebsiella pneumoniae*. Biotechnol. J. 2, 374–380.

Liu, G.Z., Shen, J.Q., 2004. Effects of culture and medium conditions on hydrogen production from starch using anaerobic bacteria. J. Biosci. Bioeng. 98, 251–256.

Liu, Y., Yu, P., Song, X., Qu, Y., 2008. Hydrogen production from cellulose by coculture of *Clostridium thermocellum* JN4 and *Thermoanaerobacterium thermosaccharolyticum* GD17. Int. J. Hydrogen Energy 33, 2927–2933.

Lo, Y.C., Chen, W.M., Hung, C.H., Chen, S.D., Chang, J.S., 2008. Dark H_2 fermentation from sucrose and xylose using H_2-producing indigenous bacteria: Feasibility and kinetic studies. Water Res. 42, 827–842.

Lo, Y.C., Saratale, G.D., Chen, W.M., Bai, M.D., Chang, J.S., 2009. Isolation of cellulose-hydrolytic bacteria and applications of the cellulolytic enzymes for cellulosic biohydrogen production. Enzyme Microb. Technol. 44, 417–425.

Lo, Y.C., Chen, C.Y., Lee, C.M., Chang, J.S., 2010. Combining dark-photo fermentation and microalgae photosynthetic processes for high-yield and CO_2-free biohydrogen production. Int. J. Hydrogen Energy 35, 10944–10953.

Logan, B.E., 2004. Extracting hydrogen and electricity from renewable resources. Environ. Sci. Technol. 38, 160A–167A.

Logan, B.E., Oh, S.E., Kim, I.S., Van Ginkel, S., 2002. Biological hydrogen production measured in batch anaerobic respirometers. Environ. Sci. Technol. 36, 2530–2535.

Luo, G., Xie, L., Zou, Z.H., Zhou, Q., Wang, J.Y., 2010. Fermentative hydrogen production from cassava stillage by mixed anaerobic microflora: Effects of temperature and pH. Appl. Energy 87, 3710–3717.

Lynd, L.R., van Zyl, W.H., McBride, J.E., Laser, M., 2005. Consolidated bioprocessing of cellulosic biomass: an update. Curr. Opin. Biotechnol. 16, 577–583.

Mackie, K.L., Brownell, H.H., West, K.L., Saddler, J.N., 1985. Effect of sulphur dioxide and sulphuric acid on steam explosion of aspenwood. J. Wood Chem. Technol. 5, 405–425.

Magnusson, L., Islam, R., Sparling, R., Levin, D., Cicek, N., 2008. Direct hydrogen production from cellulosic waste materials with a single-step dark fermentation process. Int. J. Hydrogen Energy 33, 5398–5403.

Mars, A.E., Veuskens, T., Budde, M.A.W., van Doeveren, P.F.N.M., Lips, S.J., Bakker, R.R., et al., 2010. Biohydrogen production from untreated and hydrolyzed potato steam peels by the extreme thermophiles *Caldicellulosiruptor saccharolyticus* and *Thermotoga neapolitana*. Int. J. Hydrogen Energy 35, 7730–7737.

McMillan, J.D., 1994. Pretreatment of lignocellulosic biomass. In: Himmel, M.E., Baker, J.O., Overend, R.P. (Eds.), Enzymatic Conversion of Biomass for Fuels Production. American Chemical Society, Washington, DC, pp. 292–324.

Melis, A., Zhang, L., Forestier, M., Ghirardi, M.L., Seibert, M., 2000. Sustained photohydrogen production upon reversible inactivation of oxygen evolution in the green algae *Chlamydomonas reindhardtii*. Plant Physiol. 122, 127–135.

Mes-Hartree, M., Dale, B.E., Craig, W.K., 1988. Comparison of steam and ammonia pretreatment for enzymatic hydrolysis of cellulose. Appl. Microbiol. Biotechnol. 29, 462–468.

Momirlan, M., Veziroglu, T.N., 2005. The properties of hydrogen as fuel tomorrow in sustainable energy system for a cleaner planet. Int. J. Hydrogen Energy 30, 795–802.

Mosier, N., Wyman, C., Dale, B., Elander, R., Lee, Y.Y., Holtzapple, M., et al., 2005. Features of promising technologies for pretreatment of lignocellulosic biomass. Bioresour. Technol. 96, 673–686.

Mu, Y., Yu, H.Q., Wang, G., 2007. Evaluation of three methods for enriching H_2-producing cultures from anaerobic sludge. Enzym. Microb. Technol. 40, 947–953.

Nandi, R., Sengupta, S., 1998. Microbial production of hydrogen: An overview. Crit. Rev. Microbiol. 24, 61–84.

Nath, A., Dixit, M., Bandiya, A., Chavda, S., Desai, A.J., 2008. Enhanced PHB production and scale up studies using cheese whey in fed batch culture of *Methylobacterium* sp. ZP24. Bioresour. Technol. 99, 5749–5755.

Nguyen, T.A.D., Han, S.J., Kim, J.P., Kim, M.S., Sim, S.J., 2010. Hydrogen production of the hyperthermophilic eubacterium, *Thermotoga neapolitana* under N_2 sparging condition. Bioresour. Technol. 101, S38–S41.

Noike, T., Mizuno, O., 2000. Hydrogen fermentation of organic municipal wastes. Water Sci. Technol. 42, 155–162.

Noike, T., Ko, I.B., Lee, D.Y., Yokoyama, S., 2003. Continuous hydrogen production from organic municipal wastes. In: Proceedings of the 1st NRL International Workshop on Innovative Anaerobic Technology. pp. 53–60.

Ntaikou, I., Gavala, H.N., Kornaros, M., Lyberatos, G., 2008. Hydrogen production from sugars and sweet sorghum biomass using *Ruminococcus albus*. Int. J. Hydrogen Energy 33, 1153–1163.

Ntaikou, I., Kourmentza, C., Koutrouli, E.C., Stamatelatou, K., Zampraka, A., Kornaros, M., et al., 2009. Exploitation of olive oil mill wastewater for combined biohydrogen and biopolymers production. Bioresour. Technol. http://dx. doi.org/10.1016/j.biortech. 2008.12.001.

Oh, S., Logan, B.E., 2005. Hydrogen and electricity production from a food processing wastewater using fermentation and microbial fuel cell technologies. Water Res. 39, 4673–4682.

Oh, Y.K., Park, M.S., Seol, E.H., Lee, S.J., Park, S., 2003. Isolation of hydrogen-producing bacteria from granular sludge of an upflow anaerobic sludge blanket reactor. Biotechnol. Bioprocess Eng. 8, 54–57.

Palazzi, E., Fabino, B., Perego, P., 2000. Process development of continuous hydrogen production by *Enterobacter aerogenes* in a packed column reactor. Bioprocess Eng. 22, 205–213.

Pan, C.M., Fan, Y.T., Xing, Y., Hou, H.W., Zhang, M.L., 2008. Statistical optimization of process parameters on biohydrogen production from glucose by *Clostridium* sp. Fanp2. Bioresour. Technol. 99, 3146–3154.

Panagiotopoulos, I.A., Bakker, R.R., Budde, M.A.W., de Vrije, T., Claassen, P.A.M., Koukios, E.G., 2009. Fermentative hydrogen production from pretreated biomass: a comparative study. Bioresour. Technol. 100, 6331–6338.

Panagiotopoulos, I.A., Bakker, R.R., de Vrije, T., Koukios, E.G., Claassen, P.A.M., 2010. Pretreatment of sweet sorghum bagasse for hydrogen production by *Caldicellulosiruptor saccharolyticus*. Int. J. Hydrogen Energy 35, 7738–7747.

Pattra, S., Sangyoka, S., Boonmee, M., Reungsang, A., 2008. Bio-hydrogen production from the fermentation of sugarcane bagasse hydrolysate by *Clostridium butyricum*. Int. J. Hydrogen Energy 33, 5256–5265.

Puppan, D., 2002. Environmental evaluation of biofuels. Period Polytech. Ser. Soc. Man Sci. 10, 95–116.

Rabinovich, M.L., Melnik, M.S., Bolobova, A.V., 2002. Microbial cellulases: A review. Appl. Biochem. Microbiol. 38, 305–321.

Rachman, M.A., Nakashimada, Y., Kakizono, T., Nishio, N., 1998. Hydrogen production with high yield and high evolution rate by self-flocculated cells of *Enterobacter aerogenes* in a packed-bed reactor. Appl. Microbiol. Biotechnol. 49, 450–454.

Ragauskas, A.J., Williams, C.K., Davison, B.H., Britovsek, G., Cairney, J., Eckert, C.A., et al., 2006. The path forward for biofuels and biomaterials. Science 311, 484–489.

Rajaram, S., Verma, A., 1990. Production and characterization of xylanase from *Bacillus thermoalkalophilus* growth on agricultural wastes. Appl. Microbiol. Biotechnol. 34, 141–144.

Ren, N.Q., Chua, H., Chan, S.Y., Tsang, Y.F., Wang, Y.J., Sin, N., 2007. Assessing optimal fermentation type for bio-hydrogen production in continuous flow acidogenic reactors. Bioresour. Technol. 98, 1774–1780.

Sakai, S., Yagishita, T., 2007. Microbial production of hydrogen and ethanol from glycerol-containing wastes discharged from a biodiesel fuel production plant in a bioelectrochemical reactor with thionine. J. Biosci. Bioeng. 98, 340–348.

Saratale, G.D., Chen, S.D., Lo, Y.C., Saratale, R.G., Chang, J.S., 2008. Outlook of biohydrogen production from ligno-cellulosic feedstock using dark fermentation: A review. J. Sci. Ind. Res. 67, 962–979.

Saratale, G.D., Saratale, R.G., Lo, Y.C., Chang, J.S., 2010. Multicomponent cellulase production by *Cellulomonas biazotea* NCIM-2550 and their applications for cellulosic biohydrogen production. Biotechnol. Prog. 26, 406–416.

Saratale, G.D., Chien, I.J., Chang, J.S., 2011. Enzymatic pretreatment of cellulosic wastes for anaerobic treatment and bioenergy production. In: Fang, H.H.P. (Ed.), Environmental Anaerobic Technology Applications and New Developments. Imperial College Press, London, pp. 279–308.

Schubert, C., 2006. Can biofuels finally take center stage? Nat. Biotechnol. 24, 777–784.

Schroder, C., Selig, M., Schonheit, P., 1994. Glucose fermentation to acetate, CO_2 and H_2 in the anaerobic hyperther-mophilic eubacterium *Thermotoga maritime*: Involvement of the Embden-Meyerhof pathway. Arch. Microbiol. 161, 460–470.

Shi, X., Jung, K.W., Kim, D.H., Ahn, Y.T., Shin, H.S., 2011. Direct fermentation of *Laminaria japonica* for biohydrogen production by anaerobic mixed cultures. Int. J. Hydrogen Energy 36, 5857–5864.

Shin, H.S., Youn, J.H., Kim, S.H., 2004. Hydrogen production from food waste in anaerobic mesophilic and thermo-philic acidogenesis. Int. J. Hydrogen Energy 29, 1355–1363.

Shizas, I., Bagley, D.M., 2005. Fermentative hydrogen production in a system using anaerobic digester sludge without heat treatment as a biomass source. Water Sci. Technol. 52, 139–144.

Show, K.Y., Zhang, Z.P., Lee, D.J., 2008. Design of bioreactors for biohydrogen production. J. Sci. Ind. Res. 67, 941–949.

Singh, J., Cu, S., 2010. Commercialization potential of microalgae for biofuels production. Renew. Sustain. Energy Rev. 14, 2596–2610.

Sivers, M.V., Zacchi, G., 1995. A techno-economical comparison of three processes for the production of ethanol from pine. Bioresour. Technol. 51, 43–52.

Slade, R., Bauen, A., Shah, N., 2009. The commercial performance of cellulosic ethanol supply-chains in Europe. Biotechnol. Biofuels 2 (1), 3.

Somerville, C., Youngs, H., Taylor, C., Davis, S.C., Long, S.P., 2010. Feedstocks for lignocellulosic biofuels. Science 329, 790–792.

Sparling, R., Risbey, D., Poggi-Varaldo, H.M., 1997. Hydrogen production from inhibited anaerobic composters. Int. J. Hydrogen Energy 22, 563–566.

Tanisho, S., Ishiwata, W., 1994. Continuous hydrogen production from molasses by the bacterium *Enterobacter aerogenes*. Int. J. Hydrogen Energy 19, 807–812.

Tao, H., Cornish, V.W., 2002. Milestones in directed enzyme evolution. Curr. Opin. Chem. Biol. 6, 858–864.

Thong, S.O., Prasertsan, P., Intrasungkha, N., Dhamwichukorn, S., Birkeland, N.-K., 2007. Improvement of biohydrogen production and treatment efficiency on palm oil mill effluent with nutrient supplementation at ther-mophilic condition using an anaerobic sequencing batch reactor. Enzyme Microbial. Technol. 41, 583–590.

Thong, S., Prasertsan, O.P., Karakashev, D., Angelidaki, I., 2008. Thermophilic fermentative hydrogen production by the newly isolated *Thermoanaerobacterium thermosaccharolyticum* PSU-2. Int. J. Hydrogen Energy 33, 1204–1214.

Ting, C.H., Lin, K.R., Lee, D.J., Tay, J.H., 2004. Production of hydrogen and methane from wastewater sludge using anaerobic fermentation. Water Sci. Technol. 50, 223–228.

Turner, P., Gashaw, M., Karlsson, E.N., 2007. Potential and utilization of thermophiles and thermostable enzymes in biorefining. Microbial Cell Fact. 6, 9.

Ueno, Y., Kawai, T., Sato, S., Otsuka, S., Morimoto, M., 1995. Biological production of hydrogen from cellulose by natural anaerobic microflora. J. Ferment. Bioeng. 79, 395–397.

Ueno, Y., Otsuka, S., Morimoto, M., 1996. Hydrogen production from industrial wastewater by anaerobic microflora in chemostate culture. J. Ferment. Bioeng. 82, 194–197.

Valdez-Vazquez, I., Poggi-Varaldo, H.M., 2009. Hydrogen production by fermentative consortia. Renew. Sustain. Energy Rev. 13, 1000–1013.

Valdez-Vazquez, I., Rios-Leal, E., Esparza-Garcia, F., Cecchi, F., Poggi-Varaldo, H.M., 2005. Semi-continuous solid substrate anaerobic reactors for H_2 production from organic waste: Mesophilic versus thermophilic regime. Int. J. Hydrogen Energy 30, 1383–1391.

Van Ginkel, S., Sung, S., Lay, J.J., 2001. Biohydrogen production as a function of pH and substrate concentration. Environ. Sci. Technol. 35, 4726–4730.

Van Ginkel, S.W., Oh, S., Logan, B.E., 2005. Biohydrogen gas production from food processing and domestic wastewaters. Int. J. Hydrogen Energy 30, 1535–1542.

van Niel, E.W.J., Budde, M.A.W., de Haas, G.G., van der Wal, F.J., Claassen, P.A.M., Stams, A.J.M., 2002. Distinctive properties of high hydrogen producing extreme thermophiles, *Caldicellulosiruptor saccharolyticus* and *Thermotoga elfii*. Int. J. Hydrogen Energy 27, 1391–1398.

van Wyk, J.P.H., Mohulatsi, M., 2003. Biodegradation of wastepaper by cellulase from *Trichoderma viride*. Bioresour. Technol. 86, 21–23.

Venetsaneas, N., Antonopoulou, G., Stamatelatou, K., Kornaros, M., Lyberatos, G., 2009. Using cheese whey for hydrogen and methane generation in a two-stage continuous process with alternative pH controlling approaches. Bioresour. Technnol. 100, 3713–3717.

Venkata Mohan, S., Bhaskar, Y.V., Sarma, P.N., 2007. Biohydrogen production from chemical wastewater treatment by selectively enriched anaerobic mixed consortia in biofilm conFig d reactor operated in periodic discontinuous batch mode. Water Res. 41, 2652–2664.

Venkata Mohan, S., Mohanakrishna, G., Goud, R.K., Sarma, P.N., 2009. Acidogenic fermentation of vegetable based market waste to harness biohydrogen with simultaneous stabilization. Bioresour. Technol. http://dx.doi.org/10.1016/j.biortech. 2008.12.059.

Vijayaraghavan, K., Soom, M.A.M., 2007. Trends in biological hydrogen production: A review. Int. J. Hydrogen Energy.

Vijayaraghavan, K., Ahmad, D., Ibrahim, M.K.B., 2006. Biohydrogen generation from jackfruit peel using anaerobic contact filter. Int. J. Hydrogen Energy 31, 569–579.

Wang, J.L., Wan, W., 2009. Factors influencing fermentative hydrogen production: A review. Int. J. Hydrogen Energy 34, 799–811.

Wang, C.C., Chang, C.W., Chu, C.P., Lee, D.J., Chang, B.V., Liao, C.S., 2003a. Producing hydrogen from wastewater sludge by *Clostridum bifermentans*. J. Biotechnol. 102, 83–92.

Wang, C.C., Chang, C.W., Chu, C.P., Lee, D.J., Chang, B.V., Liao, C.S., et al., 2003b. Using filtrate of waste biosolids to effectively produce bio-hydrogen by anaerobic fermentation. Water Res. 37, 2789–2793.

Wang, C.H., Lu, W.B., Chang, J.S., 2007. Feasibility study on fermentative conversion of raw and hydrolyzed starch to hydrogen using anaerobic mixed microflora. Int. J. Hydrogen Energy 32, 3849–3859.

Wang, J.P., Chen, Y.Z., Yuan, S.J., Sheng, G.P., Yu, H.Q., 2009. Synthesis and characterization of a novel cationic chitosan-based flocculant with a high water-solubility for pulp mill wastewater treatment. Water Res. 43, 5267–5275.

Wayman, M., Parekh, S.R., 1990. Biotechnology of biomass conversion; Fuels and chemicals from renewable resources. Open University Press, Milton, Keynes.

Wijffels, R.H., Barbosa, M.J., 2010. An outlook on microalgal biofuels. Science 329 (5993), 796–799.

Williams, P.J.L., Laurens, L.M.L., 2010. Microalgae as biodiesel and biomass feedstocks: Review and analysis of the biochemistry, energetics and economics. Energy Environ. Sci. 3, 554–590.

Wood, N.P., Jungermann, K.A., 1972. Inactivation of the pyruvate formate lyase reaction of *Clostridium butiricum*. FEBS Lett. 27, 49–52.

Wu, K.J., Saratale, G.D., Lo, Y.C., Chen, S.D., Chen, W.M., Tseng, Z.J., et al., 2008. Fermentative production of 2, 3 butanediol, ethanol and hydrogen with *Klebsiella* sp. isolated from sewage sludge. Bioresour. Technol. 99, 7966–7970.

Yang, H., Shao, P., Lu, T., Shen, J., Wang, D., Xu, Z., et al., 2006. Continuous bio-hydrogen production from citric acid wastewater via facultative anaerobic bacteria. Int. J. Hydrogen Energy 31, 1306–1313.

Yang, P., Zhang, R., McGarvey, J.A., Benemann, J.R., 2007. Biohydrogen production from cheese processing wastewater by anaerobic fermentation using mixed microbial communities. Int. J. Hydrogen Energy 32, 4761–4771.

Yang, Z.M., Guo, R.B., Xu, X.H., Fan, X.L., Li, X.P., 2010. Enhanced hydrogen production from lipid-extracted microalgal biomass residues through pretreatment. Int. J. Hydrogen Energy 35, 9618–9623.

Yokoi, H., Tokushige, T., Hirose, J., Hayashi, S., Takasaki, Y., 1997. Hydrogen production by immobilized cells of aciduric Enterobacter aerogenes strain HO-39. J. Ferment. Bioeng. 83, 481–484.

Yokoi, H., Saitsu, A.S., Uchida, H., Hirose, J., Hayashi, S., Takasaki, Y., 2001. Microbial hydrogen production from sweet potato starch residue. J. Biosci. Bioeng. 91, 58–63.

Yu, H.Q., Mu, Y., 2006. Biological hydrogen production in a UASB reactor with granules. II. Reactor performance in 3-year opration. Biotechnol. Bioeng. 94, 988–995.

Yu, H.Q., Zhu, Z.H., Hu, W.R., Zhang, H.S., 2002. Hydrogen production from rice winery wastewater in an upflow anaerobic reactor by using mixed anaerobic cultures. Int. J. Hydrogen Energy 27, 1359–1365.

Zhang, T., Liu, H., Fang, H.H.P., 2003. Biohydrogen production from starch in wastewater under thermophilic condition. J. Environ. Management 69, 149–156.

Zhang, P.Y.H., Himmel, M.E., Mielenz, J.R., 2006. Outlook for cellulase improvement: Screening and selection strategies. Biotechnol. Adv. 24, 452–481.

Zhang, Z.P., Tay, J.H., Show, K.Y., Yan, R., Liang, D.T., Lee, D.J., et al., 2007. Biohydrogen production in a granular activated carbon anaerobic fluidized bed reactor. Int. J. Hydrogen Energy 32, 185–191.

Zhang, P., Berson, Y.H., Sarkanen, E., Dale, S., 2009. Pretreatment and biomass recalcitrance: Fundamentals and progress. Appl. Biochem. Biotechnol. 153, 80–83.

Biohydrogen Production from Wastewater

S. Venkata Mohan, K. Chandrasekhar, P. Chiranjeevi,
P. Suresh Babu

Bioengineering and Environmental Centre, CSIR–Indian Institute of
Chemical Technology, Hyderabad, India

INTRODUCTION

In the regime of alternative, renewable, carbon-neutral, and eco-friendly fuels to accomplish the burgeoning energy demands, biohydrogen (H_2) is deemed to have a major futuristic role. H_2 production through biological roots is considered to be one of the sustainable routes envisaged to meet future energy demands. Scientific fraternity in the interdisciplinary area of environment and energy are apparently shifting their efforts from "pollution control" to "resource exploitation from waste" more recently. Wastewater at present is being considered as an important and renewable commodity for resource recovery. Wastewater treatment/remediation is an energy-intensive process that increases the economic burden on the effluent treatment plant (ETP) operators, especially pertaining to the industry. Finding ways to produce/recover useful products or value addition from wastewater remediation are gaining significance in recent times. In the perspective of environmental sustainability, negative valued wastewater can be considered as a potential substrate/feedstock for biological H_2 production by simultaneously achieving pollution control. Reducing the wastewater treatment cost by generating bioenergy, such as H_2 gas, from the organic matter present in wastewater is a sustainable opportunity. The foremost essence of sustainable water management is to transform wastewater into a usable form, which is true with H_2 production from its treatment.

WASTEWATER AS RENEWABLE SUBSTRATE FOR BIOHYDROGEN PRODUCTION

Rapid industrialization and population exploitation are obviously generating enormous magnitudes of wastewater. The regulatory requirement for their treatment prior to disposal perceptibly makes wastewater an ideal commodity to produce renewable energy in the form of H_2 by anaerobic treatment (Venkata Mohan, 2010). The intrinsic advantage of wastewater is its biodegradable organic fraction associated with its inherent net positive energy. Wastewater contains enough energy to meet a major fraction of the world's energy demand—if it could be converted to economically useful energy forms (Rittmann, 2008). Utilization of wastewater as a potential substrate for H_2 generation through biological routes documented considerable interest due to its sustainable nature. Harnessing of H_2 from wastewater will significantly reduce the cost of overall wastewater treatment process.

The Last decade witnessed considerable efforts on the application of various wastewater from domestic and industrial as potential substrates for the production of H_2 through biological machinery, mainly through light-driven (Table 1) and light-independent (Table 2) fermentation processes. Simple sugars to complex effluents, agricultural and food industry wastes rich in carbohydrates, were also evaluated for H_2 production. Theoretically, 1 kg of glucose ($C_6H_{12}O_6$) contains 1.066 kg of chemical oxygen demand (COD) (937.5 g or 5.2 mol of glucose equal to 1 kg of COD). By dark fermentation, 1 mol (180 g) of glucose can produce 4 and 2 mol of molecular H_2 based on acetate and butyrate pathways, respectively. By the photofermentation pathway, 1 mol of glucose can produce 12 mol of H_2. Theoretically, 1 kg COD can produce 20.83 mol of H_2. According to ideal gas law ($PV = RT$) at STP [standard temperature ($300\,^\circ K/27\,^\circ C$) and pressure (1 atmosphere)], 1 mol of H_2 occupies 22.4 liters volume. Accordingly, 1 mol glucose (192 g COD) can produce 89.6 liters of H_2. Therefore, 1 kg COD (5.2 mol glucose) can produce 20.83 mol of H_2 (466.6 liters of H_2/41.6 g of H_2). When 40% of COD removal efficiency was considered for H_2 production, the dark fermentation process can produce 125 g of H_2 and photofermentation conversion can yield 16.6 g of H_2. The food-processing industry in India is producing $\approx 3{,}000{,}000 \times 10^5$ liters of wastewater per year. It contains an average COD of 20 g/liter, accounting for a total of 6000×10^6 g of COD per year. Photofermentative process can produce about 300×10^6 kg of H_2 per year (with 40% removal rate) which accounts for $ 1,200 million per year (at a rate of $ 4 per kg H_2). Similarly, the same wastewater can generate revenue of $ 80 million (5×10^6 kg of H_2) per year by the dark fermentative process (on 40% removal basis). However, there are many limiting factors which influence the efficiency of photofermentation process compared to the dark fermentation.

Therefore, exploitation of wastewater as a substrate for H_2 production with simultaneous treatment will lead to a new avenue for the utilization of renewable and inexhaustible energy sources. Futuristic application of "waste from hydrogen" processes can be visualized in ETP where an anaerobic treatment process is a prerequisite.

Light-Driven and Light-Independent Fermentation for Biohydrogen Production from Wastewater

Based on the mechanism, biological H_2 production through wastewater remediation can be classified into light-driven photosynthetic and light-independent dark fermentation processes. Hydrogen gas generated by proton-reducing reactions is a common fermentation by-product generated during electron-acceptor-limited microbial processes (Madsen, 2008). The biochemistry and metabolism involved in these biological routes vary significantly based on the biocatalyst nature, operating conditions adapted, microenvironment employed, and substrate used. Photosynthetic bacteria (PSB) and few microalgae manifest H_2 production through fermentation of a wide variety of substrates, including wastewater in the presence of light. The light-driven process is of two types Viz., oxygenic and anoxygenic, where oxygen will be generated during water photolysis in the presence of light during the oxygenic

TABLE 1 Wastewater Used as Substrate for H_2 Production Through Photofermentation

Type of wastewater	Reference
Beet molasses and black satrape	Keskin and Hallenbeck, 2012; Ozgur et al., 2010b
Dark fermented effulents of sugar beet thick juice	Özkan et al., 2012
Chemical wastewater	Venkata Mohan et al., 2008f
Dark fermentation effluent of ground wheat solution	Argun and Kargi, 2010
Dairy wastewater	Venkata Mohan et al., 2008f; Seifert et al., 2010a
Tofu wastewater	Zheng et al., 2010; Zhu et al., 2002
Lignocellulose-derived organic acids	Zhu et al., 2010
Product of starch fermentation	Laurinavichene et al., 2008
Acid-hydrolyzed wheat starch	Kapdan et al., 2009
Dark fermentation effluent	Sagnak and Kargi, 2011; Ozmihci and Kargi, 2010; Chen et al., 2010; Ozgur et al., 2010a; Srikanth et al., 2009b; Chandra and Venkata Mohan, 2011; Lee et al., 2011
Soluble metabolites from dark fermentation	Lo et al., 2011
Bagasse	Wu et al., 2010
Oil palm waste hydrolysate	Pattanamanee et al., 2012
Olive mill wastewater	Eroglu et al., 2004, 2006, 2010, 2011
Pretreated olive mill waste	Ena et al., 2010
Synthetic soybean wastewater	Lu et al., 2011
Potato homogenate	Laurinavichene et al., 2010
Brewery wastewaters	Seifert et al., 2010b
Domestic wastewater and synthetic wastewater	Venkata Mohan et al., 2008f
Synthetic fatty acids	Srikanth et al., 2009a,b

TABLE 2   Source and Type of Wastewater Utilized for H_2 Production Through Dark Fermentation

Wastewater category	Type of wastewater	Reference
Food-processing industry	Food processing	Van Ginkel et al., 2005; Sentürk et al., 2010; Zhu et al., 2009
	Raw food waste	Kobayashi et al., 2012; Kim et al., 2011a, 2012; Chakkrit et al., 2011; Venkata Mohan et al., 2012; Venkateswar Reddy et al., 2011; Kima et al., 2012
	Codigested municipal food waste and sewage sludge	Zhu et al., 2011
	Sewage sludge and industrial food waste	Siddiqui et al., 2011
	Mushroom farm waste	Li et al., 2011
	Yeast waste	Chou et al., 2011
	Organic wastes composed of food waste and sewage sludge	Im et al., 2012
	Fruit and vegetables from unsold stocks	Licata et al., 2011
	Organic residues obtained from a bioethanol fermentation process using rice straw	Cheng et al., 2012
	Coffee drink manufacturing	Jung et al., 2010
	Tofu wastewater	Kim and Lee, 2010; Kim et al., 2011b
	Starch-based wastewater	Sen and Suttar, 2012; Sompong et al., 2011
	Citric acid wastewater	Yang et al., 2006
	Slaughterhouse waste	Venkata Mohan et al., 2012; Sittijunda and Reungsang, 2012
	Sweet sorghum syrup/extract	Saraphirom and Reungsang, 2010, 2011
	Potato steam peel hydrolysate	Zhu et al., 2008; Yokoi et al., 2001
Dairy-based industries	Dairy processing	Gustavo et al., 2008; Ren et al., 2007; Venkata Mohan et al., 2007c, 2008a; Julia et al., 2012; Kargi et al., 2012a,b
	Dairy waste permeate/waste lactose	Wang et al., 2009
	Cheese processing	Ferchichi et al., 2005; Yang et al., 2007; Rai et al., 2012
	Cattle wastewater	Tang et al., 2008
Alcohol-based industries	Brewery wastewater	Chang et al., 2008; Shi et al., 2010
	Wine process wastewater	Yu et al., 2002; Froylán et al., 2009

Continued

TABLE 2 Source and Type of Wastewater Utilized for H_2 Production Through Dark Fermentation—Cont'd

Wastewater category	Type of wastewater	Reference
	Molasses-based wastewater	Vatsala et al., 2008; Venkata Mohan et al., 2008c, 2011a; Han et al., 2012a,b
	Acid-hydrolyzed molasses	Morsy, 2011; Pawinee et al., 2011
	Alcohol distillery	Qiu et al., 2011
	Vinasse	Buitrón and Carvajal, 2010; Fernandes et al., 2010
	Sugar refinery	Liu et al., 2006
Plant/ agricultural-based waste	Vegetable-based waste	Rozendal et al., 2006; Venkata Mohan et al., 2009b
	Paper mill	Idania et al., 2005; Lakshmidevi and Muthukumar, 2010
	Lignocellulose-derived organic acids	Suriyamongkol et al., 2007; Zhu et al., 2010
	Wheat straw hydrolysate	Kongjan and Angelidaki, 2010
	Cassava stillage	Luo et al., 2010; Cao et al., 2012
	Waste ground wheat	Sagnak et al., 2011
	Citrus peelings	Venkata Mohan et al., 2009a
	Mixture of swine manure and fruit and vegetable market waste	Tenca et al., 2011; Mohankrishna et al., 2010a
Organic-based industries	Chemical wastewater	Saratale et al., 2008; Venkata Mohan et al., 2007a,b,d; Vijaya Bhaskar et al., 2008; Wang et al., 2009
	Glycerin from biodiesel production	Fernandes et al., 2010; Liu and Fang, 2007
	Phenol-containing wastewater	Tai et al., 2010
	Biodiesel solid residues	Gopalakrishnan et al., 2012
Oil-based industries	Palm oil mill effluent	Vijayaraghavan and Ahmad, 2006; Wu et al., 2009; Yossan et al., 2012
	Olive mill wastewater	Ntaikou et al., 2009; Eroğlua et al., 2006; Kargi and Catalkaya, 2011
Others	Landfill leachate	Liu et al., 2010; Hafez et al., 2010
	Fresh leachate and glucose	Liu et al., 2011, 2012
	Textile wastewater	Li et al., 2012
	Sludge	Guo et al., 2012
	Filtrate of activated sludge	Yang et al., 2006
	Probiotic wastewater	Sivaramkrishna et al., 2009

process, while water splitting and oxygen generation will be absent, even in the presence of light, during the anoxygenic process. Light energy will be utilized for the conversion of organic acids into storage material during the anoxygenic process, which associates with H_2 production. During the oxygenic process, the generated oxygen acts as an electron sink and will scavenge the protons and electrons without resulting in H_2 evolution. Generally, green algae follow the oxygenic process, while PSB follow the anoxygenic process. However, few species from both the classes follow both the mechanisms based on the available sources and hence can be termed as mixotrophic. PSB and a few microalgae such as *Chlorella* sp. will consider H_2 as an electron sink under anoxygenic condition by utilizing simple organic acids or even dihydrogen sulfide as an electron donor (Akkerman et al., 2002).

Purple nonsulfur (PNS) bacteria by photoheterotrophic mechanism produce H_2 in the presence of light by utilizing organic substrates through photofermentation machinery. Unlike the biophotolysis process, PNS bacteria under anoxygenic conditions generate H_2 utilizing organic substrates, including organic acids (Keskin and Hallenbeck, 2012). The phototrophic fermentative pathway by PSB bacteria is a promising approach for biohydrogen production due to its high theoretical H_2 yield from various volatile organic acids (Chen et al., 2011; Wang et al., 2010). Photofermentation is considered to be a potentially promising process because of its inherent high conversion yields of the substrate to product, lack of oxygen-evolving activity, desirable since oxygen inhibits H_2 producing enzymes, ability to use a wide spectrum of light, and capacity to generate H_2 using organic substrates derived from wastes, thereby carrying out simultaneous wastewater treatment (Keskin et al., 2011). PSB are resistant to high light intensity and changing environmental conditions. Wastewaters derived from a variety of domestic and industry processes have been used for photofermentative H_2 production (Table 1). The light-driven photofermentation mechanism seems particularly well suited for the conversion of some particular wastes to H_2, as well as being an attractive option for permitting the complete conversion of some feedstocks suited for dark fermentation, which, however, can only produce H_2 from them at low yields (Hallenbeck and Ghosh, 2009; Keskin et al., 2011). The substrate affects cell growth and H_2 production during the photofermentation process (Kim et al., 2012).

The light-independent dark fermentation process associates with the anaerobic metabolism of acidogenic bacteria (mostly) to generate H_2, along with volatile fatty acids (VFA) and CO_2. Syntrophic association and effective interspecies electron exchange is essential in the anaerobic fermentation process, especially with mixed consortia, for the conversion of wastewater organics to bioenergy (methane or H_2) (McInerney, 2008; McInerney et al., 2009; Morita et al., 2011; Stams and Plugge, 2009). Photofermentation differs from dark fermentation. Photosynthetically derived energy permits the use of substrate (organic acids) for H_2 conversion which is thermodynamically not feasible with dark fermentation process (Keskin et al., 2011). The effective conversion of organics to end products needs an electron sink, which decides the terminal product formation, especially during dark fermentation. Methanogenic microorganisms (MB) can function as a sink, consuming electrons in the reduction of carbon dioxide to methane. However, H_2 serves as an electron shuttle between the microorganisms degrading organic compounds and methanogens. Methanogens utilize the formed H_2 as an electron donor to produce methane in the presence of CO_2 that acts as an electron acceptor (syntrophic association). If the methanogenic function is restricted during fermentation, acidogenic hydrogenptrophs can function as an electron sink, resulting in H_2 evolution as a terminal product along with VFA and CO_2. Selective enrichment of the biocatalyst will help in making the H_2 an electron sink during fermentation without leading to methanogenesis. The protons get reduced to H_2 in the presence of electrons derived from ferredoxin reduction manifested by hydrogenase or nitrogenase enzymes.

Interconversion of metabolites, namely, formate and succinate, during substrate degradation also increases the availability of reducing equivalents, which further manifest H_2 production. Fermentative conversion of the organic substrate to its end products involves a series of interrelated biochemical reactions—hydrolysis, acidogenesis, acetogenesis, and methanogenesis—manifested by five physiologically distinct groups of microorganisms. Both obligate and facultative bacteria can catalyze H_2 production from organic substrates.

The dark fermentation process is rapidly gaining importance as a practically viable method among the other biological routes of H_2 production, especially with the application of wastewater as a substrate associated with the usage of mixed consortia as a biocatalyst. The process simplicity, relatively less energy intensiveness, fewer footprints, possibility of utilizing a wide spectrum of organics, and feasibility to operate at ambient temperatures and pressures are some of the inherent striking features of the dark fermentation process (Venkata Mohan et al., 2011b). The dark fermentation process is technically much simpler, requires low operating costs, and is more stable and robust (Gustavo et al., 2008; Hallenbeck and Benemann, 2002; Idania et al., 2005; Kraemer and Bagley, 2007; Venkata Mohan et al., 2009a,b, 2010a), which makes it practically more feasible for the mass production of H_2. A wide spectrum of wastewaters from industrial and domestic origin was evaluated as substrates for H_2 production with the dark fermentation process along with simultaneous remediation (Table 2).

The heterotrophic photosynthetic mechanism is of much interest in the context of wastewater utilization. At present, the photofermentation process application is very visible with the utility of acid-rich dark fermentative effluents from H_2 production as a primary substrate due to its inherent ability to utilize simple organic acids as electron donors. PSB can utilize the organic acids readily that are generated from the dark fermentation process to produce additional H_2 (Chandra and Venkata Mohan, 2011; Srikanth et al., 2009a,b; Venkata Mohan et al., 2008f). Sequential integration of dark fermentation with photofermentation is being considered an efficient and economically viable step towards further enhancement of H_2 recovery and wastewater treatment. PSB-based wastewater treatment can realize pollutants removal and resource recovery simultaneously (Liu et al., 2010).

Scientometric analysis performed with ISI Web of Knowledge [Thomson Reuters; SCI-Expanded (since 1987)], CPCI-S (since 1990), and CPCI-SSH, (since 1990) documented significant increments in biohydrogen publications since 2004. About 340 publications were documented in the year 2011 with citations of more than 6800. Out of the total records on biohydrogen, 63% relate to the fermentation process (citations, 14146) and 10% relate to the photosynthesis process (citations, 3211). Application of waste as a substrate gained prominence since 2004, registering maximum documentation in the year 2011 (records, 180; citations, 4000; H-index, 53). Records segregated based on the biocatalyst nature showed 23% pertaining to the application of mixed culture as a biocatalyst for H_2 production.

Wastewater Treatment

When wastewater is being used as a substrate, substrate degradation efficiency is also important, along with H_2 production, when process efficiency is considered. A trade-off exists between technical efficiency based on H_2 production and substrate removal in association with operating conditions (Venkata Mohan, 2010). A neutral pH is ideal for substrate degradation, while an acidic pH helps H_2 production (Venkata Mohan et al., 2007a–d, 2008b, 2011b). Balancing the conditions for combined performance is especially important in sustaining the economic viability and environmental acceptability of the process. Taguchi's design of experimental methodology was employed to enumerate the role of selected factors on both H_2 production and substrate degradation (Venkata Mohan et al., 2009c). Data enveloping analysis was used to study the role of some important factors—nature of inoculum, pretreatment method, pH, cosubstrate addition, and feed composition on combined process efficiency (Venkata Mohan et al., 2009c, 2011b). Analysis showed that the untreated anaerobic inoculum under acidic conditions utilizing simple wastewater as a substrate showed good process (combined) efficiency.

MIXED CONSORTIA AS BIOCATALYST

Both anaerobic and photosynthetic organisms are capable of generating H_2 under defined conditions. Obligate anaerobes, thermophiles, methanogens, and a few facultative anaerobes can produce H_2 through dark fermentation mechanism. Biohydrogen research during the preliminary phase was mostly associated with pure cultures employing a defined substrate. Subsequent to wastewater coming into existence as a substrate, particularly in the last decade, the application of mixed consortia as biocatalyst showed a great deal of promise. A suitable biocatalyst/inoculum selection significantly influences the H_2 production efficiency, especially when wastewater is used as the substrate. Application of mixed consortia as biocatalyst is considered to be the promising, as well as practical, approach in the context of wastewater usage, especially for upscaling purposes. Operational flexibility and stability, diverse biochemical functions, possibility of using a broad range of substrates, and nonsterile conditions are some of the impending beneficial features of mixed culture usage as a biocatalyst (Angenent et al., 2004; Venkata Mohan, 2008, 2010; Wang and Wan, 2009). It also offers a lower operational cost and ease of control (Leano and Babel, 2012; Venkata Mohan et al., 2008a, 2011b; Wang et al., 2011). Usage of mixed cultures is also gaining interest due to their abilities to utilize a variety of complex substrates in an ever-changing and complex wastewater environment. However, a mixed culture comprises various physiological groups of microbes that lead to diverse metabolic functions that are not only specific to H_2 production. Henceforth, a selective enrichment of H_2 producers by application of pretreatment to parent culture facilitates enhanced process efficiency (Goud and Venkata Mohan, 2012; Srikanth et al., 2010; Venkata Mohan et al., 2008a). Using defined mixed cultures is expected to enhance the sustainability of the process by enhancing their chances of survival in comparison to single bacterial cultures (Patel et al., 2012).

Pretreatment of Biocatalyst

Biocatalyst function is important, especially in a wastewater-mixed culture microenvironment (Venkata Mohan and Goud, 2012). H_2 production with typical anaerobic consortia gets consumed rapidly by the methanogens (Goud and Venkata Mohan, 2012; Ren et al., 2008b; Sparling et al., 1997; Venkata Mohan, 2008; Venkata Mohan et al., 2008a). Untreated consortia support proton reduction during methanogenesis rather than its shutting between intermediates during the interconversion of metabolites, which is presumed to be necessary for H_2 to form as an end product (Srikanth et al., 2010). When mixed culture is used as a biocatalyst/inoculum, pretreatment plays an important role (Kim et al., 2003; Kraemer and Bagley, 2007; Venkata Mohan, 2008; Venkata Mohan and Goud, 2012; Venkata Mohan et al., 2008a,b,c,e,f; Zhu and Beland, 2006). Applying pretreatment to the parent inoculum facilitates selective enrichment of acidogenic H_2-producing bacteria (AB) with a simultenous prevention of hydrogenotrophic methanogens (MB). Suppression or termination of methanogenic activity of anaerobic cultures allows H_2 to become the metabolic end product (Venkata Mohan, 2008; Venkata Mohan and Goud, 2012). Physiological differences between H_2-producing bacteria (AB) and H_2 uptake bacteria (MB) form the main fundamental basis for the pretreatment method (Zhu et al., 2006). H_2-producing bacteria can form spores that protect them in adverse environmental conditions (high temperature, extreme acidity and alkalinity), but methanogens lack such a capability (Venkata Mohan et al., 2008a; Zhu and Beland, 2006). Different pretreatment methods—heat shock, chemical, acid shock, alkaline shock, oxygen shock, load shock, infrared irradiation, freezing and thawing, and microwave irradiation—having different functional properties were evaluated (Table 3). Pretreatment also prevents competitive growth and coexistence of other bacteria, which are generally H_2 consumers.

The heat-shock pretreatment method is one of the most widely applied methods for preparing H_2-producing consortia. The heat-shock method relies on the thermal suppression of methanogenic *Archaea* and nonsporulating bacteria, thereby enriching the culture with sporulating H_2-producing bacteria such as *Clostridium* sp. (Lay et al., 2004). An acid-shock application suppresses methanogenic activity by simultaneously protecting the spore-forming bacteria (Liu et al., 2009b; Ren et al., 2008b). Methanogenic activity is limited to a relatively narrow pH range (6.0–7.5), while most H_2-producing acidogens can grow over a broader pH range (4.5–6.5). Chemical pretreatment methods facilitate exposure of the parent culture to inhibitory chemicals, namely, 2-bromoethanesulfonic acid (BESA), iodopropane, and acetylene, to suppress/inhibit specific metabolic function. BESA, a structural analog of coenzyme M found specifically in methanogens, selectively inhibits methanogenic activity (Chang et al., 2011; Luo et al., 2010; Venkata Mohan et al., 2008b, 2011b; Zhu et al., 2006). Acetylene is considered as nonspecific inhibitor of methanogenesis (Sparling et al., 1997; Vazquez et al., 2005). Load-shock treatment facilitates the direct cultivation of inoculums at a higher substrate load without any chemical application. Overloading or a shock-load-treated culture leads to the accumulation of organic acids, resulting in a decrease of system pH (Fang and Liu, 2002; van Ginkel et al., 2001), which also inhibits methanogens. This method can also be applied to a bioreactor as and when required during operation to regain performance by applying physical stress without adding any chemicals. Oxygen shock (forced aeration) inhibits MB and also ensures higher microbial diversity than heat shock and chemical methods. A combination of two to three pretreatment methods applied sequentially also showed a positive

TABLE 3   Pretreatment Methods Used to Selectively Enrich Dark Fermentative H_2-Producing Biocatalyst

Pretreatment method	Condition	Reference
Acid shock	pH <4	Ren et al., 2008b; Wang et al., 2011; Cai et al., 2009; Liu et al., 2009b; Chang et al., 2011; Luo et al., 2010; Moreno-Davila et al., 2010; Wang et al., 2009; Venkata Mohan et al., 2007b,c, 2008a,b, 2009b,c; Srikanth et al., 2010; Goud and Venkata Mohan, 2012
Alkaline shock	pH >9	Ren et al., 2008b; Wang et al., 2008, 2011; Liu et al., 2009b; Zhu and Béland, 2006; Chang et al., 2011; Luo et al., 2010; Wang and Wan, 2008a; Goud and Venkata Mohan, 2012
Heat shock	Extreme temperature; >80°C	Ren et al., 2008b; Leano and Babel, 2012; Wang et al., 2008, 2011; Cai et al., 2009; Liu et al., 2009b; Zhu and Béland; Chang et al., 2011; Luo et al., 2010; Srikanth et al., 2010; Moreno et al., 2010; Venkata Mohan et al., 2008a,b, 2009b,c; O-Thong et al., 2009
Load shock	In the presence of higher/toxic substrate concentration	Kongjan et al., 2011; Luo et al., 2010
Oxygen shock	In the presence of oxygen/air (0.5 mg/liter	Chang et al., 2011; Zhu and Béland, 2006; Srikanth et al., 2010; Wang and Wan, 2008a
Chemical	BESA (0.2–20 g/liter)	Zhu and Béland, 2006; Srikanth et al., 2010; Chang et al., 2011; Luo et al., 2010; Venkata Mohan et al., 2008a,b, 2009a–c
	Iodopropane (10 mmol)	Zhu and Béland, 2006
	Na_2SO_4 (0.05–0.3 mg SO_4^{2-}/ml)	Wimonsong and Nitisoravut, 2009
	Fluvastatin (1–4 µg/ml)	Wimonsong and Nitisoravut, 2009
	Chloroform (2% for 25 h)	Wang and Wan, 2008a
Freezing and thawing	−25°C for 24 h followed by 5-h thaw at room temperature	Liu et al., 2009b
Ozone treatment	Ozone bubbling (1–5 mg O_3/ml)	Wimonsong and Nitisoravut, 2009
Combined	Combination of two or more pretreatment methods	Venkata Mohan et al., 2007b,c, 2008a,b, 2009c; Srikanth et al., 2010; Liu et al., 2009b; Ren et al., 2008b; Wang and Wan, 2008a

effect on the H_2 evolution rate (Srikanth et al., 2010; Venkata Mohan et al., 2007b,c, 2008a,b, 2009c; Vijaya Bhaskar et al., 2008). Efficacy of the pretreatment method depends on the nature and composition of the parent inoculum, conditions adopted for pretreatment, and nature of substrate apart from other operating conditions (Srikanth et al., 2010; Venkata Mohan et al., 2008a,b, 2009c). Contrary to H_2 production, the pretreatment application negatively affects substrate degradation to a large extent. Application of pretreatment specifically inhibits the function of MB, which is essential in metabolizing the acid intermediates generated along with H_2 production.

Microbial Diversity Analysis

Generally, untreated (parent) anaerobic consortia have a higher bacterial population with a wide variety of biochemical functions facilitating diverse metabolic activities. On the contrary, pretreatment facilitates selective enrichment of a specific microbial population specific to acidogenic function and therefore leads to relatively less diversity. Application of pretreatment significantly influences the species composition of the microbial communities (Goud and Venkata Mohan, 2012; Ren et al., 2008b; Venkata Mohan et al., 2011a). *Clostridium* sp. became the dominant genus after acid-shock pretreatment (Lee et al., 2009). The composition of the microbial community survived in a long-term operated acidogenic reactor with combined pretreated (applied repeatedly) consortium producing H$_2$ from various types of wastewater and showed dominance of Clostridia followed by Bacteroidetes, Deltaproteobacteria, and Flavobacteria (Venkata Mohan et al., 2010c). Another acidogenic bioreactor operated for 1435 days producing H$_2$ under variable experimental conditions with a combined pretreated culture illustrating the presence of five dominant operational taxonomic units: Bacteroidetes, Clostridia, Flavobacteria, Aquificales, and Firmicutes (Goud et al., 2012). Clostridia and Bacilli were found to be the dominant classes in both the acidogenic bioreactors. The operating conditions adopted and the substrate used showed a marked influence on the microbial community structure.

The microbial population and its dynamics with the function of operating time and pretreatment methods applied help visualize the metabolic changes. The microbial community of long-term operated selectively enriched mixed culture by acid-shock and alkaline-shock pretreatment methods showed significant variation in comparison with the parent culture (Goud and Venkata Mohan, 2012). The parent culture showed dominance of the proteobacteria class followed by Firmicutes, uncultured bacteria, actionobacteria, and bacteroidia accounting majorly for non-H$_2$-producing microorganisms. The acid-pretreated culture showed dominance of Firmicutes followed by proteobacteria, uncultured bacterium, bacteroidia, and actinobacteria. Firmicutes were composed of the Clostridia class (*Clostridium cellulosi*, uncultured Ruminococcaceae) and Bacilli class (*Bacillus cereus, Lysinibacillus xylanilyticus, Bacillus thuringiensis*). The abundance of Firmicutes was found to increase with every additional application of pretreatment event or operating time. The abundance in the proteobacteria class (*Delta proteobacterium, Enterobactre* sp., *Geobacteraceae bacterium*) was also increased with time. Firmicutes became the dominant phylum after application of acid-shock pretreatment with *Clostridium* sp., *Bacillus* sp., and *Enterobacter* sp. The alkaline-pretreated culture showed dominance of the Proteobacteria class followed by Firmicutes, uncultured bacteria, and Actinobacteria/Bacteroidia, which was more or less similar to the parent inoculum with the additional presence of Clostridium (uncultured Ruminococcaceae) and Bacillus (*Lysinibacillus xylanilyticus*, uncultured Ruminococcaceae, *Bacillus boroniphilus, Bacillus cereus, Bacillus thuringiensis*, uncultured *Bacillus* sp.). The fraction of the Firmicutes class showed a decrement with every additional application of alkaline-shock pretreatment.

FACTORS INFLUENCING FERMENTATIVE H$_2$ PRODUCTION

Operating Temperature

The operating temperature plays an important role in shifting the metabolic pathways toward H$_2$ production in both light-dependent and -independent fermentation processes.

Temperature considerably influences H_2 production as well as metabolite distribution, substrate degradation, and bacterial growth. H_2 production by dark fermentation was reported under ambient (15–27°C), mesophilic (30–45°C), moderate thermophilic (50–60°C), and extreme thermophilic (over 60°C) conditions (Yokoyama et al., 2009). The optimal temperature for pure cultures was reported in the range of 37–45°C, whereas diverse optimum temperatures were reported for mixed microflora (Tang et al., 2008). Thermodynamics facilitates higher reaction rates and decreases problems associated with the contaminating H_2-consuming microorganisms (Tang et al., 2008), making thermophilic operation advantageous. Despite higher temperatures, reaction kinetics and resulting rapid changes in the system redox microenvironment inhibit the specific function of H_2-producing acidogenic bacteria (Wang and Wan, 2009). For photosynthetic bacteria, the optimum growth temperature is in the range of 30–35°C (Kapdan and Kargi, 2006; Sasikala et al., 1993). The temperature optimum for H_2 production usually depends on the nature of the biocatalyst and the type of wastewater used. Temperature control may not be a feasible option for process control in all instances. However, temperature control is especially important in regaining the spore-forming acidogenic culture during the reactor operation (Venkata Mohan and Goud, 2012).

Redox Condition

Biohydrogen production by the dark fermentation or photosynthetic process depends on the redox microenvironment. An external pH change can alter several physiological parameters, including internal pH, concentration of other ions, membrane potential, and proton motive force based on the organisms and their growth condition. The redox condition also influences the substrate metabolism, protein and storage materials synthesis, and release of metabolic by-products. An acidic microenvironment facilitates pyruvate conversion to fatty acids associated with H_2 production by AB. A neutral operation facilitates CH_4 formation by MB. On the contrary, a basic operation leads to solventogenesis. The activity of AB is crucial and rate limiting during the dark fermentation H_2 production process (Fan et al., 2004; Venkata Mohan, 2008; Venkata Mohan et al., 2008a, 2010b). AB functions well below pH 6, while the optimum range for MB is between 6.0 and 7.5. Good H_2 production was observed by maintaining the system pH in and around 6.0 (van Ginkel et al., 2001; Venkata Mohan et al., 2008b). However, a highly acidic pH (<4.5) is considered to be detrimental to H_2 production as it inactivates AB (Venkata Mohan, 2010; Zhu and Beland, 2006). Hydrogenase enzyme activity gets inhibited by maintaining low or high pH beyond the optimum range (Fan et al., 2006). Dehydrogenase-catalyzed redox reactions are higher under an acidophilic operation, which enhances proton shuttling between metabolic intermediates and redox mediators instead of getting reduced to end products (Srikanth et al., 2010). This H^+ shuttling provides a higher availability of protons to make H_2, while a neutral operation leads to a proton reduction to methane. An alkaline operation also provides higher redox reactions, where the H^+ shuttling between metabolic intermediates results in the formation of reduced compounds such as aldehydes, alcohols, and reducing sugars. Purple nonsulfur PSB can grow optimally between pH 6 and 9 depending on the substrate source (Pandey et al., 2012; Sasikala et al., 1993). A high feed pH prevents the cell from maintaining its membrane potential, therefore affecting the cellular metabolism and eventually hindering cell growth (Androga et al., 2012).

Bacterial cells maintain their membrane potential by the efflux of negatively charged OH⁻ ions from the cell to counteract the effect of taking in negatively charged organic acids (Ozgur et al., 2010a). The optimum growth pH for photosynthetic bacteria is near pH 7.0 (Kapdan and Kargi, 2006).

Accumulation of acid metabolites (VFA) during dark fermentation causes a marked drop in system pH, which reduces the buffering capacity, thereby inhibiting H$_2$ production (Lin and Lay, 2004b; Devi et al., 2010). VFA and pH express the acid–base condition of a system. System redox conditions can be controlled by increasing the buffering capacity. Stable pH can be maintained with increased buffering capacity as it maintains stable ionic strength against changes in acidic and basic constituents. Supplementation of buffer at various pH values increased the H$_2$ production time by maintaining favorable pH conditions (Zhu et al., 2009). In order to improve the system-buffering capacity, CO$_2$ was supplemented to the bioreactor to enhance H$_2$ production (Devi et al., 2010). Acidophilic conditions in association with pretreatment of biocatalyst will facilitate effective H$_2$ production during the treatment of various wastewaters (Venkata Mohan et al., 2007a–d, 2008a,b). Repression of MB indirectly promotes the H$_2$ producers within the system (Srikanth et al., 2010; Venkata Mohan et al., 2011b; Zhu and Beland, 2006). The pH range of 5.5–6.0 is optimum to avoid both methanogenesis and solventogenesis and can be considered as manipulated variable for process control, especially for the dark fermentation process (Venkata Mohan et al., 2007a, 2010b).

Reactor and Mode of Operation

Diverse reactor configurations—suspended growth, biofilm/packed bed/fixed bed, fluidized bed, expanded bed, upflow anaerobic sludge blanket, granular sludge, membrane-based systems, and immobilized systems—were reported for biohydrogen production from various types of wastewater. Biofilm/attached-growth systems showed good H$_2$ production and substrate degradation efficiency (Lalit Babu et al., 2009; Venkata Mohan et al., 2007b,c). A comparative study on the performance of a biofilm reactor (formed due to cell immobilization) and self-flocculated granules in an anaerobic-fluidized bed reactor at various loading rates showed a better retention biomass without subjection to washout of support carriers in the later case (Zhang et al., 2008). Biofilm systems are robust to shock loads, facilitate an improved reaction potential, leading to a stable and robust system, and are more resilient to changes in the process parameters. Biofilm systems are especially well suited for treating highly variable wastewater.

The operation mode of the reactor, along with its configuration, influences the reactor microenvironment, hydrodynamic behavior, wastewater–biocatalyst contact, survivability of microbial population, etc. (Venkata Mohan, 2009, 2010). Batch, fed-batch, semibatch/continuous, periodic discontinuous batch (sequencing batch operation), and continuous modes of reactor operation have been evaluated for H$_2$ production. The fed-batch mode operation reduces poor biomass retention/cell washout (Yokoi et al., 1997) and accumulation of soluble metabolic intermediates due to the fill-draw mode operation (Venkata Mohan et al., 2007b,c, 2011b). The batch mode operation, coupled with biofilm configuration, helps maintain stable and robust cultures suitable for treating highly variable wastewater due to the dual operational advantages of both systems (Lalit Babu et al., 2009; Luo et al., 2010; Venkata Mohan et al., 2007a–d, 2008a,b,d,f; Yokoi et al., 1997).

Fermentation Time

Fermentation period or hydraulic retention time (HRT) is one of the key process control variables influencing H_2 production. A longer fermentation period induces a metabolic shift from acidogenesis to methanogenesis, which is considered unfavorable for H_2 production. Maintaining a shorter HRT helps restrict MB growth as well as activity (Hawkes et al., 2007). For good H_2 production, optimum HRTs between 8.0 and 14 h were reported (Hawkes et al., 2007). Good H_2 yields are reported between 0 and 14 h with various types of wastewater as substrates (Venkata Mohan, 2010; Venkata Mohan et al., 2007a–d, 2008c, d, 2011b; Vijaya Bhaskar et al., 2008). Methanogens can be suppressed by maintaining short retention times (2–10 h) as AB grow faster (Fang and Liu, 2002; Nakamura et al., 1993; Ren et al., 2005; Venkata Mohan, 2009; Zhu et al., 2006). The nature and composition of the substrate, nature of biocatalyst, organic loading rate, and redox condition influence the HRT. Contrary to the dark fermentation process, a short retention time reduces the efficiency of substrate utilization by PSB and therefore lowers the overall process efficiency (Chen et al., 2009).

Partial Pressure of H_2

The partial pressure of H_2 is one of the important and rate-limiting factors affecting H_2 production, especially during the dark fermentation process, by influencing the hydrogenase enzyme activity due to end product inhibition. H_2 production is limited by the thermodynamics of the hydrogenase reaction, which involves the enzyme-catalyzed transfer of an electron from an intracellular electron carrier molecule to a proton (Angenent et al., 2004). When the concentration of H_2 in the liquid phase increases, the oxidation of reduced ferredoxin is less favorable and results in a reduction of oxidized ferredoxin. Hydrogenase reversibly oxidizes and reduces ferredoxin (Nicolet et al., 2002). Liquid-phase H_2 oxidizes to protons, therefore reducing the yield of H_2. At higher H_2 concentrations, the metabolic pathways shift from acidogenesis to solventogeneis with production of more reduced substrates, namely, lactate, ethanol, acetone, butanol, or alanine, which in turn decrease H_2 production (Fan et al., 2004; Nath and Das, 2004; Niel et al., 2003). H_2 production becomes thermodynamically unfavorable at H_2 partial pressures greater than 60 Pa. Operating bioreactors at a low H_2 partial pressure by stripping H_2 from the solution as it gets generated will accomplish good H_2 production (Abreu et al., 2012; Angenent et al., 2004; Hawkes et al., 2007). Prevention of H_2 consumption by propionic acid-producing bacteria, ethanol-producing bacteria, and homoacetogens and those that channel more reducing equivalents toward reduction of protons by hydrogenases can also maximize H_2 production (Angenent et al., 2004). Sparging of inert gases such as nitrogen and argon lowers the dissolved H_2 concentration, resulting in increased yields (Mizuno et al., 2000; Tanisho et al., 1998). Gas stripping and membrane-absorption technologies also help to remove H_2 from the reactor headspace (van Groenestijn et al., 2002).

Nutrients

Nitrogen is an essential component and is second only to carbon as a requirement for bacterial growth (Lin and Lay, 2004a; Wang et al., 2009). Nitrogen at an optimal concentration is beneficial to H$_2$ production, while at higher concentrations can inhibit the process performance by affecting the intracellular pH of the bacteria or inhibiting specific enzymes related to H$_2$ production (Bisaillon et al., 2006; Chen et al., 2008; Salerno et al., 2006). At elevated nitrogen levels, the metabolic path might lead towards ammonification, where the protons get consumed instead of forming H$_2$ (Salerno et al., 2006). Optimal nitrogen concentrations of 0.1/0.01 g N/liter were found to have a positive effect on H$_2$ production and substrate degradation, respectively (Wang et al., 2009). An optimum carbon-to-nitrogen (C/N) ratio helps in bacterial growth and affects H$_2$ production by mixed or pure cultures. A C/N ratio of 47 has shown to affect fermentative H$_2$ production by mixed microorganisms (Lin and Lay, 2004a). Wastewater low in carbon content can be combined with materials high in N to attain the desired C:N ratio of 30:1 (Yadvika et al., 2004). Wastewater with excess nitrogenous compounds and ammonia inhibits nitrogenase activity (Redwood and Macaskie, 2006). The major role of phosphate is energy generation to the cell in the form of ATP. Phosphates also help in maintaining the system-buffering capacity alternative to the carbonate.

REGULATORY FUNCTION OF TRACE METALS ON FERMENTATIVE H$_2$ PRODUCTION

Trace metals are required for the activation or function of many enzymes and coenzymes related to energy metabolism and are also essential for the cell growth of many microorganisms. Each of the trace metals has a specific biochemical function in the cell during metabolism, and a change in their concentration may alter that metabolic function. The functional role of some trace metals in the microbial metabolic activities is depicted in Figure 1. Several studies reported the influence of different trace metals on the fermentation process, including H$_2$ production (Garcin et al., 1999; Jakubovics and Jenkinson, 2001; Lin and Lay, 2005; Lin and Shei, 2008; Strocchi and Levitt, 1992; Wu et al., 1994; Zhang and Shen, 2006). Iron (Fe^{2+}) is very important for the function of hydrogenase and also acts as an active site component for the ferredoxin protein, which is a carrier of electrons to the hydrogenase. Fe^{2+} also acts as a mediator for intracellular electron transfer either individually or as a component of prosthetic groups. Higher concentrations of Fe^{2+} showed an enhancement in H$_2$ production efficiency due to its role as a component of hydrogenase and ferredoxin (Karadag and Puhakka, 2010; Lee et al. 2001; Srikanth and Venkata Mohan, 2012; Wang and Wan, 2008c; Zhang et al., 2005). However, at elevated levels of Fe^{2+}, it acts as an electron acceptor from metabolism to defend from its toxic influence. Optimum Fe^{2+} concentration varied from 25 to 100 mg/liter (Srikanth and Venkata Mohan, 2012; Zhang and Shen, 2006).

Nickel (Ni^{2+}) is an essential trace element for many microorganisms as an active center of four metalloenzymes and plays an important role in three major anaerobic processes: uptake and production of H$_2$, methanogenesis, and acetogenesis (Karadag and Puhakka, 2010; Wang and Wan, 2008b; Yang and Shen, 2006; Yang et al., 1989). Ni^{2+} is a crucial component of

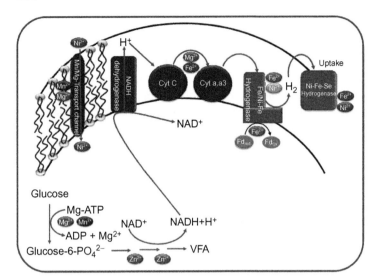

FIGURE 1 Schematic representation of role of trace metals in acidogenic fermentation for H_2 production *(redrawn after Srikanth and Venkata Mohan, 2012).*

[NiFe]- and [NiFeSe]-hydrogenases, which are the other classes of hydrogenases. [NiFe]-hydrogenases catalyze both the production and the utilization of H_2 and maintain the proper balance between electron donors or acceptors (Wang and Wan, 2008b). On the contrary, [NiFeSe]-hydrogenases are meant for H_2 uptake, especially present in methanogenic archea (Garcin et al., 1999). An optimum Ni^{2+} concentration enhances the H_2 production efficiency of the biocatalyst by activating the [NiFe]-hydrogenases, whereas higher concentrations result in cessation of H_2. This might be due to the function of Ni^{2+} in activating the uptake function of [NiFe]-hydrogenases in H_2-producing consortia or activating the [NiFeSe]-hydrogenases in methanogenic consortia at higher concentrations, shifting the metabolic pathway distant from H_2 production.

Magnesium (Mg^{2+}) plays a pivotal role in the metabolic processes for substrate utilization, especially for the activation of substrate molecules without which they cannot enter in to the metabolic pathways. The ATP molecule first gets attached with the Mg^{2+} ion (Mg-ATP), which can react further with glucose, delivering the phosphate group to glucose and forming an active glucose-X-phosphate (X may be the first or sixth position based on the necessity of the cell metabolism) that can enter the metabolic pathway. Many electron carriers, including cytochromes and protein complexes, contain Mg^{2+} as a central atom (Moncrief and Maguire, 1999). The Mg^{2+} ion is also an activator of many enzymes and a component of cell walls and cell membranes. Mg^{2+} also helps in the transport of different molecules in the cell (ATP transport) and from the medium into the cell (Ni^{2+} transport) (Moncrief and Maguire, 1999). Higher concentrations of Mg^{2+} help in higher substrate degradation that can facilitate a higher availability of protons and electrons, which can result in higher H_2 production. A higher availability of protons also activates the function of hydrogenases, which results in higher H_2 production (Srikanth and Venkata Mohan, 2012; Venkata Mohan et al., 2011). Mg^{2+} is required in higher quantities (150–600 mg/liter) than other trace metals due to its role in diverse metabolic functions (Moncrief and Maguire, 1999; Srikanth and Venkata Mohan, 2012).

Manganese (Mn^{2+}) also shows almost similar functions to Mg^{2+}. Mn^{2+} has a multifaceted role in biological systems because it can exist in 11 oxidation states ($+7$ to -3); however, it is largely restricted to Mn^{2+} complexes. Because the ionic radius of Mn^{2+} is in the middle of Mg^{2+} and Ca^{2+} ionic radii, it can be possible for an overlapping function of Mn^{2+} with these metal ions in providing the structural stabilization of several enzymes (Kehres and Maguire, 2003). The redox function of Mn^{2+} is relevant to the oxygen as the substrate or product in many enzymes, although it is the active site component. Mn^{2+} is also involved in the scavenging of superoxide radicals and hydrogen peroxide as a metal complex or as a component of enzymes such as superoxide dismutase and pseudocatalase to protect the biocatalyst from the reactive oxygen species (Horsburgh et al., 2002). Endospore formation and regermination in several bacterial species require Mn^{2+} (Jakubovics and Jenkinson, 2001). It also functions as an absolute growth factor in some species such as *Lactobacillus*, and its relative levels are found to influence substrate utilization along with the other trace metals (Imbert and Blondeau, 1998; Zhang et al., 2000). Higher concentrations of Mn^{2+} help in the uptake of several trace metals because of its function as a membrane transporter protein. However, it is required in smaller quantities (≈ 15 mg/liter) due to its complementary role to Mg^{2+}, and higher quantities may show toxic effects (Srikanth and Venkata Mohan, 2012). The uptake of other metal ions such as Ni^{2+} may also be increased under high Mn^{2+} concentrations, resulting in a shift in the metabolic pathway from the biohydrogen production process to other processes (Srikanth and Venkata Mohan, 2012). Ni^{2+} can enter the cells via the Mg^{2+}/Mn^{2+} transport system, which means that the presence of higher concentrations of any one of the three metal ions (Ni^{2+}/Mg^{2+}/Mn^{2+}) can increase the Ni^{2+} uptake into the cell (Srikanth and Venkata Mohan, 2012; Tripathi and Srivastava, 2006).

Zinc (Zn^{2+}) is involved in a wide variety of cellular processes, and maintenance of cellular Zn^{2+} is essential for growth and survival, which necessitated an improved understanding of acquisition, assimilation, and metabolism of Zn^{2+}. Zn^{2+} was reported as an important factor that can enhance H$_2$ production using anaerobic microflora, despite its toxic effects on the biocatalyst at higher concentrations (Lin and Shei, 2008; Srikanth and Venkata Mohan, 2012). Small amounts of Zn^{2+}, along with optimum concentrations of other trace metals, will enhance the H$_2$ production capability of the biocatalyst. Zn^{2+} is an essential micronutrient that plays a critical role in various physiological processes as a cofactor in all the six functional classes of enzymes, especially as an important constituent for the maintenance of protein structure (Blencowe and Morby, 2003; Srikanth and Venkata Mohan, 2012). Zn^{2+} is also related with the reactions and transformations of dehydrogenase, dismutase, hydrogenase, and methyltransferase based on its concentration (Lin and Shei, 2008). However, the Zn^{2+} role is nonspecific in H$_2$ production, as well as substrate degradation, but may cause toxic effects at higher concentrations (Srikanth and Venkata Mohan, 2012). Apart from these metal ions, other trace metals, such as calcium and copper, and a few heavy metals were also studied for their influence on the H$_2$ production process but they do not have any positive impact on the process (Chang and Lin, 2006; Oleszkiewicz and Sharma, 1990; Yu and Fang, 2001).

The influence of trace metals also plays an important function in photosynthetic machinery to regulate H$_2$ production. Molybdenum and Fe^{2+} are the major active site components of nitrogenase and hydrogenase and play significant roles in photo H$_2$ production. The presence of molybdenum and iron is necessary, as they are found in the structure of

Mo-nitrogenase (Dixon and Kahn, 2004; Koku et al., 2002); moreover, Fe is found in several electron carriers of the photosynthetic electron transport system, such as ferredoxin (Zhu et al., 2007). Reduced nitrogenase activity was reported in the absence of molybdenum and iron, indicating their role as cofactors (Jacobson et al., 1986; Kars et al., 2006). Enhancement in H_2 production was also reported by supplementing the substrate with molybdenum and Fe^{2+}. However, the range of molybdenum requirement varied between 0.1 and 16.5 mM, but most of the studies were carried out with 0.16 mM due to marginal differences in the H_2 yield at higher molybdenum concentrations (Afsar et al., 2009; Kars et al., 2006; Ozgur et al., 2010a; Uyar et al., 2009). Iron supplementation in the range of 0.1 to 10 mM is optimal for photo H_2 production similar to the dark fermentation requirement. Most studies pertaining to metal ion supplementation were carried out using defined media or single cultures (Kars et al., 2006; Zhu et al., 2007); in a few studies, real field wastewater and dark fermentation effluents (Afsar et al., 2009; Eroglu et al., 2011; Ozgur et al., 2010a; Uyar et al., 2009) were also used.

PROCESS LIMITATIONS

The main challenges observed especially with the dark fermentative H_2 production process are low substrate conversion efficiency and residual substrates present in acid-rich wastewater generated from the H_2 production process (Venkata Mohan, 2008, 2010). The main intricacy in the practical application of fermentative H_2 production from wastewater is that conversion yields by known metabolic pathways emerge to be limited to a maximum yield of 4 mol of hydrogen per mole of glucose, representing a maximum conversion efficiency of 33% (Cheng et al., 2012). About 60–70% residual organic carbon remains in the effluent after dark fermentation, even under optimal operating conditions, and it requires further treatment prior to disposal. The persistent accumulation of acidogenic by-products such as VFA causes a sharp drop in the pH, resulting in process inhibition (Venkata Mohan et al., 2011b; Wang and Wan, 2009). Biological limitations such as H_2 end product inhibition, acid or solvent accumulation, and H_2 partial pressure limit the process efficiency. The H_2 yield is lower when more reduced organic compounds, such as lactic acid, propionic acid, and ethanol, are produced as fermentation products, as these represent the end products of metabolic pathways that bypass the major H_2-producing reaction. Apart from a lower conversion efficiency, the nonutilized organic fraction usually remaining as a soluble fermentation product from the acidogenic process is another major concern that needs to be resolved. Environmental and economic concerns suggest that it is advisable to use the residual carbon fraction of the acidogenic outlet for additional energy generation in the process of its treatment (Mohanakrishna et al., 2010b). One of the major problems associated with the photofermentation process is the high operation cost associated due to the artificial light requirement (Chen et al., 2010). Light utilization is the key factor in the photofermentation process, and developing effective photobioreactors with efficient internal light distribution characteristics is critical, especially for the industrial scale where self-shading inside the fermenter has to be prevented to ensure maximum H_2 production (Akkerman et al., 2002; Gadhamshetty et al., 2011; Keskin et al., 2011).

STRATEGIES TO ENHANCE PROCESS EFFICIENCY

Integration Approaches

Where as an integrated approaches were studied extensively to overcome the persistent limitation of the acidogenic process. Utilization of the residual acid-rich organic fraction originating from acidogenic bioreactor can be made possible by integrating two-stage energy-producing (hybrid) processes (Table 4). Some of the possible process integrations used in this context are also illustrated in Figure 2. Various secondary processes Viz., methanogenesis (AF) for methane production, acidogenic fermentation (DF) for additional H_2 production, photobiological process (PF) for additional H_2 production, microbial electrolysis (MEC) for additional H_2 production, anoxygenic nutrient-limiting process for bioplastics production, hetrotrophic algae cultivation for lipid accumulation, and microbial fuel cell (MFCEF) for bioelectricity generation—were integrated with the primary dark fermentative (DF) H_2 production process. Integration approaches facilitate a reduction in the wastewater load with the advantage of value addition to the existing process in the form of product recovery, making the whole process viable economically and environmentally. The efficiency of H_2 production as well as substrate degradation was found to depend on the process used in the first stage, along with the composition of the substrate (Venkata Mohan et al., 2008f; Mohanakrishna and Venkata Mohan, 2013).

Photobiological Process

Photosynthetic bacteria can readily utilize the organic acids generated from the dark fermentation process to produce additional H_2 (Chandra and Venkata Mohan, 2011; Rai et al., 2012; Srikanth et al., 2009a,b). Some PNS bacteria can utilize metabolic intermediates of dark fermentation such as short-chain VFA. Henceforth, integration of anoxygenic photofermentation with dark fermentation will have dual advantages of increased H_2 production along with substrate removal (Chandra and Venkata Mohan, 2011; Srikanth et al., 2009b; Venkata Mohan et al., 2008f; Mohanakrishna and Venkata Mohan, 2013). Green algae such as Chlorella also have the capability to utilize organic acids produced during dark fermentation for H_2 production, especially when acetate is used as a substrate (Amutha and Murugesan, 2011). Photofermentation of acid-rich effluents from the H_2 production process is considered to be more complicated than dark fermentation due to light penetration problems, complex nutritional requirements of the bacteria, strict control of environmental conditions, substrate (VFA and ammonia) inhibitions, and susceptibility for contamination (Ozmihci and Kargi, 2010; Liu et al., 2009a; Ozkan et al., 2012). Substrate inhibition is the main obstacle observed when using a dark fermentation effluent as the sole substrate for the photofermentation process (Chen et al., 2010). The formation of ammonium ions from a dark-fermented feed inhibits the photofermentation process (Androga et al., 2012; Guwy et al., 2011). Ammonium ions repress the nitrogenase enzyme, which is responsible for catalyzing H_2 production in the photosynthetic bacteria (Akkose et al., 2009) either directly via feedback inhibition (reversible) or at the genetic level by the repression of nif genes that encode subunits of the nitrogenase enzyme (Pekgoz et al., 2011).

Biodegradable Plastics

The effluents generated from dark-fermentative biohydrogen production process are rich in VFA. These VFA rich effluents can be used as potential substrate for synthesis of bioplastic

TABLE 4   Details of Various Process Integration Routes Studied with Fermentative H$_2$ Production

Primary substrate	Biogas/product recovered[a]		Reference
	First stage	Second stage	
Acidified sorghum extract	Hydrogen (DF)	Methane (AF)	Antonopoulou et al., 2012
Molasses effluent with high ammonium concentration	Hydrogen (DF)	Hydrogen (PF)	Androga et al., 2012
Rice straw bioethanol residues	Hydrogen (DF)	Methane (AF)	Cheng et al., 2012
Water hyacinth	Hydrogen (DF)	Methane (AF)	Chuang et al., 2011
Molasses	Hydrogen (DF)	Ethanol (DF)	Han et al., 2012b; Ozgur et al., 2010b
Thin stillage (by-product from ethanol production)	Hydrogen (DF)	Methane (DF)	Nasr et al., 2012
Laminaria japonica	Hydrogen (DF)	Methane (AF)	Jung et al., 2012
Cornstalk waste	Hydrogen (DF)	Methane (AF)	Cheng et al., 2012
Synthetic fruit juice wastewater	Hydrogen (DF)	Electricity (EF)	Campo et al., 2012
Cornstalks	Hydrogen (DF)	Hydrogen (DF) and methane (AF)	Cheng and Liu, 2012
Food waste	Hydrogen (DF)	Methane (AF)	Kim et al., 2012
Recalcitrant lignocellulosic materials	Hydrogen (DF)	Hydrogen (MEC)	Lalaurette et al., 2009
Synthetic wastewater	Hydrogen (DF)	Bioplastic	Venkateswar Reddy et al., 2011; Venkata Mohan et al., 2010a
Food waste	Hydrogen (DF)	Bioplastic	Venkateswar Reddy and Venkata Mohan, 2011
Acid-rich effluents from H$_2$ and CH$_4$ producing reactor	Hydrogen (DF)	Hydrogen (PF)	Srikanth et al., 2009b
Synthetic waste and food waste	Hydrogen (DF)	Hydrogen (PF)	Chandra and Venkata Mohan, 2011
Synthetic wastewater	Hydrogen (DF)	Methane (AF)	Venkata Mohan et al., 2008e
Cheese whey	Hydrogen (DF)	Hydrogen (PF)	Rai et al., 2012
Synthetic wastewater	Hydrogen (DF)	Bioelectricity (EF)	Mohankrishna et al., 2010b
Molasses	Hydrogen (DF)	Hydrogen (PF)	Androga et al., 2012
Sugar beet thick juice	Hydrogen (DF)	Hydrogen (PF)	Özkan et al., 2012
Synthetic wastewater and fermentation effluents	Hydrogen (DF)	Hydrogen (DF)	Lee et al., 2011
Acid-hydrolyzed wheat starch	Hydrogen (DF)	Hydrogen (PF)	Sagnak and Kargi, 2011
Ground wheat starch	Hydrogen (DF)	Hydrogen (PF)	Ozmihci and Kargi, 2010
Bagasse	Hydrogen (DF)	Hydrogen (PF)	Wu et al., 2010
Ground wheat solution	Hydrogen (DF)	Hydrogen (PF)	Argun and Kargi, 2010

Continued

TABLE 4 Details of Various Process Integration Routes Studied with Fermentative H_2 Production—Cont'd

Primary substrate	Biogas/product recovered[a]		Reference
	First stage	Second stage	
Potato homogenate	Hydrogen (DF)	Hydrogen (PF)	Laurinavichene et al., 2010
Wheat straw	Hydrogen (DF)	Methane (DF)	Prasad et al., 2009
Wheat straw	Bioethanol (DF)	Methane (DF)	Prasad et al., 2009
Olive oil mill wastewater	Hydrogen (DF)	Bopolymer	Ntaikou et al., 2009
Pea shells slurry	Hydrogen (DF)	Bioplastic	Patel et al., 2012
Food waste	Hydrogen (DF)	Lipids	Venkata Mohan and Devi, 2012
Acidogenic effluents	Hydrogen (DF)	Hydrogen (PF)	Ozgur et al., 2010a
Soluble metabolites from dark fermentation effluents	Hydrogen (DF)	Hydrogen (PF)	Lo et al., 2011
Starch fermentation	Hydrogen (DF)	Hydrogen (PF)	Laurinavichene et al., 2008

[a]*DF, dark fermetation; AF, anaerobic fermentation; PF, photofermentation; EF, electrogenic fermentation; MEC, microbial electrolysis cell.*

viz., polyhydroxyalkanoates (PHA) which are considered as sustainable alternatives to the conventional synthetic plastics. Bioplastics in the form of polyhydroxyalkanoates (PHA) are alternatives to these plastics. PHA represent a group of biologically derived biodegradable biopolyesters of hydroxyalkanoates that accumulate as cellular reserve storage materials produced under excess carbon and nutrient-deprived conditions in microbial cells (Madsen, 2008; Van Loosdrecht et al., 1997). In the presence of an excessive carbon source and when other nutrients are growth limiting, biopolyesters are deposited as water-insoluble, cytoplasmic microsized inclusions. Conventionally, PHA are produced using commercial-grade substrates with pure cultures, which is expensive due to their production costs. Compared to carbohydrates, lipids, and amino acids, VFA has a simple structure with a lower number of carbon atoms [acetate (2C), propionate (3C), and butyrate (4C)], which facilitates easy synthesis to PHA without the involvement of glycolysis and β-oxidation pathways by using a fewer number of metabolic reactions and enzymes (Venkata Mohan et al., 2010a). PHB production from individual fatty acids and acid-rich effluents from a H_2-producing reactor was reported under an anoxic microenvironment using a mixed culture (Venkata Mohan et al., 2010a; Venkateswar Reddy et al., 2011). The coupling of H_2 and PHB production processes and utilization of their effluents for methanogenesis are likely to make the whole process economical and sustainable (Patel et al., 2012). Higher volumetric productivity and product purity can be achieved by separating the PHA-producing process from acidification (Li and Yu, 2011).

Electrically Driven Biohydrogenesis

Using microbial electrolysis cells as an alternative electrically driven H_2 production process facilitates the conversion of biodegradable material into H_2 gas under applied external voltages (Fig. 3). The MEC process is also called a bioelectrochemically assisted microbial reactor, electrohydrogenesis, electrofermentation, or biocatalyzed electrolysis cell

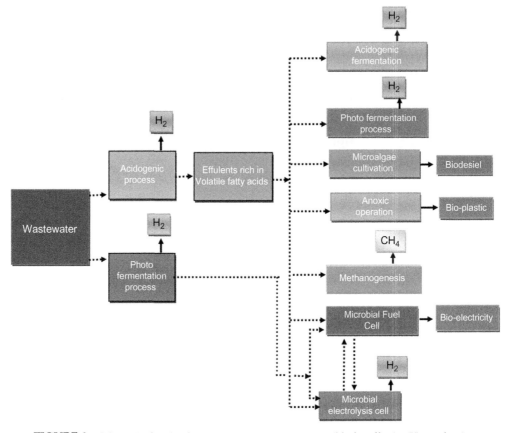

FIGURE 2 Schematic details of various integration routes possible for effective H_2 production.

FIGURE 3 Schematic details of a single-chamber microbial electrolysis cell.

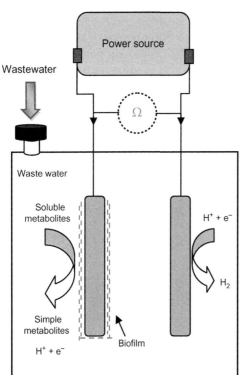

(Cheng and Logan, 2007; Liu et al., 2005; Rozendal et al., 2006). The MEC process more or less resembles a microbial fuel cell (MFC), where the basic difference exists with the requirement of a small input of external potential. Based on thermodynamics, a potential greater than 0.11 V in addition to that generated by bacteria (-0.3 V) will yield H_2 gas at the cathode (Cheng and Logan, 2007). The standard redox potential for the reduction of protons to H_2 is -0.414 V. This mechanism extends the H_2 production to past the endothermic barrier imposed due to the formation of fermentation end products, such as acetic acid. The potential required is relatively low compared to the theoretically applied voltage of 1.23 V for water electrolysis (Rozendal et al., 2007). In practice, a relatively higher voltage than this is required due to overpotentials created by physicochemical and microbial factors (Hallenbeck, 2011). Compared to the dark fermentation process where 33% of energy recovery is generally possible, MEC can achieve more than 90% of H_2 recovery (Cheng and Logan, 2007). Low-energy consumption compared to conventional water electrolysis, high product (H_2) recovery, and substrate degradation than the dark fermentation process are some of the benefits that make MEC an alternate process.

The performance of MEC depends on the nature of the biocatalyst, electrode materials, membrane, applied potential, nature of the substrate, and its loading rate and cell. MEC was initially operated in a dual chamber and later shifted to a single chamber. A single chamber configured MEC operation significantly reduces the internal resistance, acidification of the anode chamber due to the production of protons, and basification of the cathode chamber from proton consumption (Hallenbeck, 2011; Hu et al 2008). Membrane elimination will reduce both pH energy and ohmic energy losses (Clauwaert and Verstraete, 2009; Lee et al., 2009), which are significant with dual-chamber MEC (Rozendal et al., 2007). MEC process showed feasibility in utilizing a wide variety of soluble organic matter, including wastewater, to generate H_2 or methane in association with simultaneous treatment of wastewater (Wagner et al., 2009). A two-stage process was studied to convert acid-rich dark fermentation effluents as substrates for additional H_2 production (Lalaurette et al. 2009; Venkata Mohan and Lenin Babu, 2011; Wagner et al., 2009). Integrating MEC with other processes is also gaining much attention (Cusick et al. 2010; Lalaurette et al., 2009; Wang et al, 2011). Integrating the wastewater treatment with MEC to produce economically feasible H_2 production attaining recent interest among the research fraternity.

Bioaugmentation

The anaerobic bioreactor (producing methane) was shifted metabolically from methanogenesis to acidogenesis to produce H_2 as an end product by applying bioaugmentation strategy (Venkata Mohan et al., 2007d). The H_2 production rate improved significantly after augmenting with selectively enriched AB (in immobilized form). Co-cultures of *Clostridium acetobutylicum* X9 and *Ethanoigenens harbinense* B49 augmentation showed significant improvement in both hydrolysis and subsequent H_2 production (Ren et al., 2008a). Augmentation of a constructed microbial consortium (*Enterobacter cloacae* IIT-BT 08: *Citrobacter freundii* IIT-BT L139: *Bacillus coagulans* IIT-BT S1) showed an improvement in H_2 production (Kotay and Das, 2010). A bioaugmentation strategy can be applied to the full-scale anaerobic reactors producing CH_4 to shift toward H_2 production. Application of this strategy can improve process efficiency within a short period of time. Application of bioaugmentation improves the H_2 producing capacity of the system and system stability during operation. However, success of the bioaugmentation strategy depends on a number of factors, such as survivability and persistence of augmented biocatalysts in the system, operating conditions, and substrate composition and nature, as well as diversity of native microflora of the system.

FUTURE OUTLOOK

Focused and multidisciplinary research on biohydrogen production integrated with wastewater remediation is underway. This process possesses certain inherent limitations, namely, low substrate conversion efficiency, accumulation of carbon-rich acid intermediates, dynamic buffering, and redox change, which need considerable attention. Fundamental understanding of the potential limiting factors is essential to overcome these limitations in the direction of enhancing process efficiency. Process engineering and optimization of operational factors govern the performance of any biological system and also have considerable influence on fermentative H_2 production. Understanding the biochemistry and microbiological aspects based on the functional role of membrane components and mechanism of proton reduction, community analysis, culture development aspects, and design and development of efficient bioreactors for both dark and photofermentation operations are some of the key areas where considerable focus is required. Optimization of process parameters is essential for upscaling of the technology. Nonutilized residual organic fraction remaining as a soluble fermentation product after the acidogenic process is one of the key limitations that need significant attention. Integration approaches towards the utilization of acid-rich wastewater with simultaneous bioenergy recovery are to be effectively and completely established for economic viability of the process towards commercialization. Light utilization is the key factor in the photofermentation process, and developing effective photobioreactors with efficient internal light distribution characteristics is especially important for the industrial scale application. Metabolic engineering is one of the promising areas that can be used advantageously to enhance the H_2 production rate. Both basic and applied research is on the way to gain more insight into the process for understanding and establishing optimized conditions. Several novel approaches have been proposed in recent years to surpass some of the persistent drawbacks. Biohydrogen technology requires multidisciplinary research to make the process environmentally sustainable and economically viable.

Acknowledgments

The authors thank the Director, CSIR-IICT, for his kind encouragement. The authors also wish to acknowledge the financial support from CSIR in the form of XII five year network project on 'Sustainable Waste Management Technologies for Chemical and Allied Industries-SETCA' and Ministry of New and Renewable Energy (MNRE, Project No. 103/131/2008-NT), Government of India on 'Biohydrogen'.

References

Abreu, A.A., Karakashev, D., Angelidaki, I., Sousa, D.Z., Alves, M.M., 2012. Biohydrogen production from arabinose and glucose using extreme thermophilic anaerobic mixed cultures. Biotechnol. Biofuels 5, 6.
Afsar, N., Ozgur, E., Gurgan, M., de Vrije, T., Yucel, M., Gunduz, U., 2009. Hydrogen production by *R. capsulatus* on dark fermenter effluent of potato steam peel hydrolysate. Chem. Eng. Trans. 18, 385–390.

Akkerman, I., Janssen, M., Rocha, J., Wijffels, R.H., 2002. Photobiological hydrogen production: Photochemical efficiency and bioreactor design. Int. J. Hydrogen Energy 27, 1195–1208.

Akkose, S., Gunduz, U., Yucel, M., Eroglu, I., 2009. Effects of ammonium ion, acetate and aerobic conditions on hydrogen production and expression levels of nitrogenase genes in *Rhodobacter sphaeroides* O.U.001. Int. J. Hydrogen Energy 34, 8818–8827.

Amutha, K.B., Murugesan, A.G., 2011. Biological hydrogen production by the algal biomass *Chlorella vulgaris* MSU 01 strain isolated from pond sediment. Bioresour. Technol. 102, 194–199.

Androga, D.D., Özgür, E., Eroglu, I., Gündüz, U., Yücel, M., 2012. Amelioration of photofermentative hydrogen production from molasses dark fermenter effluent by zeolite-based removal of ammonium ion. Int. J. Hydrogen Energy 37 (21), 16421–16429.

Angenent, L.T., Karim, K., Al-Dahhan, M.H., Wrenn, B.A., Domíguez-Espinosa, R., 2004. Production of bioenergy and biochemicals from industrial and agricultural wastewater. Trends Biotechnol. 22, 477–485.

Antonopoulou, G., Gavala, H.N., Skiadas, I.V., Lyberatos, G., 2012. ADM1-based modeling of methane production from acidified sweet sorghum extract in a two stage process. Bioresour. Technol. 106, 10–19.

Argun, H., Kargi, F., 2010. Photo-fermentative hydrogen gas production from dark fermentation effluent of ground wheat solution: Effects of light source and light intensity. Int. J. Hydrogen Energy 35 (4), 1595–1603.

Bisaillon, A., Turcot, J., Hallenbeck, P.C., 2006. The effect of nutrient limitation on hydrogen production by batch cultures of *Escherichia coli*. Int. J. Hydrogen Energy 31, 1504–1508.

Blencowe, D.K., Morby, A.P., 2003. Zn(II) metabolism in prokaryotes. FEMS Microbiol. Rev. 27, 291–311.

Buitrón, G., Carvajal, C., 2010. Biohydrogen production from Tequila vinasses in an anaerobic sequencing batch reactor: effect of initial substrate concentration, temperature and hydraulic retention time. Bioresour. Technol. 101, 9071–9077.

Cai, J.L., Wang, G.C., Li, Y.C., Zhu, D.L., Pan, G.H., 2009. Enrichment and hydrogen production by marine anaerobic hydrogen-producing microflora. Chin. Sci. Bull. 54 (15), 2656–2661.

Campo, A.G., Canizares, P., Lobato, J., Rodrigo, M.A., Fernandez, F.J., 2012. Electricity production by integration of acidogenic fermentation of fruit juice wastewater and fuel cells. Int. J. Hydrogen Energy 37 (11), 9028–9037.

Cao, G.L., Guo, W.Q., Wang, A.J., Zhao, L., Xu, C.J., Zhao, Q., et al., 2012. Enhanced cellulosic hydrogen production from lime-treated cornstalk wastes using thermophilic anaerobic microflora. Int. J. Hydrogen Energy 37 (17), 13161–13166.

Chakkrit, S., Plangklang, P., Imai, T., Reungsang, A., 2011. Co-digestion of food waste and sludge for hydrogen production by anaerobic mixed cultures: Statistical key factors optimization. Int. J. Hydrogen Energy 36 (21), 14227–14237.

Chandra, R., Venkata Mohan, S., 2011. Microalgal community and their growth conditions influence biohydrogen production during integration of dark-fermentation and photo-fermentation processes. Int. J. Hydrogen Energy 36 (19), 12211–12219.

Chang, F.Y., Lin, C.Y., 2006. Calcium effect on fermentative hydrogen production in an anaerobic up-flow sludge blanket system. Water Sci. Technol. 54, 105–112.

Chang, J., Chou, C., Ho, C., Chen, W., Lay, J., Huang, C., 2008. Syntrophic co-culture of aerobic Bacillus and anaerobic Clostridium for bio-fuels and bio-hydrogen production. Int. J. Hydrogen Energy 33, 5137–5146.

Chang, S., Zheng, J., Li, L.F., 2011. Evaluation of different pretreatment methods for preparing hydrogen-producing seed inocula from waste activated sludge. Renew. Energy 36, 1517–1522.

Chen, S.D., Lee, K.S., Lo, Y.C., Chen, W.M., Wu, J.F., Lin, C.Y., 2008. Batch and continuous biohydrogen production from starch hydrolysate by Clostridium species. Int. J. Hydrogen Energy 33 (7), 1803–1812.

Chen, W., Sung, S., Chen, S., 2009. Biological hydrogen production in an anaerobic sequencing batch reactor: pH and cyclic duration effects. Int. J. Hydrogen Energy 34, 227–234.

Chen, C.Y., Yeh, K.L., Lo, Y.C., Wang, H.M., Chang, J.S., 2010. Engineering strategies for the enhanced photo-H_2 production using effluents of dark fermentation processes as substrate. Int. J. Hydrogen Energy 35 (24), 13356–13364.

Chen, C.Y., Liu, C.H., Lo, Y.C., Chang, J.S., 2011. Perspectives on cultivation strategies and photobioreactor designs for photo-fermentative hydrogen production. Bioresour. Technol. 102, 8484–8492.

Cheng, H.H., Whang, L.M., Wu, C.W., Chung, M.C., 2012. A two-stage bioprocess for hydrogen and methane production from rice straw bioethanol residues. Bioresour. Technol. 113, 23–29.

Cheng, X.Y., Liu, C.Z., 2012. Enhanced coproduction of hydrogen and methane from cornstalks by a three-stage anaerobic fermentation process integrated with alkaline hydrolysis. Bioresour. Technol. 104, 373–379.

Cheng, S., Logan, B.E., 2007. Sustainable and efficient biohydrogen production via electrohydrogenesis. Proc. Natl. Acad. Sci. USA 104, 18875–118873.

Cheng, X.Y., Li, Q., Liu, C.Z., 2012. Coproduction of hydrogen and methane via anaerobic fermentation of cornstalk waste in continuous stirred tank reactor integrated with up-flow anaerobic sludge bed. Bioresour. Technol. 114, 327–333.

Chou, C.H., Han, C.L., Chang, J.J., Lay, J.J., 2011. Co-culture of *Clostridium beijerinckii* L9, *Clostridium butyricum* M1 and *Bacillus thermoamylovorans* B5 for converting yeast waste into hydrogen. Int. J. Hydrogen Energy 36 (21), 13972–13983.

Chuang, Y.S., Lay, C.H., Sen, B., Chen, C.C., Gopalakrishnan, K., Wu, J.H., et al., 2011. Biohydrogen and biomethane from water hyacinth (*Eichhornia crassipes*) fermentation: Effects of substrate concentration and incubation temperature. Int. J. Hydrogen Energy 36 (21), 14195–14203.

Clauwaert, P., Verstraete, W., 2009. Methanogenesis in membraneless microbial electrolysis cell. Appl. Microbiol. Biotechnol. 82 (5), 829–836.

Cusick, R.D., Kiely, P.D., Logan, B.E., 2010. A monetary comparison of energy recovered from microbial fuel cells and microbial electrolysis cells fed winery or domestic wastewaters. Int. J. Hydrogen Energy 35 (17), 8855–8861.

Devi, M.P., Venkata Mohan, S., Mohanakrishna, G., Sarma, P.N., 2010. Regulatory influence of CO_2 sparging on fermentative hydrogen production. Int. J. Hydrogen Energy 35, 10701–10709.

Dixon, R., Kahn, D., 2004. Genetic regulation of biological nitrogen fixation. Nat. Rev. Microbiol. 2 (8), 621–631.

Ena, A., Pintucci, C., Carlozzi, P., 2010. Production of $bioH_2$ by *Rhodopseudomonas palustris* (strain 6A) grown in pretreated olive mill waste, under batch or semi- continuous regime. J. Biotechnol. 150, 180.

Eroglu, E., Gündüz, U., Yücel, M., Türker, L., Eroglu, I., 2004. Photobiological hydrogen production from olive mill wastewater as sole substrate sources. Int. J. Hydrogen Energy 29 (2), 163–171.

Eroglu, E., Eroglu, I., Gunduz, U., Turker, L., Yucel, M., 2006. Biological hydrogen production from olive mill wastewater with two-stage processes. Int. J. Hydrogen Energy 31 (11), 1527–1535.

Eroglu, E., Gunduz, U., Yucel, M., Eroglu, I., 2010. Photosynthetic bacterial growth and productivity under continuous illumination or diurnal cycles with olive mill wastewater as feedstock. Int. J. Hydrogen Energy 35 (11), 5293–5300.

Eroglu, E., Gunduz, U., Yucel, M., Eroglu, I., 2011. Effect of iron and molybdenum addition on photofermentative hydrogen production from olive mill wastewater. Int. J. Hydrogen Energy 36, 5895–5903.

Eroğlua, E., Eroğlua, I., Gündüzb, U., Türkerc, L., Yücelb, M., 2006. Biological hydrogen production from olive mill wastewater with two-stage processes. Int. J. Hydrogen Energy 31, 1527–1535.

Fan, Y.T., Li, C.L., Lay, J.J., Hou, H.W., Zhang, G.S., 2004. Optimization of initial substrate and pH levels for germination of sporing hydrogen-producing anaerobes in cow dung compost. Bioresour. Technol. 91, 189–193.

Fang, H.H.P., Liu, H., 2002. Effect of pH on hydrogen production from glucose by a mixed culture. Bioresour. Technol. 82, 87–93.

Ferchichi, M., Crabbe, V., Gil, G.H., Hintz, W., Almadidy, A., 2005. Influence of initial pH on hydrogen production from cheese whey. J. Biotechnol. 120, 402–409.

Fernandes, B.S., Peixoto, G., Albrecht, F.R., Katia, N., Aguila, S.D., Zaiat, M., 2010. Potential to produce biohydrogen from various wastewaters. Energy Sustain. Dev. 14 (2), 143–148.

Froylán, M.E.E., Carlos, P.O., Jose, N.C., Yolanda, G.G., André, B., Humberto, G.P., 2009. Anaerobic digestion of the vinasses from the fermentation of Agave tequilanaWeber to tequila: The effect of pH, temperature and hydraulic retention time on the production of hydrogen and methane. Biomass Bioenergy 33, 14–20.

Gadhamshetty, V., Sukumaran, A., Khandan, N.N., 2011. Review on photoparameters in photofermentative biohydrogen production. Crit. Rev. Environ. Sci. Technol. 41 (1), 1–51.

Garcin, E., Vernede, X., Hatchikian, E.C., Volbeda, A., Frey, M., Fontecilla-Camps, J.C., 1999. The crystal structure of a reduced [NiFeSe] hydrogenase provides an image of the activated catalytic center. Structure 7, 557–566.

Gopalakrishnan, K., Lay, C.H., Chu, C.Y., Wu, J.H., Lee, S.C., Lin, C.Y., 2012. Seed inocula for biohydrogen production from biodiesel solid residues. Int. J. Hydrogen Energy 37 (20), 15489–15495.

Goud, R.K., Venkata Mohan, S., 2012. Acidic and alkaline shock pretreatment to enrich acidogenic biohydrogen producing mixed culture: Long term synergetic evaluation of microbial inventory, dehydrogenase activity and bio-electro kinetics. RSC Adv. 2, 6336–6353.

Goud, R.K., Veer Raghavulu, S., Mohanakrishna, G., Naresh, K., Venkata Mohan, S., 2012. Predominance of Bacilli and Clostridia in microbial community of biohydrogen producing biofilm sustained under diverse acidogenic operating conditions. Int. J. Hydrogen Energy 37 (5), 4068–4076.

Guo, L., Zhao, J., She, Z., Lu, M., Zong, Y., 2012. Effect of S-TE (solubilization by thermophilic enzyme) digestion conditions on hydrogen production from waste sludge. Bioresour. Technol. 117, 368–372.

Gustavo, D.V., Felipe, A.M., Antonio de, L.R., Elías, R.F., 2008. Fermentative hydrogen production in batch experiments using lactose, cheese whey and glucose: Influence of initial substrate concentration and pH. Int. J. Hydrogen Energy 33, 4989–4997.

Guwy, A.J., Dinsdale, R.M., Kim, J.R., Massanet-Nicolau, J., Premier, G., 2011. Fermentative biohydrogen production systems integration. Bioresour. Technol. 102, 8534–8542.

Hafez, H., Nakhla, G., Naggar, H.E.l., 2010. An integrated system for hydrogen and methane production during landfill leachate treatment. Int. J. Hydrogen Energy 35, 5010–5014.

Hallenbeck, P.C., 2011. Microbial paths to renewable hydrogen production. Biofuels 2, 285–302.

Hallenbeck, P.C., Benemann, J.R., 2002. Biological hydrogen production; fundamental and limiting processes. Int. J. Hydrogen Energy 27, 1185–1193.

Hallenbeck, P.C., Ghosh, D., 2009. Advances in fermentative biohydrogen production: The way forward? Trends Biotechnol. 27, 287–297.

Han, W., Chen, H., Jiao, A., Wang, Z., Li, Y., Ren, N., 2012a. Biological fermentative hydrogen and ethanol production using continuous stirred tank reactor. Int. J. Hydrogen Energy 37 (1), 843–847.

Han, W., Wang, B., Zhou, Y., Wang, D.X., Wang, Y., Yue, L., et al., 2012b. Fermentative hydrogen production from molasses wastewater in a continuous mixed immobilized sludge reactor. Bioresour. Technol. 110, 219–223.

Hawkes, F.R., Hussy, I., Kyazze, G., Dinsdale, R., Hawkes, D.L., 2007. Continuous dark fermentative hydrogen production by mesophilic microflora: Principles and progress. Int. J. Hydrogen Energy 32, 172–184.

Horsburgh, M.J., Wharton, S.J., Karavolos, M., Foster, S.J., 2002. Manganese: Elemental defence for a life with oxygen. Trends Microbiol. 10, 496–501.

Hu, Q., Sommerfeld, M., Jarvis, E., Ghirardi, M., Posewitz, M., Seibert, M., et al., 2008. Microalgal triacylglycerols as feedstocks for biofuel production: Perspectives and advances. Plant J. 54, 621–639.

Idania, V.V., Richard, S., Derek, R., Noemi, R.S., Poggi-Varaldo, H.M., 2005. Hydrogen generation via anaerobic fermentation of paper mill wastes. Bioresour. Technol. 96, 1907–1913.

Im, W.T., Kim, D.H., Kim, K.H., Kim, M.S., 2012. Bacterial community analyses by pyrosequencing in dark fermentative H$_2$-producing reactor using organic wastes as a feedstock. Int. J. Hydrogen Energy 37 (10), 8330–8337.

Imbert, M., Blondeau, R., 1998. On the iron requirement of lactobacilli grown in chemically defined medium. Curr. Microbiol 37, 64–66.

Jacobson, M.R., Premakumar, R., Bishop, P.E., 1986. Transcriptional regulation of nitrogen fixation by molybdenum in *Azotobacter vinelandii*. J. Bacteriol. 167 (2), 480–486.

Jakubovics, N.S., Jenkinson, H.F., 2001. Out of the iron age: New insights into the critical role of manganese homeostasis in bacteria. Microbiology 147, 1709–1718.

Julia, C.R., Celis, L.B., Felipe, A.M., Elías, R.F., 2012. Different start-up strategies to enhance biohydrogen production from cheese whey in UASB reactors. Int. J. Hydrogen Energy 37 (7), 5591–5601.

Jung, K.W., Kim, D.H., Shin, S.H., 2010. Continuous fermentative hydrogen production from coffee drink manufacturing wastewater by applying UASB reactor. Int. J. Hydrogen Energy 35, 13370–13378.

Jung, K.W., Kim, D.H., Shin, H.S., 2012. Continuous fermentative hydrogen and methane production from *Laminaria japonica* using a two-stage fermentation system with recycling of methane fermented effluent. Int. J. Hydrogen Energy 37 (20), 15648–15657.

Kapdan, I.K., Kargi, F., 2006. Bio-hydrogen production from waste materials. Enzyme Microbial Technol. 38, 569–582.

Kapdan, I.K., Kargi, F., Oztekin, R., Argun, H., 2009. Bio-hydrogen production from acid hydrolyzed wheat starch by photo-fermentation using different *Rhodobacter* sp. Int. J. Hydrogen Energy 34 (5), 2201–2207.

Karadag, D., Puhakka, J.A., 2010. Enhancement of anaerobic hydrogen production by iron and nickel. Int. J. Hydrogen Energy 35, 8554–8560.

Kargi, F., Catalkaya, E.C., 2011. Hydrogen gas production from olive mill wastewater by electrohydrolysis with simultaneous COD removal. Int. J. Hydrogen Energy 36 (5), 3457–3464.

Kargi, F., Eren, N.S., Ozmihci, S., 2012a. Bio-hydrogen production from cheese whey powder (CWP) solution: Comparison of thermophilic and mesophilic dark fermentations. Int. J. Hydrogen Energy 37 (10), 8338–8342.

Kargi, F., Eren, N.S., Ozmihci, S., 2012b. Hydrogen gas production from cheese whey powder (CWP) solution by thermophilic dark fermentation. Int. J. Hydrogen Energy 37 (3), 2260–2266.

Kars, G., Gunduz, U., Yucel, M., Turker, L., Eroglu, I., 2006. Hydrogen production and transcriptional analysis of nifD, nifK and hupS genes in *Rhodobacter sphaeroides* O.U.001 grown in media with different concentrations of molybdenum and iron. Int. J. Hydrogen Energy 31, 1536–1544.

Kehres, D.G., Maguire, M.E., 2003. Emerging themes in manganese transport, biochemistry and pathogenesis in bacteria. FEMS Microbiol. Rev. 27, 263–290.

Keskin, T., Hallenbeck, P.C., 2012. Hydrogen production from sugar industry wastes using single-stage photofermentation. Bioresour. Technol. 112, 131–136.

Keskin, T., Abo-Hashesh, M., Hallenbeck, P.C., 2011. Photofermentative hydrogen production from wastes. Bioresour. Technol. 102, 8557–8568.

Kim, M.S., Lee, D.Y., 2010. Fermentative hydrogen production from tofu-processing waste and anaerobic digester sludge using microbial consortium. Bioresour. Technol. 101, S48–S52.

Kim, J., Park, C., Tak-Hyun, K., Lee, M., Kim, S., Seung-Wook, K., et al., 2003. Effects of various pretreatments for enhanced anaerobic digestion with waste activated sludge. J. Biosci. Bioeng. 95, 271–275.

Kim, D.H., Kim, S.H., Jung, K.W., Kim, M.S., Shin, H.S., 2011a. Effect of initial pH independent of operational pH on hydrogen fermentation of food waste. Bioresour. Technol. 102 (18), 8646–8652.

Kim, M.S., Lee, D.Y., Kim, D.H., 2011b. Continuous hydrogen production from tofu processing waste using anaerobic mixed microflora under thermophilic conditions. Int. J. Hydrogen Energy 36 (14), 8712–8718.

Kim, M.S., Kim, D.H., Cha, J., Lee, J.K., 2012. Effect of carbon and nitrogen sources on photo-fermentative H_2 production associated with nitrogenase, uptake hydrogenase activity, and PHB accumulation in *Rhodobacter sphaeroides* KD131. Bioresour. Technol. 116, 179–183.

Kima, S.H., Cheona, H.C., Lee, C.Y., 2012. Enhancement of hydrogen production by recycling of methanogenic effluent in two-phase fermentation of food waste. Int. J. Hydrogen Energy 37 (18), 13777–13782.

Kobayashi, T., Xu, K.Q., Li, Y.Y., Inamori, Y., 2012. Effect of sludge recirculation on characteristics of hydrogen production in a two-stage hydrogenemethane fermentation process treating food wastes. Int. J. Hydrogen Energy 37 (7), 5602–5611.

Koku, H., Eroglu, I., Gunduz, U., Yucel, M., Turker, L., 2002. Aspects of the metabolism of hydrogen production by *Rhodobacter sphaeroides*. Int. J. Hydrogen Energy 28 (4), 1315–1329.

Kongjan, P., Angelidaki, I., 2010. Extreme thermophilic biohydrogen production from wheat straw hydrolysate using mixed culture fermentation: Effect of reactor configuration. Bioresour. Technol. 101, 7789–7796.

Kongjan, P., O-Thong, O., Angelidaki, I., 2011. Biohydrogen production from desugared molasses (DM) using thermophilic mixed cultures immobilized on heat treated anaerobic sludge granules. Int. J. Hydrogen Energy 36 (21), 14261–14269.

Kotay, S.M., Das, D., 2010. Microbial hydrogen production from sewage sludge bioaugmented with a constructed microbial consortium. Int. J. Hydrogen Energy 35, 10653–10659.

Kraemer, J.T., Bagley, D.M., 2007. Improving the yield from fermentative hydrogen production. Biotechnol. Lett. 29, 685–695.

Lakshmidevi, R., Muthukumar, K., 2010. Enzymatic saccharification and fermentation of paper and pulp industry effluent for biohydrogen production. Int. J. Hydrogen Energy 35, 3389–3400.

Lalaurette, E., Thammannagowda, S., Mohagheghi, A., Maness, P.C., Logan, B.E., 2009. Hydrogen production from cellulose in a two-stage process combining fermentation and electrohydrogenesis. Int. J. Hydrogen Energy 34 (15), 6201–6210.

Lalit Babu, V., Venkata Mohan, S., Sarma, P.N., 2009. Influence of reactor configuration on fermentative hydrogen production during wastewater. Int. J. Hydrogen Energy 34, 3305–3312.

Laurinavichene, T.V., Tekucheva, D.N., Laurinavichius, K.S., Ghirardi, M.L., Seibert, M., Tsygankov, A.A., 2008. Towards the integration of dark and photo fermentative waste treatment. 1. Hydrogen photoproduction by purple bacterium *Rhodobacter capsulatus* using potential products of starch fermentation. Int. J. Hydrogen Energy 33, 7020–7026.

Laurinavichene, T.V., Belokopytov, B.F., Laurinavichius, K.S., Tekucheva, D.N., Seibert, D.N., Tsygankov, A.A., 2010. Towards the integration of dark- and photo-fermentative waste treatment. 3. Potato as substrate for sequential dark fermentation and light-driven H_2 production. Int. J. Hydrogen Energy 35 (16), 8536–8543.

Lay, J.J., Fan, K.S., Chang, J., Ku, C.H., 2004. Influence of chemical nature of organic wastes on their conversion to hydrogen by heat-shock digested sludge. Int. J. Hydrogen Energy 28, 1361–1367.

Leano, E.P., Babel, S., 2012. Effects of pretreatment methods on cassava wastewater for biohydrogen production optimization. Renew. Energy 39 (1), 339–346.

Lee, Y.J., Miyahara, T., Noike, T., 2001. Effect of iron concentration on hydrogen fermentation. Bioresour. Technol. 80, 227–231.

Lee, D., Li, Y., Oh, Y., Kim, M., Noike, T., 2009. Effect of iron concentration on continuous H_2 production using membrane bioreactor. Int. J. Hydrogen Energy 34, 1244–1252.

Lee, C.M., Hung, G.J., Yang, C.F., 2011. Hydrogen production by Rhodopseudomonas palustris WP 3-5 in a serial photobioreactor fed with hydrogen fermentation effluent. Bioresour. Technol. 102 (18), 8350–8356.

Li, W.W., Yu, H.Q., 2011. From wastewater to bioenergy and biochemicals via two-stage bioconversion processes: A future paradigm. Biotechnol. Adv. 29 (6), 972–982.

Li, Y.C., Wu, S.Y., Chu, C.Y., Huang, H.C., 2011. Hydrogen production from mushroom farm waste with a two-step acid hydrolysis process. Int. J. Hydrogen Energy 36 (21), 14245–14251.

Li, Y.C., Chu, C.Y., Wu, S.Y., Tsai, C.Y., Wang, C.C., Hung, C.H., et al., 2012. Feasible pretreatment of textile wastewater for dark fermentative hydrogen production. Int. J. Hydrogen Energy 37 (20), 15511–15517.

Licata, B.L., Sagnelli, F., Boulanger, A., Lanzini, A., Leone, P., Zitella, P., et al., 2011. Bio-hydrogen production from organic wastes in a pilot plant reactor and its use in a SOFC. Int. J. Hydrogen Energy 36 (13), 7861–7865.

Lin, C.Y., Lay, C.H., 2004a. Carbon/nitrogen-ratio effect on fermentative hydrogen production by mixed microflora. Int. J. Hydrogen Energy 29, 41–45.

Lin, C.Y., Lay, C.H., 2004b. Effects of carbonate and phosphate concentrations on hydrogen production using anaerobic sewage sludge microflora. Int. J. Hydrogen Energy 29, 275–281.

Lin, C.Y., Lay, C.H., 2005. A nutrient formulation for fermentative hydrogen production using anaerobic sewage sludge microflora. Int. J. Hydrogen Energy 30, 285–292.

Lin, C.Y., Shei, S.H., 2008. Heavy metal effects on fermentative hydrogen production using natural mixed microflora. Int. J. Hydrogen Energy 33, 587–593.

Liu, F., Fang, B., 2007. Optimization of biohydrogen production from biodiesel wastes by Klebsiella pneumoniae. Biotech. J. 2, 374–380.

Liu, H., Grot, S., Logan, B.E., 2005. Electrochemically assisted microbial production of hydrogen from acetate. Environ. Sci. Technol. 39, 4317–4320.

Liu, X.H., Hai-ping, X.U., Bin, H.E., Guo-lin, Y., Yun-tao, G., 2006. Conversion of High Concentration Organic Wastewater from Sugar Works. Environ. Sci. Technol. 03.

Liu, B.F., Ren, N.Q., Ding, J., Xie, G.J., Cao, G.L., 2009a. Enhanced photo-H_2 production of R. faecalis RLD-53 by separation of CO_2 from reaction system. Bioresour. Technol 100 (3), 1501–1504.

Liu, H., Wanga, G., Zhu, D., Pan, V., 2009b. Enrichment of the hydrogen-producing microbial community from marine intertidal sludge by different pretreatment methods. Int. J. Hydrogen Energy 34, 9696–9701.

Liu, B.F., Ren, N.Q., Xie, G.J., Ding, J., Guo, W.Q., Xing, D.F., 2010. Enhanced biohydrogen production by the combination of dark and photo-fermentation in batch culture. Bioresour. Technol. 101, 5325–5329.

Liu, Q., Zhang, X., Yu, L., Zhao, A., Tai, J., Liu, J., et al., 2011. Fermentative hydrogen production from fresh leachate in batch and continuous bioreactors. Bioresour. Technol. 102 (9), 5411–5417.

Liu, Q., Zhang, X.L., Jun, Z., Zhao, A.H., Chen, S.P., Liu, F., et al., 2012. Effect of carbonate on anaerobic acidogenesis and fermentative hydrogen production from glucose using leachate as supplementary culture under alkaline conditions. Bioresour. Technol. 113, 37–43.

Lo, Y.C., Chen, C.Y., Lee, C.M., Chang, J.S., 2011. Photo fermentative hydrogen production using dominant components (acetate, lactate, and butyrate) in dark fermentation effluents. Int. J. Hydrogen Energy 36 (21), 14059–14068.

Lu, H., Zhang, G., Wan, T., Lu, Y., 2011. Influences of light and oxygen conditions on photosynthetic bacteria macromolecule degradation: Different metabolic pathways. Bioresour. Technol 102, 9503–9508.

Luo, G., Xie, L., Zou, Z., Wang, W., Zhou, Q., 2010. Evaluation of pretreatment methods on mixed inoculum for both batch and continuous thermophilic biohydrogen production from cassava stillage. Bioresour. Technol. 101, 959–964.

Madsen, E.L., 2008. Environmental Microbiology from Genomes to Biogeochemistry. Blackwell Publishers.

McInerney, M.J., 2008. Physiology, ecology, phylogeny, and genomics of microorganisms capable of syntrophic metabolism. Ann. N.Y. Acad. Sci. 1125, 58–72.

McInerney, M.J., Sieber, J.R., Gunsalus, R.P., 2009. Syntrophy in anaerobic global carbon cycles. Curr. Opin. Biotechnol. 20, 623–632.

Mizuno, O., Dinsdale, R., Hawkes, F.R., Hawkes, D.L., Noike, T., 2000. Enhancement of hydrogen production from glucose by nitrogen gas sparging. Bioresour. Technol. 73, 59–65.

Mohanakrishna, G., Venkata Mohan, S., 2013. Multiple process integrations for broad perspective analysis of fermentative H_2 production from wastewater treatment: Technical and environmental considerations. Appl. Energy 107, 244–254.

Mohanakrishna, G., Goud, R.K., Venkata Mohan, S., Sarma, P.N., 2010a. Enhancing biohydrogen production through sewage supplementation of composite vegetable based market waste. Int. J. Hydrogen Energy 35 (2), 533–541.

Mohanakrishna, G., Venkata Mohan, S., Sarma, P.N., 2010b. Utilizing acid-rich effluents of fermentative hydrogen production process as substrate for harnessing bioelectricity: An integrative approach. Int. J. Hydrogen Energy 35 (8), 3440–3449.

Moncrief, M.B.C., Maguire, M.E., 1999. Magnesium transport in prokaryotes. J. Biol. Inorg. Chem. 4, 523–527.

Moreno-Davila, I.M.M., Rios-GonZalez, L.J., Gaona-Llozano, J.G., Garza-Garcia, Y., Rodriguez-Martinez, G.J., 2010. Biohydrogen production by anaerobic biofilms from a pretreated mixed microflora. Appl. Sci. 5 (6), 376–382.

Morita, M., Malvankara, N.S., Franksa, A.E., Summersa, Z.M., Giloteauxa, L., Rotarua, A.E., et al., 2011. Potential for Direct Interspecies Electron Transfer in Methanogenic Wastewater Digester Aggregates. mBio 2 (4), e00159–11.

Morsy, F.M., 2011. Hydrogen production from acid hydrolyzed molasses by the hydrogen overproducing *Escherichia coli* strain HD701 and subsequent use of the waste bacterial biomass for biosorption of Cd(II) and Zn(II). Int. J. Hydrogen Energy 36 (22), 14381–14390.

Nakamura, M., Kanbe, H., Matsumoto, J., 1993. Fundamental studies on hydrogen production in the acid-forming phase and its bacteria in anaerobic treatment processes-the effects of solids retention time. Water Sci. Technol. 28, 81.

Nasr, N., Elbeshbishy, E., Hafez, H., Nakhl, G., Hesham, M., Naggar, E., 2012. Comparative assessment of single-stage and two-stage anaerobic digestion for the treatment of thin stillage. Bioresour. Technol. 111, 122–126.

Nath, K., Das, D., 2004. Improvement of fermentative hydrogen production: Various approaches. Appl. Microbiol. Biotechnol. 65, 520–529.

Nicolet, Y., Cavazza, C., Fontecilla-Camps, J.C., 2002. Fe-only hydrogenases: Structure, future and evolution. J. Inorg. Biochem. 91, 1–8.

Niel, E.W.J.V., Claassen, P.A.M., Stams, A.J.M., 2003. Substrate and production inhibition of hydrogen production by the extreme thermophile *Caldicellulosiruptor saccharolyticus*. Biotechnol. Bioeng. 81, 255.

Ntaikou, I., Kourmentza, C., Koutrouli, E.C., Stamatelatou, K., Zampraka, A., Kornaros, M., et al., 2009. Exploitation of olive oil mill wastewater for combined biohydrogen and biopolymers production. Bioresour. Technol. 100 (15), 3724–3730.

Oleszkiewicz, J.A., Sharma, V.K., 1990. Stimulation and inhibition of anaerobic processes by heavy metals. Biol. Wastes 31, 45–67.

O-Thong, S., Prasertsan, P., Birkeland, N.K., 2009. Evaluation of methods for preparing hydrogen-producing seed inocula under thermophilic condition by process performance and microbial community analysis. Bioresour. Technol. 100, 909–918.

Ozgur, E., Afsar, N., Vrije, T., Yucel, M., Gunduz, U., Claassen, P., et al., 2010a. Potential use of thermophilic dark fermentation effluents in photofermentative hydrogen production by *Rhodobacter capsulatus*. J. Cleaner Prod. 18, S23–S28.

Ozgur, E., Mars, A., Peksel, B., Lowerse, A., Afsar, N., Vrije, T., et al., 2010b. Biohydrogen production from beet molasses by sequential dark and photofermentation. Int. J. Hydrogen Energy 35, 511–517.

Ozkan, E., Uyar, B., Özgür, E., Yücel, M., Eroglu, I., Gündüz, U., 2012. Photofermentative hydrogen production using dark fermentation effluent of sugar beet thick juice in outdoor conditions. Int. J. Hydrogen Energy 37 (2), 2044–2049.

Ozmihci, S., Kargi, F., 2010. Bio-hydrogen production by photo-fermentation of dark fermentation effluent with intermittent feeding and effluent removal. Int. J. Hydrogen Energy 35 (13), 6674–6680.

Pandey, A., Srivastava, N., Sinha, P., 2012. Optimization of hydrogen production by Rhodobacter sphaeroides NMBL-01. Biomass Bioenergy 37, 251–256.

Patel, S.K.S., Singh, M., Kumar, P., Purohit, H.J., Kalia, V.C., 2012. Exploitation of defined bacterial cultures for production of hydrogen and polyhydroxybutyrate from pea-shells. Biomass Bioenergy 36, 218–225.

Pattanamanee, W., Choorit, W., Deesan, C., Sirisansaneeyakul, S., Chisti, Y., 2012. Photofermentive production of biohydrogen from oil palm waste hydrolysate. Int. J. Hydrogen Energy 37 (5), 4077–4087.

Pawinee, S., Rangsunvigit, P., Leethochawalit, M., Chavadej, S., 2011. Hydrogen production from alcohol distillery wastewater containing high potassium and sulfate using an anaerobic sequencing batch reactor. Int. J. Hydrogen Energy 36 (20), 12810–12821.

Pekgoz, G., Gunduz, U., Eroglu, I., Yucel, M., Kovacs, K., Rakhely, G., 2011. Effect of inactivation of genes involved in ammonium regulation on the biohydrogen production of *Rhodobacter capsulatus*. Int. J. Hydrogen Energy 36, 13536–13546.

Prasad, K., Serrano, M., Thomsen, A.B., Prawit, K., Angelidaki, I., 2009. Bioethanol, biohydrogen and biogas production from wheat straw in a biorefinery concept. Biores. Technol. 100 (9), 2562–2568.

Qiu, C., Wen, J., Jia, X., 2011. Extreme-thermophilic biohydrogen production from lignocellulosic bioethanol distillery wastewater with community analysis of hydrogen-producing microflora. Int. J. Hydrogen Energy 36 (14), 8243–8251.

Rai, K.P., Singh, S.P., Asthana, R.K., 2012. Biohydrogen production from cheese whey wastewater in a two-step anaerobic process. Appl. Biochem. Biotechnol. 167, 1540–1549.

Redwood, M.D., Macaskie, L.E., 2006. A two-stage, two-organism process for biohydrogen from glucose. Int. J. Hydrogen Energy 31, 1514–1521.

Ren, N., Chen, Z., Wang, A., Hu, D., 2005. Removal of organic pollutants and analysis of MLSS-COD removal relationship at different HRTs in a submerged membrane bioreactor. Int. Biodeter. Biodegrad. 55, 279–284.

Ren, N.Q., Chua, H., Chan, S.Y., Tsang, Y.F., Wang, Y.J., Sin, N., 2007. Assessing optimal fermentation type for bio-hydrogen production in continuous flow acidogenic reactors. Bioresour. Technol. 98, 1774–1780.

Ren, N., Wang, A., Gao, L., Xin, L., Lee, D-J, Su, A., 2008a. Bioaugmented hydrogen production from carboxymethyl cellulose and partially delignified corn stalks using isolated cultures. Int. J. Hydrogen Energy 33, 5250–5255.

Ren, N-Q., Guo, W-Q., Wang, X-J., Xiang, W-S., Liu, B-F., Wang, X-Z., et al., 2008b. Effects of different pretreatment methods on fermentation types and dominant bacteria for hydrogen production. Int. J. Hydrogen Energy 33, 4318–4324.

Rittmann, B.E., 2008. Opportunities for renewable bioenergy using microorganisms. Biotechnol. Bioeng. 100 (2), 203–212.

Rozendal, R.A., Hamelers, H.V.M., Euverink, G.J.W., Metz, S.J., Buisman, C.J.N., 2006. Principle and perspectives of hydrogen production through biocatalyzed electrolysis. Int. J. Hydrogen Energy 31, 1632–1640.

Rozendal, R.A., Hamelers, H.V.M., Molenkamp, R.J., Buisman, C.J.N., 2007. Performance of single chamber biocatalyzed electrolysis with different types of ion exchange membranes. Water Res 41 (9), 1984–1994.

Sagnak, R., Kargi, F., 2011. Photo-fermentative hydrogen gas production from dark fermentation effluent of acid hydrolyzed wheat starch with periodic feeding. Int. J. Hydrogen Energy 36 (7), 4348–4353.

Sagnak, R., Kargi, F., Kapdan, I.K., 2011. Bio-hydrogen production from acid hydrolyzed waste ground wheat by dark fermentation. Int. J. Hydrogen Energy 36 (20), 12803–12809.

Salerno, M.B., Park, W., Zuo, Y., Logan, B.E., 2006. Inhibition of biohydrogen production by ammonia. Water Res 40, 1167–1172.

Saraphirom, P., Reungsang, A., 2010. Optimization of biohydrogen production from sweet sorghum syrup using statistical methods. Int. J. Hydrogen Energy 35 (24), 13435–13444.

Saraphirom, P., Reungsang, A., 2011. Biological hydrogen production from sweet sorghum syrup by mixed cultures using an anaerobic sequencing batch reactor (ASBR). Int. J. Hydrogen Energy 36 (14), 8765–8773.

Saratale, G.D., Chen, S.D., Lo, Y.C., Saratale, R.G., Chang, J.S., 2008. Outlook of biohydrogen production from lignocellulosic feedstock using dark fermentation -a review. J. Sci. Ind. Res. 67, 962–979.

Sasikala, K., Ramana, C.V., Rao, P.R., Kovacs, K.L., 1993. Anoxygenic phototropic bacteria: Physiology and advances in hydrogen production technology. Adv. Appl. Microbiol. 38, 211–295.

Seifert, K., Waligorska, M., Laniecki, M., 2010a. Hydrogen generation in photobiological process from dairy wastewater. Int. J. Hydrogen Energy 35 (18), 9624–9629.

Seifert, K., Waligorska, M., Laniecki, M., 2010b. Brewery wastewaters in photobiological hydrogen generation in presence of *Rhodobacter sphaeroides* O.U.001. Int. J. Hydrogen Energy 35, 4085–4091.

Sen, B., Suttar, R.R., 2012. Mesophilic fermentative hydrogen production from sago starch-processing wastewater using enriched mixed cultures. Int. J. Hydrogen Energy 37 (20), 15588–15597.

Sentürk, E., Ince, M., Engin, O.G., 2010. Treatment efficiency and VFA composition of a thermophilic anaerobic contact reactor treating food industry wastewater. J. Haz. Mat. 176 (1–3), 843–848.

Shi, X.Y., Jin, D.W., Sun, Q.Y., Li, W.W., 2010. Optimization of conditions for hydrogen production from brewery wastewater by anaerobic sludge using desirability function approach. Renewable Energy 35, 1493–1498.

Siddiqui, Z., Horan, N.J., Salter, M., 2011. Energy optimisation from co-digested waste using a two-phase process to generate hydrogen and methane. Int. J. Hydrogen Energy 36 (8), 4792–4799.

Sittijunda, S., Reungsang, A., 2012. Biohydrogen production from waste glycerol and sludge by anaerobic mixed cultures. Int. J. Hydrogen Energy 37 (18), 13789–13796.

Sivaramakrishna, D., Sreekanth, D., Himabindu, V., Anjaneyulu, Y., 2009. Biological hydrogen production from probiotic wastewater as substrate by selectively enriched anaerobic mixed microflora. Renew. Energy 34, 937–940.

Sompong, O.T., Hniman, A., Prasertsan, P., Imai, T., 2011. Biohydrogen production from cassava starch processing wastewater by thermophilic mixed cultures. Int. J. Hydrogen Energy 36 (5), 3409–3416.

Sparling, R., Risbey, D., Poggi-Varaldo, H.M., 1997. Hydrogen production from inhibited anaerobic composters. Int. J. Hydrogen Energy 22, 563–566.

Srikanth, S., Venkata Mohan, S., 2012. Regulatory function of divalent cations in controlling the acidogenic biohydrogen production process. RSC Adv. 2, 6576–6589.

Srikanth, S., Venkata Mohan, S., Devi, M.P., Dinikar, P., Sarma, P.N., 2009a. Acetate and butyrate as substrates for hydrogen production through photo-fermentation using mixed culture: Optimization of process parameters and combined process evaluation. Int. J. Hydrogen Energy 34, 7513–7522.

Srikanth, S., Venkata Mohan, S., Devi, M.P., Lenin Babu, M., Sarma, P.N., 2009b. Effluents with soluble metabolites generated from acidogenic and methanogenic processes as substrate for additional hydrogen production through photo-biological process. Int. J. Hydrogen Energy 34, 1771–1779.

Srikanth, S., Venkata Mohan, S., Lalit Babu, V., Sarma, P.N., 2010. Metabolic shift and electron discharge pattern of anaerobic consortia as a function of pretreatment method applied during fermentative hydrogen Production. Int. J. Hydrogen Energy 35, 10693–10700.

Stams, A.J.M., Plugge, C.M., 2009. Electron transfer in syntrophic communities of anaerobic bacteria and archaea. Nat. Rev. Microbiol. 7, 568–577.

Strocchi, A., Levitt, M.D., 1992. Factors affecting hydrogen production and consumption by human fecal flora: The critical roles of hydrogen tension and methanogenesis. J. Clin. Invest. 89, 1304–1311.

Tai, J., Adav, S.S., Su, A., Lee, D.J., 2010. Biological hydrogen production from phenol-containing wastewater using *Clostridium butyricum*. Int. J. Hydrogen energy 35, 13345–13349.

Tang, G.L., Huang, J., Sun, Z.J., Tang, Q.Q., Yan, C.H., Liu, G.Q., 2008. Biohydrogen production from cattle wastewater by enriched anaerobic mixed consortia: Influence of fermentation temperature and pH. J. Biosci. Bioeng. 106, 80–87.

Tanisho, S., Kuromoto, M., Kadokura, N., 1998. Effect of CO_2 removal on hydrogen production by fermentation. Int. J. Hydrogen Energy 23, 559–563.

Tenca, A., Schievano, A., Perazzolo, F., Adani, F., Oberti, R., 2011. Biohydrogen from thermophilic co-fermentation of swine manure with fruit and vegetable waste: Maximizing stable production without pH control. Bioresour. Technol. 102 (18), 8582–8588.

Tripathi, V.N., Srivastava, S., 2006. Ni^{2+}-uptake in *Pseudomonas putida* strain S4: A possible role of Mg^{2+}-uptake pump. J. Biosci 31 (1), 61–67.

Uyar, B., Schumacher, M., Gebicki, J., Modigell, M., 2009. Photoproduction of hydrogen by *Rhodobacter capsulatus* from thermophilic fermentation effluent. Bioprocess Biosyst. Eng. 32, 603–606.

Van Ginkel, S., Sung, S.W., Lay, J.J., 2001. Biohydrogen production as a function of pH and substrate concentration. Environ. Sci. Technol. 35, 4726–4730.

Van Ginkel, S.W., Oh, S., Logan, B.E., 2005. Biohydrogen gas production from food processing and domestic wastewaters. Int. J. Hydrogen Energy 30, 1535–1542.

Van Groenestijn, J., Hazewinkel, J., Nienoord, M., Bussmann, P., 2002. Energy aspects of biological hydrogen production in high rate bioreactors operated in the thermophilic temperature range. Int. J. Hydrogen Energy 27 (11), 1141–1147.

Van Loosdrecht, M.C.M., Pot, M.A., Heijnen, J.J., 1997. Importance of bacterial storage polymers in bioprocesses. Water Sci. Technol. 35 (1), 41–47.

Vatsala, T.M., Mohan Raj, S., Manimaran, A., 2008. A pilot-scale study of biohydrogen production from distillery effluent using defined bacterial co-culture. Int. J. Hydrogen Energy 33, 5404–5415.

Vazquez, I., Sparling, R., Rinderknecht-Seijas, N., Risbey, D., Poggi-Varaldo, H.M., 2005. Hydrogen from the anaerobic fermentation of industrial solids waste. Bioresour. Technol 96, 1907–1913.

Venkata Mohan, S., 2008. Fermentative hydrogen production with simultaneous wastewater treatment: Influence of pretreatment and system operating conditions. J. Sci. Ind. Res 67 (11), 950–961.

Venkata Mohan, S., 2009. Harnessing of biohydrogen from wastewater treatment using mixed fermentative consortia: Process evaluation towards optimization. Int. J. Hydrogen Energy 34, 7460–7474.

Venkata Mohan, S., 2010. Waste to renewable energy: A sustainable and green approach towards production of biohydrogen by acidogenic fermentation. In: Om, S., Steve, H. (Eds.), Sustainable Biotechnology: Renewable Resources and New Perspectives. Springer, New York, pp. 129–164.

Venkata Mohan, S., Devi, M.P., 2012. Fatty acid rich effluents from acidogenic biohydrogen reactor as substrates for lipid accumulation in heterotrophic microalgae with simultaneous treatment. Bioresour. Technol. 123, 627–635.

Venkata Mohan, S., Goud, R.K., 2012. Pretreatment of biocatalyst as viable option for sustained production of biohydrogen from wastewater treatment. In: Mudhoo, A. (Ed.), Biogas Production: Pretreatment Methods in Anaerobic Digestion. John Wiley & Sons, Inc., Hoboken, NJ.

Venkata Mohan, S., Lenin Babu, M., 2011. Dehydrogenase activity in association with poised potential during biohydrogen production in single chamber microbial electrolysis cell. Bioresour. Technol. 102 (12), 6637–6782.

Venkata Mohan, S., Bhaskar, Y.V., Krishna, T.M., Chandrasekhara Rao, N., Lalit Babu, V., Sarma, P.N., 2007a. Biohydrogen production from chemical wastewater as substrate by selectively enriched anaerobic mixed consortia: Influence of fermentation pH and substrate composition. Int. J. Hydrogen Energy 32, 2286–2295.

Venkata Mohan, S., Bhaskar, Y.V., Sarma, P.N., 2007b. Biohydrogen production from chemical wastewater treatment by selectively enriched anaerobic mixed consortia in biofilm configured reactor operated in periodic discontinuous batch mode. Water Res. 41, 2652–2664.

Venkata Mohan, S., Lalit Babu, V., Sarma, P.N., 2007c. Anaerobic biohydrogen production from dairy wastewater treatment in sequencing batch reactor (AnSBR): Effect of organic loading rate. Enzyme Microbial Technol 41 (4), 506–515.

Venkata Mohan, S., Mohanakrishna, G., Veer Raghuvulu, S., Sarma, P.N., 2007d. Enhancing biohydrogen production from chemical wastewater treatment in anaerobic sequencing batch biofilm reactor (AnSBBR) by bioaugmenting with selectively enriched kanamycin resistant anaerobic mixed consortia. Int. J. Hydrogen Energy 32, 3284–3292.

Venkata Mohan, S., Lalit Babu, V., Sarma, P.N., 2008a. Effect of various pretreatment methods on anaerobic mixed microflora to enhance biohydrogen production utilizing dairy wastewater as substrate. Bioresour. Technol. 99, 59–67.

Venkata Mohan, S., Lalit Babu, V., Srikanth, S., Sarma, P.N., 2008b. Bio-electrochemical evaluation of fermentative hydrogen production process with the function of feeding pH. Int. J. Hydrogen Energy 33 (17), 4533–4546.

Venkata Mohan, S., Mohanakrishna, G., Ramanaiah, S.V., Sarma, P.N., 2008c. Simultaneous biohydrogen production and wastewater treatment in biofilm configured anaerobic periodic discontinuous batch reactor using distillery wastewater. Int. J. Hydrogen Energy 33 (2), 550–558.

Venkata Mohan, S., Mohanakrishna, G., Reddy, S.S., Raju, B.D., Rao, K.S., Sarma, P.N., 2008d. Self-immobilization of acidogenic mixed consortia on mesoporous material (SBA-15) and activated carbon to enhance fermentative hydrogen production. Int. J. Hydrogen Energy 33, 6133–6142.

Venkata Mohan, S., Mohankrishna, G., Sarma, P.N., 2008e. Integration of acidogenic and methanogenic processes for simultaneous production of biohydrogen and methane from wastewater treatment. Int. J. Hydrogen Energy 33, 2156–2166.

Venkata Mohan, S., Srikanth, S., Dinakar, P., Sarma, P.N., 2008f. Photo-biological hydrogen production by the adopted mixed culture: Data enveloping analysis. Int. J. Hydrogen Energy 33 (2), 559–569.

Venkata Mohan, S., Lenin Babu, M., Mohanakrishna, G., Sarma, P.N., 2009a. Harnessing of biohydrogen by acidogenic fermentation of *Citrus limetta* peelings: Effect of extraction procedure and pretreatment of biocatalyst. Int. J. Hydrogen Energy 34 (15), 6149–6156.

Venkata Mohan, S., Mohanakrishna, G., Goud, R.K., Shrma, P.N., 2009b. Acidogenic fermentation of vegetable based market waste to harness the biohydrogen. Bioresour. Technol. 100 (12), 3061–3068.

Venkata Mohan, S., Raghuvulu, S.V., Mohanakrishna, G., Srikanth, S., Sarma, P.N., 2009c. Optimization and evaluation of fermentative hydrogen production and wastewater treatment processes using data enveloping analysis (DEA) and Taguchi design of experimental (DOE) methodology. Int. J. Hydrogen Energy 34, 216–226.

Venkata Mohan, S., Reddy, M.V., Subhash, G.V., Sarma, P.N., 2010a. Fermentative effluents from hydrogen producing bioreactor as substrate for poly (β-OH) butyrate production with simultaneous treatment: An integrated approach. Bioresour. Technol. 101 (23), 9382–9386.

Venkata Mohan, S., Srikanth, S., Lenin Babu, M., Sarma, P.N., 2010b. Insight into the dehydrogenase catalyzed redox reactions and electron discharge pattern during fermentative hydrogen production. Bioresour. Technol. 101, 1826–1833.

Venkata Mohan, S., Veer Raghavulu, S., Goud, R.K., Srikanth, S., Lalit Babu, V., Sarma, P.N., 2010c. Microbial diversity analysis of long term operated biofilm configured anaerobic reactor producing hydrogen from wastewater under diverse conditions. Int. J. Hydrogen Energy 35, 1208–12215.

Venkata Mohan, S., Agarwal, L., Mohanakrishna, G., Srikanth, S., Kapley, A., Purohit, H.J., et al., 2011a. Firmicutes with iron dependent hydrogenase drive hydrogen production in anaerobic bioreactor using distillery wastewater. Int. J. Hydrogen Energy 36, 8234–8242.

Venkata Mohan, S., Mohanakrishna, G., Srikanth, S., 2011b. Biohydrogen production from industrial effluents. In: Pandey, A., Larroche, C., Ricke, S.C., Dussap, C.G., Gnansounou, E. (Eds.), Biofuels: Alternative Feedstocks and Conversion Processes. Academic Press, New York, pp. 499–524.

Venkata Mohan, S., Chiranjeevi, P., Mohanakrishna, G., 2012. A rapid and simple protocol for evaluating biohydrogen production potential (BHP) of wastewater with simultaneous process optimization. Int. J. Hydrogen Energy 37, 3130–3141.

Venkateswar Reddy, M., Chandrasekhar, K., Venkata Mohan, S., 2011. Influence of carbohydrates and proteins concentration on fermentative hydrogen production using canteen based waste under acidophilic microenvironment. J. Biotechnol. 155, 387–395.

Vijaya Bhaskar, Y., Venkata Mohan, S., Sarma, P.N., 2008. Effect of substrate loading rate of chemical wastewater on fermentative biohydrogen production in biofilm configured sequencing batch reactor. Bioresour. Technol. 99, 6941–6948.

Vijayaraghavan, K., Ahmad, D., 2006. Biohydrogen generation from palm oil mill effluent using anaerobic contact filter. Int. J. Hydrogen Energy 31, 1284–1291.

Wagner, R.C., Regan, J.M., Oh, S.E., Zuo, Y., Logan, B.E., 2009. Hydrogen and methane production from swine wastewater using microbial electrolysis cells. Water Res. 43, 1480–1488.

Wang, J., Wan, W., 2008a. Comparison of different pretreatment methods for enriching hydrogen-producing bacteria from digested sludge. Int. J. Hydrogen Energy 33 (12), 2934–2941.

Wang, J., Wan, W., 2008b. Influence of Ni^{2+} concentration on biohydrogen production. Bioresour. Technol. 99, 8864–8868.

Wang, J.L., Wan, W., 2008c. Effect of Fe^{2+} concentrations on fermentative hydrogen production by mixed cultures. Int. J. Hydrogen Energy 33, 1215–1220.

Wang, J., Wan, W., 2009. Factors influencing fermentative hydrogen production: A review. Int. J. Hydrogen Energy 34, 799–811.

Wang, B., Wan, W., Wang, J., 2009. Effect of ammonia concentration on fermentative hydrogen production by mixed cultures. Bioresour. Technol. 100, 1211–1213.

Wang, Y.Z., Liao, Q., Zhu, X., Tian, X., Zhang, C., 2010. Characteristics of hydrogen production and substrate consumption of Rhodopseudomonas palustris-CQK01 in an immobilized-cell photobioreactor. Bioresour. Technol. 101 (11), 4034–4041.

Wang, Y.Y., Ai, P., Hu, C.X., Zhang, Y.L., 2011. Effects of various pretreatment methods of anaerobic mixed microflora on biohydrogen production and the fermentation pathway of glucose. Int. J. Hydrogen Energy 36, 390–396.

Wimonsong, P., Nitisoravut, S., 2009. Pretreatment evaluation and its application on palam oil mill effluent for biohydrogen enhancement and methanogenic activity repression. Pakistan J. Biol. Sci. 12 (16), 1127–1133.

Wu, L.F., Navarro, C., Pina, K., de Quenard, M., Mandrand, M.A., 1994. Antagonistic effect of nickel on the fermentative growth of Escherichia coli K-12 and comparison of nickel and cobalt toxicity on the aerobic and anaerobic growth. Environ. Health Perspect. 102, 297–300.

Wu, T.Y., Mohammad, A.W., Md Jahim, J., Anuar, N., 2009. A holistic approach to managing palm oil mill effluent (POME): Biotechnological advances in the sustainable reuse of POME. Biotechnol. Adv. 27, 40–52.

Wu, X., Li, Q., Dieudonne, M., Cong, Y., Zhou, J., Long, M., 2010. Enhanced H$_2$ gas production from bagasse using adhE inactivated *Klebsiella oxytoca* HP1 by sequential dark-photo fermentations. Bioresour. Technol. 101 (24), 9605–9611.

Yadvika, S., Sreekrishnan, T.R., Kohli, S., Rana, V., 2004. Enhancement of biogas production from solid substrates using different techniques: A review. Bioresour. Technol. 95 (1), 1–10.

Yang, H.J., Shen, J.Q., 2006. Effect of ferrous iron concentration on anaerobic bio-hydrogen production from soluble starch. Int. J. Hydrogen Energy 31, 2137–2146.

Yang, H.C., Daniel, S.L., Hsu, T.D., Drake, H.L., 1989. Nickel transport by the thermophilic acetogen *Acetogenium kivui*. Appl. Environ. Microbiol. 55, 1078.

Yang, H., Shao, P., Lu, T., Shen, J., Wang, D., Xu, Z., et al., 2006. Continuous biohydrogen production from citric acid wastewater via facultative anaerobic bacteria. Int. J. Hydrogen Energy 31, 1306–1313.

Yang, P., Zhang, R., McGarvey, J.A., Benemann, J.R., 2007. Biohydrogen production from cheese processing wastewater by anaerobic fermentation using mixed microbial communities. Int. J. Hydrogen Energy 32, 4761–4771.

Yokoi, H., Tokushige, T., Hirose, J., Hayashi, S., Takasaki, Y., 1997. Hydrogen production by immobilized cells of aciduric *Enterobacter aerogenes* strain HO-39. J. Ferment. Bioeng. 83, 481–484.

Yokoi, H., Saitsu, A.S., Uchida, H., Hirose, J., Hayashi, S., Takasaki, Y., 2001. Microbial hydrogen production from sweet potato starch residue. J. Biosci. Bioeng. 91, 58–63.

Yokoyama, H., Ohmori, H., Waki, M., Ogino, A., Tanaka, Y., 2009. Continuous hydrogen production from glucose by using extreme thermophilic anaerobic microflora. J. Biosci. Bioeng. 107 (1), 64–66.

Yossan, S., Sompong, O.T., Prasertsan, P., 2012. Effect of initial pH, nutrients and temperature on hydrogen production from palm oil mill effluent using thermotolerant consortia and corresponding microbial communities. Int. J. Hydrogen Energy 37 (18), 13806–13814.

Yu, H.Q., Fang, H.H.P., 2001. Inhibition on acidogenesis of dairy waster by zinc and copper. Environ. Technol. 22, 1459–1465.

Yu, H., Zhu, Z., Hu, W., Zhang, H., 2002. Hydrogen production from rice winery wastewater in an upflow anaerobic reactor by using mixed anaerobic cultures. Int. J. Hydrogen Energy 27, 1359–1365.

Zhang, Y., Shen, J., 2006. Effect of temperature and iron concentration on the growth and hydrogen production of mixed bacteria. Int. J. Hydrogen Energy 31, 441–446.

Zhang, Y.M., Wong, T.Y., Chen, L.Y., Lin, C.S., Liu, J.K., 2000. Induction of a futile Embden-Meyerhof-Parnas pathway in *Deinococcus radiodurans, Deinococcus radiodurans* by Mn: Possible role of the pentose phosphate pathway in cell survival. Appl. Environ. Microbiol. 66, 105–112.

Zhang, Y.F., Liu, G.Z., Shen, J.Q., 2005. Hydrogen production in batch culture of mixed bacteria with sucrose under different iron concentrations. Int. J. Hydrogen Energy 30, 855–860.

Zhang, Z.P., Show, K.Y., Tay, J.H., Liang, D.T., Lee, D.J., 2008. Biohydrogen production with anaerobic fluidized bed reactors: A comparison of biofilm-based and granule-based systems. Int. J. Hydrogen Energy 33 (5), 1559–1564.

Zheng, G.H., Wang, L., Kang, Z.H., 2010. Feasibility of biohydrogen production from tofu wastewater with glutamine auxotrophic mutant of *Rhodobacter sphaeroides*. Renew. Energy 35, 2910–2913.

Zhu, H., Beland, M., 2006. Evaluation of alternative methods of preparing hydrogen producing seeds from digested wastewater sludge. Int. J. Hydrogen Energy 31, 1980–1988.

Zhu, H., Ueda, S., Asada, Y., Miyake, J., 2002. Hydrogen production as a novel process of wastewater treatment-studies on tofu wastewater with entrapped *R. sphaeroides* and mutagenesis. Int. J. Hydrogen Energy 27, 1349–1357.

Zhu, H., Fang, H.H.P., Zhang, T., Beaudette, L.A., 2007. Effect of ferrous ion on photo heterotrophic hydrogen production by Rhodobacter sphaeroides. Int. J. Hydrogen Energy 32, 4112–4118.

Zhu, H., Stadnyk, A., Beland, M., Seto, P., 2008. Co-production of hydrogen and methane from potato waste using a two-stage anaerobic digestion process. Bioresour. Technol. 99, 5078–5084.

Zhu, H., Parker, W., Basnar, R., Proracki, A., Falletta, P., Béland, M., et al., 2009. Buffer requirements for enhanced hydrogen production in acidogenic digestion of food wastes. Bioresour. Technol 100 (21), 5097–5102.

Zhu, Z., Shi, J., Zhou, Z., Fengxian, H.u., Bao, J., 2010. Photo-fermentation of *Rhodobacter sphaeroides* for hydrogen production using lignocellulose-derived organic acids. Proc. Biochem. 45 (12), 1894–1898.

Zhu, H., Parker, W., Conidi, D., Basnar, R., Seto, P., 2011. Eliminating methanogenic activity in hydrogen reactor to improve biogas production in a two-stage anaerobic digestion process co-digesting municipal food waste and sewage sludge. Bioresour. Technol. 102 (14), 7086–7092.

11

Fermentative Biohydrogen Production from Solid Wastes

Mi-Sun Kim, Jaehwan Cha, Dong-Hoon Kim

Clean Fuel Center, Korea Institute of Energy Research, Daejeon 305-343, Republic of Korea

INTRODUCTION

Around the world, massive organic solid wastes are discharged from municipal, agricultural, and industrial sources. These wastes cause serious environmental contamination and their treatment presents an economic burden. The conversion of organic solid wastes to energy has thus been considered a sustainable waste management strategy. Various waste-to-energy technologies such as incineration, gasification, pyrolysis, and anaerobic digestion have been developed to manage waste. Among them, the bioconversion of anaerobic digestion has the lowest environmental impact (Finnveden et al., 2005; Ozeler et al., 2006). In particular, as hydrogen produces only water upon combustion with a high-energy yield, fermentative hydrogen production has attracted much attention. Hydrogen can be produced by fermentative bacteria, grown in the dark or light on carbohydrate-rich substrates. Compared with photofermentation, dark fermentative hydrogen production has several advantages, including a fast reaction rate, no need of light, and a smaller footprint. In addition, because a wide range of organic substrates can be used as a substrate, dark fermentation is the most suitable approach for practical biohydrogen production from organic solid wastes (Levin et al., 2007).

In anaerobic mineralization, organic matter is converted into methane, hydrogen, carbon dioxide, ammonia, hydrogen sulfide, and new biomass. These conversions are conducted via a series of interrelated microbial metabolic pathways, including hydrolysis, acidogenesis, acetogenesis, and methanogenesis. Fermentative bacteria hydrolyze and ferment carbohydrates, proteins, and lipids to volatile fatty acids, which are further converted to acetate, CO_2, and H_2 by acidogenic and acetogenic bacteria. The concept of fermentative hydrogen production involves blocking methanogenesis by eliminating

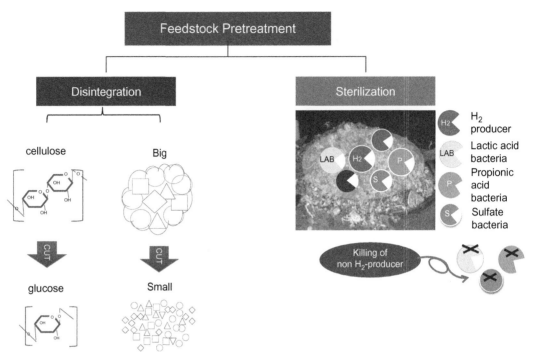

FIGURE 1   The role of feedstock pretreatment on hydrogen production from solid wastes.

or inhibiting hydrogen-consuming bacteria (e.g., methanogens and homoacetogens). There-fore, when mixed cultures are used, it is very important to retain favorable environment conditions for hydrogen-producing bacteria. In general, hydrogen-producing bacteria can form spores, allowing them to better survive under harsh conditions than nonspore-forming hydrogen-consuming bacteria. To date, many studies have been conducted to in-vestigate various pretreatment methods for enriching hydrogen-producing bacteria, such as heat, pH, aeration, chemical, and freezing and thawing treatments (Cheong and Hansen, 2006; Mu et al., 2006; Zhu and Beland, 2006; Wang and Wan, 2008).

Organic solid wastes are a plentiful source of readily available and inexpensive substrates for fermentative hydrogen production. The global generation of solid wastes is estimated to range from 2.75 to 4 kg per capita per day in high-income countries and below 0.5 kg per capita per day in low-income countries (Thitame et al., 2010). While the composition of solid wastes depends on the topography of the area, food habits, seasons, and so on, organic solid wastes occupy a considerable part of total solid wastes. Organic solid wastes contain different amounts of lipids, proteins, and carbohydrates, which determine their biodegradability and hydrogen production potential. Many researchers have investigated fermentative hydrogen from various organic solid wastes such as municipal solid wastes, agricultural wastes, and livestock wastes (Gilroyed et al., 2008; Gomez et al., 2006; Kim and Lee, 2010; Tawfik and El-Qelish, 2012; Wang et al., 2007).

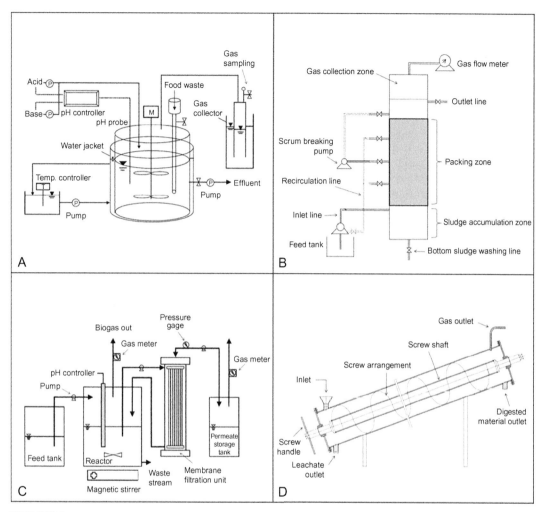

FIGURE 2 Different continuous bioreactor configurations for hydrogen production. (a) Continuous stirred tank reactor (Shin and Youn, 2005), (b) upflow anaerobic contact filter (Vijayaraghavan et al., 2006), (c) membrane bioreactor (Kim et al., 2011b), and (d) inclined plug-flow reactor (Jayalakshmi et al., 2009).

This chapter presents an up-to-date overview of fermentative hydrogen production from organic solid wastes. It first covers characterization of feedstock and its hydrogen production potential. Several case studies of fermentative hydrogen production in batch or continuous operation are also introduced. Effective strategies, including pretreatment of feedstock and optimization of operational conditions, are then presented. Finally, reactor configurations for enhancing hydrogen production are discussed.

CHARACTERISTICS OF ORGANIC SOLID WASTES FOR HYDROGEN PRODUCTION

Fermentative hydrogen production is one of the feasible options for the effective management of organic solid wastes as it can achieve two goals simultaneously: reduction of environmental burden and production of clean energy. However, the performance of hydrogen production highly depends on the composition of organic matters involved. Carbohydrates, such as glucose, xylose, and sucrose, are the most favorable substrates for hydrogen production due to their high biodegradability (Li and Fang, 2007). In contrast, proteins are hydrolyzed to amino acids, which can produce only a small amount of hydrogen (Okamoto et al., 2000). Lipids, which consist mainly of glycerol, are also not suitable for hydrogen production (Heyndrickx et al., 1991). Lay et al. (2003) showed that the hydrogen production potential of carbohydrate-rich organic wastes was approximately 20 times greater than that of fat-rich and of protein-rich organic wastes.

The most common organic solid wastes are food waste and lignocellulosic biomass. In Asia, food wastes account for a considerable part of municipal solid wastes and present significant problems, including groundwater contamination, odor, and toxic gas emission (Kim et al., 2004b; Yasin et al., 2011). They include kitchen refuse, municipal waste, and food industry coproducts. In general, the composition of food wastes is 85–95% volatile solids and 75–85% moisture (Li et al., 2008). Food wastes are a suitable substrate for fermentative hydrogen production because they are readily hydrolysable and carbohydrate rich. Studies have reported that the performance of hydrogen production from food wastes ranges from 50 to 194 ml-H_2/g-VS (Dong et al., 2009; Li et al., 2008).

Lignocellulosic wastes are also an abundant and readily available organic waste that can be used as the feedstock for fermentative hydrogen production. They are generated mainly from agricultural residues, dedicated energy crops, wood residues, and municipal paper wastes. Ren et al. (2009) estimated that the annual yields of lignocellulosic biomass residues worldwide exceed 220 billion tons, equivalent to 60–80 billion tons of crude oil. Lignocellulosic biomass is composed mainly of cellulose (32–47%), hemicellulose (19–27%), and lignin (5–24%). Both hemicelluloses and lignin provide a protective sheath around the cellulose and must be pretreated. Pretreatment entails removing lignin, partly hydrolyzing hemicellulose, and altering the lignocellulosic structure so that enzymes can convert the carbohydrate polymers into fermentable sugars. In order to achieve a higher H_2 yield, various pretreatment methods have been applied (Fan et al., 2006b; Wang et al., 2010a; Zhang et al., 2007). In a study on biohydrogen production from cornstalk, the substrate was pretreated with 0.2% HCl and the highest hydrogen production yield was 149.69 ml-H_2/g-VS, which was 47 times higher than that without acid pretreatment (Zhang et al., 2007).

Livestock wastes are discharged in various forms such as slurry, liquid, or solid manure. Because they are mainly protein rich, biohydrogen yields from livestock are much lower than those from food waste or lignocellulosic biomass, with values ranging from 4 to 29 ml-H_2/g-VS (Kotsopoulos et al., 2009; Yokoyama et al., 2007). The addition of glucose to livestock wastes may increase hydrogen production, but is impractical (Wu et al., 2009; Zhu et al., 2009b). Thus, it is better to treat livestock waste at an ambient temperature for traditional methane production or for acid generation in a conventional two-stage process. However, the addition of livestock wastes as a supplement to carbohydrate-rich wastes seems feasible, as they are rich in nutrients and in buffering capacity. The pH of food-processing wastes is generally below pH 5.0, whereas the pH of livestock wastes is above pH 7.5. Therefore, the addition of livestock wastes may be beneficial in keeping the operation at the optimal range of pH 6–8 for fermentative hydrogen production (Cheng et al., 2002; Valdez-Vazquez et al., 2005).

BIOHYDROGEN PRODUCTION FROM ORGANIC SOLID WASTES

Hydrogen Production from Organic Solid Wastes in Batch Operation

Many kinds of organic solid wastes, such as food wastes, agricultural wastes, livestock wastes, and sewage wastes, can potentially be exploited as a plentiful and inexpensive source of feedstock for hydrogen production. Numerous studies have been conducted to evaluate the biohydrogen production potential from various organic solid wastes in batch operation, and the results are summarized in Table 1. The maximum hydrogen production yield reported in recent studies varied widely from 16 ml-H_2/g-VS (Xiao and Liu, 2009) to 194 ml-H_2/g-VS (Li et al., 2008). Food wastes have been the most extensively studied substrate, and the average hydrogen production yield is substantially higher than values obtained from livestock wastes and sewage sludge. Dong et al. (2009) estimated the hydrogen production potential from seven typical components of the organic fraction of municipal solid waste, including rice, potato, lettuce, lean meat, oil, fat, and banyan leaves. In their study, rice, a carbohydrate-rich substrate, showed a very high hydrogen yield of 134 ml-H_2/g-VS, while protein- and lipid-rich substrates showed very low or even no hydrogen production (Dong et al., 2009). Food waste can serve not only as a substrate but also as a source of hydrogen-producing bacteria when pretreatment is applied to the waste. Kim et al. (2009) showed that lactic acid bacteria were the most abundant species in untreated food waste, while hydrogen-producing bacteria were dominant in heat-, acid-, and alkali-pretreated food wastes. When using heat pretreatment, they achieved a hydrogen yield of 154 ml-H_2/g-VS (corresponding to 2.05 mol-H_2/mol-hexose$_{cosumed}$).

Even though food waste is a suitable substrate for producing hydrogen due to its high organic content and easily hydrolysable nature, it may be short of nitrogen, which is a vital nutrient for the growth of hydrogen-producing bacteria. Sewage sludge is a good candidate as a cosubstrate to meet the nitrogen source requirement. Sreela-or et al. (2011b) found that the addition of sludge to food waste supplied a more balanced carbon-to-nitrogen ratio, thereby enhancing hydrogen production. Lay et al. (1999) and Li et al. (2008) also showed a very high hydrogen production yield from codigestion of food waste and sewage sludge. Along with food wastes, agricultural wastes could also be a suitable substrate for fermentative hydrogen production. Because agricultural wastes often contain appreciable amounts of cellulose, hemicellulose, and lignin, most studies on biohydrogen production from agricultural wastes have used various pretreatments, such as acid, heat, and biopretreatment. Fan et al. (2008) achieved a high hydrogen production yield of 176 ml-H_2/g-TS from cornstalk waste by biopretreatment using microbe additives.

Many studies expressed hydrogen production amount per volatile solid (ml-H_2/g-VS) as a yield unit to directly show the hydrogen production capability of organic solid wastes.

TABLE 1 H$_2$ Production from Organic Solid Wastes in Batch Operation

Substrate		Substrate pretreatment	pH	Temperature (°C)	Maximum H$_2$ production yield	Reference
Food wastes	Food waste from household	None	6.5–7	50	57 ml-H$_2$/g-VS	Pan et al., 2008
	Food waste from cafeteria	Heat pretreatment	7	35	154 ml-H$_2$/g-VS	Kim et al., 2009
	Food waste from cafeteria	None	6.5	35	120 ml-H$_2$/g-VS	Zhu et al., 2009b
	Municipal solid waste (rice)	None	5.5	37	134 ml-H$_2$/g-VS	Dong et al., 2009
	Municipal solid waste (potato)	None	5.5	37	106 ml-H$_2$/g-VS	
	Municipal solid waste (lettuce)	None	5.5	37	50 ml-H$_2$/g-VS	
	Apple pomace	Ammonia pretreatment	7	37	101 ml-H$_2$/g-TS	Feng et al., 2010
	Food waste from cafeteria	None	7.2	30	105 ml-H$_2$/g-VS	Sreela-or et al., 2011a
	Food waste from household	Ultrasonication	5.5	37	80 ml-H$_2$/g-VS	Elbeshbishy et al., 2012
Agricultural wastes	Wheat straw waste	Acid pretreatment	7	36	68 ml-H$_2$/g-TVS	Fan et al., 2006c
	Cornstalk waste	Acid pretreatment	7	36	150 ml-H$_2$/g-TVS	Zhang et al., 2007
	Corn stover biomass	Acid + steam explosion	5.5	35	66 ml-H$_2$/g-TS	Datar et al., 2007
	Cornstalk waste	Biopretreatment	5.5	36	176 ml-H$_2$/g-TS	Fan et al., 2008
	Cassava stillage	None	6	60	68 ml-H$_2$/g-VS	Luo et al., 2010
	Wheat straw waste	Acid + heat pretreatment		36	141 ml-H$_2$/g-VS	Nasirian, 2012
Livestock wastes	Cow waste slurry	None		60	29 ml-H$_2$/g-VS	Yokoyama et al., 2007
	Dairy manure	Acid pretreatment	7	36	28 ml-H$_2$/g-VS	Xing et al., 2010

Continued

TABLE 1 H_2 Production from Organic Solid Wastes in Batch Operation—Cont'd

Substrate		Substrate pretreatment	pH	Temperature (°C)	Maximum H_2 production yield	Reference
Sewage sludge	Waste-activated sludge	Sterilization using autoclave	7	37	16 ml-H_2/g-VS	Xiao and Liu, 2009
	Waste-activated sludge	Heat pretreatment	10	37	27 ml-H_2/g-VSS	Zhao et al., 2010
	Waste-activated sludge	Enzymatic pretreatment	6	36	43 ml-H_2/g-VSS	Guo et al., 2012
Codigestion	Food waste + night solid sludge + sewage sludge	None	5	37	180 ml-H_2/g-TVS	Lay et al., 1999
	Food waste + sewage sludge	None	6	35	60 ml-H_2/g-VS	Kim et al., 2004b
	Food waste + sewage sludge	None	5	36	194 ml-H_2/g-VS	Li et al., 2008
	Food waste + sewage sludge	None		30	101 ml-H_2/g-VS	Sreela-or et al., 2011b

However, it is difficult to compare hydrogen production efficiencies based solely on VS, as carbohydrates have a much higher hydrogen production potential than other components (i.e., fats and proteins) and each organic solid waste has a different carbohydrate content. Therefore, it is recommended that substrate characteristics, including carbohydrate content, are considered to compare the hydrogen production potential with a yield unit of hydrogen production amount per carbohydrate or hexose.

Hydrogen Production from Organic Solid Wastes in Continuous Operation

Even though batch-mode bioreactors are easy to operate and have been used to determine the hydrogen production potential from organic solid wastes, continuous processes are strongly recommended for practical application with the management of organic solid wastes. Continuous hydrogen production from organic solid wastes can be divided into mesophilic and thermophilic regimes. As shown in Table 2, the hydrogen production rate is relatively higher in a thermophilic condition (2.2–19.9 liter-H_2/liter/day) than in a mesophilic condition (2.0–4.8 liter-H_2/liter/day). This might be attributable to a faster biochemical reaction, enhancement of hydrolysis, and/or faster change of the microbial

TABLE 2 H_2 Production from Organic Solid Wastes in Continuous Operation

Temperature	Substrate	Reactor[a]	Organic loading rate	HRT	Maximum H_2 production yield	Maximum H_2 production rate (liter-H_2/ liter/day)	Reference
Mesophilic	Municipal solid waste		11 g-VS/ kg-wet mass reactor/day		165 ml-H_2/g-VS		Valdez-Vazquez et al., 2005
	Food waste	SCRD	22.65 kg-VS/ m³/day		65 ml-H_2/ g-VS		Wang and Zhao, 2009
	Molasses	PBR		6 h		2.8	Park et al., 2010
	Food waste	SBHR	45.9 g-COD/ liter/day	2 h		4.8	Elbeshbishy and Nakhla, 2011
	Food waste	ABR	35 g-COD/ liter/day	26 h	370 ml-H_2/ g-VS		Tawfik et al., 2011
	Kitchen waste	I-CSTR	46 g-COD/ liter/day	2 h	96 ml-H_2/ g-VS	2	Li et al., 2011
			100 g-COD/ liter/day	1.3 h	86 ml-H_2/ g-VS	3.4	
Thermophilic	Municipal solid waste		11 g-VS/ kg-wet mass reactor/day		360 ml-H_2/ g-VS		Valdez-Vazquez et al., 2005
	Garbage slurry+paper	CSTR	97 g-COD/ liter/day	1.2 days	2.4 mol-H_2/ mol-hexose	5.4	Ueno et al., 2007
	Pig slurry	CSTR	24.9 g-VS/ day	1 day	3.65 ml-H_2/ g-VS		Kotsopoulos et al., 2009
	Wheat straw hydrolysate	CSTR		3 days	178 ml-H_2/ g-sugars		Kongjan et al., 2010
	Food waste	CSTR	39 g-COD/ liter/day	1.9 days		11.1	Lee et al., 2010
	Starch-rich kitchen waste	I-CSTR	39 g-COD/ liter/day	8 days	2.1 mmol-H_2/g-COD	2.2	Wang et al., 2010b
	Tofu-processing waste	MBR		8 h	1.87 mol-H_2/ mol-hexose	12.8	Kim et al., 2011b
				4 h	1.45 mol-H_2/ mol-hexose	19.9	

[a] SCRD, semicontinuous rotating drum; PBR, packed bed reactor; SBHR, sonicated biological hydrogen reactor; ABR, anaerobic baffled reactor; I-CSTR, intermittent-continuous stirred tank reactor; CSTR, continuous stirred tank reactor; and MBR, membrane bioreactor, wherein the membrane filtration unit was coupled to the CSTR.

population dynamics favorable for hydrogen production. Using a continuous mesophilic fermenter, Li et al. (2011) investigated the effects of the volumetric loading rate on hydrogen production from kitchen wastes. The highest hydrogen production yield they observed was 96 ml-H_2/g-VS with a loading rate of 46 g-COD/liter/day, while the highest specific hydrogen production rate was 3.4 liter-H_2/liter/day with a loading rate of 100 g-COD/liter/day. Tawfik et al. (2011) operated two identical anaerobic baffle reactors connected in series under a mesophilic condition. In the first reactor, the hydrogen production yield was 250 ml-H_2/g-VS with a loading rate of 58 g-COD/liter/day, while the hydrogen production in the second reactor further increased to 370 ml-H_2/g-VS with a loading rate of 35 g-COD/liter/day. They achieved a total hydrogen yield of 4.9 mol-H_2/mol-hexose using the two-step process.

The hydrogen production rate in a thermophilic condition is often higher than in a mesophilic condition. Valdez-Vazquez et al. (2005) compared the performance of semicontinuous solid substrate anaerobic reactors under mesophilic (35°C) and thermophilic (50°C) conditions. The H_2 yield of a thermophilic bioreactor (360 ml-H_2/g-VS) was significantly higher than that in a mesophilic bioreactor (165 ml-H_2/g-VS). Kim et al. (2011b) conducted biohydrogen production from tofu-processing waste using a membrane bioreactor (MBR). They obtained a hydrogen production rate of 19.9 liter-H_2/liter/day at 4 h hydraulic retention time (HRT) under a thermophilic condition (60°C).

Several studies have reported on large-scale bioreactors for fermentative hydrogen production (Kim et al., 2010; Lee and Chung, 2010; Wang and Zhao, 2009). Wang and Zhao (2009) tested a pilot-scale two-stage fermentation process consisting of a hydrolysis/acetogenesis rotating drum for hydrogen production (200 liter of working volume) and a methane fermentation reactor (800 liter of working volume). They demonstrated that indigenous microorganisms in food waste could be applied for hydrogen and methane production in the pilot-scale integrated fermentation process. Similarly, Lee and Chung (2010) tested a pilot-scale combined hydrogen/methane fermentation process consisting of a hydrogen fermentation tank (500 liter of working volume) and an upflow anaerobic sludge blanket (UASB, 2300 liter of working volume). They obtained a high hydrogen yield of 1.82 mol-H_2/mol-glucose from food waste under mesophilic conditions.

STRATEGIES FOR ENHANCING HYDROGEN PRODUCTION FROM SOLID WASTES

Pretreatment of Feedstock

Organic solid wastes are plentiful and cost-effective substrates for biohydrogen production. However, the productivity and the conversion yield of the hydrogen-producing process using organic solid wastes are generally low due to the complex structure of organic solid wastes and the numerous kinds of indigenous hydrogen-consuming and nonhydrogen-producing bacteria. Therefore, appropriate pretreatment is required to use organic solid wastes as a feedstock for effective hydrogen production.

Hydrolysis Improvement

Organic solid waste, especially lignocellulosic biomass, is composed mainly of complex carbohydrate polymers (cellulose, hemicellulose) and lignin. The amounts of these constituents vary with different types of lignocellulosic materials, as shown in Table 3. Both hemicellulose and lignin provide a protective barrier around cellulose, which prevents plant cell degradation for efficient biohydrogen production. Therefore, before the conversion of biomass to hydrogen, the lignocellulosic material should be hydrolyzed to fermentable sugars. In the hydrolysis step, the biomass is primarily hydrolyzed by removing lignin and partly hydrolyzing hemicellulose. The cellulose is then hydrolyzed by breaking down the carbohydrate chains with the release of their constituent sugars. Although the hydrolysis step is usually an expensive process in biohydrogen production, it can contribute greatly to improved hydrogen yield from organic solid wastes.

There are various pretreatment methods for the hydrolysis of lignocellulosic biomass, including physical (mechanical comminution, ultrasonication), chemical (ozonolysis, acid hydrolysis, alkali hydrolysis), physicochemical (steam explosion, ammonia fiber explosion), biological (white rot fungi), and electrical techniques (Kumar et al., 2009; Ren et al., 2009). When selecting a pretreatment method, the following requirements should be considered: (1) achieving high yield of fermentable sugar, (2) avoiding degradation or loss of carbohydrate, (3) avoiding the formation of inhibitory by-products, and (4) cost-effectiveness.

Physical pretreatments are effective in breaking down cellulose crystallinity. Lignocellulosic biomass can be comminuted by various chipping, grinding, and milling processes. Mechanical comminution usually reduces the particle size to 10–30 mm via chipping and 0.2–2 mm via grinding or milling (Sun and Cheng, 2002). It destroys cellulose fiber crystallinity and makes biomass more amenable to subsequent enzymatic hydrolysis by increasing the accessible surface area of solid particles. However, mechanical comminution is energy intensive and expensive. Furthermore, it does not result in lignin removal, which can limit the accessibility of cellulose and inhibit cellulases (Zheng et al., 2009).

Ultrasonication has been widely used as a mechanical disintegration pretreatment method for anaerobic digestion to enhance the solubilization of organic matter. Ultrasonication produces alternating low- and high-pressure waves in the aqueous phase, causing the formation and vigorous collapse of microbubbles. It leads to so-called cavitation, resulting in strong hydrodynamic shear forces that can erode solid particles (Tiehm et al., 1997). Elbeshbishy et al. (2012) reported that ultrasonication promoted the release of carbohydrates and proteins into the liquid phase and enhanced biohydrogen production from food pulp waste. In their study, the highest hydrogen yield was 141 ml-H_2/g-VS at a sonication time of 30 min, while unsonicated food waste showed a yield of only 80 ml-H_2/g-VS.

Chemical pretreatments are used to improve the biodegradability of cellulose by removing lignin and/or hemicellulose and decreasing the degree of polymerization and crystallinity of cellulose. Ozonolysis is one of the chemical pretreatments available to increase the digestibility of lignocellulosic biomass. Ozone can mainly degrade lignin and partly affects hemicellulose but cannot degrade cellulose (Kumar et al., 2009). Ozonolysis is an environmentally friendly technique because it does not produce toxic residues. Furthermore, ozone can be decomposed easily using a catalytic bed or increasing the temperature. However, ozonolysis is an expensive process due to the large amount of ozone required.

TABLE 3 Contents of Cellulose, Hemicellulose, and Lignin in Common Agricultural Residues and Wastes[a]

Lignocellulosic materials	Cellullose (%)	Hemicellulose (%)	Lignin (%)
Hardwoods stems	40–55	24–40	18–25
Softwood stems	45–50	25–35	25–35
Nut shells	25–30	25–30	30–40
Corn cobs	45	35	15
Grasses	25–40	35–50	10–30
Paper	85–99	0	0–15
Wheat straw	30	50	15
Sorted refuse	60	20	20
Leaves	15–20	80–85	0
Cotton seed hairs	80–95	5–20	0
Newspaper	40–55	25–40	18–30
Waste papers from chemical pulps	60–70	10–20	5–10
Primary wastewater solids	8–15	NA[b]	24–29
Swine waste	6	28	NA
Solid cattle manure	1.6–4.7	1.4–3.3	2.7–5.7
Coastal Bermuda grass	25	35.7	6.4
Switch grass	45	31.4	12

[a] From Sun and Cheng (2002).
[b] Not available.

Acid pretreatment methods have been widely used to improve the enzymatic hydrolysis of lignocellulosic biomass to release fermentable sugars. These methods can be classified as concentrated acid hydrolysis and diluted acid hydrolysis. Concentrated acids such as H_2SO_4 and HCl are powerful agents for cellulose hydrolysis, but they are toxic, corrosive, and hazardous. Concentrated acid pretreatment thus requires expensive corrosion-resistant reactors. Recovery of the concentrated acids after hydrolysis can make the process economically feasible (Sivers and Zacchi, 1995). Alternatively, dilute acid pretreatment has been developed successfully. It can be applied to a wide range of feedstock, including softwood, agricultural waste, and municipal solid waste. Dilute acid (typically sulfuric acid) is mixed with biomass at 160–220°C for a few minutes to hydrolyze hemicellulose to xylose and other sugars. Subsequently, xylose is broken down to form furfural. In general, there are two types of dilute acid hydrolysis: (1) high temperature (T >160°C) and continuous flow process for low solid loading (5–10% w/w) and (2) low temperature (T <160°C) and batch process for high solid loading (10–40% w/w). Cao et al. (2009) applied dilute sulfuric acid pretreatment to hydrolyze corn stover for biohydrogen production. They investigated the effects of H_2SO_4

concentration and reaction time on the composition of hemicellulose hydrolysate and sugar concentration. The highest hydrogen yield of 2.24 mol-H_2/mol-sugar was achieved using 1.69% sulfuric acid and a 117-min reaction time, where these conditions were determined by response surface methodology.

Alkaline pretreatment is generally effective on lignocellulosic biomass with a low lignin content (Bjerre et al., 1996). It leads to the removal of lignin from the biomass, thus enhancing the reactivity of the polysaccharides. Furthermore, it can remove acetyl and various uronic acid substitutions on hemicellulose. Compared with other pretreatments, alkaline treatment uses lower temperatures and pressures, even ambient conditions, but requires a much longer reaction time (on the order of hours or days). Unlike acid pretreatment, some alkali is converted to irrecoverable salts or incorporated as salts into the biomass during alkaline hydrolysis (Mosier et al., 2005). Sodium, potassium, calcium, and ammonium hydroxides are suitable alkaline treatment agents. In a study on biohydrogen production from bagasse with thermally preheated sludge, the bagasse substrate was hydrolyzed through an alkaline pretreatment (Chairattanamanokorn et al., 2009). A mixture of bagasse and a NaOH solution was boiled at 100°C for 2 h and was subsequently neutralized by washing with distilled water. Compared with the untreated case, the alkaline pretreatment of bagasse significantly improved the cellulose hydrolysis efficiency from 0.25 to 0.79 g-reducing sugar/g-bagasse, and the hydrogen production yield was thereby enhanced 13-fold higher than that under the neutral condition.

Steam explosion is the most commonly used pretreatment method, which uses both physical and chemical techniques to break down the lignocellulosic structure. In this hydrothermal pretreatment, high-pressure saturated steam is injected into a reactor filled with biomass. The steam injection is typically conducted for several seconds to a few minutes with an increase of temperature ranging from 160 to 260°C. The reactor is then depressurized rapidly, which makes the biomass undergo an explosive decompression with hemicellulose degradation and lignin matrix disruption (Kumar et al., 2009). It exposes the cellulose surface and increases the accessibility of enzymes to the cellulose microfibrils. Datar et al. (2007) demonstrated that steam explosion of corn stover resulted in hemicellulose-enriched hydrolyzate fractions suitable to support fermentative hydrogen production. They found that most of the sugars in the liquid fractions obtained by steam explosion of corn stover were oligomeric. The highest hydrogen yield was 2.84 mol-H_2/mol-hexose (corresponding to 71% of theoretical conversion efficiency) when steam treatment was carried out for 3 min at 220°C. However, the total hydrogen production decreased when corn stover was pretreated for 5 min at 220°C. They contended that more severe conditions would cause a loss of total sugars or the formation of inhibitory products such as furfural. Ren et al. (2008) compared hydrolysis and biohydrogen production from cornstalks with the strain *Clostridium acetobutylicum* X$_9$ when the substrate was pretreated with H_2SO_4, NaOH, and NH_3 soaking and steam explosion. In their study, steam explosion was the most effective method among the four tested pretreatments for hydrolyzing cellulose and yielding hydrogen.

Ammonia fiber explosion is another physicochemical process similar to steam explosion pretreatment. In ammonia fiber explosion, the biomass material is exposed to liquid ammonia under high pressures and moderate temperatures ranging from 60 to 100°C for a few minutes. Much like steam explosion, the ammonia and biomass mixture is saturated in a pressurized reactor, and then the pressure is reduced suddenly. This leads to rapid expansion of the ammonia gas, which causes swelling of the lignocellulosic materials, disrupting the lignin–carbohydrate linkage, and hydrolyzing hemicellulose.

Suppression of Indigenous Nonhydrogen-Producing Bacteria

Although biohydrogen can be obtained from a variety of organic solid wastes, hydrogen production was often deteriorated by indigenous microorganisms. Organic solid wastes itself can be an unwanted inoculum source because they have many kinds of hydrogen-consuming bacteria and nonhydrogen-producing acidogens (Li and Fang, 2007). Therefore, in order to enhance hydrogen production, organic solid wastes require pretreatment to suppress the activities of the undesirable organisms, such as methanogens, homoacetogens, and lactic acid bacteria. In principle, hydrogen-producing bacteria such as *Clostridium* are physiologically different from nonhydrogen-producing bacteria (Das and Veziroglu, 2001). *Clostridium* can form protective spores under harsh conditions such as high temperature, extreme acidity, and alkalinity, whereas methanogens cannot. Thus, researchers have applied several pasteurizing methods for eliminating or suppressing the activity of nonhydrogen-producing bacteria from organic solid wastes, as shown in Table 4.

TABLE 4 Pasteurizing Methods for Eliminating or Suppressing the Activity of Nonhydrogen-Producing Bacteria

Substrate	Substrate pretreatment	Seed sludge	Operation	H_2 production yield or rate	Reference
Food waste	Base (pH 12.5, 1 day)	Heat-treated digested sludge	Sequencing batch	62.6 ml-H_2/ g-VS	Kim and Shin, 2008
Food waste	Heat (90°C, 20 min)	No inoculum addition	Batch	96.9 ml-H_2/ g-VS	Kim et al., 2009
	Acid (pH 1, 1 day)	No inoculum addition	Batch	89.5 ml-H_2/ g-VS	
	Base (pH 13, 1 day)	No inoculum addition	Batch	50.9 ml-H_2/ g-VS	
Food waste	Low temperature (4°C)	Heat-treated digested sludge	Continuous	2000 ml-H_2/ day	Jo et al., 2007
Food waste + sewage sludge	Heat (100°C, 10 min)	Digested sludge	Continuous	48 ml-H_2/ g-VS	Zhu et al., 2011
Beer lees waste	Acid + heat (30 min)	Heat-treated cow dung compost	Batch	60.2 ml-H_2/ g-VS	Fan et al., 2006b
	Base + heat (30 min)	Heat-treated cow dung compost	Batch	11.5 ml-H_2/ g-VS	
Cornstalk waste	Acid + heat (30 min)	Cow dung compost	Batch	149.7 ml-H_2/ g-VS	Zhang et al., 2007

Thermal pretreatment has been widely used to facilitate the suppression of nonspore-forming bacteria. Noike et al. (2002) found that continuous hydrogen production from bean curd manufacturing waste was impossible due to the existence of lactic acid bacteria in the substrate, but could be overcome by heat treatment of the feedstock at 50–90°C for 30 min. Similarly, a study on hydrogen production from a mixture of food waste and sewage sludge also reported that heat treatment of feedstock (100°C, 10 min) was effective in eliminating methanogenic activity in a semicontinuous flow reactor (Zhu et al., 2011). Hydrogen from food waste without inoculum addition is also feasible. Kim et al. (2009) demonstrated that food waste successfully served not only as a substrate, but also as a source of hydrogen-producing bacteria when heat (90°C for 20 min), acid (pH 1.0 for 1 day), or alkali (pH 13.0 for 1 day) treatment was applied. These pretreatment methods inhibited lactate production and increased hydrogen and butyrate production. Among three pretreatments, heat treatment showed the highest hydrogen yield (97 ml-H_2/g-VS), followed by acid treatment (90 ml-H_2/g-VS) and alkali treatment (51 ml-H_2/g-VS).

However, a low temperature could repress lactic bacteria activity. Jo et al. (2007) observed shifts in the microbial community from *Clostridium* spp. to *Lactobacillus* spp. within a food waste-fed anaerobic reactor, which resulted in the conversion of hydrogen fermentation to lactic acid fermentation. It was found that the substrate competition between hydrogen producers and nonhydrogen producers caused an instability of continuous hydrogen production. However, when the feed solution storage tank was controlled at a low temperature (4°C), significant shifts in the microbial community did not occur and hydrogen production was maintained stably.

Most methanogens are neutrophilic with a narrow pH range of 6–8, while hydrogen-producing bacteria can grow over a relatively broad pH range. Acid/base treatments are thus efficient to repress methanogenic activity and to allow growth of spore-forming bacteria. Kim and Shin (2008) investigated the effects of acid and base pretreatments on the reduction of indigenous bacteria in food waste and the microbial population in hydrogen fermentation. They observed that base pretreatment reduced indigenous anaerobic bacteria in food waste by 4.9 log and enabled stable long-term operation over 90 days with a hydrogen yield of 62.6 ml-H_2/g-VS, whereas acid pretreatment of the feedstock showed no positive effect on hydrogen production.

Optimization of Operational Parameters

Organic solid wastes are plentiful and cost-effective substrates for biohydrogen production. However, the productivity and the conversion yield of hydrogen-producing processes using organic solid wastes are generally quite low due to the presence of complex organic substances and many kinds of indigenous hydrogen-consuming and nonhydrogen-producing bacteria. Therefore, feedstock is often pretreated by thermal, chemical, or physical methods. Although these pretreatments were effective in terms of improving hydrogen production in a laboratory-scale reactor, they might be impractical in full-scale applications because a large amount of feedstock must be pretreated. However, optimization of the operational conditions of biological reactors could also help improve hydrogen production. The following factors should be considered to achieve successful hydrogen production from organic solid wastes.

pH

pH is one of the most important parameters for fermentative hydrogen production because it can affect hydrogenase activity, metabolic pathways, and the dominant bacterial population (Dabrock et al., 1992; Fang and Liu, 2002; Lay, 2000). Generally, the optimum pH for hydrogen production has been reported to be in a range of 5.0–7.0, which might favor the activity of hydrogenase (Mohan, 2008). The acidophilic pH range of 5.0–6.5 is also suitable in suppressing methanogenesis and solventogenesis. However, some studies have found that hydrogen-consuming methanogenic activity was detected even at pH 5.0 and was totally inhibited at pH 4.5 (Hwang et al., 2004; Kim et al., 2004a).

For biohydrogen production from organic solid wastes, pH levels influence the rate of hydrolysis, the types and amount of acidogenic metabolites, and the rate and amount of hydrogen production (Li and Fang, 2007). Zhang et al. (2005) reported that both hydrolysis and acidogenesis of kitchen wastes in anaerobic digestion were improved by adjusting the pH to 7.0. At pH 7, 86% of the total organic carbon and 82% of chemical oxygen demand (COD) were solubilized while the total volatile fatty acid (VFA) concentration yield was 0.27 g/g-total solid (TS), which was two times higher than the yield without pH adjustment. Shin and Youn (2005) found that pH change within a narrow range (5.0, 5.5, and 6.0) significantly affected the metabolic pathway and hydrogen production from food waste with thermophilic acidogenesis. As pH was decreased from 5.5 to 5.0, the amount of butyrate decreased from 62 % to 38%, while the amount of lactate increased from 2% to 39%. However, at pH 6.0, the butyrate content increased to 57%, while the lactate content decreased to 3%. In their study, the highest hydrogen yield (2.2 mol-H_2/mol-hexose) was found at pH 5.5, followed by pH 6.0 (2.0 mol-H_2/mol-hexose). In contrast, the hydrogen yield at pH 5.0 was only 1.0 mol-H_2/mol-hexose. Fan et al. (2006a) observed that acetate and butyrate were the major by-products at pH 6.0 or below during fermentative hydrogen production from brewery waste. However, when pH was higher than 6.5, solventogenesis (propanol, butanol, and ethanol) occurred.

Many studies have indicated that the final pH level dropped to 4.0–5.0, regardless of initial pH (Liu and Shen, 2004; Pattra et al., 2008; Yokoi et al., 2001; Zhang et al., 2003). The decrease in pH is due to organic acid formation during fermentation, which depletes the buffering capacity of the medium. The pH drop can alter the metabolic pathway to nonhydrogenic processes or inhibit substrate utilization (Kim et al., 2011a). In addition, it can also limit hydrogen production directly, as hydrogenase activity declines at excessively low pH (Kapdan and Kargi, 2006). Hence, especially in a batch operation, the initial pH can be separated from operational pH. Even if the active microbial consortium is prepared at a certain initial pH, it cannot ensure high performance unless operational pH is controlled (Kim et al., 2011a). Two approaches can be used to maintain operational pH at the optimal level. Online pH monitoring with the addition of an acid and base can be used for pH control, but it is challenging to implement at a field scale. An alternative is to supplement the feedstock with a sufficient buffer to compensate the pH decrease. A study on batch acidogenic digestion of food waste reported that hydrogen production was improved considerably with a hydrogen yield of 28 ml-H_2/g-VS by adding a 0.1 M phosphate buffer solution, whereas a negligible quantity of hydrogen was produced without buffer addition (Zhu et al., 2009a). In their study, with the addition of the buffer solution, the pH decreased slightly from 7.0 to 6.4 in the initial 3 h and was maintained at over 6.0 for 25 h. However, without buffer addition, pH dropped from 7.0 to 4.0 in the initial 3 h.

Temperature

Temperature is the most important operational parameter affecting both biohydrogen production and microbial metabolisms, along with pH. The optimal temperature for hydrogen production from organic solid wastes varies with the feedstock because of complex microbial communities and the constituent materials (Guo et al., 2010). Anaerobic digestion is commonly performed under mesophilic conditions (30–40°C) or thermophilic conditions (50–60°C). Economically, mesophilic biohydrogen production is preferred due to less requirement of external heating, whereas thermophilic biohydrogen production often shows a high hydrogen yield due to the suppression of hydrogen-consuming bacteria and the enhancement of substrate utilization, that is, hydrolysis (Ren et al., 2009).

Lin et al. (2011) investigated the effects of temperature on biohydrogen production from food-processing wastewater by changing the temperature from 35 to 55°C at an interval of 10°C. With operation at 55°C, the highest hydrogen production yields (189 ml-H_2/g-COD) and hydrogen production rate (32 ml-H_2/g-VSS/h) were obtained, which was twofold higher than that at 35°C. Similarly, in a study on biohydrogen production from grass silage, hydrogen yield increased from 3.2 to 7.2 and 16 ml-H_2/g-VS when the temperature was increased from 33 to 55 and 70°C, respectively (Pakarinen et al., 2008). However, the time taken to reach the maximum hydrogen yield was longer at 70°C (25 days) compared to at 55°C (10 days) and 35°C (3–4 days).

The thermophilic condition can have inhibitory effects on methane and propionate production. In a study on hydrogen production from food waste, Shin et al. (2004) found that the biogas produced from a thermophilic acidogenic culture (55°C) was free of methane, whereas methane was detected from a mesophilic acidogenic culture (35°C). Also, in the thermophilic test, butyrate was the main acid product, while propionate was negligible. In contrast, propionate was one of the major acid products in the mesophilic test. They concluded that higher hydrogen production from the thermophilic condition than the mesophilic condition was caused by free of methane and negligible production of propionate, which were hydrogen consumers (Shin et al., 2004). Several studies have reported that the formation of acetate and butyrate accompanied hydrogen production, whereas hydrogen was consumed during the production of propionate (Horiuchi et al., 2002; Ueno et al., 2001).

Temperature in anaerobic fermentation can change the microbial community structure, thereby affecting the hydrogen production yield. Yokoyama et al. (2007) examined hydrogen production from dairy cow waste slurry at different temperatures (37, 50, 55, 60, 67, and 75°C) to investigate the effects of fermentation temperature on hydrogen production and microbial communities. Broadly, hydrogen production showed a tendency to increase with an increase in temperature, but two optima hydrogen production were observed at 60 and 75°C (392 and 248 ml-H_2/L-slurry, respectively). A denaturing gradient gel electrophoresis analysis showed that predominant bacteria at 60°C were affiliated to hydrogen-producing bacteria *Bacteroides xylanolyticus*, *Clostridium stercorarium*, and *Clostridium thermocellum*, while three strains of the extreme thermophiles, *Caldanaerobacter subterraneus*, were dominant at 75°C (Yokoyama et al., 2007).

High temperature can promote hydrolysis and simplify microbial diversity favorable to hydrogen production. However, especially in treating organic solid wastes, monotonous microbial diversity at a high temperature can result in incomplete substrate degradation. Furthermore, hydrogen fermentation at a high temperature is energy-intensive and expensive. Therefore, in order to select the optimum temperature, not only the hydrogen production performance but also substrate degradation and an economic analysis should be considered.

Nutrient Supplementation

Organic solid wastes contain a wide variety of constituents, which vary with sources, collection method, weather, etc. In order to produce hydrogen successfully, the feedstock components should meet the carbon and nitrogen needed for the involved microorganisms. In addition, phosphorous, ferrous, and other trace metals are also necessary for microbial metabolisms. Among various nutrients, nitrogen is the most essential for bacterial growth. Nitrogen is usually supplied by various forms, such as ammonium, yeast extract, and peptone. For biohydrogen production from steam-exploded corn straw using *Clostridium butyricum* AS1.209, NH_4HCO_3, urea, and yeast extract were added as a nitrogen source in batch experiments (Li and Chen, 2007). In their study, hydrogen yields by organic nitrogen sources (urea and yeast extract) were much higher than that by an inorganic nitrogen source (NH_4HCO_3). When changing the concentration of urea added (0.06–0.3 g/g-substrate), the maximum hydrogen yield and the shortest lag stage were obtained at a urea concentration of 0.15 g/g-substrate. At higher or lower urea concentrations, the hydrogen yield decreased. They confirmed that a proper C/N (carbon/nitrogen) ratio is needed to enhance biohydrogen production from corn straw using the *C. butyricum* strain. Similarly, Yokoi et al. (2001) obtained the highest hydrogen yield from sweet potato starch residue using *C. butyricum* upon supplementation of a nitrogen source (0.1% polypeptone). However, no hydrogen production was observed when polypeptone was not added. Yokoi et al. (2002) reported that corn-steep liquor was a suitable nitrogen source for hydrogen production. Because corn-steep liquor is a waste discharged from the corn starch-manufacturing process, use of corn-steep liquor as a nitrogen source can reduce operating costs for biohydrogen production.

The external iron concentration can affect the fermentative hydrogen production, as hydrogenase present in anaerobic bacteria oxidizes ferredoxin to produce molecular hydrogen. Liu and Shen (2004) reported that hydrogen production from corn starch increased with increasing the concentration of iron added. At an iron concentration of 10 mg/liter, the maximum specific hydrogen rate and hydrogen yield were 171 ml-H_2/g-VSS/day and 140 ml-H_2/g-starch, respectively. However, hydrogen production was inhibited at a high iron concentration over 20 mg/liter. Lay et al. (2005) found the optimal nutrient concentration for hydrogen production from food wastes by *Clostridium*-rich composts. Using response surface methodology, the highest level of hydrogen production was obtained in the presence of 132 mg-Fe^{2+}/liter, 537 mg $- NH_4^+$/liter, and 1331 mg $- PO_4^{3-}$/liter. They suggested that iron had a synergistic effect on biohydrogen production by Clostridia.

Specific nutrients, termed germinants, have been reported to affect spore germination of *Clostridium* and *Bacillus* genera, and consequently hydrogen production (Valdez-Vazquez et al., 2009). These germinants are generally amino acids, sugars, purine nucleosides, or a mixture of nutrients. Valdez-Vazquez et al. (2009) investigated the effects of representative germinants on batch hydrogen production from the organic fraction of municipal solid waste. The highest hydrogen production was obtained in the reactor supplemented with L-alanine, whereas the fermenter without germinant addition showed very poor hydrogen production.

Hydraulic Retention Time

The hydraulic retention time is closely related to the amount of substrate that can be handled per unit time, and thereby has a direct impact on the economic feasibility of a bioprocess. A short HRT yields a higher hydrogen production rate and lowers capital outlay by reducing the size of the bioreactor. However, if the HRT is too short, the system will fail due to washout of the slowly growing bacteria essential for hydrogen production (Zaher et al., 2009). Furthermore, a short HRT can result in reduced efficiency of substrate degradation. In general, short HRT is preferred for hydrogen-producing bacteria, as the growth rate of hydrogen-producing bacteria is much higher than that of hydrogen-consuming bacteria such as methanogens. Zhang et al. (2006) found that the increase of the hydrogen yield after changing the HRT from 8 to 6 h was caused by washout of the propionate producing population at a short HRT. However, the optimal HRT for biohydrogen production from organic solid wastes is usually longer, as the feedstock requires sufficient hydrolysis. In a study on biohydrogen production from food waste, when the HRT was shortened to 5, 3, and 2 days, the hydrogen production rate and yield decreased from 1.0 to 0.5 liter-H_2/liter/day and from 2.2 to 1.0 mol-H_2/mol-hexose, respectively (Shin and Youn, 2005). They also showed that the efficiencies of carbohydrate decomposition in food waste reduced from 90 to 76% with a decrease of the HRT from 5 to 2 days. Ueno et al. (1996) also reported that more carbohydrate in sugary wastewater was decomposed at longer HRT, but the hydrogen yield decreased. They speculated that homoacetogenic fermentation, which yields no hydrogen, could gradually become the dominant process at longer HRT, and/or the metabolism of hydrogen-producing acidogens could shift to no formation of hydrogen due to the decrease in their growth rate. Li and Fang (2007) showed that optimal HRTs for biohydrogen production from organic solid wastes varied considerably from 6–9 h for bean curd waste to 5 days for food waste. It depends on substrate type, operational temperature, process stage, and reactor type.

REACTOR CONFIGURATION

Reactor configuration is an important factor in biohydrogen production from organic solid wastes, as it influences the contact between the substrate and microorganisms, substrate hydrolysis, biomass retention time, etc. Generally, according to feeding modes, fermentative hydrogen production has been conducted in batch or continuous bioreactors. The batch systems are simple to operate and easy to control. Therefore, batch bioreactors have been widely used to determine the hydrogen production potential of various organic solid wastes and to investigate the effects of key parameters such as temperature, pH, and inoculum concentration. Dong et al. (2009) carried out batch experiments to estimate the hydrogen potential of seven different components of the organic fraction of municipal solid waste, including rice, potato, lettuce, lean meat, oil, fat, and banyan leaves. Their experiment results clearly showed that carbohydrates were the most optimal substrate for fermentative hydrogen production compared with proteins, lipids, and lignocelluloses. Sreela-or et al. (2011a) operated a batch bioreactor to optimize several key factors for enhancing hydrogen production from food waste. Using response surface methodology, they achieved the highest hydrogen yield (105 ml-H_2/g-VS) under the optimal condition of 2.30 g-VSS/liter of inoculum concentration, 2.54 g-VS/liter of substrate concentration, and 0.11 M of citrate buffer concentration.

Although most studies on biohydrogen from organic wastes have been performed in batch reactors, the continuous hydrogen producing processes are highly recommended for practical applications and economic considerations (Guo et al., 2010). In a single-stage system, a continuous stirred tank reactor (CSTR) is the most commonly used system for continuous hydrogen production (Kotsopoulos et al., 2009; Li et al., 2011; Shin and Youn, 2005). In the CSTR, hydrogen-producing bacteria are usually suspended and well mixed with the feedstock, but biomass may be washed out easily at a low hydraulic retention time. Ren et al. (2010) directly compared a conventional suspended-sludge CSTR with an attached-sludge CSTR under identical operational conditions. In their study, the attached-sludge CSTR was shown to be more stable than the suspended-sludge CSTR in terms of hydrogen production and substrate utilization efficiency. This indicates that a reactor configuration capable of maintaining a high biomass concentration is required to enhance hydrogen production. Similarly, comparisons between immobilized cell systems and suspended cell systems based on biomass growth in the form of granules, biofilms, and flocs were carried out by Show et al. (2010). Their experimental results showed that the formation of granules or biofilms enhanced biomass retention substantially, which was found to be proportional to the hydrogen production rate. So far, several studies have shown that a reactor configuration capable of maintaining a high biomass concentration resulted in high performance of hydrogen production from organic solid wastes. Vijayaraghavan et al. (2006) introduced an upflow anaerobic contact filter for treating solid waste generated from the fruit-processing industry (Fig. 2b). Rigid porous plastic balls were used as packing material for supporting microorganisms. They showed that high hydrogen production (396 ml-H_2/g VS) remained stable in the anaerobic contact filter under continuous operation. An upflow anaerobic sludge bed reactor and anaerobic filter (AF) reactor have also been used to improve biomass retention and hydrogen production rates. Kongjan and Angelidaki (2010) reported that higher hydrogen production from wheat straw hydrolysate was achieved from the UASB and AF reactors compared to the CSTR. This was mainly due to a higher and more stable biomass concentration on the granules (UASB) and plastic carriers (AF). In their study, both UASB and AF reactors exhibited stable performance at a low HRT of 1 h, while biomass washout was observed in the CSTR when the HRT was lowered to 2.5 days. Kim et al. (2011b) carried out continuous biohydrogen production from tofu-processing waste using a membrane bioreactor composed of a continuous stirred tank reactor coupled with a hollow fiber membrane unit (Fig. 2c). The MBR increased the solid retention time and the biomass concentration, resulting in considerable improvement of hydrogen production.

In a single reactor, it is difficult to achieve high efficiencies in both substrate hydrolysis and hydrogen production simultaneously. Chen et al. (2009) developed a two-stage process combining starch hydrolysis and hydrogen fermentation, where starch was hydrolyzed in a sequencing batch reactor (SBR), and then the starch hydrolysate was used for hydrogen production in a CSTR. They achieved a stable starch hydrolysis rate (1.86 g-starch/h/liter) from the SBR and a high hydrogen production rate (0.52 liter-H_2/h/liter) from the CSTR. Jayalakshmi et al. (2009) developed a novel reactor configuration to treat kitchen waste for biohydrogen production. They fabricated an inclined cylindrical-shaped plug-flow reactor, which was kept at a 20°C angle with the horizontal to facilitate easy movement of the wastes within the reactor (Fig. 2d). A screw arrangement was provided inside the reactor to achieve mixing and to push the feedstock from the inlet at the bottom side to the outlet at the top side.

The screw was designed with 14 leads to maintain a 7-day retention time by rotating two turns of the screw per day. This ensured sufficient hydrolysis time for organic solid waste.

CONCLUSION

Biohydrogen production from organic solid wastes seems to be the most promising and environmentally friendly option for the future energy economy, as it can solve the problems of energy production and waste management at the same time. Many studies have already shown the hydrogen production potential of various kinds of organic solid wastes. Among different solid wastes, food wastes are the most favorable substrate for hydrogen production due to their high carbohydrate content and high biodegradability. However, indigenous microorganisms such as lactic acid bacteria in food wastes can limit hydrogen production, and thus appropriate pasteurizing methods are recommended. Heat and acid pretreatment are effective methods to suppress the activity of these nonhydrogen-producing bacteria. Along with food wastes, lignocellulosic wastes generated from agricultural resides have also been the most extensively studied substrate favorable for hydrogen production. Since these wastes are mainly composed of cellulose, hemicellulose, and lignin, proper pretreatments to remove lignin or to hydrolyze hemicellulose are necessary. Hydrogen production is not only restricted by the composition of the organic wastes, but also is dependent on the operational conditions. In general, a pH ranging from 5.0 to 6.0 and thermophilic conditions are suitable for hydrogen production, while the optimal level of HRT depends on the substrate type, operational temperature, process stage, and reactor type. Although most studies on biohydrogen from organic wastes have been performed in batch reactors, continuous hydrogen producing processes are highly recommended for practical applications and economic considerations. In the continuous processes, suspended biomass may be washed out easily at a low HRT. Thus, attached cell systems using supporting materials, granules, or biofilms are preferable for enhancing hydrogen production.

References

Bjerre, A.B., Olesen, A.B., Fernqvist, T., Ploger, A., Schmidt, A.S., 1996. Pretreatment of wheat straw using combined wet oxidation and alkaline hydrolysis resulting in convertible cellulose and hemicellulose. Biotechnol. Bioeng. 49, 568–577.

Cao, G., Ren, N., Wang, A., Lee, D.J., Guo, W., Liu, B., et al., 2009. Acid hydrolysis of corn stover for biohydrogen production using *Thermoanaerobacterium thermosaccharolyticum* W16. Int. J. Hydrogen Energy 34, 7182–7188.

Chairattanamanokorn, P., Penthamkeerati, P., Reungsang, A., Lo, Y.C., Lu, W.B., Chang, J.S., 2009. Production of biohydrogen from hydrolyzed bagasse with thermally preheated sludge. Int. J. Hydrogen Energy 34, 7612–7617.

Chen, S.D., Lo, Y.C., Lee, K.S., Huang, T.I., Chang, J.S., 2009. Sequencing batch reactor enhances bacterial hydrolysis of starch promoting continuous bio-hydrogen production from starch feedstock. Int. J. Hydrogen Energy 34, 8549–8557.

Cheng, S.S., Chang, S.M., Chen, S.T., 2002. Effects of volatile fatty acids on a thermophilic anaerobic hydrogen fermentation process degrading peptone. Water Sci. Technol. 46, 209–214.

Cheong, D.Y., Hansen, C.L., 2006. Bacterial stress enrichment enhances anaerobic hydrogen production in cattle manure sludge. Appl. Microbiol. Biotechnol. 72, 635–643.

Dabrock, B., Bahl, H., Gottschalk, G., 1992. Parameters affecting solvent production by *Clostridium pasteurianum*. Appl. Environ. Microbiol. 58, 1233–1239.

Das, D., Veziroglu, T.N., 2001. Hydrogen production by biological processes: A survey of literature. Int. J. Hydrogen Energy 26, 13–28.

Datar, R., Huang, J., Maness, P.C., Mohagheghi, A., Czernik, S., Chornet, E., 2007. Hydrogen production from the fermentation of corn stover biomass pretreated with a steam-explosion process. Int. J. Hydrogen Energy 32, 932–939.

Dong, L., Zhenhong, Y., Yongming, S., Xiaoying, K., Yu, Z., 2009. Hydrogen production characteristics of the organic fraction of municipal solid wastes by anaerobic mixed culture fermentation. Int. J. Hydrogen Energy 34, 812–820.

Elbeshbishy, E., Nakhla, G., 2011. Comparative study of the effect of ultrasonication on the anaerobic biodegradability of food waste in single and two-stage systems. Bioresour. Technol. 102, 6449–6457.

Elbeshbishy, E., Hafez, H., Nakhla, G., 2012. Viability of ultrasonication of food waste for hydrogen production. Int. J. Hydrogen Energy 37, 2960–2964.

Fan, K.S., Kan, N.R., Lay, J.J., 2006a. Effect of hydraulic retention time on anaerobic hydrogenesis in CSTR. Bioresour. Technol. 97, 84–89.

Fan, Y.T., Zhang, G.S., Guo, X.Y., Xing, Y., Fan, M.H., 2006b. Biohydrogen-production from beer lees biomass by cow dung compost. Biomass Bioenergy 30, 493–496.

Fan, Y.T., Zhang, Y.H., Zhang, S.F., Hou, H.W., Ren, B.Z., 2006c. Efficient conversion of wheat straw wastes into biohydrogen gas by cow dung compost. Bioresour. Technol. 97, 500–505.

Fan, Y.T., Xing, Y., Ma, H.C., Pan, C.M., Hou, H.W., 2008. Enhanced cellulose-hydrogen production from corn stalk by lesser panda manure. Int. J. Hydrogen Energy 33, 6058–6065.

Fang, H.H.P., Liu, H., 2002. Effect of pH on hydrogen production from glucose by a mixed culture. Bioresour. Technol. 82, 87–93.

Feng, X., Wang, H., Wang, Y., Wang, X., Huang, J., 2010. Biohydrogen production from apple pomace by anaerobic fermentation with river sludge. Int. J. Hydrogen Energy 35, 3058–3064.

Finnveden, G., Johansson, J., Lind, P., Moberg, A., 2005. Life cycle assessment of energy from solid waste. 1. General methodology and results. J. Clean. Prod. 13, 213–229.

Gilroyed, B.H., Chang, C., Chu, A., Hao, X., 2008. Effect of temperature on anaerobic fermentative hydrogen gas production from feedlot cattle manure using mixed microflora. Int. J. Hydrogen Energy 33, 4301–4308.

Gomez, X., Moran, A., Cuetos, M.J., Sanchez, M.E., 2006. The production of hydrogen by dark fermentation of municipal solid wastes and slaughterhouse waste: A two-phase process. J. Power Sour. 157, 727–732.

Guo, X.M., Trably, E., Latrille, E., Carrere, H., Steyer, J.P., 2010. Hydrogen production from agricultural waste by dark fermentation: A review. Int. J. Hydrogen Energy 35, 10660–10673.

Guo, L., Zhao, J., She, Z., Lu, M., Zong, Y., 2012. Effect of S-TE (solubilization by thermophilic enzyme) digestion conditions on hydrogen production from waste sludge. Bioresour. Technol. 117, 368–372.

Heyndrickx, M., De Vos, P., Vancanneyt, M., De Ley, J., 1991. The fermentation of glycerol by *Clostridium butyricum* LMG 1212 t$_2$ and 1213t$_l$ and *C. pasteurianum* LMG 3285. Appl. Microbiol. Biotechnol. 34, 637–642.

Horiuchi, J.I., Shimizu, T., Tada, K., Kanno, T., Kobayashi, M., 2002. Selective production of organic acids in anaerobic acid reactor by pH control. Bioresour. Technol. 82, 209–213.

Hwang, M.H., Jang, N.J., Hyun, S.H., Kim, I.S., 2004. Anaerobic bio-hydrogen production from ethanol fermentation: The role of pH. J. Biotechnol. 111, 297–309.

Jayalakshmi, S., Joseph, K., Sukumaran, V., 2009. Bio hydrogen generation from kitchen waste in an inclined plug flow reactor. Int. J. Hydrogen Energy 34, 8854–8858.

Jo, J.H., Jeon, C.O., Lee, D.S., Park, J.M., 2007. Process stability and microbial community structure in anaerobic hydrogen-producing microflora from food waste containing kimchi. J. Biotechnol. 131, 300–308.

Kapdan, I.K., Kargi, F., 2006. Bio-hydrogen production from waste materials. Enzyme Microb. Technol. 38, 569–582.

Kim, M.S., Lee, D.Y., 2010. Fermentative hydrogen production from tofu-processing waste and anaerobic digester sludge using microbial consortium. Bioresour. Technol. 101, S48–S52.

Kim, S.H., Shin, H.S., 2008. Effects of base-pretreatment on continuous enriched culture for hydrogen production from food waste. Int. J. Hydrogen Energy 33, 5266–5274.

Kim, I.S., Hwang, M.H., Jang, N.J., Hyun, S.H., Lee, S.T., 2004a. Effect of low pH on the activity of hydrogen utilizing methanogen in bio-hydrogen process. Int. J. Hydrogen Energy 29, 1133–1140.

Kim, S.H., Han, S.K., Shin, H.S., 2004b. Feasibility of biohydrogen production by anaerobic co-digestion of food waste and sewage sludge. Int. J. Hydrogen Energy 29, 1607–1616.

Kim, D.H., Kim, S.H., Shin, H.S., 2009. Hydrogen fermentation of food waste without inoculum addition. Enzyme Microb. Technol. 45, 181–187.

Kim, D.H., Kim, S.H., Kim, K.Y., Shin, H.S., 2010. Experience of a pilot-scale hydrogen-producing anaerobic sequencing batch reactor (ASBR) treating food waste. Int. J. Hydrogen Energy 32, 1590–1594.

Kim, D.H., Kim, S.H., Jung, K.W., Kim, M.S., Shin, H.S., 2011a. Effect of initial pH independent of operational pH on hydrogen fermentation of food waste. Bioresour. Technol. 102, 8646–8652.

Kim, M.S., Lee, D.Y., Kim, D.H., 2011b. Continuous hydrogen production from tofu processing waste using anaerobic mixed microflora under thermophilic conditions. Int. J. Hydrogen Energy 36, 8712–8718.

Kongjan, P., Angelidaki, I., 2010. Extreme thermophilic biohydrogen production from wheat straw hydrolysate using mixed culture fermentation: Effect of reactor configuration. Bioresour. Technol. 101, 7789–7796.

Kongjan, P., O-Thong, S., Kotay, M., Min, B., Angelidaki, I., 2010. Biohydrogen production from wheat straw hydrolysate by dark fermentation using extreme thermophilic mixed culture. Biotechnol. Bioeng. 105, 899–908.

Kotsopoulos, T.A., Fotidis, I.A., Tsolakis, N., Martzopoulos, G.G., 2009. Biohydrogen production from pig slurry in a CSTR reactor system with mixed cultures under hyper-thermophilic temperature (70°C). Biomass Bioenergy 33, 1168–1174.

Kumar, P., Barrett, D.M., Delwiche, M.J., Stroeve, P., 2009. Methods for pretreatment of lignocellulosic biomass for efficient hydrolysis and biofuel production. Ind. Eng. Chem. Res. 48, 3713–3729.

Lay, J.J., 2000. Modeling and optimization of anaerobic digested sludge converting starch to hydrogen. Biotechnol. Bioeng. 68, 269–278.

Lay, J.J., Lee, Y.J., Noike, T., 1999. Feasibility of biological hydrogen production from organic fraction of municipal solid waste. Water Res. 33, 2579–2586.

Lay, J.J., Fan, K.S., Chang, J., Ku, C.H., 2003. Influence of chemical nature of organic wastes on their conversion to hydrogen by heat-shock digested sludge. Int. J. Hydrogen Energy 28, 1361–1367.

Lay, J.J., Fan, K.S., Hwang, J.I., Chang, J.I., Hsu, P.C., 2005. Factors affecting hydrogen production from food wastes by Clostridium-rich composts. J. Environ. Eng. ASCE 131, 595–602.

Lee, Y.W., Chung, J., 2010. Bioproduction of hydrogen from food waste by pilot-scale combined hydrogen/methane fermentation. Int. J. Hydrogen Energy 35, 11746–11755.

Lee, D.Y., Ebie, Y., Xu, K.Q., Li, Y.Y., Inamori, Y., 2010. Continuous H_2 and CH_4 production from high-solid food waste in the two-stage thermophilic fermentation process with the recirculation of digester sludge. Bioresour. Technol. 101, S42–S47.

Levin, D.B., Zhu, H., Beland, M., Cicek, N., Holbein, B.E., 2007. Potential for hydrogen and methane production from biomass residues in Canada. Bioresour. Technol. 98, 654–660.

Li, D., Chen, H., 2007. Biological hydrogen production from steam-exploded straw by simultaneous saccharification and fermentation. Int. J. Hydrogen Energy 32, 1742–1748.

Li, C., Fang, H.H.P., 2007. Fermentative hydrogen production from wastewater and solid wastes by mixed cultures. Crit. Rev. Environ. Sci. Technol. 37, 1–39.

Li, M., Zhao, Y., Guo, Q., Qian, X., Niu, D., 2008. Bio-hydrogen production from food waste and sewage sludge in the presence of aged refuse excavated from refuse landfill. Renew. Energy 33, 2573–2579.

Li, S.L., Lin, J.S., Wang, Y.H., Lee, Z.K., Kuo, S.C., Tseng, I.C., et al., 2011. Strategy of controlling the volumetric loading rate to promote hydrogen-production performance in a mesophilic-kitchen-waste fermentor and the microbial ecology analyses. Bioresour. Technol. 102, 8682–8687.

Lin, Y.H., Juan, M.L., Hsien, H.J., 2011. Effects of temperature and initial pH on biohydrogen production from food-processing wastewater using anaerobic mixed cultures. Biodegrad 22, 551–563.

Liu, G., Shen, J., 2004. Effects of culture and medium conditions on hydrogen production from starch using anaerobic bacteria. J. Biosci. Bioeng. 98, 251–256.

Luo, G., Xie, L., Zou, Z., Zhou, Q., Wang, J.Y., 2010. Fermentative hydrogen production from cassava stillage by mixed anaerobic microflora: Effects of temperature and pH. Appl. Energy 87, 3710–3717.

Mohan, S.V., 2008. Fermentative hydrogen production with simultaneous wastewater treatment: Influence of pretreatment and system operating conditions. J. Sci. Ind. Res. 67, 950–961.

Mosier, N., Wyman, C., Dale, B., Elander, R., Lee, Y.Y., Holtzapple, M., et al., 2005. Features of promising technologies for pretreatment of lignocellulosic biomass. Bioresour. Technol. 96, 673–686.

Mu, Y., Zheng, X.J., Yu, H.Q., Zhu, R.F., 2006. Biological hydrogen production by anaerobic sludge at various temperatures. Int. J. Hydrogen Energy 31, 780–785.

Nasirian, N., 2012. Biological hydrogen production from acid-pretreated straw by simultaneous saccharification and fermentation. Afr. J. Agric. Res. 7, 876–882.

Noike, T., Takabatake, H., Mizuno, O., Ohba, M., 2002. Inhibition of hydrogen fermentation of organic wastes by lactic acid bacteria. Int. J. Hydrogen Energy 27, 1367–1371.

Okamoto, M., Miyahara, T., Mizuno, O., Noike, T., 2000. Biological hydrogen potential of materials characteristic of the organic fraction of municipal solid wastes. Water Sci. Technol. 41, 25–32.

Ozeler, D., Yetis, U., Demirer, G.N., 2006. Life cycle assesment of municipal solid waste management methods: Ankara case study. Environ. Int. 32, 405–411.

Pakarinen, O., Lehtomaki, A., Rintala, J., 2008. Batch dark fermentative hydrogen production from grass silage: The effect of inoculum, pH, temperature and VS ratio. Int. J. Hydrogen Energy 33, 594–601.

Pan, J., Zhang, R., El-Mashad, H.M., Sun, H., Ying, Y., 2008. Effect of food to microorganism ratio on biohydrogen production from food waste via anaerobic fermentation. Int. J. Hydrogen Energy 33, 6968–6975.

Park, M.J., Jo, J.H., Park, D., Lee, D.S., Park, J.M., 2010. Comprehensive study on a two-stage anaerobic digestion process for the sequential production of hydrogen and methane from cost-effective molasses. Int. J. Hydrogen Energy 35, 6194–6202.

Pattra, S., Sangyoka, S., Boonmee, M., Reungsang, A., 2008. Bio-hydrogen production from the fermentation of sugarcane bagasse hydrolysate by *Clostridium butyricum*. Int. J. Hydrogen Energy 33, 5256–5265.

Ren, N., Wang, A., Gao, L., Xin, L., Lee, D.J., Su, A., 2008. Bioaugmented hydrogen production from carboxymethyl cellulose and partially delignified corn stalks using isolated cultures. Int. J. Hydrogen Energy 33, 5250–5255.

Ren, N., Wang, A., Cao, G., Xu, J., Gao, L., 2009. Bioconversion of lignocellulosic biomass to hydrogen: Potential and challenges. Biotechnol. Adv. 27, 1051–1060.

Ren, N.Q., Tang, J., Liu, B.F., Guo, W.Q., 2010. Biological hydrogen production in continuous stirred tank reactor systems with suspended and attached microbial growth. Int. J. Hydrogen Energy 35, 2807–2813.

Shin, H.S., Youn, J.H., 2005. Conversion of food waste into hydrogen by thermophilic acidogenesis. Biodegrad. 16, 33–44.

Shin, H.S., Youn, J.H., Kim, S.H., 2004. Hydrogen production from food waste in anaerobic mesophilic and thermophilic acidogenesis. Int. J. Hydrogen Energy 29, 1355–1363.

Show, K.Y., Zhang, Z.P., Tay, J.H., Liang, D.T., Lee, D.J., Ren, N., et al., 2010. Critical assessment of anaerobic processes for continuous biohydrogen production from organic wastewater. Int. J. Hydrogen Energy 35, 13350–13355.

Sivers, M.V., Zacchi, G., 1995. A techno-economical comparison of three processes for the production of ethanol from pine. Bioresour. Technol. 51, 43–52.

Sreela-or, C., Imai, T., Plangklang, P., Reungsang, A., 2011a. Optimization of key factors affecting hydrogen production from food waste by anaerobic mixed cultures. Int. J. Hydrogen Energy 36, 14120–14133.

Sreela-or, C., Plangklang, P., Imai, T., Reungsang, A., 2011b. Co-digestion of food waste and sludge for hydrogen production by anaerobic mixed cultures: Statistical key factors optimization. Int. J. Hydrogen Energy 36, 14227–14237.

Sun, Y., Cheng, J., 2002. Hydrolysis of lignocellulosic materials for ethanol production: A review. Bioresour. Technol. 83, 1–11.

Tawfik, A., El-Qelish, M., 2012. Continuous hydrogen production from co-digestion of municipal food waste and kitchen wastewater in mesophilic anaerobic baffled reactor. Bioresour. Technol. 114, 270–274.

Tawfik, A., Salem, A., El-Qelish, M., 2011. Two stage anaerobic baffled reactors for bio-hydrogen production from municipal food waste. Bioresour. Technol. 102, 8723–8726.

Thitame, S.N., Pondhe, G.M., Meshram, D.C., 2010. Characterization and composition of municipal solid waste (MSW) generated in Sangamner city, district Ahmednagar, Maharashtra, India. Environ. Monit. Assess. 170, 1–5.

Tiehm, A., Nickel, K., Nies, U., 1997. The use of ultrasound to accelerate the anaerobic digestion of sewage sludge. Water Sci. Technol. 36, 121–128.

Ueno, Y., Otsuka, S., Morimoto, M., 1996. Hydrogen production from industrial wastewater by anaerobic microflora in chemostat culture. J. Ferment. Bioeng. 82, 194–197.

Ueno, Y., Haruta, S., Ishii, M., Igarashi, Y., 2001. Microbial community in anaerobic hydrogen-producing microflora enriched from sludge compost. Appl. Microbiol. Biotechnol. 57, 555–562.

Ueno, Y., Fukui, H., Goto, M., 2007. Operation of a two-stage fermentation process producing hydrogen and methane from organic waste. Environ. Sci. Technol. 41, 1413–1419.

Valdez-Vazquez, I., Rios-Leal, E., Esparza-Garcia, F., Cecchi, F., Poggi-Varaldo, H.M., 2005. Semi-continuous solid substrate anaerobic reactors for H_2 production from organic waste: Mesophilic versus thermophilic regime. Int. J. Hydrogen Energy 30, 1383–1391.

Valdez-Vazquez, I., Ponce-Noyola, M.T., Poggi-Varaldo, H.M., 2009. Nutrients related to spore germination improve H_2 production from heat-shock-treated consortia. Int. J. Hydrogen Energy 34, 4291–4295.

Vijayaraghavan, K., Ahmad, D., Ibrahim, M.K.B., 2006. Biohydrogen generation from jackfruit peel using anaerobic contact filter. Int. J. Hydrogen Energy 31, 569–579.

Wang, J.L., Wan, W., 2008. Comparison of different pretreatment methods for enriching hydrogen-producing cultures from digested sludge. Int. J. Hydrogen Energy 33, 2934–2941.

Wang, X., Zhao, Y.C., 2009. A bench scale study of fermentative hydrogen and methane production from food waste in integrated two-stage process. Int. J. Hydrogen Energy 34, 245–254.

Wang, C.H., Lu, W.B., Chang, J.S., 2007. Feasibility study on fermentative conversion of raw and hydrolyzed starch to hydrogen using anaerobic mixed microflora. Int. J. Hydrogen Energy 32, 3849–3859.

Wang, Y., Wang, H., Feng, X., Wang, X., Huang, J., 2010a. Biohydrogen production from cornstalk wastes by anaerobic fermentation with activated sludge. Int. J. Hydrogen Energy 35, 3092–3099.

Wang, Y.H., Li, S.L., Chen, I.C., Tseng, I.C., Cheng, S.S., 2010b. A study of the process control and hydrolytic characteristics in a thermophilic hydrogen fermentor fed with starch-rich kitchen waste by using molecular-biological methods and amylase assay. Int. J. Hydrogen Energy 35, 13004–13012.

Wu, X., Zhu, J., Dong, C., Miller, C., Li, Y., Wang, L., et al., 2009. Continuous biohydrogen production from liquid swine manure supplemented with glucose using an anaerobic sequencing batch reactor. Int. J. Hydrogen Energy 34, 6636–6645.

Xiao, B., Liu, J., 2009. Biological hydrogen production from sterilized sewage sludge by anaerobic self-fermentation. J. Hazard. Mater. 168, 163–167.

Xing, Y., Li, Z., Fan, Y., Hou, H., 2010. Biohydrogen production from dairy manures with acidification pretreatment by anaerobic fermentation. Environ. Sci. Pollut. Res. 17, 392–399.

Yasin, N.H.M., Rahman, N.A., Man, H.C., Yusoff, M.Z.M., Hassan, M.A., 2011. Microbial characterization of hydrogen-producing bacteria in fermented food waste at different pH values. Int. J. Hydrogen Energy 36, 9571–9580.

Yokoi, H., Saitsu, A., Uchida, H., Hirose, J., Hayashi, S., Takasaki, Y., 2001. Microbial Hydrogen production from sweet potato starch residue. J. Biosci. Bioeng. 91, 58–63.

Yokoi, H., Maki, R., Hirose, J., Hayashi, S., 2002. Microbial production of hydrogen from starch-manufacturing wastes. Biomass Bioenergy 22, 389–395.

Yokoyama, H., Waki, M., Moriya, N., Yasuda, T., Tanaka, Y., Haga, K., 2007. Effect of fermentation temperature on hydrogen production from cow waste slurry by using anaerobic microflora within the slurry. Appl. Microbiol. Biotechnol. 74, 474–483.

Zaher, U., Li, R., Jeppsson, U., Steyer, J.P., Chen, S., 2009. GISCOD: General integrated solid waste co-digestion model. Water Res. 43, 2717–2727.

Zhang, T., Liu, H., Fang, H.H.P., 2003. Biohydrogen production from starch in wastewater under thermophilic conditions. J. Environ. Manag. 69, 149–156.

Zhang, B., Zhang, L.L., Zhang, S.C., Shi, H.Z., Cai, W.M., 2005. The influence of pH on hydrolysis and acidogenesis of kitchen wastes in two-phase anaerobic digestion. Environ. Technol. 26, 329–339.

Zhang, Z.P., Show, K.Y., Tay, J.H., Liang, D.T., Lee, D.J., Jiang, W.J., 2006. Effect of hydraulic retention time on biohydrogen production and anaerobic microbial community. Process Biochem. 41, 2118–2123.

Zhang, M.L., Fan, Y.T., Xing, Y., Pan, C.M., Zhang, G.S., Lay, J.J., 2007. Enhanced biohydrogen production from cornstalk wastes with acidification pretreatment by mixed anaerobic cultures. Biomass Bioenergy 31, 250–254.

Zhao, Y., Chen, Y., Zhang, D., Zhu, X., 2010. Waste activated sludge fermentation for hydrogen production enhanced by anaerobic process improvement and acetobacteria inhibition: The role of fermentation pH. Environ. Sci. Technol. 44, 3317–3323.

Zheng, Y., Pan, Z., Zhang, R., 2009. Overview of biomass pretreatment for cellulosic ethanol production. Int. J. Agric. Biol. Eng. 2, 51–68.

Zhu, H.G., Beland, M., 2006. Evaluation of alternative methods of preparing hydrogen producing seeds from digested wastewater sludge. Int. J. Hydrogen Energy 31, 1980–1988.

Zhu, H., Parker, W., Basnar, R., Proracki, A., Falletta, P., Beland, M., et al., 2009a. Buffer requirements for enhanced hydrogen production in acidogenic digestion of food wastes. Bioresour. Technol. 100, 5097–5102.

Zhu, J., Li, Y., Wu, X., Miller, C., Chen, P., Ruan, R., 2009b. Swine manure fermentation for hydrogen production. Bioresour. Technol. 100, 5472–5477.

Zhu, H., Parker, W., Conidi, D., Basnar, R., Seto, P., 2011. Eliminating methanogenic activity in hydrogen reactor to improve biogas production in a two-stage anaerobic digestion process co-digesting municipal food waste and sewage sludge. Bioresour. Technol. 102, 7086–7092.

12

Thermochemical Route for Biohydrogen Production

Thallada Bhaskar, Bhavya Balagurumurthy,
Rawel Singh, Mukesh Kumar Poddar

Council of Scientific and Industrial Research–Indian Institute of Petroleum,
Bio-Fuels Division, Dehradun, India

INTRODUCTION

Fossil fuels account for most of the 6.5 billion tons (gigatons) of carbon—the amount present in 25 gigatons of CO_2 that people around the world vent into the atmosphere every year. As the amount of greenhouse gas increases, so does the likelihood of triggering a debilitating change in the Earth's climate (Robert, 2004). The depletion of fossil fuels and the worldwide spread of environmental problems due to fossil fuels have induced the development of alternative energy production. Hydrogen (H_2) is the most common element on earth and does not exist mostly in elemental form. It is present in water, biomass, and hydrocarbons. H_2 is called a clean energy fuel, as the chemical energy stored in the H—H bond is released when it combines with oxygen, yielding only water as a reaction product. Accordingly, a future energy infrastructure based on H_2 has been perceived as an ideal long-term solution to energy-related environmental problems (Bockris, 2002). In addition, H_2 is a flammable, odorless, colorless gas. Any significant H_2 explosion accident could prevent the public from accepting H_2 as transportation fuel H_2. There is no doubt that H_2 has the potential to provide a clean and affordable energy supply that can minimize our dependence on oil and therefore enhance the global economy and reduce environmental pollution. H_2 economy is one of the most discussed topics in recent years. It is thought to have the highest potential to be the future energy carrier and could play a very significant role in the reduction of emissions of greenhouse gases (Christopher and Dimitrios, 2012).

Distinct from first-generation (e.g., solid coal) and second-generation (e.g., liquid gasoline, diesel) fuels, third-generation transportation fuels include hydrogen and electricity, both of which work as energy carriers that can be converted to kinetic work efficiently without the restriction of the second law of thermodynamics. Both H_2 and electricity can be generated from various primary energy sources, such as biomass, solar energy, wind energy, geothermal energy, and tidal energy. The H_2-fuel cell-electricity system will play a predominant role because of (1) very high energy conversion efficiency through fuel cells, (2) minimal pollutants generated, (3) much higher energy storage densities than rechargeable batteries alone, and (4) diverse H_2-producing means from primary energy resources. However, large-scale implementation of the H_2 economy must break four technological hurdles: low-cost H_2 production from any primary energy resources, high H_2 density storage means (>9 mass%), affordable fuel distribution infrastructure, and affordable fuel cells throughout the whole life cycle (Schlapbach and Zuttel, 2001; Steele and Heinzel, 2001).

H_2, a small and energetic molecule, can diffuse through container materials or react with materials. For example, H_2 cannot be simply delivered by today's natural gas pipeline systems because of steel embrittlement, accompanied with increased maintenance costs, leakage rates, and material replacement costs. H_2 pipelines are much more expensive than electric transmission lines and natural gas pipelines (Zhang, 2009).

It is generally understood that the renewable energy-based processes of H_2 production (solar photochemical and photobiological water decomposition, electrolysis of water coupled with photovoltaic cells or wind turbines, etc.) are unlikely to involve significant reductions in H_2 costs over the next 10 to 20 years. Industry generates some 48 million metric tons of H_2 globally each year from fossil fuel. Almost half of this H_2 goes into making ammonia; refineries use the second largest volume of H_2 for chemical processes, such as removing sulfur from gasoline and converting heavy hydrocarbons into gasoline and diesel fuel. Food producers add a small percentage of H_2 to some edible oils through a catalytic hydrogenation process. The demand for H_2 is expected to grow over the next 10 years, for both traditional uses, such as making ammonia and running fuel cells (Navarro et al., 2009).

Even if the hydrogen economy were technically and economically feasible today, weaning the world off carbon-based fossil fuels would still take decades. During that time, carbon combustion will continue to pour greenhouse gases into the atmosphere—unless scientists find a way to reroute them (Robert, 2004).

HYDROGEN PRODUCTION FROM FOSSIL FUELS

The effectiveness of biomass as a H_2 source depends critically on the production efficiency and the use of residuals from the biomass conversion process. Currently, H_2 is mainly produced from fossil fuels (96%). None of the routes can compete with the well-established H_2 production technology, that is, the steam reforming (SR) of natural gas (Chen and He, 2011). SR of natural gas is the most frequently employed process (48%), which is obviously not a sustainable and CO_2-neutral route (Fermoso et al., 2012).

There will likely be an extended period of time when the new technologies consume more energy than they produce. With this in mind, we need to think carefully about how many intermediate technology steps we introduce and how long (and at what cost) we must operate them in order to make the energy payback positive. The energy required to sustain a growth rate must also be taken into account (Turner, 2004).

METHODS OF HYDROGEN PRODUCTION FROM RENEWABLE SOURCES

Production of H_2 from renewable sources derived from agricultural or other waste streams offers the possibility to contribute to the production capacity with lower or no net greenhouse gas emissions (without carbon sequestration technologies), increasing the flexibility and improving the economics of distributed and semicentralized reforming. Electrolysis, thermocatalytic, and biological production can be adapted easily to on-site decentralized production of H_2, circumventing the need to establish a large and costly distribution infrastructure. Each of these H_2 production technologies, however, faces technical challenges, including conversion efficiencies, feedstock type, and the need to safely integrate H_2 production systems with H_2 purification and storage technologies.

Renewable Electrolysis

Renewable electrolysis uses an electric current to split water into H_2 and oxygen. The electricity required can be generated using any of a number of resources. Renewable energy sources such as photovoltaics, wind, biomass, hydropower, and geothermal can provide clean, sustainable electricity. The hydrogen produced from renewable electrolysis can be used in fuel cells or internal combustion engines to produce electricity during peak demand or low-power production. This hydrogen can also be used as transportation fuel.

Solar Routes

Solar energy is the most abundant renewable energy on Earth, but its low-energy density, dispersivity, and discontinuity make it difficult to store. Solar photovoltaic technology has entered the commercial application. However, the shortage and high cost of the raw material have become main obstacles for the large-scale application of solar photovoltaic technology. For the industrialization of solar thermal power generation, great breakthroughs must be made in key technologies in the future, such as improving the efficiency of solar concentrators, reducing the cost of a solar tracking system, and developing more advanced thermal storage technology. The thermochemical, photochemical, and biological processes, in which solar energy will be converted directly into H_2 or other energy products, are ideal ways to solving the aforementioned problems of solar utilization. The possible one-step ways for solar H_2 production include solar thermochemical decomposition of water and biomass, solar photocatalytic/photoelectricchemical water splitting, photobiological H_2 production from water and biomass, and so on. In the short to medium term, the solar thermochemical decomposition of water and biomass for H_2 is most likely to achieve industrialization's mean of large-scale H_2 production from renewable energy, and in the long term, solar photocatalytic/photoelectricchemical and photobiological H_2 production from water and biomass are the most inexpensive and ideal ways to H_2 production (Lu et al., 2011). Another water-splitting method, called high-temperature thermochemical water splitting, uses high temperatures generated by solar concentrators (mirrors that focus and intensify sunlight) or nuclear reactors to drive a series of chemical reactions to split water into H_2 and oxygen through a series of chemical reactions. All of the intermediate process chemicals are recycled within the process.

Photolysis

Different approaches to solar water splitting include semiconductor particles as photocatalysts and photoelectrodes, molecular donor–acceptor systems linked to catalysts for H_2, and oxygen evolution and photovoltaic cells coupled directly or indirectly to electrocatalysts. Despite several decades of research, solar H_2 generation is efficient only in systems that use expensive photovoltaic cells to power water electrolysis. Direct photocatalytic water splitting is a challenging problem because the reaction is thermodynamically uphill. Light absorption results in the formation of energetic charge-separated states in both molecular donor–acceptor systems and semiconductor particles. Unfortunately, energetically favorable charge recombination reactions tend to be much faster than the slow multielectron processes of water oxidation and reduction. Consequently, visible light water splitting has been achieved only recently in semiconductor-based photocatalytic systems and remains an inefficient process (Youngblood et al., 2009).

Nuclear High-Temperature Electrolysis

Heat from a nuclear reactor can be used to improve the efficiency of water electrolysis to produce H_2. By increasing the temperature of the water, less electricity is required to split it into H_2 and oxygen, which reduces the total energy required. Three methods are actively being researched to produce H_2 from nuclear power. One method is conventional electrolysis of water using electricity generated from nuclear power plants. The inefficiencies of several energy conversions from nuclear heat to the end product of H_2 limit the viability of electrolysis for large-scale use and have prompted research into ways to use nuclear heat directly, including thermochemical water splitting. Sulfur–iodine, hybrid sulfur, and calcium–bromine cycles are being researched, as the efficiency of thermochemical water-splitting processes is much higher than that for electrolysis. The third method being researched is high-temperature electrolysis of steam, which has a potential efficiency higher than conventional electrolysis (Bartels et al., 2010; Henderson and Taylor, 2006).

Biotechnological Routes

Biohydrogen production could be achieved by either photosynthetic or anaerobic microorganisms. Photosynthetic microorganisms basically use CO_2 and water for H_2 gas production or photoheterophic microorganisms use carbohydrates or organic acids to produce H_2 and CO_2. Even though H_2 production through this process is high, it is not applicable, as light energy needs to be supplied and it is difficult to design an efficient photoreactor. During the anaerobic acidogenesis process, H_2 is produced as a by-product with organic acid production. Naturally, H_2 will be produced by H_2 producers in an anaerobic treatment system. However, H_2 in the liquid phase will be consumed readily by methanogenic H_2 consumers (de Vrije and Claassen, 2003). Therefore, in order to produce H_2 gas in bulk, methanogenesis must be eliminated from the anaerobic treatment system. This could be achieved by the elimination of H_2

consumers from the system and operating at a shorter retention time (Cheng et al., 2002; Chonga et al., 2009).

Direct biophotolysis of H_2 production is a biological process using microalgae photosynthetic systems to convert solar energy into chemical energy in the form of H_2. Two photosynthetic systems are responsible for photosynthesis process: (i) photosystem I (PSI) producing reductant for CO_2 reduction and (ii) photosystem II (PSII) splitting water and evolving oxygen. In the biophotolysis process, two photons from water can yield either CO_2 reduction by PSI or H_2 formation with the presence of hydrogenase. The concept of indirect biophotolysis involves the following four steps: (i) biomass production by photosynthesis; (ii) biomass concentration; (iii) aerobic dark fermentation yielding 4 mol H_2/mol glucose in the algae cell, along with 2 mol of acetates; and (iv) conversion of 2 mol of acetates into H_2 (Ni et al., 2006).

FEEDSTOCKS FOR THERMOCHEMICAL METHODS OF HYDROGEN PRODUCTION

Biomass, as a product of photosynthesis, is an abundant and carbon-neutral renewable energy resource for the sustainable production of H_2. Energy production from biomass has the advantage of forming smaller amounts of greenhouse gases compared to the conversion of fossil fuels, as the CO_2 generated during the energy conversion is consumed during subsequent biomass regrowth (Stocker, 2008). Biomass is any organic matter wood, crops, seaweed, or animal wastes that can be used as an energy source. First-generation biofuels use edible biomass for producing biofuels. In contrast, second-generation biofuels are produced from nonedible feedstocks such as lignocellulosic feedstocks, which include agro residue (stalk, husk), forest residue (branch, twigs, bark, leaves), and several others. Lignocellulosic biomass is composed of cellulose, hemicelluloses, lignin, and other inorganic compounds. Although fairly straightforward, the conversion of biomass to H_2 has low conversion efficiency from sunlight to H_2, and any system designed to generate substantial amounts of H_2 must be rather large. Nonetheless, if the biomass used is a waste by-product, then this is perhaps the least expensive of the H_2 generation technologies (Turner, 1999). Hence, agro wastes, forest residues, and defatted algae will be best sources for hydrogen production by thermochemical methods of conversion.

THERMOCHEMICAL METHODS OF CONVERSION

The main advantage of the thermochemical method of conversion is that it utilizes the entire biomass as such. This eliminates the cost and energy input in the pretreatment steps, whether in the form of acid hydrolysis or enzyme hydrolysis or various other methods. At present, the removal of lignin is an essential step in enabling fermentation by microorganisms. The other advantage is the speed at which products can be obtained from biomass. Depending on the type of method chosen, the product is obtained in few seconds to an hour or two. The biochemical route at present requires many hours for fermentation to occur. The microorganisms are feed specific, and even the slightest change could lead to failure of fermentation. This poses a major risk in the commercialization of the process at an industrial level, as the biomass availability in terms of quality and quantity is never constant throughout the year.

The base of thermochemical conversion is the pyrolysis process in most cases. Pyrolysis is the fundamental chemical reaction process and is simply defined as the chemical changes occurring when heat is applied to a material in the absence of oxygen. The various types of thermochemical methods of conversion are combustion, gasification, pyrolysis, liquefaction, and carbonization. The products of thermochemical methods of conversion include water, charcoal (carbonaceous solid), biocrude, tars, and permanent gases, including methane, hydrogen, carbon monoxide, and carbon dioxide depending on the reaction parameters, such as environment, reactors used, final temperature, rate of heating, and source of heat. The following sections explain the various methods of hydrogen production from biomass using various thermochemical methods of conversion. The representative thermochemical methods of hydrogen production from biomass are illustrated in Figure 1.

Gasification

Biomass gasification generally refers to the thermochemical conversion of solid biomass fuels using a gasifying agent [e.g., steam, air (partial oxidation), or CO_2] to a mixture of combustible product gases, including H_2, CH_4, CO, and CO_2. Gasification is one of the most efficient ways to extract energy from fuel sources and convert it into a usable form by partial or total transformation of solids to gases. It is the energy conversion process that has been studied as an alternative solution to environmental issues associated with energy production (Boerrigter and Rauch, 2005). Biomass gasification offers the earliest and most economical route for the production of renewable H_2.

The chemistry of biomass gasification is complex. Biomass gasification proceeds primarily via a two-step process—pyrolysis followed by gasification. Pyrolysis is decomposition of the biomass feedstock by heat. This step, also known as devolatilization, is endothermic and produces 75 to 90% volatile materials in the form of gaseous and liquid hydrocarbons. The remaining nonvolatile material, containing a high carbon content, is referred to as char (Bridgwater and Evans, 1993). The volatile hydrocarbons and char are subsequently converted to syngas in the second step—gasification. A few of the major reactions involved in this step are listed here (Bridgwater and Evans, 1993).
Exothermic reactions:

$$\text{Combustion } \{biomass\ volatiles/char\} + O_2 \rightarrow CO_2 \tag{1}$$

$$\text{Partial oxidation } \{biomass\ volatiles/char\} + O_2 \rightarrow CO \tag{2}$$

$$\text{Methanation } \{biomass\ volatiles/char\} + H_2 \rightarrow CH_4 \tag{3}$$

$$\text{Water}-\text{gas shift } CO + H_2O \rightarrow CO_2 + H_2 \tag{4}$$

$$\text{CO methanation } CO + 3H_2 \rightarrow CH_4 + H_2O \tag{5}$$

Endothermic reactions:

$$\text{Steam}-\text{carbon reaction } \{biomass\ volatiles/char\} + H_2O \rightarrow CO + H_2 \tag{6}$$

$$\text{Boudouard reaction } \{biomass\ volatiles/char\} + CO_2 \rightarrow 2CO \tag{7}$$

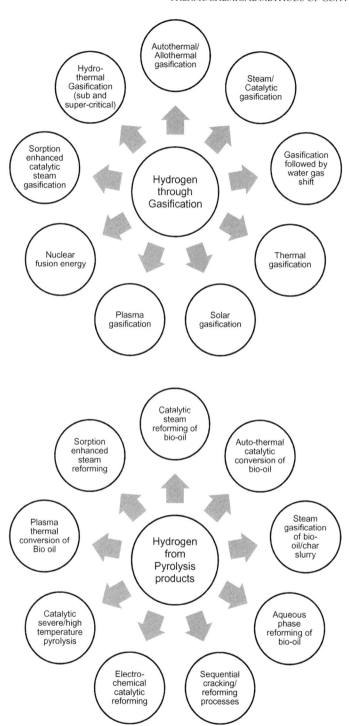

FIGURE 1 Representative thermo-chemical methods of biohydrogen production from biomass.

In the presence of an oxidizing agent at high temperature, the large polymeric molecules of biomass decompose into lighter molecules and eventually to permanent gases (CO, H_2, CH_4, and lighter hydrocarbons), ash, char and tar, and minor contaminants. Char and tar are the result of an incomplete conversion of biomass. Several methods of biomass gasification are described in the following sections.

Autothermal gasification

Gasification is an endothermic process; therefore, heat is needed to sustain the gasification process. The process could be either autothermal or allothermal, depending on how this heat is provided. In the case of autothermal gasification, the necessary heat is generated directly by partial oxidation inside the gasifier, while, during indirect heating, heat is provided by combusting some of the feedstock, char, or clean syngas in a separate reactor and transferring heat through exchangers using preheated bed material (Rezaiyan and Cheremisinoff, 2005).

Autothermal and allothermal gasifications are the main technologies used for syngas production. Low-lower heating value (LHV) syngas can be produced with autothermal processes using air as the oxidizing agent and medium-LHV syngas by allothermal or steam/oxygen-blown autothermal gasification.

In these systems, the reactor temperature is controlled by the oxidant feed rate. If air is used as the oxidant, the product gas has a low heating value of 4 to 5 MJ/m^3 (107–134 Btu/ft^3) due to nitrogen dilution. Autothermal processes generate enough heat to sustain themselves in the reaction temperature. These kinds of reactors are relatively inexpensive to construct because no heat transfer area has to be mounted inside the gasifier, but they can also be inconvenient to operate due to the lack of temperature control by means of cooling medium. Thus, the only way to control the temperature inside the reactor is to feed excess of one of the components or use a diluent. The Foster Wheeler-pressurized steam/oxygen-blown circulating fluidized bed gasifier is an example of an autothermal reactor suitable for biomass gasification. The gasifier is refractory lined and complemented with a uniflow cyclone. No heat transfer surfaces are mounted inside the gasifier, but an air preheater has been integrated into the gas duct just below the cyclone. The gasifier can be operated flexibly according to different fuels and applications. Normal operating temperatures for woody biomass are in the range of 850–900 °C. Examples of this technology are the Gas Technology Institute (GTI) and the SynGas gasifiers.

Allothermal gasification

In allothermal (heated indirectly) gasifiers, heat required for gasification reactions is produced outside the reactor and then transferred inside. One of the major advantages of allothermal gasification is the ability to produce medium-heating value gas without needing expensive oxygen separation technology. Allothermal gasifiers are usually either heated by heat exchangers inside the gasifier or circulated hot bed material is used to transfer the heat between the combustion and the gasification zone. Examples of this kind of technology include the Battelle and fast internally circulating fluidized bed gasifiers. Both these processes use two physically separate reactors: (1) a gasifier where biomass is converted to medium-heating value gas and residual char and (2) a combustor where the residual char is combusted to provide heat for the gasification. The heat transfer is carried out by circulating solid bed material (sand) between these two reactors. An example of indirectly heated gasification technology is the BCL/FERCO gasifier (Turn, 1999). It utilizes a bed of hot particles (sand), which is fluidized using steam. Solids (sand and char) are separated from the syngas via a cyclone and then transported to a second fluidized bed reactor. The second bed is air blown and acts as a char combustor, generating a flue gas exhaust stream and a stream of hot particles. The hot (sand) particles are separated from the flue gas and recirculated to the gasifier to provide the heat required for pyrolysis. This approach separates the combustion reaction from the remaining gasification reactions, producing a product gas that is practically nitrogen free with a heating value of 15 MJ/m^3 (403 Btu/ft^3) (Turn, 1999).

H$_2$ production by gasification of biomass is a complex process influenced by a number of factors, such as feedstock composition, moisture content, gasifier temperature, gasifier pressure, amount of oxidant present, gasifier geometry, and mode of gas–solid contact. Contact between the solid fuel particles and gases takes place in a reactor or gasifier (Seitarides et al., 2008).

Thermal gasification

As the development of thermal fuel gasification processes was going on, it faced two main disadvantages: low gas yields and corrosion of downstream equipment caused by the high concentration of tar vapor. Product gas exits a gasifier containing some particles, alkali compounds, tars, and nitrogen-containing components. The formation of tar (complex mixture of organic liquid constituents) and char (solid carbonaceous materials) during the gasification process is the most severe of all problems and, because of this problem, none of the processes currently available are universally accepted for commercialization (Blasi et al., 1999).

Steam gasification

When the thermochemical conversion of biomass is undertaken, using steam as the gasifying agent, the resulting product gas is rich in H$_2$. The use of steam, instead of air or CO$_2$, leads to higher H$_2$ yields due to the additional H$_2$ produced from the decomposition of H$_2$O. In addition, compared with partial oxidation using air, the product gas has a higher heating value because dilution with N$_2$ is avoided (Gil et al., 1999).

In the absence of O$_2$, conventional steam gasification is endothermic, which means that an additional heat source is required to drive the reaction system. This is a challenge because, obviously, this input of energy reduces the maximum efficiency of the process. A further challenge is the provision of additional heat without compromising the quality of the product gas. Methods to meet this energy shortfall involve the (i) combustion of a fraction of the biomass fuel or the unconverted biomass residue to generate heat, (ii) use of a fraction of the produced H$_2$ or other combustible product gases to generate energy via thermochemical or chemical pathways, and (iii) use of alternative renewable energy resources, for example, solar thermal energy. A range of different operating configurations, including dual fluidized bed gasifiers and circulating fluidized bed gasifiers, have been investigated for transferring heat from an exothermic process to an endothermic gasification process.

The process is believed to involve three main steps, delineated by reaction temperature (Pinto et al., 2003) (i) devolatilization, which occurs at relatively low temperatures, between 300 and 500°C, during which 70–90%-wt of the biomass is converted to volatile matter and solid char (Antal et al., 1984); (ii) cracking and reforming of the volatile matter and tars, typically defined as condensable organic contaminants (with molecular weights greater than benzene) (Abatzoglou et al., 2000), which occurs at temperatures greater than 600 °C; and (iii) char gasification, which occurs at high temperatures (>800 °C). The reaction mechanism is the summation of a complex series of competing reactions, including (i) gas–solid reactions between the biomass fuel particles and the steam and (ii) gas–gas reactions between the steam and the evolved gas species. These reactions result in a distribution of products, including tar, char, and a product gas composed mainly of H_2, CO, CO_2, CH_4, C_2H_4, and C_2H_6. The cracking and reforming reactions are dominant in influencing the final product gas composition during biomass gasification. The final composition of the product gas in any gasification system can therefore be manipulated by controlling the reaction parameters, including the temperature and steam-to-biomass ratio. A high concentration of tar (>3 /Nm^3) in the product gas is a significant problem in gasification reactors. High tar levels disrupt the operation of the process (e.g., due to blockages), as well as limiting utilization of the product gas for most applications, including gas engines and turbines (Sutton et al., 2001). Aside from operational issues, the destruction of tar results in an increase in gas yield and thus an enhanced output of H_2. The main technical challenges associated with steam gasification are that it requires an indirect or external heat supply for gasification, yields high tar content in syngas, and requires catalytic tar reforming steps (de Lasa et al., 2011).

The composition of products in the pyrolysis zone varies greatly, depending on temperature. It is generally accepted that a low temperature favors the formation of tar and char, whereas a high temperature favors the formation of gas products, which may not be favorable/valuable due to various reasons. Warnecke (2000) showed that fluidized bed gasification has been found to be advantageous. However, these processes still provide significant amounts of tar (a complex mixture of higher hydrocarbons) in the product gas even if operated at 800–1000 °C. Thus, product gas cleaning is a challenging point for most of its application. Catalytic cleaning and upgrading of hot dry gas is nowadays the best solution. Nickel-based catalyst is very effective for tar conversion in the secondary reactor at around 700–800 °C, resulting in about 98% tar removal from the product gas (Caballero et al., 2000). Using a catalyst in the primary reactor simplifies the overall process; however, a limited number of works focused on the direct use of a catalyst in the primary bed. Although nickel-based catalysts have been found to be effective in the primary reactor to reduce tar content in the product gas at above 750 °C (Arauzo et al., 1997; Rapagna et al., 1998), the catalysts deactivate suddenly because of carbon deposition on the surface (Baker et al., 1987). Asadullah et al. (2002, 2003) demonstrated the low-temperature catalytic steam gasification of biomass model compounds (cellulose) and biomass. The gasification process is operated at considerably low temperatures and thus formed tar and char as major products in the pyrolysis zone. The pyrolytic products then are exposed to the cracking zone where tar, char, and gases take part in the secondary reactions in the presence of Rh-based catalysts. The $Rh/CeO_2/SiO_2$ catalyst with 35 wt% of CeO_2 exhibited excellent performance for cellulose gasification at 500–600 °C. Almost complete C-conversion was achieved when at least 1111 µmol/min of steam was introduced with 51 cm^3/min of air for the gasification of 85 mg/min at 600 °C. Catalyst deactivation was not observed even after 8 h of reaction time (Asadullah et al., 2002, 2003). The $Rh/CeO_2/SiO_2(60)$ catalyst appeared to be an effective catalyst for biomass gasification, even in the absence of any gasifying agent within the temperature range of 873–973 K. Secondary char or coke, as well as primary char, was remarkably reduced on the $Rh/CeO_2/SiO_2(60)$ catalyst in the presence of any type of gasifying agent, such as O_2, CO_2, and steam. A great deal of char or coke conversion with CO_2 on this catalyst within this low-temperature range emphasized the recycling of CO_2 from the product gas in order to increase carbon utilization from the biomass. The external steam supply with a rate of 1666 μmol/min increased both char and tar conversion, which suggested that biomass with a high moisture content is also feasible for the gasification on the $Rh/CeO_2/SiO_2(60)$ catalyst (Asadullah et al., 2003).

Gasification followed by water gas shift reaction

Oxygen or air gasification coupled with a water–gas shift is the most widely practiced process route for biomass to H_2. Thermal, steam, and partial oxidation gasification technologies are under development all around the world. Thermal gasification is essentially high-severity pyrolysis, although steam is generally present. An example of this is the Sylvagas (BCL/FERCO) low-pressure, indirectly heated circulating fluid bed.

$$Biomass + Energy \rightarrow CO + H_2 + CH_4 + \ldots \ldots \tag{8}$$

By including oxygen in the reaction gas, a separate supply of energy is not required, but the product gas is diluted with CO_2 and, if air is used to provide the oxygen, then nitrogen is also present. Examples of this are the GTI (formerly IGT) high-pressure oxygen-blown gasifier, as well as the circulating fluid bed by TPS Termiska.

$$Biomass + O_2 \rightarrow \quad CO + H_2 + CO_2 + Energy \tag{9}$$

Other relevant gasifier types are bubbling fluid beds being tested by Enerkem and the high-pressure, high-temperature slurry-fed entrained flow Texaco gasifier.

All of these gasifier examples will need to include significant gas conditioning, including the removal of tars and inorganic impurities and the conversion of CO to H_2 by the water–gas shift (WGS) reaction:

$$CO + H_2O \rightarrow \quad CO_2 + H_2 \tag{10}$$

Relatively pure H_2 can be obtained from syngas produced from biomass gasification by steam reforming followed by a water–gas shift reactor. During the WGS reaction, CO and H_2O react in a 1:1 molar ratio on a catalytically active metal site to form CO_2 and H_2. This is a reversible reaction and therefore steam is added in excess to shift the equilibrium toward the product side. The WGS reaction can be carried out at two temperature ranges: (1) the high-temperature reaction is carried out using Fe and/or Cr-based oxide catalysts at temperatures between 350 and 500 °C and (2) the low-temperature reaction is carried out over Cu–Zn oxide catalysts (Bartholomew and Farrauto, 2006) at 200–250 °C. Low-temperature WGS reactions have also been carried out on metal catalysts supported on partially reducible metal oxides, such as transition metal catalysts (Barbier and Duprez, 1994) and Au (Jacobs et al., 2005) supported on Al_2O_3, CeO_2 (Schlatter and Mitchell, 1980; Trovarelli, 1996), and CeO_2–ZrO_2.

CeO_2 is by far the most studied partially reducible support because of its ability to readily reduce from Ce^{4+} to Ce^{3+} over the temperature range where a low-temperature WGS reaction is favorable. Gold supported on ceria has shown great promise and has been reported to be stable up to 300 °C (Liu et al., 2005). However, the higher temperatures upstream in a plant require more active and stable catalysts. Traditionally, high-temperature shift catalysts based on oxides of Fe and Cr are used in the industry. However, Lei et al. (2005, 2006) reported that promotion of Fe_2O_3–Cr_2O_3 catalysts with small amounts of Rh can increase the activity by up to a factor of 4. It is believed that the presence of Rh enhances the reduction of Fe_2O_3 to $Fe_3O_4$and also increases the H_2 release rate during the reoxidation by water. Examples of Fe and Cr oxide catalyst promotion by Cu, Hg, and Ag can also be found in the literature (Rhodes et al., 2002; Rhodes and Hutchings, 2003).

The CO concentration in the gas stream decreases to approximately 2% after the high-temperature WGS reactor and finally reduces to less than 0.2% after the low-temperature WGS reactor. However, CO is a poison for the anode catalyst used in proton exchange membrane fuel cells (Cheng et al., 2007). Therefore, generally it is expected that the CO concentration of the feed stream in the fuel cell is at a sub-ppm level (Farrauto et al., 2003). One of the ways to achieve this economically is to oxidize CO preferentially (CO-PROX) (Navarro et al., 2007).

Recent improvements in biomass gasification to produce syngas

Reaction swing methodology by Southern Illinois University, chemical looping processes, and so on have been developed for H_2 production involving the use of CaO as the CO_2 acceptor. In an effort to increase the purity of H_2 in the product stream by separation of the product gases from gasification, a new process to produce a Ca-based CO_2 sorbent with enhanced sorption capacities and life has been developed (Dasgupta et al., 2008).

Improvements in gasification technology include H_2 Production by Reaction Integrated Novel Gasification (HyPr-RING) technology. In this technology, the gasification and WGS reaction is combined in one reactor with simultaneous absorption of CO_2 and other pollutants to increase the H_2 yield, while maintaining a relatively low temperature of 650 °C.

The main goal of this process is to produce tar-free syngas for H_2 or liquid fuel production. This process takes place in two steps. In the first step, biomass is converted to tar-containing gas and charcoal in a pyrolysis unit that operates at 500 °C. The tar-containing gas is combusted in a high-temperature gasifier by cofeeding oxygen. Here, the charcoal from the first reactor is gasified completely to syngas at 1500 °C in an entrained flow gasifier (Carbo V-Process).

In this process, lignocellulosic biomass such as straw and other nonwoody biomass is first liquefied by fast pyrolysis at many local small plants. The resulting bio-oil and char slurry is transported to a central facility where large pressurized-entrained flow gasifiers combust the slurry to produce tar-free syngas. The gasifier is operated at 26 bar pressure and at temperatures (typically above the ash melting point) ranging between 1200 and 1600 °C (BIOLIQ process).

Solar gasification

The use of concentrated solar energy as a heat source for pyrolysis and gasification of biomass is an efficient means for producing H_2-rich synthesis gas. Utilizing molten alkali carbonate salts as a reaction and heat transfer media promises enhanced stability to solar transients and faster reaction rates. The molten salt (eutectic blend of lithium, potassium, and sodium carbonate salts) increases the rate of pyrolysis by 74% and increases gasification rates by more than an order of magnitude while promoting a product gas composition nearer to thermodynamic equilibrium predictions (Hathaway et al., 2011).

Supercritical water gasification of biomass driven by concentrated solar energy is a novel way for solar H_2 production. Proof of concept for the novel system of H_2 production by biomass gasification in supercritical water using concentrated solar energy has been demonstrated (Chen et al., 2010) using biomass model compounds (glucose) and biomass (corn meal, wheat stalk).

Plasma gasification

Studies have shown that there is potential to improve conventional biomass processing by coupling a plasma reactor to a pyrolysis cyclone reactor. The role of the plasma is twofold: it acts as a purifying stage by reducing the production of tars and aerosols and simultaneously produces a rich H_2 syngas. Biomass is converted into syngas at temperature levels close to 1400 °C to reach gas characteristics compatible with a downstream Fischer–Tropsch synthesis. Technological tools able to heat an industrial gasifier with an external source are electric arc or plasma torches. With regard to the gases conversion/treatment, nontransferred plasma torches seem a mature technology with several industrial applications even if an optimization is necessary to adapt the torch to a special plasmagen gas (that means gas ionized by the arc). Oxygen and CO_2 were added in experiments as oxidizing mediums to reduce the production of solid carbon from the surplus of carbon in a treated biomass. Syngas with a high content of H_2 and carbon monoxide and a very low content of CO_2 was produced. A very low content of complex hydrocarbons and tar was detected (Hrabovsky et al., 2009).

H_2 production from biomass using nuclear fusion energy

This process potentially provides H_2 at significantly better efficiency than other proposed reactions, for example, electrolysis, from renewable sources with little or no CO_2 emission. Experimental results at 1000 °C showed complete gasification of cellulose, and H_2 was obtained at a conversion efficiency of 40% of material and 50% of heat (concept of a reactor and fusion-H_2 plant) (Kimura et al., 2006).

Sorption-enhanced catalytic steam gasification process

Single-stage production of high-purity H_2 from raw solid lignocellulosic biomass by sorption-enhanced catalytic steam gasification (CSG) has been demonstrated in a combined downdraft flow fluidized bed and fixed bed reactor. A Pd/Co–Ni catalyst derived from a hydrotalcite-like material (HT) and dolomite as a CO_2 acceptor are fed together with a biomass (chestnut wood sawdust) as the reactor feed. Almost pure H_2 (>99.9 vol%) and high H_2 yields (up to 90%) can be achieved by this process. The *in situ* removal of CO_2 by the carbonation reaction of dolomite shifts the equilibrium of steam reforming and WGS reactions toward H_2 production. The catalyst has shown high activity in the cleavage of C—C and C—H bonds during the conversion of tars and hydrocarbons resulting from fast pyrolysis and CSG, as a result of which further char formation is reduced (Fermoso et al., 2012).

Hydrothermal gasification

The main goal of hydrothermal biomass gasification is to benefit from the special properties of near (sub)- and supercritical water as a solvent and its presence as a reaction partner. Relatively fast hydrolysis of biomass in sub- and supercritical water leads to a rapid degradation of the polymeric structure of biomass. Consecutive reactions also are rather fast, as a result of which the gas is formed at low temperatures compared to "dry processes." The high solubility of the intermediates in the reaction medium, especially in water under supercritical conditions, inhibits tar and coke formation significantly: the reactive species originating from biomass are "diluted" by solvation in water and, consequently, the reaction rate of polymerization to unwanted products such as tar and coke is reduced. Altogether, this leads to high gas yields of hydrothermal gasification processes at relatively low temperatures (Kruse, 2009).

GASIFICATION UNDER SUBCRITICAL CONDITIONS

Elliott et al. (1994), at the Pacific Northwest National Laboratory, developed a process to gasify biomass under subcritical conditions (350 °C, 20 MPa) using a variety of catalysts. Rhodium, ruthenium, and nickel phases deposited on ZrO_2 (monoclinic), α-Al_2O_3, TiO_2 (rutile), and carbon. In general, the product gas consisted of more than 50% vol% CH_4, 40–50 vol% CO_2, less than 10 vol% H_2, and light hydrocarbons at trace levels.

Osada et al. (2004, 2008) investigated several supported noble metal catalysts for the gasification of lignin and propyl phenols. They suggested that the metal catalyzes the decomposition of lignin to lower molecular weight products, that is, alkylated phenols, and also causes the gasification of these phenolics. Usui et al. (2000) carried out a study on cellulose gasification at 200-400 °C using nickel catalysts in a stirred autoclave. The gas yield was found to be a strong function of the amount of nickel catalyst (Usui et al., 2000).

Subcritical water gasification by partial oxidation of glucose was carried out in the presence of various alkali catalysts: NaOH, KOH, $Ca(OH)_2$, Na_2CO_3, K_2CO_3, and $NaHCO_3$ in a closed batch reactor under 330 °C temperature and 13.5 MPa pressure. The H_2 gas yield in relation to the alkali catalyst was in the following order: NaOH > KOH > Ca$(OH)_2$ > K_2CO_3 > Na_2CO_3 > $NaHCO_3$. Results indicated that NaOH, KOH, and $Ca(OH)_2$ could promote biomass decomposition and improve the amount of H_2 product and inhibited and suppressed tar and char formation. In addition, results showed that hydrothermal gasification at a low temperature depended not only on the metal ion (Na^+, K^+, and Ca^{2+}), but also on their species and concentrations (Muangrat et al., 2010). The rate of H_2 production for the biomass samples was in the following order: glucose > cellulose, starch, rice straw > potato > rice husk (Onwudili and Williams, 2009).

The effect of additives, such as an inorganic alkali and a nickel catalyst, on the hydrothermal process was examined to generate H_2 from biomass with high selectivity at relatively low temperatures around 400 °C. At first, a cellulose sample as a model biomass was subjected to the hydrothermal process at 400 °C less than 25 MPa in the presence of an alkali (Na_2CO_3) and a nickel catalyst (Ni/SiO_2). Furthermore, the hydrothermal process of real biomass, such as wood chips, organic fertilizer, and food waste, in the presence of both the two additives, resulted in a highly selective production of H_2 even at 400 °C (Ishida et al., 2009).

GASIFICATION UNDER SUPERCRITICAL CONDITIONS

Supercritical water (SCW) gasification is the process in which water having a pressure of over 22 MPa and a temperature over 374 °C (i.e., supercritical conditions) is used as the gasifying agent. Modell (1985) was the first to demonstrate that wood could be gasified in supercritical water without the formation of char and tars at low conversions. In SCW technology, the biomass will be gasified almost completely, and the produced gas usually contains mainly H_2 and CO_2 with a very small fraction of CH_4 and CO, which are separated easily from the water phase by cooling to ambient temperature. The overall chemical conversion can be presented by the simplified reaction.

$$C_xH_yO_z + (2x - z)\,H_2O \rightarrow xCO_2 + (2x - z + y/2)H_2 \tag{11}$$

Thermodynamic calculations indicate that at temperatures above 600 °C only a gas rich in H_2, CH_4, CO, and CO_2, with no solid carbon, is formed. Cellulose is the most stable component of biomass but suffers rapid decomposition at temperatures somewhat below the critical temperature of water. At temperatures above 190 °C, a fraction of lignin and hemicelluloses reacts via solvolysis after only a few minutes of exposure to hot water. The initial products generated in solvolysis undergo several reactions, such as dehydration, isomerization, fragmentation, and condensation, finally forming gas and tars (Navarro et al., 2009).

Differing from the conventional process, SCW gasification only requires a single reactor and has the advantage that wet biomass can be used. It has a minimal dryness requirement and is particularly suitable for the high moisture biomass. There is high gasification efficiency. There is a high molar fraction of the H_2 in the gaseous products, meaning a lower cost for separation of the H_2 from the production gas. With SCW, the biomass decomposes rapidly into small molecules or gases; consequently, neither tar nor char occurs (Zhang, 2008).

High-temperature SCW gasification is conducted in the 500–800 °C range. Under these operating conditions, the yields of H_2 and CO_2 increase sharply, while that of CO follows the opposite trend. These results indicate a stronger water gas-shift activity at temperatures above 600 °C. The pressure of the process has little effect on either product gas composition or gasification efficiency within a wide range of pressures, including supercritical as well as subcritical pressures (60–400 bar). Additionally, the concentration of the feedstock has a major influence on the gas yield. This resulted in a drop of H_2 yield and carbon gasification efficiency when the concentration of the organic feedstock exceeded 5–10% (Peterson et al., 2008).

Yu et al. (1993) investigated supercritical gasification of wet biomass and glucose. They reported that complete gasification of glucose can occur at 600 °C, 34.5 MPa, and a 30-s residence time and also that Inconel walls of the reactor catalyze the water gas-shift reaction strongly.

Schmieder et al. (2000) and Kruse et al. (2000) on SCW gasification of a range of feedstocks demonstrated complete gasification to H_2 at 625 °C and 25 MPa in both batch and continuous tubular-flow reactors. They found that potassium compounds such as KOH and K_2CO_3 increased the yield of H_2 drastically.

In this work, influences of the $NaHCO_3$ and the $KHCO_3$ on the WGS reaction were investigated at 10 and 23 MPa and 230–300 °C. The presence of $NaHCO_3$ and $KHCO_3$ promotes the WGS reaction significantly and to the same extent (Akguel et al., 2012). In order to obtain valuable products such as monosaccharides and H_2 from rice straw (RS), two-stage processing— hydrothermal treatment in the first stage and steam gasification in the second stage—was carried out. The maximum yield of monosaccharides from the raw RS sample was 1.1 wt% (C basis). In the second stage, conversion of the hydrothermal-treated rice straw residue (HT-RSR) into H_2 was performed by steam gasification using a fixed-bed reactor, and influence of the nickel catalyst was examined. Total amount of H_2 evolved from samples loaded with nickel more than 2 wt% was 50–60 mmol/g-RSR and about three times larger than that from the HT-RSR sample without catalyst (Azadi and Farnood, 2011; Murakami et al., 2012).

Absorption-enhanced reforming (AER)

The *in situ* capture of CO_2 during gasification is an especially attractive process, as it allows for very high H_2 content with very low (near zero) CO_2 and tar contents in the gasification gas (Hausseman, 1994; Lin et al., 2004). In the AER process, because CO_2 produced during steam gasification is separated from the reactor by an adsorbent, the resulting product gas contains a high H_2 concentration and low concentration of carbon oxides. CO_2 absorption not only shifts the equilibrium toward the desired product, but also delivers heat for the endothermic reactions. A gas with increased CO_2 concentration is produced along the regeneration step, which simplifies CO_2 separation. A nearly nitrogen-free product gas with a caloric value of 12–14 MJ Nm^{-3} (dry) is produced in the AER gasifier.

Adsorbents and membranes for H_2 separation

The WGS reaction is used to increase the concentration of H_2 in syngas; large amounts of CO_2 are produced as well. Thus, for the process to achieve high yields and high purity, WGS has to be followed by some kind of H_2 separation technology. Pressure swing adsorption (PSA) is used to separate gas species from a mixture of gases. It is based on molecular sieves, manufactured so accurately that the pore diameter of the sieve allows the separation of different sized molecules of a syngas stream. The single most important feature of the PSA system is its ability to remove virtually any gas phase impurity to a required level and, as a consequence, reach very high levels of purity (99–99.99%). Membrane separation systems are based on the preferentially selective permeation of H_2 from a mixed gas stream through a polymer, metallic, or ceramic membrane. Metallic membranes—pure metals, metal alloys, or ceramic–metallic (cermet) composites—typically dissociate the H_2 molecule into atoms that pass through the metal film and the atoms recombine into H_2 molecules on the other side of the membrane. Dense ceramic membranes separate H_2 by transferring H_2 ions (i.e., protons) and electrons through the membrane and require a higher temperature operation to achieve adequate flux rates compared to other membrane technologies. Advancements in membrane separation technologies have the potential to reduce costs, improve efficiency, and simplify H_2 separation and purification systems. Many H_2 membrane separation systems are presently at the research phase and require further advancement to overcome technical challenges, such as limited durability due to material or catalyst failure in the presence of contaminants and/or thermal cycling, inability to consistently fabricate defect-free membranes on a large scale, and lack of proper high-pressure seals, among others.

Graphdiyne, a recently synthesized one-atom-thick carbon allotrope, is atomistically porous (characterized by a regular nanomesh) and suggests application as a separation membrane for H_2 purification. Graphdiyne provides a unique, chemically inert and mechanically stable platform facilitating selective gas separation at nominal pressures using a homogeneous material system, without needing chemical functionalization or the explicit introduction of molecular pores (Cranford and Buehler, 2012). It is shown that the two-stage arrangement using membranes with a H_2/CO_2 selectivity of 9 allows production of H_2 fuel with a H_2 content of 98% in vol (vol/vol) and H_2 recovery of around 75% (Makaruk et al., 2012).

Pyrolysis

Pyrolysis is the fundamental chemical reaction process that is the precursor of both the gasification and the combustion of solid fuels and is simply defined as the chemical changes occurring when heat is applied to a material in the absence of oxygen. Pyrolysis can be classified into mainly eight different categories, such as carbonization, conventional, fast, flash-liquid, flash-gas, ultra, vacuum, hydropyrolysis, and methanopyrolysis, based on process environment, parameters (residence time, temperature, heating rates), and products obtained (Demirbas, 2005). The most important reactor configurations are fluidized beds (circulating, bubbling), rotating cones, auger, vacuum, and ablative pyrolysis reactors. Fluidized beds and rotating cones are easier for scaling and possibly more cost-effective (Venderbosch and Prins, 2010). The higher yield (70–75%) of bio-oil could be possible with reactor configurations such as fluidized bed, circulating fluidized bed, and rotating cone reactor by proper selection of biomass feedstock. One of the approaches for the value addition of the pyrolysis bio-oil is to produce the bio-H_2 through various methods.

Single-step process/direct production of H_2-rich stream from biomass

Hydrogen can be produced in a single step from biomass pyrolysis using the impregnation of certain metal salts. Samples of nickel and iron metal-impregnated cellulose, hemicelluloses (beech wood xylan), and lignin were pyrolyzed to produce hydrogen. Beech wood samples with different metal contents (0.10, 0.26, and 0.65 mmol/g wood) were also tested to compare the effects of iron and nickel in the biomass. As a result of metal impregnation, significant changes in the yields of pyrolysis products, in the concentration of tar compounds (determined by GC-MS), and in the composition of the gaseous fraction were obtained. Nickel was more efficient than iron in rearrangement of the aromatic rings in the matrix, which contributed to the large increase in H_2 production observed with nickel-impregnated samples (Collard et al., 2012).

This method aims to utilize steam autogenerated from biomass moisture as a reactant to react with the intermediate products of pyrolysis to produce additional H_2. The heating rate is a key role in the process. The use of sweeping gas is unfavorable to H_2 production due to the reduced residence time of both the autogenerated steam and the volatile. Moisture content of biomass feedstock has a great effect on H_2 production. Under conditions of a fast heating rate and without the use of sweeping gas, the pyrolysis of biomass (wet biomass with a moisture content of 47.4%, wet basis) exhibits a higher H_2 yield of 495 ml/g, H_2 content of 38.1 vol%, and carbon conversion efficiency of 87.3% than those (267 ml/g, 26.9 vol%, and 68.2%) from the pyrolysis of B_{TD} (predried biomass with a moisture content of 7.9%, wet basis) (Hu et al., 2009).

Gasification of charcoal for production of H_2

Slow pyrolysis is being used for the production of charcoal, which can also be gasified to obtain hydrogen-rich gas. The short residence time pyrolysis of biomass (flash pyrolysis), at moderate temperatures, is being used to obtain a high yield of liquid products (up to 70% wt), particularly interesting as energetic vectors. Bio-oil can substitute for fuel oil or diesel fuel in many static applications, including boilers, furnaces, engines, and turbines for electricity generation. While commercial biocrudes can substitute easily for heavy fuel oils, it is necessary to improve the quality in order to consider biocrudes as a replacement for light fuel oils.

For transportation fuels, high severity chemical/catalytic processes are needed. An attractive future transportation fuel can be hydrogen, produced by steam reforming of the whole oil, or its carbohydrate-derived fraction. Pyrolysis gas containing a significant amount of carbon dioxide, along with methane, may be used as a fuel for industrial combustion. Presently, heat applications are most economically competitive, followed by combined heat and power applications; electrical applications are generally not competitive (Vamvuka, 2011).

Catalytic steam reforming (CSR) of bio-oil

Catalytic steam reforming is a widely used and thermodynamically proven process for H_2 production from biomass fast pyrolysis oil as a feedstock (Wang et al., 1998), followed by shift conversion and H_2 purification. Steam reforming can be performed using fixed or fluidized bed tubular reactor systems. Commercial catalysts used in the steam reforming of bio-oil or its fractions are designed specifically for use in a fixed bed, and attrition problems have been observed when they have been used in fluidized-bed reactors (Czernik et al., 2002). However, as a major shortcoming, this technique is accompanied by the formation of carbonaceous deposits over the catalyst surface, rendering it inactive and requiring frequent regeneration. Coke formation has been cited as the major disadvantage of bio-oil reforming, and it is more pronounced when Ni-based catalysts are used. The most important parameters in the steam-reforming process [Eq. (13)] are lower pressure, higher temperatures (700–1100 °C), steam-to-carbon molar ratio (S/C) of 5–20, and a gas hourly space velocity of 6000–32,000 h⁻1; in the presence of a metal-based catalyst (nickel), superheated steam (750 °C) reacts with bio-oil to yield carbon monoxide and H_2. High temperature and excess steam shift the equilibrium favorably and increase the rate of reforming. WGS to convert CO is carried out in a separate reactor operating at a lower temperature.

$$C_nH_mO_k + (n - k)H_2O \rightarrow nCO + (n + m/2 - k)H_2 \tag{12}$$

Additional H_2 can be recovered by a low temperature gas-shift reaction with the carbon monoxide produced:

$$nCO + nH_2O \leftrightarrow nCO_2 + nH_2 \tag{13}$$

The WGS reaction is an exothermic reaction that results in an increased reactor temperature, thereby decreasing the yield of H_2. The overall reaction can be shown as

$$C_nH_mO_k + (2n - k)H_2O \rightarrow nCO_2 + (2n + m/2 - k)H_2 \tag{14}$$

Overall, steam reforming is an endothermic reaction, and a high temperature can favor the reaction; some decomposition reactions may occur because of high temperature, which may result in coke formation that plugs the reactor and deactivates the catalysts and the Boudouard reaction can occur simultaneously:

$$C_nH_mO_k \leftrightarrow C_xH_yO_z + gases\ (H_2, H_2O, CO, CO_2, CH_4....) + coke \tag{15}$$

$$2CO \leftrightarrow CO_2 + C \tag{16}$$

The maximum H_2 yield of 68% with a fixed-bed reactor (under temperatures 750–850 °C, S/C 7–12, LHSV 0.8–1.5 h⁻¹) and 75% with a fluidized-bed reactor (under temperature 700–800 °C, S/C 15–20, LHSV 0.5–1.0 h⁻¹) could be possible (Lan et al., 2010). The S/C required in the fluidized bed was higher than that of the fixed bed, and the LHSV in the fluidized bed was lower than that of the fixed bed. The maximum H_2 yield obtained in the fluidized bed was 7% higher than that of the fixed bed.

Various Ni–Co bimetallic catalysts were employed to investigate their effects on H_2 production from steam reforming of bio-oil (from pyrolysis of sawdust at 450–500 °C). The catalyst can promote the generation of H_2 as well as the transformation of CO and CH_4 and plays an active role in steam reforming of bio-oil or gaseous products from bio-oil pyrolysis (Zhang et al., 2012).

The aqueous fraction of bio-oil, generated from fast pyrolysis, was catalytically steam reformed at 825 and 875 °C, high space velocity (up to 126,000 h^{-1}), and low residence time (26 ms). Using a fixed-bed microreactor interfaced with a molecular beam mass spectrometer, a variety of research and commercial nickel-based catalysts were tested. Since the main constraint in reforming bio-oils is catalyst deactivation caused by carbon deposition, two strategies were applied to improve the performance of the catalysts. The first approach aimed at enhancing steam adsorption to facilitate the partial oxidation, that is, gasification of coke precursors. The second one attempted to slow down the surface reactions leading to the formation of the coke precursors due to cracking, deoxygenation, and dehydration of adsorbed intermediates.

Fast pyrolysis followed by steam reforming

Fast pyrolysis of biomass followed by catalytic steam reforming of the pyrolysis oil or its fractions is another concept for the production of hydrogen. A major advantage of this route is the fact that bio-oil is easier and less expensive to transport than either biomass or hydrogen. Therefore, the processing of biomass and the production of H can be performed at separate locations, optimized with respect to feedstock supply and to H distribution infrastructure. This approach makes the process well suited for both centralized and distributed H production (Czernik et al., 2007).

Steam reforming using commercial nickel-based catalysts can efficiently convert volatile oil components to H_2 and carbon oxides. However, nonvolatile compounds such as sugars and lignin-derived oligomers tend to decompose thermally and to form carbonaceous deposits on the catalyst surface and in the reactor freeboard. To reduce this undesirable effect, we employed a fluidized bed reformer configuration with fine mist feed injection to the catalyst bed. The H_2 yields obtained from the carbohydrate-derived bio-oil fraction exceeded 80% of that possible by stoichiometric conversion. Although 90% of the feed carbon was converted to CO_2 and CO, carbonaceous deposits were formed on the catalyst surface, which resulted in gradual loss of its activity. The catalyst was regenerated easily by steam or CO_2 gasification of the deposits (Czernik et al., 1999).

H_2 production from the aqueous phase derived from fast pyrolysis of biomass was carried out by catalytic steam reforming in a fluidized-bed reactor. The H_2 yield of 64.6%, potential H_2 yield of 77.6%, and the carbon selectivity for product gases of 84.3% can be obtained at the optimized conditions of reaction temperature 800 °C, S/C 10, and WHSV 1.0 h^{-1} (Zhang et al., 2011).

Steam reforming of the aqueous fraction is an alternative process that increases the H_2 content of the syngas. Ni–Al catalysts modified with Ca or Mg used in the steam reforming of the aqueous fraction of pyrolysis liquids results in the deposition of coke on catalysts surface.

The catalyst composition influenced the quantity and type of coke deposits. Calcium improved the formation of carbonaceous products, leading to lower H_2/CO ratios, while magnesium improved the WGS reaction (Medrano et al., 2011).

The aqueous phase derived from fast pyrolysis was reformed with bio-oil to produce H_2 by using the Ni-base catalyst. The effects of reaction conditions such as steam-to-carbon ratio, temperature, and reaction time on catalytic reforming indicated that the developed process was effective and feasible (Li et al., 2009).

Autothermal catalytic conversion of bio-oil

The National Renewable Energy Laboratory developed a new route to produce H_2 from stabilized pyrolysis oil (stabilization with 10 wt% of methanol). This concept consists of consecutive volatilization ($\sim$400 °C), followed by oxidative cracking ($\sim$650 °C, addition of oxygen) and autothermal catalytic conversion to H_2 (addition of steam) to minimize secondary and tertiary tar formation. The H_2 is separated using a supported membrane from the H_2O, CO, CH_4, and CO_2, and the residual gas is combusted to supply the heat for the process. After the pyrolysis oil is evaporated, the vapors are oxidized further. These oxidized vapors are then highly reactive and converted easily by the steam-reforming catalyst at a relative low temperature. Czernik et al. (2007) studied the impact of the oxidation step of vapors and steam reforming of methanol-stabilized pyrolysis oil in the "conventional" fluidized bed setup; an 85 wt% of H_2 yield could be achieved with C11-NK catalyst and a S/C ratio $\sim$5.8.

Steam gasification of bio-oil/char slurry

Steam gasification is another route used to produce H_2 from bio-oil/char slurry. When char, a by-product of the fast pyrolysis process, is mixed with bio-oil, a bio-oil/char slurry with an even higher density is obtained. The typical procedure will follow steam gasification, followed by methanation and shift equilibria to improve the H_2 yields. Depending on the conditions, the reverse Boudouard reaction may also occur. Therefore, in the absence of solid carbon formation, the stoichiometric maximum yield of H_2 that can be obtained is $2 + (m - 2k)/2n$ moles per mole of carbon in the feed material, as mentioned in Eq. (14). From Eq. (14), 171 g of H_2 can be obtained from 1 kg of the bio-oil at the maximum yield (Sakaguchi, 2010). Dinjus et al. (2004) reported gasification of bio-oil/char slurry with oxygen carried out in a 3–6 MW entrained flow gasifier at 26 bar and temperature of 1200–1600 °C. Slurry consisted of bio-oil and 23–26 wt% of char and 3 wt% of straw ash with a char diameter of 10–1000 μm. The oxygen gas-feeding amount corresponded to $\lambda = 0.4$–0.6 (ratio of oxygen used to the stoichiometric amount for complete combustion). It is observed that complete carbon conversion (>99%) was obtained at a high gasification temperature (>1000 °C). Also, tar-free synthesis gas was obtained at 1200 °C.

The gasification reaction of biomass-derived oil was carried out at 800 °C under atmospheric pressure in an Inconel tubular fixed-bed down-flow microreactor using mixtures of CO_2 and N_2 and H_2 and N_2. Also, steam gasification was performed by feeding biomass-derived oil at a flow rate of 5 g/h along with steam (2.5–10 g/h) and nitrogen (30 ml/min) as a carrier gas. The gas product essentially consisted of H_2, CO, CO_2, CH_4, C_2, C3, and C_{4+} components. The composition of various gas components ranged as syngas ($H_2 + CO$) from 75 to 80 mol%, including 48–52 mol% H_2 and CH_4 from 12 to 18 mol%. Heating values of the product gas ranged between 460 and 510 Btu/Scf. The present study shows that there is a strong potential of making syngas, H_2, and medium-heating-value Btu gas from the steam gasification of biomass-derived oil (Panigrahi et al., 2003).

Pure H_2 production from pyrolysis oil using the steam-iron process

In the steam-iron process, relatively pure H_2 can be produced from pyrolysis oil in a redox cycle with iron oxides. Experiments in a fluidized bed showed that H_2 production from pyrolysis oil increases with increasing temperature during reduction. The experimental H_2 production at nearly 1000 °C with noncatalytic (blast furnace) and catalytic (ammonia synthesis) iron oxide was found to be 1.39 and 1.82 Nm^3 of H_2/kg of dry oil, respectively. The gasification of pyrolysis oil over an iron oxide bed results in an increased carbon-to-gas conversion compared to gasification over a sand bed. Near-complete gasification of oil is achieved when temperatures above 900 °C are applied in a fluidized-bed setup containing iron oxide (Bleeker et al., 2010).

Aqueous phase reforming (APR) of bio-oil

Aqueous phase reforming produces hydrogen from biomass-derived oxygenated compounds such as glycerol, sugars, and sugar alcohols. APR is unique in that reforming is done in the liquid phase. The process generates hydrogen without volatilizing water, which represents a major energy savings. Furthermore, it occurs at temperatures and pressures where the water–gas shift reaction is favorable, making it possible to generate hydrogen with low amounts of CO in a single chemical reactor. By taking place at low temperatures, the process also minimizes undesirable decomposition reactions typically encountered when carbohydrates are heated to elevated temperatures. In another mode, the reactor and catalysts can be altered to allow generation of high-energy hydrocarbons (propane, butane) from biomass-derived compounds (Vispute and Huber, 2009). Aqueous phase processing, developed by Davda and co-workers (2005), involves the selective conversion of sugars and polyols (ethylene glycol, 1,2-propanediol, 1,4-butanediol) to H_2. Production of H_2 and CO_2 from carbohydrates may be accomplished in a single-step, low-temperature process, in contrast to the multireactor steam-reforming system required for producing H_2 from hydrocarbons (Davda et al., 2005).

Aqueous phase reforming is conducted at pressures (typically 15–50 bar) where the H_2-rich effluent can be purified effectively using pressure-swing adsorption or membrane technologies; the CO_2 can also be separated effectively for either sequestration or use as a chemical. The catalytic pathway for the production of H_2 and CO_2 by APR of oxygenated hydrocarbons involves cleavage of C—C bonds, as well as C—H and/or O—H bonds, to form adsorbed species on the catalyst surface. Cleavage of these bonds occurs readily over group VIII metals, such as Pd and Rh (Mavrikakis and Barteau, 1998). Ethylene glycol and glycerol decompose on Pt to form adsorbed CO at room temperature (Gootzen et al., 1997).

The first step in this process is to add water to the bio-oil and separate it into aqueous and organic phases. The aqueous phase is then sent to a low-temperature hydrogenation unit where thermally unstable functionalities are hydrogenated to thermally stable compounds. In this hydrogenation step, aldehydes are converted to alcohols, sugars to sugar alcohols, and aromatics are hydrogenated. Undesired methane is also formed from this reaction. A key need in this low-temperature step is to reduce the amount of H_2 consumed. After the low-temperature hydrogenation step, either H_2 can be produced by aqueous phase reforming or aqueous phase dehydration/hydrogenation, respectively (Vispute and Huber, 2009). APR involves C—C bond cleavage (to produce CO), followed by the WGS reaction over supported metal catalysts, within the temperature range 200–250 °C and pressures of 10–50 bar. It is claimed that H_2 is produced at a H_2 selectivity of 60% from the water-soluble part of bio-oil, which is comparable to that observed for pure sorbitol at similar conversions (Vispute and Huber, 2009).

The fixed-bed reactor or high-throughput reactors can be used for aqueous phase reforming of oxygenated hydrocarbons. The reforming of oxygenated hydrocarbons (C:O=1:1) to form CO and H_2 is thermodynamically favorable at relatively lower temperatures. Side reactions such as methanation and dehydration/hydrogenation take place to produce C_1 to C_6 alkanes as by-products.

$$C_nH_{2y}O_n \rightarrow n\,CO + yH_2 \tag{17}$$

Also, the subsequent water–gas shift reaction, necessary to convert the carbon monoxide generated by reforming to CO_2, becomes increasingly favorable at lower temperature. The overall reaction for H_2 production from oxygenated hydrocarbon with C:O=1:1 by APR is the following:

$$C_xH_{2x}O_x + xH_2O \rightarrow 2xH_2 + x\,CO_2 \tag{18}$$

H_2 can be produced from sugars and alcohols at temperatures near 227 °C in a single-reactor aqueous phase-reforming process using a platinum-based catalyst. Catalytic aqueous phase reforming may prove useful for the generation of H_2-rich fuel gas from carbohydrates extracted from renewable biomass and biomass waste streams (Cortright et al., 2002). The selectivity of the reforming process depends on various factors, such as nature of the catalytically active metal, support, solution pH, feed, and process conditions.

Sequential cracking/reforming processes

The sequential cracking (Iojoiu et al., 2007) process alternates a cracking reaction step in which the bio-oil is transformed into a gas mixture containing H_2, CH_4, C_{2+}, CO, and CO_2 and a regeneration step allowing combustion of the coke deposited during the cracking step, which restores the catalytic activity of the catalysts. The advantage of this process is H_2 and CO_2 are produced in alternate steps, thereby reducing the energy requirement for H_2 purification (Vagia and Lemonidou, 2008).

It is generally acknowledged that the continuous production of rich H_2 is made using a sequential biomass pyrolysis reactor, which combines the pyrolysis of rice husks with the secondary decomposition of gaseous intermediate. The result shows that $Fe/\gamma\text{-}Al_2O_3$ could fully convert biomass pyrolysis volatile into gaseous products, such as H_2, CH_4, and CO (Xu et al., 2012).

A sequential process aiming at H_2 production was studied over two Ni-based catalysts, using crude beech wood oil as the feed. The process alternates cracking/reforming steps, during which a H_2+CO-rich stream is produced and carbon is stored on the catalyst, with regeneration steps where the carbon is combusted under oxygen. The two Ni-based catalysts exhibited good performances for H_2 production from bio-oil, with the gaseous products stream consisting in 45–50% H_2. The regeneration step was found fast and efficient, with the coke being readily combusted and the catalyst activity fully recovered (Davidian et al., 2007).

Electrochemical catalytic reforming (ECR)

In place of conventional catalytic steam reforming, a route of electrochemical catalytic reforming can be chosen to produce H_2 from bio oil. Ye et al. (2009) reported a new method for H_2 production using ECR of the bio-oil over the NiCuZn–Al_2O_3-reforming catalyst, giving well reforming performance, including a high H_2 yield and high carbon conversions >90%, even at a low reforming temperature (300–400 °C), whereas the same results were obtained with conventional catalytic steam reforming at >700 °C (Ye et al., 2009). An electrified Ni–Cr wire was used for heating the catalyst and synchronously providing the thermal electrons onto the catalyst during the bio-oil reforming, described as the ECR mode. The catalyst bed is heated supplementary by an outside furnace or cooled via a circulation water system to make a certain reforming temperature.

Chemical looping steam reforming (CLR)

The chemical looping steam reforming process is different from the conventional steam reforming process by cycling between fuel-steam feed and oxidation by air steps in order to achieve autothermality without depending on the oxygen feed for partial oxidation. The principle that works behind this process is an oxygen transfer material (OTM), which acts as a steam-reforming catalyst when in its reduced state. The advantage of CLR is it allows the steam-reforming reaction to be performed at a lower temperature than the conventional steam-reforming process and also overcomes the issue of coking of the catalyst associated with bio-oils while also using the heat generated to support the steam reforming in the subsequent fuel feed cycle. Lea-Langton et al. (2012) studied the feasibility of H_2 production from pine and pine empty fruit bunches pyrolysis oil. The reaction occurred during the fuel feed step in addition to steam reforming, and WGS is "unmixed combustion," that is, the direct reduction of NiO by the fuel.

$$C_nH_mO_k + (2n + m/2 - k)\, NiO \rightarrow nCO_2 + (m/2)\, H_2O + (2n + m/2 - k)\, Ni \qquad (19)$$

The oxidation reactions occurring during the air feed step are as follow:

$$C + O_2 \rightarrow CO_2 \qquad (20)$$

$$C + 0.5O_2 \rightarrow CO \qquad (21)$$

$$Ni + 0.5O_2 \rightarrow NiO \qquad (22)$$

Pine oil and palm empty fruit bunches (EFB) oil were investigated with a Ni/Al_2O_3 catalyst doubling as OTM. With a downward fuel feed configuration and using a H_2-reduced catalyst, maximum averaged fuel conversions of approximately 97% for pine oil and 89% for EFB oil were achieved at S/C ratios of 2.3 and 2.6, respectively (on a water-free oil basis). This produced H_2 with a yield efficiency of approximately 60% for pine oil and 80% for EFB oil notwithstanding equilibrium limitations and with little CH_4 by-product (Lea-Langton et al., 2012).

Secondary steam-reforming process

The pyrolysis of biomass was combined with the secondary decomposition of gaseous intermediates for H_2-rich gas production, with the avoidance of N_2 and CO_2 dilution to the energy density of gaseous effluents. In order to acquire the optimum conditions for H_2 generation, effects of operating parameters on this two-step decomposition of biomass were analyzed through simulation of thermodynamic equilibrium and experiments using the Ni/cordierite catalyst. A H_2 content of above 60% and a H_2 yield of around 65 g/kg biomass were achieved with optimized conditions (Zhao et al., 2010).

Catalytic severe/high temperature pyrolysis

Direct route or catalytic fast pyrolysis at higher temperature can be utilized for the production of H_2 from biomass. Demirbas (2001) reported that the largest H_2-rich gas yields of 70.3, 59.9, and 60.3% were obtained from olive husk, cotton cocoon shell, and tea waste, respectively, using about 13% $ZnCl_2$ as the catalyst at about 750 °C.

Using two different reactor configurations—a captive sample wire mesh reactor for fast pyrolysis and a fixed bed reactor for noncatalytic and catalytic pyrolysis—Zabaniotou et al. (2008) showed that pyrolysis in the semiablative-type reactor produced more H_2-rich gas than in the fixed-bed reactor, while the fixed-bed reactor configuration seemed to favor the production of liquid products. The fast pyrolysis experiments were carried out at a temperature range of 300–750 °C with a heating rate of 40–300 °C/s at atmospheric pressure in a helium environment. The yield and composition of H_2-rich gas depend mainly on the type of pyrolysis conditions and biomass composition.

Integrated gasification–pyrolysis process

The influence of biomass composition, temperature, steam/biomass mole ratio, and effect of CO_2 absorbent for the pyrolysis gasification of different waste biomasses, that is, palm oil fiber, palm oil shell, corncob, and coconut shell, has been investigated in a stainless-steel fixed-bed reactor from 327 to 927 °C. Results showed that H_2 and char production were related to the biomass composition. At higher gasification temperatures, more H_2, CO and CH_4 were formed, but the char production reduced at the same operating conditions. The addition of steam significantly promoted H_2 and CO production and also reduced the solid char product. Furthermore, H_2 production also increased as CO_2 was removed from the system using CaO as a CO_2 absorbent. It was shown that an optimum H_2 concentration was obtained at >727 °C, 1 bar, and steam/biomass mole ratio equal to 1 with *in situ* CO_2 removal for the palm oil fiber (Isha and Williams, 2011).

Plasma thermal conversion of bio-oil to H_2

Numerous processes exist or are proposed for the energetic conversion of biomass. The use of thermal plasma is proposed in the frame of the GALACSY project for the conversion of bio-oil to H_2 and carbon monoxide. For this purpose, an experimental apparatus has been built. The feasibility of this conversion at very high temperature, as encountered in thermal plasma, is examined both experimentally and numerically. This zero-dimensional study tends to show that a high temperature (around 2227 °C or above) is needed to ensure a high yield of H_2 (about 50 mol%) and about 95 mol% of CO + H_2. Predicted CO + H_2 yield and

CO/H_2 ratio are consistent with measurements. It is also expected that the formation of particles and tars is hampered (Guenadou et al., 2012).

EXERGY AND ECONOMIC ISSUES

Exergy is defined as the amount of work available from an energy source. The maximum amount of work is obtainable when matter and/or energy such as thermal energy is brought to a state of thermodynamic equilibrium with the common components of the environment in which this process takes place with the dead state by means of a reversible process, involving interaction only with the aforementioned components of nature (Szargut et al., 1988). Exergy is a measurement of how far a certain system deviates from a state of equilibrium with its environment. The quality of energy can be expressed as the quantity of exergy per unit of energy. The quality increases with the temperature (provided that the temperature is higher than that of the environment). It is clear that the exergy concept incorporates both quantitative and qualitative properties of energy. Every irreversible phenomenon causes exergy losses, leading to exergy destruction of the process or to an increased consumption of energy from whatever source the energy was derived. The objective of exergy analysis is to determine the exergy losses (thermodynamic imperfections) and to evaluate quantitatively the causes of the thermodynamic imperfection of the process under consideration. Exergy analysis can lead to all kinds of thermodynamic improvement of the process under consideration (Szargut et al., 1988). Since H_2 had to be produced, all processes in which electricity was produced were followed by electrolysis of water. Its exergetic efficiency was calculated to be 67.5%. A great impact on the final exergy efficiency of all processes was that of H_2 liquefaction. Because H_2 is transported in a liquid form, the liquefaction process is also taken into consideration in this work. The liquefaction process is very energy intensive and, as a consequence, requires a lot of exergy. H_2 liquefaction has a very low exergetic efficiency value of 13.3%. The exergy efficiency for electrolysis is taken to be the same for all processes using renewable energy sources. Process with the hydropower input has the highest exergy efficiency at 5.6% and the photovoltaic process has the lowest at 1.0%. The biomass gasification energy system has the second highest exergetic efficiency, reaching the level of 5.4%. Biomass gasification is a direct route to H_2 production, so there is no need for water electrolysis.

In the event that no liquefaction takes place, hydro process efficiency reaches the level of 41.6% and that of the photothermal process is equal to 8.6%. Exergy efficiencies before and after the liquefaction process have great fluctuations. If there was development of an efficient H_2 gas collection system, the use of renewable energy sources could be highly efficient, especially in the case of the use of hydropower, wind power, and biomass gasification. Even though there is low energy efficiency, the emissions of the processes are very low and there are no greenhouse gas emissions (Koroneos et al., 2004). It must also be noted that because renewable energy sources are free, exergy efficiency has an impact on the cost of the construction of the processes (Christopher and Dimitrios, 2012).

A superstructure for the production process of H_2 via gasification of switchgrass has been proposed embedding two technologies for the gasification (direct and indirect), two more for gas reforming (steam reforming and partial oxidation), and WGS to evaluate the trade-offs resulting from the use of either of the alternatives. The production process with the lowest cost ($0.68/kg of H_2) involves the use of indirect gasification with steam reforming of the hydrocarbon generated in the syngas. The cost of oxygen and the high consumption of steam penalize the direct gasification and partial oxidation alternatives. The solution also reveals that steam should be added to the WGS reactor to drive it to total conversion (Mariano and Ignacio, 2011).

OPPORTUNITIES AND CHALLENGES

Hydrogen offers the possibility to respond to all the major energy policy objectives (greenhouse gas emissions reduction, energy security, and reduction of local air pollution and noise) in the transport sector at the same time. If biological H_2 systems are to become commercially competitive, they must be able to synthesize H_2 at rates sufficient for practical application. Both thermocatalytic processes such as aqueous reforming of organic molecules and biological fermentation systems generate gas mixtures of H_2 and CO_2. Thus, all renewable H_2 production systems require the concomitant development and integration of H_2 purification and storage systems (David and Richard, 2010).

Gasification of biomass has been identified as a possible system for producing renewable H_2, which is beneficial to exploit biomass resources, to develop a highly efficient clean way for large-scale H_2 production, and to have less dependence on insecure fossil energy sources. Biomass gasification for hydrogen production, still at an early stage today, is expected to become the least expensive renewable hydrogen supply option in the coming decades. Biomass gasification is applied in small decentralized plants during the early phase of infrastructure rollout and in centralized plants in later periods. Compared with other biomass thermochemical gasifications, such as air gasification or steam gasification, supercritical water gasification can deal directly with the wet biomass without drying and have high gasification efficiency in a lower temperature (Yan et al., 2006). The major disadvantage of these processes is the decomposition of the biomass feedstock leading to char and tar formation (Swami et al., 2008). The major challenge for H_2 production by steam reforming of oxygenated hydrocarbons is the development of inexpensive catalysts with high conversion efficiencies. This is also the case for alkaline-enhanced reforming, with the added challenge that the carbon generated by the reforming processes is sequestered as a sodium carbonate precipitate, which creates problems with respect to catalyst fouling. Supercritical water partial oxidation generates clean H_2, but requires a significant energy input to attain the temperatures and pressures above a mixture's thermodynamic critical point. Hydrothermal technologies for the production of hydrogen from biomass may have energetic advantages, as a phase change to steam is avoided when water is heated at high pressures, which avoids large enthalpic energy penalties. Challenges include unknown or largely uncharacterized reaction pathways and kinetics, inadequate catalysts that do not withstand hydrothermal conditions, inadequate solid management practices that lead to precipitation of inorganic materials and can result in fouling and plugging issues, recovery of homogeneous catalysts, coking and deactivation of heterogeneous catalysts, wall effects, heat transfer and recovery, feedstock and impurities, and a need for specialized materials to withstand the high-temperature, high-pressure, and often corrosive environments of hydrothermal media (Peterson et al., 2008). Supercritical water gasification of biomass is safe, nontoxic, readily available, inexpensive, and environmentally benign. However, a high temperature and pressure are required to meet the minimum reaction condition. Therefore, the high operating cost has become the biggest obstacle to the development of this technology. Chars and tars formation may be the most significant technological problem. However, catalysis should be the solution to obtain higher yields of hydrogen and to decrease the amounts of chars and tars. The key point of the process is energy recovery, as the chemical reaction is endothermic and needs a high temperature and a rather large ratio of water/biomass.

H_2 production from biomass has major challenges and there are no completed technology demonstrations. It is believed that, in the future, biomass can become an important sustainable source of H_2. Due to its environmental merits, the share of H_2 from biomass in the automotive fuel market will grow fast in the next decade. Steam reformers and electrolyzers can also be scaled down and implemented on site at fueling stations (although still more expensive), while coal gasification or nuclear energy is for large-scale, central production only and therefore restricted to later phases with a high hydrogen demand.

Generally, the hydrogen production mix is very sensitive to the country-specific context and is influenced strongly by the assumed feedstock prices; resource availability and policy support also play a role, particularly for hydrogen from renewable and nuclear energy. The fossil hydrogen production option dominates the first two decades while the infrastructure is being developed and also later periods if only economic criteria are applied: initially on the basis of natural gas and subsequently with increasing gas prices more and more on the basis of coal. Renewable hydrogen is mainly an economic option in countries with a large renewable resource base and/or a lack of fossil resources, for remote and sparsely populated areas (such as islands), or for storing surplus electricity from intermittent renewable energies. Otherwise renewable hydrogen needs to be incentivized or mandated. To reduce overall H_2 cost, research is focused on reducing capital equipment, operations, and maintenance costs, as well as improving the efficiency of H_2 production technologies. Related research includes improving biomass growth, harvesting, and handling to reduce the cost of biomass resources used in H_2 production.

Catalysis plays an important role in the thermochemical hydrogen production process. New catalytic, low-cost, active components have to be explored that will make the process economically viable. Hybrid catalytic supports that can withstand high pressure and temperature can also be used. The reactor configuration should be designed to increase the heat transfer between the biomass particles. The chemical reaction engineering in the case of solid gas reactions is different from those of the liquid–liquid or liquid–gas systems.

In the present scenario, it is very difficult to decide the best candidate for all the available biomass in a nearby locality. It depends on the composition of the feedstock, end product utilization at the locality, size, and cost of process to be utilized. To decide the best process for a typical biomass feedstock, various other parameters, such as life cycle assessment, environmental impact, water, and carbon footprint, have to be considered.

Acknowledgments

The authors thank the director of the Indian Institute of Petroleum, Dehradun, for his constant encouragement and support. RS thanks the Council of Scientific and Industrial Research, New Delhi, India, for providing a Senior Research Fellowship.

References

Abatzoglou, N., Barker, N., Hasler, P., Knoef, H., 2000. The development of a draft protocol for the sampling and analysis of particulate and organic contaminants in the gas from small biomass gasifiers. Biomass Bioenergy 18, 5–17.

Akguel, G., Kruse, A., Sueleyman, D., 2012. Influence of salts on the subcritical water-gas shift reaction. J. Supercrit. Fluid 66, 207–214.

Antal Jr., J.M., Edwards, W.E., Friedman, H.L., Rogers, R.E., 1984. A study of the steam gasification of organic wastes. Project: Report to United States Environmental Protection Agency: Liberick, W.W. Project Officer.

Arauzo, J., Radlein, D., Piskorz, J., Scott, D.S., 1997. Catalytic pyrogasification of biomass. Evaluation of modified nickel catalysts. Ind. Eng. Chem. Res. 36, 67–75.

Asadullah, M., Ito, S., Kunimori, K., Yamada, M., Tomishige, K., 2002. Energy efficient production of hydrogen and syngas from biomass: Development of low-temperature catalytic process for cellulose gasification. Environ. Sci. Technol. 36, 4476–4481.

Asadullah, M., Miyazawa, T., Ito, S., Kunimori, K., Tomishige, K., 2003. Catalyst performance of $Rh/CeO_2/SiO_2$ in the pyrogasification of biomass. Energy Fuels 17, 842–849.

Azadi, P., Farnood, R., 2011. Review of heterogeneous catalysts for sub- and supercritical water gasification of biomass and wastes. Int. J Hydrogen Energy 36, 9529–9541.

Baker, E.G., Mudge, L.K., Brown, M.D., 1987. Steam gasification of biomass with nickel secondary catalysts. Ind. Eng. Chem. Res. 26, 1335–1339.

Barbier Jr., J., Duprez, D., 1994. Steam effects in three-way catalysis. Appl. Catal. B 4, 105–140.

Bartels, J.R., Pate, M.B., Olson, N.K., 2010. An economic survey of H_2 production from conventional and alternative energy sources. Int. J. Hydrogen Energy 35, 8371–8384.

Bartholomew, C.H., Farrauto, R.J., 2006. Fundamentals of Industrial Catalytic, second ed. Wiley, UK.

Blasi, C.D., Signorelli, G., Portoricco, G., 1999. Countercurrent fixed-bed gasification of biomass at laboratory scale. Ind. Eng. Chem. Res. 38, 2571–2581.

Bleeker, M.F., Veringa, H.J., Kersten, S.R.A., 2010. Pure H_2 production from pyrolysis oil using the steam-iron process: Effects of temperature and iron oxide conversion in the reduction. Ind. Eng. Chem. Res. 49, 53–64.

Bockris, J.O.M., 2002. The origin of ideas on a hydrogen economy and its solution to the decay of environment. Int. J. Hydrogen Energy 27, 731–740.

Boerrigter, H., Rauch, R., 2005. Review of applications of gases from biomass gasification. In: Knoef, H.A.M. (Ed.), Handbook Biomass Gasification. Biomass Technology Group (BTG), The Netherlands, pp. 211–230.

Bridgwater, A.V., Evans, G.D., 1993. An Assessment of Thermochemical Conversion Systems for Processing Biomass and Refuse. Energy Technology Support Unit (ETSU) on behalf of the Department of Trade, ETSU B/T1/00207/REP.

Caballero, M.A., Corella, J., Aznar, M.P., Gil, J., 2000. Biomass gasification with air in fluidized bed: Hot gas cleanup with selected commercial and full-size nickel-based catalysts. Ind. Eng. Chem. Res. 39, 1143–1154.

Chen, D., He, L., 2011. Towards an efficient H_2 production from biomass: A review of processes and materials. ChemCatChem 3, 490–511.

Chen, J.W., Lu, Y.J., Guo, L.J., Zhang, X.M., Xiao, P., 2010. H_2 production by biomass gasification in supercritical water using concentrated solar energy: System development and proof of concept. Int. J. Hydrogen Energy 35, 7134–7141.

Cheng, C.C., Lin, C.Y., Lin, M.C., 2002. Acid-base enrichment enhances anaerobic H_2 production process. Appl. Microbiol. Biotechnol. 58, 224–228.

Cheng, X., Shi, Z., Glass, N., Zhang, L., Zhang, J., Song, D., et al., 2007. A review of PEM H_2 fuel cell contamination: Impacts, mechanisms and mitigation. J. Power Sources 165, 739–756.

Chonga, M.L., Sabaratnamb, V., Shiraic, Y., Hassana, M.A., 2009. Biohydrogen production from biomass and industrial wastes by dark fermentation. Int. J. Hydrogen Energy 34, 3277–3287.

Christopher, K., Dimitrios, R., 2012. A review on exergy comparison of H_2 production methods from renewable energy sources. Energy Environ. Sci. 5, 6640–6651.

Cranford, S.W., Buehler, M.J., 2012. Biomass gsification for selective H_2 purification through graphdiyne under ambient temperature and pressure. Nanoscale 4, 4587–4593.

Collard, F.X., Blin, J., Bensakhria, A., Valette, J., 2012. Influence of impregnated metal on the pyrolysis conversion of biomass constituents. J. Anal. Appl. Pyrol. 95, 213–226.

Cortright, R.D., Davda, R.R., Dumesic, J.A., 2002. H_2 from catalytic reforming of biomass-derived hydrocarbons in liquid water. Nature 418, 964–967.

Czernik, S., French, R., Feik, C., Chornet, E., 1999. Fluidized bed catalytic steam reforming of pyrolysis oil for production of H_2. In: Overend, R.P., Chornet, E. (Eds.), Biomass: A Growth Opportunity in Green Energy and Value-Added Products. Proceedings of the Biomass Conference of the Americas, Oakland, CA, pp. 827–832.

Czernik, S., French, R., Feik, C., Chornet, E., 2002. Hydrogen by catalytic steam reforming of liquid byproducts from biomass thermoconversion processes. Ind. Eng. Chem. Res. 41, 4209–4215.

Czernik, S., Evans, R., French, R., 2007. H_2 from biomass: Production by steam reforming of biomass pyrolysis oil. Catal. Today 129, 265–268.

Dasgupta, D., Mondal, K., Wiltowski, T., 2008. Robust, high reactivity and enhanced capacity CO_2 removal agents for H_2 production applications. Int. J. Hydrogen Energy 33, 303–311.

Davda, R.R., Shabaker, J.W., Huber, G.W., Cortright, R.D., Dumesic, J.A., 2005. A review of catalytic issues and process conditions for renewable H_2 and alkanes by aqueous-phase reforming of oxygenated hydrocarbons over supported metal catalysts. Appl. Catal. B Environ. 56, 171–186.

David, B.L., Richard, C., 2010. Challenges for renewable H_2 production from biomass. Int. J. Hydrogen Energy 35, 4962–4969.

Davidian, T., Guilhaume, N., Iojoiu, E., Provendier, H., Mirodatos, C., 2007. H_2 production from crude pyrolysis oil by a sequential catalytic process. Appl. Catal. B Environ. 73, 116–127.

de Lasa, H., Salaices, E., Mazumder, J., Lucky, R., 2011. Catalytic steam gasification of biomass: Catalysts, thermodynamics and kinetics. Chem. Rev. 111, 5404–5433.

de Vrije, T., Claassen, P.A.M., 2003. Dark H_2 fermentations. In: Reith, J.H., Wijffels, R.H., Barten, H. (Eds.), Biomethanation and BioH$_2$: Status and Perspectives of Biological Methane and H_2 Production. Dutch Biological H2 Foundation, pp. 103–123.

Demirbas, A., 2001. Yields of hydrogen-rich gaseous products via pyrolysis from selected biomass samples. Fuel 80, 885–1891.

Demirbas, A., 2005. Pyrolysis of ground beech wood in irregular heating rate conditions. J. Anal. Appl. Pyrol. 73, 39–43.

Dinjus, E., Henrich, E., Weirich, F., 2004. A two stage process for synfuel from biomass. In: IEA Bioenergy Agreement Task 33: Thermal Gasification of Biomass. Vienna, May 3–5.

Elliott, D., Sealock Jr., L., Baker, E., 1994. Chemical processing in high-pressure aqueous environments. 3. Batch reactor process development experiments for organics destruction. Ind. Eng. Chem. Res. 33, 558–565.

Farrauto, R., Hwang, S., Shore, L., Ruettinger, W., Lampert, J., Giroux, T., et al., 2003. New material needs for hydrocarbon fuel processing: Generating H_2 for the PEM fuel cell. Annu. Rev. Mater. Res. 33, 1–27.

Fermoso, J., Rubiera, F., Chen, D., 2012. Sorption enhanced catalytic steam gasification process: a direct route from lignocellulosic biomass to high purity hydrogen. Energy Environ. Sci. 5, 6358–6367.

Gil, J., Corella, J., Aznar, M.P., Caballero, M.A., 1999. Biomass gasification in atmospheric and bubbling fluidized bed: Effect of the type of gasifying agent on the product distribution. Biomass Bioenergy 17, 389–403.

Gootzen, J.F.E., Wonders, A.H., Visscher, W., van Veen, J.A.R., 1997. Adsorption of C_3 alcohols, 1-butanol and ethane on platinized platinum as studied with FTIRS and DEMS. Langmuir 13, 1659–1667.

Guenadou, D., Lorcet, H., Peybernes, J., Catoire, L., Osmont, A., Gokalp, I., 2012. Plasma thermal conversion of bio oil for H_2 production. Int. J. Energy Res. 36, 409–414.

Hathaway, B.J., Davidson, J.H., Kittelson, D.B., 2011. Solar gasification of biomass: Kinetics of pyrolysis and steam gasification in molten salt. J. Solar Energy Eng. 133, 021011.

Hausseman, W.B., 1994. High-yield hydrogen production by catalytic gasification of coal or biomass. Int. J. Hydrogen Energy 19, 413–419.

Henderson, D., Taylor, A., 2006. The U.S. Department of Energy research and development programme on H_2 production using nuclear energy. Int. J. Nuclear Hydrogen Prod. Appl. 1, 51–56.

Hrabovsky, M., Hlina, M., Konrad, M., Kopecky, V., Kavka, T., Chumak, O., et al., 2009. Thermal plasma gasification of biomass for fuel gas production. High Temp. Mater. 13, 299–313.

Hu, G., Hao, H., Li, Y., 2009. H_2-rich gas production from pyrolysis of biomass in an auto generated steam atmosphere. Energy Fuel 23, 1748–1753.

Iojoiu, E.E., Domine, M.E., Nolven, T.D., Mirodatos, G.C., 2007. Hydrogen production by sequential cracking of biomass-derived pyrolysis oil over noble metal catalysts supported on ceria-zirconia. Appl. Catal. A Gen. 323, 147–161.

Isha, R., Williams, P.T., 2011. Pyrolysis-gasification of agriculture biomass wastes for H_2 production. J. Energy Inst. 84, 80–87.

Ishida, Y., Kumabe, K., Hata, K., Tanifuji, K., Hasegawa, T., Kitagawa, K., et al., 2009. Selective H_2 generation from real biomass through hydrothermal reaction at relatively low temperatures. Biomass Bioenerg. 33, 8–13.

Jacobs, G., Ricote, S., Patterson, P.M., Graham, U.M., Dozier, A., Khalid, S., et al., 2005. Low temperature water-gas shift: Examining the efficiency of Au as a promoter for ceria-based catalysts prepared by CVD of an Au precursor. Appl. Catal. A 292, 229–243.

Kimura, H., Takeuchi, Y., Yamamoto, Y., Konishi, S., 2006. H_2 production from biomass using nuclear fusion energy. In: IEEE/NPSS Symposium on Fusion Engineering 21st. pp. 496–499.

Koroneos, C., Dompros, A., Roumbas, G., Moussiopoulos, N., 2004. Life cycle assessment of H_2 fuel production processes. Int. J. Hydrogen Energy 29, 1443–1450.

Kruse, A., 2009. Hydrothermal gasification. J. Super Fluids 47, 391–399.

Kruse, A., Meier, D., Rimbretch, P., Schacht, M., 2000. Gasifcation of pyrocatecholin supercritical water in the presence of potassium hydroxide. Ind. Eng. Chem. Res. 39, 4842–4848.

Lan, P., Xu, Q., Zhou, M., Lan, L., Zhang, S., Yan, Y., 2010. Catalytic steam reforming of fast pyrolysis bio-oil in fixed bed and fluidized bed reactors. Chem. Eng. Technol. 33, 2021–2028.

Lea-Langton, A., Zin, R.M., Dupont, V., Twigg, M.V., 2012. Biomass pyrolysis oils for H_2 production using chemical looping reforming. Int. J. Hydrogen Energy 37, 2037–2043.

Lei, Y., Cant, N.W., Trimm, D.L., 2005. Activity patterns for the water gas shift reaction over supported precious metal catalysts. Catal. Lett. 103, 133–136.

Lei, Y., Cant, N.W., Trimm, D.L., 2006. The origin of rhodium promotion of Fe_3O_4–Cr_2O_3 catalysts for the high-temperature water-gas shift reaction. J. Catal. 239, 227–236.

Li, H., Xu, Q., Xue, H., Yan, Y., 2009. Catalytic reforming of the aqueous phase derived from fast-pyrolysis of biomass. Renew. Energy 34, 2872–2877.

Lin, S., Harada, M., Suzuki, Y., Hatano, H., 2004. Continuous experiment regarding hydrogen production by coal/CaO reaction with steam (1) gas products. Fuel 83, 869–874.

Liu, X., Ruettinger, W., Xu, X., Farrauto, R., 2005. Deactivation of Pt/CeO_2 water-gas shift catalysts due to shutdown/startup modes for fuel cell applications. Appl. Catal. B 56, 69–75.

Lu, Y., Zhao, L., Guo, L., 2011. Technical and economic evaluation of solar H_2 production by supercritical water gasification of biomass in China. Int. J. Hydrogen Energy 36, 14349–14359.

Makaruk, A., Miltner, M., Harasek, M., 2012. Membrane gas permeation in the upgrading of renewable H_2 from biomass steam gasification gases. Appl. Therm. Eng. 43, 134–140.

Mariano, M., Ignacio, E.G., 2011. Energy optimization of H_2 production from lignocellulosic biomass. Comput. Chem. Eng. 35, 1798–1806.

Mavrikakis, M., Barteau, M.A., 1998. Oxygenate reaction pathways on transition metal surfaces. J. Mol. Catal. A Chem. 131, 135–147.

Medrano, J.A., Oliva, M., Ruiz, J., Garcia, L., Arauzo, J., 2011. H_2 from aqueous fraction of biomass pyrolysis liquids by catalytic steam reforming in fluidized bed. Energy 36, 2215–2224.

Modell, M., 1985. In: Overend, R.P., Milne, T.A., Mudge, L.K. (Eds.), Fundamentals of Thermochemical Biomass Conversion. Elsevier Applied Science, New York, p. 95.

Muangrat, R., Onwudili, J.A., Williams, P.T., 2010. Influence of alkali catalysts on the production of H_2-rich gas from the hydrothermal gasification of food processing waste. Appl. Catal. B Environ. 100, 440–449.

Murakami, K., Kasai, K., Kato, T., Sugawara, K., 2012. Conversion of rice straw into valuable products by hydrothermal treatment and steam gasification. Fuel 93, 37–43.

Navarro, R.M., Pena, M.A., Fierro, J.L.G., 2007. H_2 production reactions from carbon feedstocks: Fossil fuels and biomass. Chem. Rev. 107, 3952–3991.

Navarro, R.M., Sanchez-Sanchez, M.C., Alvarez-Galvan, M.C., del Valle, F., Fierro, J.L.G., 2009. H_2 production from renewable sources: Biomass and photocatalytic opportunities. Energy Environ. Sci. 2, 35–54.

Ni, M., Leung, D.Y.C., Leung, M.K.H., Sumathy, K., 2006. An overview of H_2 production from biomass. Fuel Process Technol. 87, 461–472.

Onwudili, J.A., Williams, P.T., 2009. Role of sodium hydroxide in the production of H_2 gas from the hydrothermal gasification of biomass. Int. J. Hydrogen Energy 34, 5645–5656.

Osada, M., Sato, T., Watanabe, M., Adschiri, T., Arai, K., 2004. Low temperature catalytic gasification of lignin and cellulose with a ruthenium catalyst in supercritical water. Energy Fuels 18, 327–333.

Osada, M., Hiyoshi, N., Sato, O., Arai, K., Shirai, M., 2008. Subcritical water regeneration of supported ruthenium catalyst poisoned by Sulfur. Energy Fuel 22, 845–849.

Panigrahi, S., Dalai, A.K., Chaudhari, S.T., Bakhshi, N.N., 2003. Synthesis gas production from steam gasification of biomass-derived oil. Energy Fuel 17, 637–642.

Peterson, A.A., Vogel, F., Lachance, R.P., Fröling, M., Antal Jr., M.J., Tester, J.W., 2008. Thermochemical biofuels production in hydrothermal media: A review of sub- and supercritical water technologies. Energy Environ. Sci. 1, 32–65.

Pinto, C., Gulyurtlu, F., Cabrita, I., 2003. The study of reactions influencing the biomass steam gasification process. Fuel 82, 835–842.

Rapagna, S., Jand, N., Foscolo, P.U., 1998. Catalytic gasification of biomass to produce hydrogen rich gas. Int. J Hydrogen Energy 23, 551–557.

Rezaiyan, J., Cheremisinoff, N.P., 2005. Gasification Technologies a Primer for Engineers and Scientists. Taylor & Francis Group CRC Press.

Rhodes, C., Hutchings, G.J., 2003. Studies of the role of the copper promoter in the iron oxide/chromia high temperature water gas shift catalyst. Phys. Chem. Chem. Phys. 5, 2719–2723.

Rhodes, C., Peter, W.B., King, F., Hutchings, G.J., 2002. Promotion of Fe_3O_4/Cr_2O_3 high temperature water gas shift catalyst. Catal. Commun. 3, 381–384.

Robert, F.S., 2004. The carbon conundrum. Science 305, 962–963.

Sakaguchi, M., 2010. Gasification of Bio-Oil and Bio-Oil/Char Slurry. Vancouver Ph.D Thesis, University of British Colombia.

Schlapbach, L., Zuttel, A., 2001. Hydrogen-storage materials for mobile applications. Nature 414, 353–358.

Schlatter, J.C., Mitchell, P.J., 1980. Three-way catalyst response to transients. Ind. Eng. Chem. Prod. Res. Dev. 19, 288–293.

Schmieder, H., Abeln, J., Boukis, N., Dinjus, E., Kruse, A., Kluth, M., et al., 2000. Hydrothermal gasification of biomass and organic wastes. J. Supercrit. Fluids 17, 145–153.

Seitarides, T., Athanasiou, C., Zabaniotou, A., 2008. Modular biomass gasification-based solid oxide fuel cells (SOFC) for sustainable development. Renew. Sust. Energy Rev 12, 1251–1276.

Steele, B.C.H., Heinzel, A., 2001. Materials for fuel cell technologies. Nature 414, 345–352.

Stocker, M., 2008. Biofuels and biomass-to-liquid fuels in the biorefinery: Catalytic conversion of lignocellulosic biomass using porous materials. Angew. Chem. Int. Ed. 47, 9200–9211.

Sutton, D., Kelleher, B., Ross, J.R.H., 2001. Review of literature on catalysts for biomass gasification. Fuel Process. Technol. 72, 155–173.

Swami, S.M, Chaudhari, V., Kim, D.S., Sim, S.J., Abraham, M.A., 2008. Production of H_2 from glucose as a biomass simulant: Integrated biological and thermochemical approach. Ind. Eng. Chem. Res. 47, 3645–3651.

Szargut, J., Morris, D.R., Steward, F.R., 1988. Exergy Analysis of Thermal Chemical and Metallurgical Processes. Hemisphere Publishing Corporation.

Trovarelli, A., 1996. Catalytic properties of ceria and CeO_2-containing materials. Catal. Rev. Sci. Eng. 38, 439–520.

Turn, S.Q., 1999. Biomass integrated gasifier combined cycle technology: Application in the cane sugar industry. Int. Sugar J. 101, 267–272.

Turner, J.A., 1999. A realizable renewable energy future. Science 285, 687–689.

Turner, J.A., 2004. Sustainable H_2 production. Science 305, 972–974.

Usui, Y., Minowa, T., Inoue, S., Ogi, T., 2000. Selective hydrogen production from cellulose at low temperature catalyzed by supported group 10 metals. Chem. Lett. 29, 1166–1167.

Vagia, E.C., Lemonidou, A.A., 2008. Thermodynamic analysis of hydrogen production via autothermal steam reforming of selected components of aqueous bio-oil fraction. Int. J. Hydrogen Energy 33, 2489–2500.

Vamvuka, D., 2011. Bio-oil, solid and gaseous biofuels from biomass pyrolysis processes: An overview. Int. J. Energy Res. 35, 835–862.

Venderbosch, R.H., Prins, W., 2010. Fast pyrolysis technology. Dev. Biofuel Bioprod. Bioref. 4, 178–208.

Vispute, T.P., Huber, G.W., 2009. Production of hydrogen, alkanes and polyols by aqueous phase processing of wood-derived pyrolysis oils. Green Chem. 11, 1433–1445.

Wang, D., Czernik, S., Chornet, E., 1998. Production of hydrogen from biomass by catalytic steam reforming of fast pyrolysis oils. Energy Fuels 12, 19–24.

Warnecke, R., 2000. Gasification of biomass: Comparison of fixed bed and fluidized bed gasifier. Biomass Bioenerg 18, 489–497.

Xu, X., Jiang, E., Wang, M., Li, B., 2012. Rich H_2 production from crude gas secondary catalytic cracking over $Fe/\gamma-Al_2O$. Renew. Energy 39, 126–131.

Yan, Q., Guo, L., Lu, Y., 2006. Thermodynamic analysis of H_2 production from biomass gasification in supercritical water. Energy Convers. Manage. 47, 1515–1528.

Ye, T., Yuan, L., Chen, Y., Kan, T., Tu, J., Zhu, X., et al., 2009. High efficient production of hydrogen from bio-oil using low-temperature electrochemical catalytic reforming approach over NiCuZn–Al_2O_3 catalyst. Catal. Lett. 127, 323–333.

Youngblood, W.J., Lee, S.H.A., Maeda, K., Mallouk, T.E., 2009. Visible light water splitting using dye-sensitized oxide semiconductors. Accounts Chem. Res. 42, 1966–1973.

Yu, D., Aihara, M., Antal Jr., M.J., 1993. Hydrogen production by steam reforming glucose in supercritical water. Energy Fuels 7, 574–577.

Zabaniotou, A., Ioannidou, O., Antonakou, E., Lappas, A., 2008. Experimental study of pyrolysis for potential energy, hydrogen and carbon material production from lignocellulosic biomass. Int. J. Hydrogen Energy 33, 2433–2444.

Zhang, J., 2008. H_2 production by biomass gasification in supercritical water. Energeia 19, .

Zhang, Y.H.P., 2009. A sweet out-of-the-box solution to the H_2 economy: Is the sugar-powered car science fiction? Energy Environ. Sci. 2, 272–282.

Zhang, S.P., Li, X.J., Li, Q.Y., Xu, Q.L., Yan, Y.J., 2011. H_2 production from the aqueous phase derived from fast pyrolysis of biomass. J. Anal. Appl. Pyrol. 92, 158–163.

Zhang, Y., Li, W., Zhang, S., Xu, Q., Yan, Y., 2012. Steam reforming of bio-oil for H_2 production: Effect of Ni-Co bimetallic catalysts. Chem. Eng. Technol. 35, 302–308.

Zhao, B., Zhang, X., Sun, L., Meng, G., Chen, L., Yi, X., 2010. H_2 production from biomass combining pyrolysis and the secondary decomposition. Int. J. Hydrogen Energy 35, 2606–2611.

13

Bioreactor and Bioprocess Design for Biohydrogen Production

Kuan-Yeow Show*, Duu-Jong Lee[†]

*Department of Environmental Science and Engineering, Fudan University, Shanghai, China
[†]Department of Chemical Engineering, National Taiwan University, Taipei, Taiwan

INTRODUCTION

The current global economy and energy supply rely extensively on fossil fuels. This has resulted in an unprecedented increase in the atmospheric carbon dioxide concentration from the hypercombustion of fossil fuels and alarming depletion of nonrenewable fossil fuel resources. The fast escalating greenhouse gas emissions due primarily to carbon dioxide are considered the main culprit of global warming and climate change. To mitigate global warming and other environmental issues, substantial effort is being put in at the global level to explore renewable energy sources that could replace fossil fuels.

Hydrogen is a promising alternative biofuel to conventional fossil fuels because it releases explosive energy while producing water as the only by-product. It has been envisaged as the ultimate transport fuel for vehicles because of its nonpolluting nature with the use of fuel cells to convert chemical energy to electricity efficiently and gently (Forsberg, 2007). Most of the current hydrogen is produced from fossil fuels through thermochemical processes (hydrocarbon reforming, coal gasification, and partial oxidation of heavier hydrocarbons) (Das and Veziroglu, 2001; Levin et al., 2004). As the current hydrogen fuel is based on the use of nonrenewable fossil fuel resources with an escalating cost of production, a major concern of conventional hydrogen production is its sustainability.

Biohydrogen production is deemed to be a key development to a sustainable global energy supply and a promising alternative to fossil fuels, as it has the potential to overcome most of the problems and challenges that fossil fuels brought about. Using appropriate technologies, biohydrogen can become a desired green energy product of natural microbial conversion.

Current development of biological hydrogen (biohydrogen) production is focusing on biophotolysis of water, photodecomposition, and dark fermentation of organic compounds using various species of microorganisms. Dark biohydrogen fermentation using anaerobes appears to be the most favorable, as hydrogen could be generated at higher rates. In addition, the anaerobic conversion can be realized with various organic wastes and carbohydrate-enriched liquid effluents, thus achieving sustainable low-cost biohydrogen production with concomitant waste purification and renewable clean hydrogen energy.

While encouraging technological development of biohydrogen production has been demonstrated in laboratory studies, they are yet confined to laboratory- and pilot-scale studies. The technology is yet to compete with commercial hydrogen production from fossil fuels in terms of production cost, efficiency, and reliability for full-scale applications (Lee et al., 2011; Show et al., 2008, 2010, 2011). Improving the hydrogen yield and production rate are two major challenges for biohydrogen production. This chapter provides an overview of the state-of-the-art and perspectives of biohydrogen production research in the context of pathways, microorganisms, metabolic flux analysis, process design, and reactor system. Challenges and prospects of biohydrogen production are also outlined.

PATHWAYS OF BIOHYDROGEN PRODUCTION

There are a variety of biological processes in which hydrogen can be produced via microorganisms. In essence, the metabolic pathways can be differentiated into two distinct categories: light-dependent and light-independent processes. Light-dependent pathways include direct or indirect photolysis and photofermentation, whereas dark fermentation is the major light-independent process.

Direct Photolysis

Direct photolysis capitalizes on the photosynthetic capability of algae and cyanobacteria to split water directly into oxygen and hydrogen. Algae have evolved the ability to harness solar energy to extract protons and electrons from water via water-splitting reactions. Biohydrogen production takes place via direct absorption of light and transfer of electrons to two groups of enzymes—hydrogenases and nitrogenases (Manis and Banerjee, 2008). Under anaerobic conditions or when too much energy is captured in the process, some microorganisms vent excess electrons by using a hydrogenase enzyme that converts the hydrogen ions to hydrogen gas (Sorensen, 2005; Turner et al., 2008). It has been reported that the protons and electrons extracted via the water-splitting process are recombined by a chloroplast hydrogenase to form molecular hydrogen gas with a purity of up to 98% (Hankamer et al., 2007).

In addition to producing hydrogen, the microorganisms also produce oxygen, which in turn suppresses hydrogen production (Kapdan and Kargi, 2006; Kovacs et al., 2006). Research work has been carried out to engineer algae and bacteria so that the majority of the solar energy is diverted to hydrogen production, with bare energy diverted to carbohydrate production to solely maintain cells. Researchers are attempting to either identify or engineer less oxygen-sensitive microorganisms, isolate the hydrogen and oxygen cycles, or change the ratio of photosynthesis to respiration to prevent oxygen buildup (U.S. DOE, 2007). Addition of sulfate has been found to suppress oxygen production. However, the hydrogen production mechanisms are also inhibited (Sorensen, 2005; Turner et al., 2008).

The merit of direct photolysis is that the principal feed is water, which is readily available and inexpensive. While this technology has significant promise, it is also seeing tremendous challenges. Currently, this process requires a significant surface area to collect sufficient light. Another challenge is achieving continuous hydrogen production under aerobic conditions (U.S. DOE, 2007).

Indirect Photolysis

Alternatively, biohydrogen can be prepared through the use of some microorganisms (algae) that can directly produce hydrogen under certain conditions (Manis and Banerjee, 2008). Anaerobic cultures of green algae with energy derived from light under deprivation of sulfur would induce the "hydrogenase pathway" to produce hydrogen photosynthetically (Melis, 2002). In the course of such a hydrogen production condition (sulfur deprivation), cells consumed significant amounts of internal starch and protein (Zhang et al., 2002). Such catabolic reactions apparently sustain the hydrogen production process indirectly.

Indirect photolysis biohydrogen production via algae is deemed feasible if photon conversion efficiency can be improved via large-scale algal bioreactors. Algal bioreactors can provide an engineering approach to regulate light inputs to the culture to improve the photon conversion efficiency of algal cells. The improvement of photosynthesis efficiency is too difficult to achieve for conventional crop plants (Hankamer et al., 2007). Research has reported a substantial increase in light utilization efficiency of up to 15% compared with the previous utilization of around 5% (Laurinavichene et al., 2008; Tetali et al., 2007). Some researchers claim that efficiency between 10 and 13% is attainable by engineering the microorganisms to better utilize the solar energy (Turner et al. 2008).

Biohydrogen production can be improved via bioengineering (Hankamer et al., 2007). Several algae strains have been isolated and manipulated. Mutant algae with less chlorophyll can allow more light penetrating into deeper algae layers in the bioreactor. Hence sunlight is made available for more algal cells to generate hydrogen, thus improving the production rate.

It was estimated that if the entire capacity of the photosynthesis of algae could be diverted toward hydrogen production, 80 kg of hydrogen could be produced commercially per acre per day. With a realistic 50% capacity, the cost of producing hydrogen comes close to $2.80 a kilogram (Melis and Happe, 2001). Biohydrogen could compete with gasoline at this price, assuming 1 kg of hydrogen is equivalent to a gallon of gasoline. Currently, less than 10% of the algae photosynthetic capacity was utilized for biohydrogen production. Research is underway in improving further algae photosynthetic capacity using a molecular engineering approach. With technology advancement, biohydrogen production via algae bioreactors will offer a sustainable alternative energy resource in the future.

Photofermentation

In the photofermentation approach, photosynthetic microorganisms are able to convert solar energy to hydrogen directly from organic substrates. Despite the relatively lower yields of hydrogen photodecomposition of organic compounds by photosynthetic bacteria, the fermentative route is a promising method of biohydrogen production due to its higher rate of hydrogen evolution in the absence of any light source, as well as the versatility of the substrates used.

The photosynthetic system of purple bacteria is relatively simpler compared with green algae. It consists of only one photosystem, which is fixed in the intracellular membrane. The photosystem itself is not powerful enough to split water. Under anaerobic circumstances, however, these bacteria are able to use simple organic acids, such as acetic acid, or even dihydrogen sulfide as an electron donor (Akkerman et al., 2003).

Electrons released from the organic carbon are driven through substantial electron carriers, during which protons are pumped through the membrane. A proton gradient is developed and the ATP synthase enzyme generates ATP through the gradient. Extra energy in the form of ATP can be used to transport the electrons further to the electron acceptor ferredoxin. Under nitrogen-deficient circumstances, these electrons can be used by the nitrogenase enzyme with the extra energy from ATP to reduce molecular nitrogen into ammonium. In the absence of molecular nitrogen, this enzyme can, again with the extra ATP energy, reduce protons into hydrogen gas with the electrons derived from the ferredoxin (Akkerman et al., 2003). In this manner, various organic acids can be transformed into hydrogen and carbon dioxide.

Photofermentation by purple nonsulfur (PNS) bacteria is a major field of research through which the overall yield for biological hydrogen production can be improved significantly by the optimization of growth conditions and immobilization of active cells. Various processes of biohydrogen production using PNS bacteria along with several current developments had been examined. However, suitable process parameters, such as carbon and nitrogen ratio, illumination intensity, bioreactor configuration, and inoculum age, may lead to higher yields of hydrogen generation using PNS bacteria (Basak and Das, 2007).

Dark Fermentation

Among the biological biohydrogen production processes, dark fermentation or heterotrophic fermentation under anaerobic conditions seems to be more favorable. Hydrogen is yielded at a high rate, and various organic compounds and wastewaters are enriched with carbohydrates as the substrate results in low-cost hydrogen production (Hallenbeck and Ghosh, 2009).

The conventional anaerobic fermentation process has been well known since the production of methane, a useful energy carrier, and is associated with the degradation of organic pollutants in two distinct stages: acidification and methane production. Each stage is carried out by specific microorganisms through syntrophic interactions. Hydrogen is produced in the first stage as an intermediate metabolite, which in turn is used as an electron donor by many methanogens at the second stage of the process. It might be feasible to harvest hydrogen at the acidification stage of anaerobic fermentation, leaving the remaining acidification products for further methanogenic treatment. Therefore, it is possible that only acidogens are left to produce hydrogen gas, CO_2, and volatile fat acids (VFAs) if the final stage or methanogenesis and other hydrogen-consuming biochemical reactions are inhibited during the dark fermentation. Such a purpose is able to achieve through regulating biohydrogen cultures at a low pH and/or short hydraulic retention time (HRT) conditions (Kim et al., 2005; Mizuno et al., 2000) or through inactivating hydrogen consumers by heat treatment (Lay, 2001; Logan et al., 2002) and chemical inhibitors (Sparling et al., 1997; Wang et al., 2003b).

Currently, dark fermentation technologies are under the development at a laboratory scale to produce biohydrogen from wet biomass (e.g., molasses, organic wastes, sewage sludge) using anaerobic hydrogen-fermenting bacteria. Wastes and biomass rich in sugars and/or complex carbohydrates turn out to be the most suitable feedstocks for biohydrogen generation (Ntaikou et al., 2010). During dark fermentation, various organic acids are also produced. These compounds can subsequently be converted to hydrogen by a process denoted as photofermentation.

During dark fermentation, carbohydrates are converted into hydrogen gas and VFAs and alcohols, which are organic pollutants and energy carriers. For the purpose of energy production and protection of the water bodies, a second-stage process is necessary to recover the energy residues remaining in the effluent in the form of VFAs and alcohols. Thus the fermentative reactor becomes part of a process wherein the effluent post-treatment process and hydrogen utilization should also be included. A possible second-stage process is photofermentation, anaerobic digestion, or microbial fuel cells, which has been assessed elsewhere (Hawkes et al., 2007).

For hydrogen produced from dark fermentation to be used alone in an internal combustion engine or a fuel cell, some issues, such as biohydrogen purification, storage, and transport, need to be addressed. The gas produced is a mixture of primary hydrogen (generally less than 70%) and CO_2, but may also contain other gases such as CH_4, H_2S, ammonia (NH_4), and/or moisture. Purification of the hydrogen is essential before the hydrogen utilization can be practical. Relevant suggestions on hydrogen separation and purification can be found in the review literature (Das and Veziroglu, 2001; Hallenbeck and Benemann, 2002).

Although positive features of dark fermentation technology, such as high production rate, low energy demand, easy operation, and sustainability, have been demonstrated in laboratory studies (Claassen et al., 1999; Hallenbeck and Benemann, 2002; Hawkes et al., 2002; Kapdan and Kargi, 2006; Nandi and Sengupta, 1998; Nath and Das, 2004,), this technology is yet to compete with commercial hydrogen production processes from fossil fuels in terms of cost, efficiency, and reliability. Enhancing the hydrogen production efficiency is thus a major challenge to dark hydrogen fermentation. To achieve such a purpose, numerous studies have been conducted, and several hundreds of public reports were published during the past decade.

HYDROGEN-PRODUCING MICROORGANISMS

Biological processes are carried out largely at ambient temperatures and pressures, and hence are less energy-intensive than chemical or electrochemical ones. A large number of microbial species, including significantly different taxonomic and physiological types, can produce hydrogen. Biological processes use the enzyme hydrogenase or nitrogenase as hydrogen-producing protein. This enzyme regulates the hydrogen metabolism of uncountable prokaryotes and some eukaryotic organisms, including green algae. The function of nitrogenase, as well as hydrogenase, is linked with utilization of the products of photosynthetic reactions that generate reductants from water. In fact, hydrogen is utilized by the consortia as it is produced, mainly by methanogenic archaea, acetogenic bacteria, and sulfate-reducing bacteria (Adams et al., 1980).

Several algae mutant strains related to indirect photolysis biohydrogen production have been isolated, for example, the *Chlamydomonas reinhardtii* strain can increase biohydrogen production under high starch conditions (Hankamer et al., 2007). It is also reported that an algae mutant with less chlorophyll was manipulated for a large-scale commercial bioreactor. The algae mutant will absorb less light to allow more light penetrating into deeper algae layers. Hence sunlight is made available for more algal cells to generate hydrogen, thus improving the overall hydrogen production rate.

Research studies on dark fermentative organisms have been developed intensively in recent years, and some new or efficient bacterial species and strains for dark hydrogen fermentation have been isolated and recognized. A wide variety of dark fermentative bacteria are capable of hydrogen production, which are distributed across 10 of the 35 bacterial groups (Fang et al., 2002b). In a review on hydrogen-producing microorganisms (Nandi and Sengupta, 1998), such bacterial groups include strict anaerobes (*Clostridia*, *Methylotrophs*, methanogenic bacteria, Rumen bacteria, archaea), facultative anaerobes (*Escherichia coli* and *Enterobacter*), aerobes (*Alcaligenes* and *Bacillus*), photosysnthetic bacteria, and *Cyanobacteria*. In general, the isolated and identified mesophiles are mainly affiliated with two genera: facultative *Enterobactericeae* (Kumar and Das, 2000; Penfold et al., 2003; Tanisho and Ishiwata, 1995; Yokoi et al., 1995) and strictly anaerobic *Clostridiaceae* (Collet et al., 2004; Evvyernie et al., 2001; Wang et al., 2003a), whereas most thermophiles belong to genus *Thermoanaerobacterium* (Ahn et al., 2005; Ueno et al., 2001; Zhang et al., 2003).

In addition to the use of anaerobes in dark fermentation, aerobes such as *Bacillus* (Kalia et al., 1994; Kumar et al., 1995; Shin et al., 2004), *Aeromonons* spp., *Pseudomonos* spp., and *Vibrio* spp. (Oh et al., 2003) have been cultivated or isolated, but their hydrogen yields are generally less than 1.2 mol-H_2/mol-glucose under anaerobic conditions. Pure cultures of seed have been used to produce hydrogen from various substrates. *Clostridium* and *Enterobacter* were used most widely as inoculum for fermentative hydrogen production (Wang and Wan, 2009; Zhang et al., 2006b, 2008c). Most of the studies using pure cultures of bacteria for fermentative hydrogen production were conducted in batch mode and used glucose as the substrate (Wang and Wan, 2009).

Because facultative anaerobes can be grown more easily than obligate ones, attempts have been made to work with pure cultures of facultative anaerobes of genus *Enterobactericeae* (Kumar and Das, 2001; Kumar et al., 2000; Rachman et al., 1997; Tanisho and Ishiwata, 1995). Among the fermentative anaerobes, *Clostridia* have been well known and studied extensively, not for their hydrogen production capability, but for their role in the industrial solvent production from various carbohydrates. Some species such as *C. acetobutylicum* (Andersch et al., 1983; Holt et al., 1984; Kim et al., 1984) and *C. beijerinckii* (Chen and Blaschek, 1999) were used previously as industrial solvent producers. Use of *Clostridia* for hydrogen production attracts much attention. Hydrogen yields of pure cultures belonging to the genus *Clostridiaceae* have been examined, including *C. paraputrificum* (Evvyernie et al., 2001), *C. lentocellum* (Ravinder et al., 2000), *C. thermosuccinogenes* (Sridhar and Eiteman, 2001), *C. bifermentans*

(Wang et al., 2003a), *C. thermolacticum* (Collet et al., 2004), *C. butyricum* (Chen et al., 2005), *C. saccharoperbutylacetonicum* (Ferchichi et al., 2005), *C. acetobutylicum* (Zhang et al., 2006a), and *C. pasteurianum* (Dabrock et al., 1992). The optimum hydrogen yields observed of each species vary between 1.1 and 2.6 mol-H_2/mol-hexose, dependent on the organism per se as well as environmental conditions.

Use of undefined mixed cultures in nonsterile conditions would be more attractive for anaerobic hydrogen production, as the pure culture is contaminated easily by other bacteria, including hydrogen consumers such as methane-producing bacteria and sulfate-reducing bacteria (Adams et al., 1980; Ueno et al., 1995). There is a risk of hydrogen consumers coexisting in the mixed hydrogen-producing cultures. Several types of anaerobic bacteria are able to use hydrogen as a source of energy by coupling its oxidation to the reduction of a variety of electron acceptors. This group also includes facultative anaerobes such as *E. coli*, which is capable of hydrogen-dependent fumarate reduction, and *Paracoccus denitrificans*, which, although classified as an aerobic hydrogen bacterium, uses either O_2 or nitrate as a terminal electron acceptor during respiratory hydrogen oxidation. All these reactions are characterized by the uptake of hydrogen via a hydrogenase (Adams et al., 1980). Mixed culture derived from anaerobic reactor treating molasses-based wastewater subjected to multiple pretreatment processes showed better hydrogen production than the inoculum collected from a laboratory-scale anaerobic biofilm reactor (Venkata Mohan et al., 2009; Zhang et al., 2007a,c, 2008b).

Hydrogen yields by mixed thermophilic cultures are reported in a range of 0.7–3.2 mol-H_2/mol-hexose (Liu et al., 2003; Monmoto et al., 2004; Shin et al., 2004; Ueno et al., 1995, 1996; Valdez-Vazquez et al., 2005). The highest yield was obtained from organic solid wastes (60% fruit/vegetable waste and 40% office paper waste, based on dried weight) by a mixed culture at 55°C (Valdez-Vazquez et al., 2005), whereas the lowest one was achieved on fermenting starch-containing wastewater at 55°C (Zhang et al., 2003). Several reports on microbial hydrogen production using hyperthemophiles have been reported. For example, van Niel et al. (2002) obtained a hydrogen yield of 3.33 mol-H_2/mol-hexose using a pure culture of extremophiles—either *Caldicellulosiruptor saccharolyticus* on sucrose (70°C) or *Thermotoga elfi* on glucose (65°C)—and the same yield was reported by Kanai et al. (2005) from starch fermentation with the hyperthermophilic archaeon *Thermococcus kodakaraensis* KOD1 at 85°C. Moreover, it is stressed that a yield of 4 mol-H_2/mol-hexose, the maximal theoretical value, was reported by Schröder et al. (1994) with a batch eubacterial culture of *Thermotoga maritima* on glucose at 80°C. However, the authors pointed out that both microbial growth of the culture and glucose utilization are low. In addition, compared with mesophilic cultures, hyperthermophiles exhibit much lower production rates of hydrogen—generally ranging from 0.01 to 0.2 liter/liter·h—which is largely attributed to the slow-growing characteristics of hyperthermophiles.

Aerobic *Alcaligenes eutrophus* has been shown to grow heterotrophically on gluconate and fructose and produced hydrogen when exposed to anaerobic conditions (Kuhn et al., 1984). It contains a soluble NAD-reducing hydrogenase (Pinchukova et al., 1979). Aerobic hydrogen-producing *Bacillus licheniformis* was isolated from cattle dung (Kalia et al., 1994). It produced 0.5 mol H_2/mol glucose (Kumar et al., 1995). Immobilized cells had an average hydrogen yield of 1.5 mol per mole glucose and cells were stable for 60 days.

PROCESS DESIGN OF BIOHYDROGEN PRODUCTION

Batch Processes

In general, hydrogen-producing cultures were found to perform inefficiently when conducted under batch mode laboratory conditions (Li and Fang, 2007; Zhang et al., 2005) with relatively inferior hydrogen production rates (0.06–0.66 liter/liter·h). Enhancing hydrogen production efficiency is thus a major challenge to batch biohydrogen production systems. A semicontinuous mode for anaerobic hydrogen production from food waste was developed (Han and Shin, 2004). Pretreated seed sludge and food waste were fed into an anaerobic-leaching bed reactor, and dilution water was pumped continuously into the reactor at different dilution rates. Microbial conversion was deemed accomplished when biogas production ceased in about a week. It was found that appropriate control of the dilution rate could enhance fermentation efficiency by improving the degradation of nondegradable fractions of the food waste. The dilution rate may also delay the shift of predominant metabolic flow from a hydrogen- and acid-forming pathway to a solvent-forming pathway. In another study involving municipal solid wastes, it was reported that hydrogen can be produced steadily in reactors fed with substrate twice a week in a draw-and-fill mode in an anaerobic chamber and operated continuously at 35 and 55 °C for 40 days (Valdez-Vazquez et al., 2005).

A bioreactor operated under sequencing batch mode was used to evaluate the hydrogen productivity of an acid-enriched sewage sludge microflora at 35 °C (Lin and Chou, 2004). A 4-h cycle, including feed, reaction, settle, and decant stages, was employed to operate the 5-liter anaerobic sequencing batch reactor. The sucrose substrate concentration was kept at 20 g chemical oxygen demand (COD)/liter, and HRT was maintained initially at 12–120 h and thereafter at 4–12 h. The reaction to settling time ratio was maintained at 1.7. The hydrogenic activity of biomass microflora was HRT dependent, with peak activity recorded at an HRT of 8 h and an organic loading rate (OLR) of 80 kg COD/m^3 reactor liquid per day. Each mole of sucrose converted produced 2.8 mol of hydrogen, whereas each gram of biomass produced 39 mmol of hydrogen per day. An excessively low HRT might deteriorate hydrogen productivity. The concentration ratios of butyric acid to acetic acid, as well as volatile fatty acids and soluble microbial products to alkalinity, can be used as monitoring indicators for hydrogenic activity. Proper pH control was necessary for stable operation of the sequencing batch reactor.

Continuous Processes

Continuous stirred tank reactors (CSTRs) are used frequently for continuous hydrogen production (Chen and Lin, 2003; Majizat et al., 1997; Yu et al., 2002). Compared with batch reactor systems, hydrogen-producing bacteria in a CSTR are well suspended in mixed liquor and suffer less from mass transfer resistance. Because of its intrinsic structure and operating pattern, a CSTR is unable to maintain high levels of biomass inventory. Depending on the operating HRT, biomass inventory in a range between 1 and 4 g-VSS/liter is commonly reported (Chen and Lin, 2003; Horiuchi et al., 2002; Lin and Chang, 1999; Zhang et al., 2006b). Biomass washout is likely at short HRTs, resulting in deterioration in hydrogen production rates. Hydrogen production rates are thus restricted considerably by a low CSTR biomass retention and low hydraulic loading (Lay et al., 1999; Yu et al., 2002). The highest hydrogen production rate of CSTR culture fermenting sucrose with a mixed hydrogen-producing microflora was reported as 1.12 liter/liter·h (Chen and Lin, 2003).

Spontaneous granulation of hydrogen-producing bacteria can occur with shortened HRT in CSTR (Fang et al., 2002a; Oh et al., 2004b; Yu and Mu, 2006). Show et al. (2007) and Zhang et al. (2007c) found that the formation of granular sludge significantly increased overall biomass inventory to as much as 16.0 g-VSS per liter of reactor volume, enabling the CSTR to be operated at higher OLRs of up to 20 g-glucose/liter·h, which hence enhanced hydrogen production. In another study, it was reported that granular sludge disappeared within 3 weeks when CSTRs were incubated statically instead of being shaken (Vanderhaegen et al., 1992).

A reactor coupled with a membrane has been used to increase biomass concentration in the system. At a HRT of 3.3 h, Oh et al. (2004a) demonstrated that biomass concentration increased from 2.2 g/liter in a control reactor (without membrane fitted) to 5.8 g/liter in an anaerobic membrane bioreactor (MBR). This was achieved by controlling sludge retention time (SRT) at 12 h, corresponding to a marginal increase in the hydrogen production rate from 0.50 to 0.64 liter/literh. Increasing SRT can enhance biomass retention, which favors substrate utilization, but may result in a decrease in the hydrogen production rate. Li and Fang (2007) summarized that hydrogen production rates ranged between 0.25 and 0.69 liter/liter·h in the MBR systems reviewed. While the biomass concentration can be enhanced in a MBR, there is no clear advantage of a MBR over other types of efficient hydrogen production systems. In addition, membrane fouling and a high operating cost may limit the use of the MBR process in hydrogen fermentation.

Immobilized Cell Processes

In order to achieve higher biomass concentration and dilution rates without biomass wash-out, immobilized cell systems in the form of granules, biofilm, or gel-entrapped biomass particles have been developed (Show et al., 2007; Zhang et al., 2008c). Immobilized pure or mixed cultures of hydrogen-producing bacteria can be achieved by entrapment as biogels such as *C. butyricum* strain IFO13949 in agar gel (Yokoi et al., 1997a), *E. aerogenes* strain HO-39 in k-carrageenan, calcium alginate or agar gel, sewage sludge in calcium alginate beads, or alginate beads with added activated carbon powder, polyurethane, and acrylic latex/silicone (Wu et al., 2002), and sewage sludge and activated carbon powder fixed by eth-ylene vinyl acetate copolymer (Wu et al., 2003). Peak hydrogen production rates of continu-ous gel-immobilized sludge ranged from 0.090 to 0.93 liter/liter·h in a stirred chemostat and a fluidized bed reactor, respectively (Wu et al., 2005).

In comparison with gel-entrapped bioparticles, biofilm attached on solid and porous sup-port carriers seems to be superior for continuous hydrogen production. Hydrogen produc-tion and glucose consumption rates (Yokoi et al., 1997a) with *C. butyricum* immobilized on porous glass beads were higher than corresponding values with cells immobilized in agar gel at HRTs of 3 and 5 h in continuous cultures without pH control. In another study (Yokoi et al., 1997b), the hydrogen production rate of agar gel and porous glass beads devel-oped with *E. aerogenes* strain HO-39 culture increased with decreasing HRT, which was not observed with free cells culture. In the same study, the hydrogen production rate of glass beads was superior to that of agar gels at HRTs below 3 h, reaching the maximum rate at 0.85 liter/liter·h with a HRT of 1 h. Biofilm culture had an advantage in hydrogen production rate compared to that of gel-entrapped bioparticles with lower mass transfer efficiency and cell stability. It can be deduced that gel-entrapped biomass particles may not be the technology of choice for fermentative hydrogen production. However, there are also several operational and economic drawbacks of biofilm systems. Washout of support carriers might be an intrinsic problem of biofilm processes, especially when operating under low HRTs. Moreover, a large amount of support carriers is normally required to support microbial growth in biofilm processes. This leads to a considerable reduction in reactor effective volume for efficient biomass–substrate interactions, thereby resulting in lower reactor performance. A system upset might occur once interstitial void spaces in a pack bed reactor are clogged with biomass. The process stability of such a biofilm-based system may be challenged by long-term operation. As support carriers need to be replaced periodically due to wear and tear, the cost of carrier replacement can be a major economic consideration in maintenance.

Microbial self-aggregation, in which microbial cells are organized into dense and fast-settling granules, is studied extensively due to its practical importance in biohydrogen production. Granular sludge has some advantages over biofilm sludge in continuous dark hydrogen fermentation. Possible advantages of microorganisms in anaerobic granule in comparison with flocculated or suspended counterparts are a higher retention of granulated biomass in a reactor, diversity of physiological functions of microorganisms in the granule, and resistance of the microorganisms within the granule to toxic substances. It is favorable for the microorganisms to be very close to each other in the granule in order to achieve a high substrate conversion to hydrogen. Aggregation leads to a heterogeneous community and facilitate syntrophic relationships, especially interspecies metabolic transfer, and under unfavorable conditions for growth (e.g., toxicity), a more favorable microenvironment can be maintained within the aggregates so that metabolism can be sustained.

Biogranulation is a gradual process from seed sludge to compact aggregates, then to granular sludge, and finally to mature granules. A major drawback of granulation is its long startup period and development of active biogranules for efficient functioning of the process. A complete development of hydrogen-producing granules may take several months (Fang et al., 2002a; Yu and Mu, 2006). During startup of an UASB hydrogen-producing reactor, Mu and Yu (2006) found that small granules (diameter 400–500 µm) were formed at the reactor bottom after 140 days of operation. Granules developed further to sizes larger than 2.0 mm upon 200 days. Although the reactor reached steady-state hydrogen production and substrate degradation after 5 months of startup operation, development and accumulation of mature and stable granular sludge were only completed beyond 8 months of operation. Chang and Lin (2004) noted that a UASB reactor took 39 days to achieve constant gas production at a HRT of 24 h and granules became visible after 120 days of operation. A longer period (180 days), however, was required for further development of granules.

Granulation of hydrogen-producing cultures can be accelerated markedly. Packing of a small quantity of carrier matrices significantly accomplished sludge granulation within 80–290 h in a novel carrier-induced granular sludge bed bioreactor (Lee et al., 2006). Granulation of seed sludge could take place in all carrier-packed reactors as HRTs were shortened to 4–8 h, dependent on carrier type. By adding cationic polymer (cationic polyacrylamide) and anionic polymer (silica sol), rapid granulation of hydrogen-producing culture could be accomplished within 5 min (Kim et al., 2005). Zhang et al. (2008a) developed an approach of acid incubation to initiating formation of hydrogen-producing granules rapidly in a CSTR. Hydrogen-producing granules were formed rapidly within 114 h as the seed microbial culture was subjected to a 24-h period of acid incubation at a pH of 2.0. Changing culture pH would result in an improvement in surface physicochemical properties of culture favoring microbial granulation.

BIOHYDROGEN REACTOR SYSTEMS

Photobioreactors

Different types of photobioreactors have been reviewed by Janssen (2002). Small-scale flat panel reactors consisting of a rectangular transparent box were mixed with gas introduced via a perforated tube at the bottom of the reactor. In order to create a high degree of turbulence, 3 to 4 liter of air per liter of reactor volume per minute has to be provided. The panels were illuminated from one side by direct sunlight and the panels were placed vertically, or inclined toward the sun. Light/dark cycles were short in flat panel reactors, which is probably the key factor leading to the high photosynthetic efficiency. A disadvantage of flat panel reactors systems is that the power consumption of aeration (or mixing with another gas) is high, although mixing is always necessary in any reactor. The large-scale flat plate reactor consisted of a rectangular air-lift photo bioreactor with a large number of light redistributing plates fixed a few centimeters from each other. Mixing was provided by air injected between adjacent plates and the culture liquid rises in between.

Tubular photo bioreactors consist of long transparent tubes with diameters ranging from 3 to 6 cm and lengths ranging from 10 to 100 m (Janssen 2002). The culture liquid is pumped through these tubes by means of mechanical or air-lift pumps. The tubes can be positioned in many different ways: in a horizontal plane as straight tubes with a small or large number of U bends; vertical, coiled as a cylinder or a cone; in a vertical plane, positioned in a fence-like structure using U bends or connected by manifolds; or horizontal or inclined parallel tubes connected by manifolds; in addition, horizontal tubes can be placed on different reflective surfaces with a certain distance between the tubes. Although the tubular reactor design is rather diverse, the predominant effect of the specific designs on the light regime is a difference in the photon flux density incident on the reactor surface. The shape of the light gradient in the tubes is similar in most designs. Also with respect to liquid mixing, the circumstances in most designs are similar. The length of the tubes is limited because of accumulation of gas, although this might not be so important for nitrogenase-based processes, as they may be less inhibited by hydrogen. The way to scale up is to connect a number of tubes via manifolds. Flat panel reactors normally show a high photochemical efficiency or biomass yield on light energy, while biomass density is also high. Tubular reactors in theory should show better efficiencies because of the shorter average light/dark cycles.

Dark Fermentation Bioreactors

The influence of reactor configuration on hydrogen production was evaluated by examining CSTR and anaerobic fluidized-bed reactor (AFBR) systems with granular sludge for hydrogen production (Show et al., 2007, 2010; Zhang et al., 2007b, 2008d). Maximum hydrogen production rates of the suspended sludge, CSTR granular sludge, AFBR granular sludge, and biofilm sludge were essentially correlated to the biomass concentration, indicating that the system performance is largely influenced by reactor biomass retention. Compared to the CSTR, much more biomass retention was retained in the AFBR reactor (35 g-VSS/liter vs 16 g-VSS/liter), which might be attributed to the reactor intrinsic structure. A larger height/section diameter ratio of the column reactor, such as AFBR, may be an advantage favoring solid/liquid separation and biomass retention, and hence hydrogen production. It should be stressed that Wu et al. (2006) reported the highest hydrogen production rate of up to 15 liter/literh, which was achieved by a granule-based CSTR system at a hydraulic retention time of 0.5 h. The biomass concentration of the CSTR culture maintained at a level of 32.5 g-VSS/liter, which was comparable to that of the present AFBR system, but was two times higher than that of the present CSTR.

Additionally, because the recirculation flow rate was kept consistently at 200 ml/min, its effect on mass transfer and reactor performance was not examined in the present study. Nevertheless, it is possible that an increasing liquid recirculation ratio might improve mass transfer of the immobilized culture and substrate in a column reactor. A fixed- or packed-bed reactor is operated under conditions with a lesser extent of hydraulic turbulence, thus its immobilized cultures usually encounter mass transfer resistance, which would result in inferior rates of substrate conversion and hydrogen production. Rachman et al. (1998) found that a high hydrogen molar yield could not be maintained consistently in a packed-bed reactor, although the pH in the effluent was controlled at more than 6.0. This was due to pH gradient distribution along the reactor column, resulting in a heterogeneous distribution of microbial activity. In order to overcome the mass transfer resistance and pH heterogeneous distribution, they recommended that a fluidized-bed or an expanded bed reactor system with recirculation flow is more appropriate for further enhancing the hydrogen production rate and yield. In a later study, Kumar and Das (2001) observed that hydrogen production and substrate conversion rates of a packed-bed reactor both increased with the recycling ratio and obtained the maximum hydrogen production rate of 1.69 l/liter·h at a recirculation ratio of 6.4. This means that the mass transfer resistance in a packed-bed reactor can be reduced by increasing the slurry recycle ratio.

An appropriate reactor configuration would help increase hydraulic turbulence of the column reactors and result in an increase in hydrogen production. Kumar and Das (2001) investigated hydrogen production by attaching *E. cloacae* on coir in packed bed reactors with different configurations, that is, tubular, tapered, or rhomboid shape at a HRT of 1.08 h. The comparative study indicated that the rhomboid bioreactor with a convergent–divergent configuration has a maximum hydrogen production rate of 1.60 liter/liter·h as compared with 1.46 liter/liter·h in a tapered reactor and 1.40 liter/liter·h in a tubular reactor. The enhancement in the hydrogen production rate could be attributed to higher turbulence caused by reactor geometry favoring mass transfer and reduced gas hold-up.

Microbial Electrolysis Cells

Microbial-aided electrolysis cells (MEC), also called a bioelectrochemically assisted microbial reactor, use electrohydrogenesis to directly convert biodegradable material into hydrogen (Call and Logan, 2008; Cheng and Logan, 2007; Ditzig et al., 2007). The MEC is a modified microbial fuel cell. In a microbial fuel cell, special microbial exoelectrogens decompose organic material and transfer electrons to the anode. The electrons combine at the cathode, after traveling through an external load, with protons and oxygen forming water (Holladay et al., 2009). An MEC operates in anaerobic state, and an external voltage is applied to the cell rather than generated by it. The added energy is required, as acetate substrate decomposition is not spontaneous under standard conditions (Call and Logan, 2008; Cheng and Logan, 2007; Ditzig et al., 2007). Hydrogen gas production occurs at the cathode via reaction of hydrogen ion with electrons.

Operation of MECs is, in fact, versatile in the sense that suitable microbial cultures can be selected during operation, and a variety of substrates such as wastewater can be metabolized. Because of the versatility of the microbial community, MECs are not confined to using fermentation metabolites as substrate but can also decompose glucose or glucose-containing substrates such as cellulose to hydrogen (Ditzig et al., 2007). Achieving satisfactory efficiency of MEC with nominal electrical inputs, however, is still a major challenge. Hydrogen production rates are still substantially lower than those derived from dark fermentations at relatively high voltages (800 mV). There remain substantial challenges to the practical implementation of this technology, including the replacement of expensive platinum electrode, high current densities, and reduction of the electrical input requirement.

Hybrid Bioreactors

Hybrid two-stage systems are operated on the notion that conversion of substrate to hydrogen and organic acids takes place in a conventional reactor in a first stage of process and additional gaseous energy in the form of methane or hydrogen is extracted in the second stage (Koutrouli et al., 2009). In optimizing gas production, the second-stage reactor is operated under different conditions, such as higher pH and longer HRT (Hwang et al., 2011). Applications of photofermentation (Kapdan and Kargi, 2006) or fuel cell (Garcia-Pena et al., 2009) for maximizing the overall energy extraction in the second stage have been reported. Additional hydrogen was recovered from metabolites of the dark fermentation in the studies. In another approach, MECs were incorporated into the second stage, in which electricity applied to a microbial fuel cell provided the energy needed in converting organic acids to hydrogen (Call and Logan, 2008; Wang et al., 2011). In principle, a second-stage MEC could produce 12 H_2/glucose with only a small input of energy.

A hybrid two-stage photofermentation system in which the second stage was fed with effluent from a hydrogen-producing reactor has been investigated (Nath and Das, 2006). Hydrogen yields well below the hypothetical values were reported. Conversely, it was found that a combination of dark and photofermentation in a hybrid two-stage system enhanced the overall yield of hydrogen (Nath and Das, 2006). In the first stage, the biomass is fermented to acetate, carbon dioxide, and hydrogen in a thermophilic dark fermentation. In the second-stage photobioreactor, acetate is converted to hydrogen and carbon dioxide. Enhancement of the hybrid system was attributable to maximal utilization of the substrate with improved thermodynamics. The combination of dark and photofermentation has the potential to almost attaining the theoretical maximum production of 12 mol of hydrogen per mol glucose equivalent. Theoretically, complete conversion of organics into hydrogen via photofermentations is possible due to the fact that hydrogen production is driven by ATP-dependent nitrogenase. The required ATP is formed with energy derived from sunlight. While research work has so far indicated encouraging findings, the main technical barrier to practical application of photofermentation lies with low photosynthetic efficiencies with concomitant low hydrogen yields under moderate- to high-light intensities.

The hybrid two-stage system has been scaled up to the pilot plant stage (Ueno et al., 2007). The pilot system offers several merits over conventional methanogenesis-enhanced substrate solubilization, and tolerance to high OLRs is among the merits reported. Successful operation of such pilot systems using actual wastes has also been documented. A pilot-scale fermentation of kitchen waste generated relatively high production rates of $5.4 \, m^3 \, H_2 \, m^{-3} \, day^{-1}$ and $6.1 \, m^3 \, CH_4 \, m^{-3} \, day^{-1}$ while achieving a COD removal of 80% (Ueno et al., 2007). The superior performance of this hybrid system is illustrated from the methane yields, which were twofold higher than a single-stage process (Ueno et al., 2007).

Multistage Bioreactors

Multistage bioreactors involving three or even four stages have been reported in biohydrogen production (Show et al., 2011; Wang et al., 2011; U.S. DOE, 2007). Sunlight is first filtered through first-stage direct photolysis in which visible light is utilized by blue-green algae, and an unfiltered infrared ray is used by photosynthetic microbes in the second-stage photofermentative reactor. Effluent from the second-stage photofermentation, together with the biomass feedstock, is fed into a third-stage dark fermentation reactor where bacteria convert the substrate into hydrogen and organic acids. As the effluent is enriched with organic acids, a supply of external organic acids for the photofermentative process can be eliminated. The fourth stage involves use of a MEC to convert the organic acids generated from dark fermentation into hydrogen under the light-independent process. It thus can be operated during the night or low-light conditions.

Integration of multiple biochemical conversion processes poses significant challenges for multistage reactor engineering, system design, process control, operation, and maintenance. Major challenges with the simultaneous production of hydrogen and oxygen from photolytic hydrogen production include respiration to photosynthetic capacity ratio, coculture balance, and concentration and processing of cell biomass (Holladay et al., 2009).

CHALLENGES AND THE WAY FORWARD

Low hydrogen yields and production rates are two major challenges for practical application of biohydrogen production. Major efforts are underway to increase the hydrogen yield substantially, considering that less than 15% of the energy from the organic source can typically be obtained in the form of hydrogen (Logan, 2004). The U.S. DOE (2007) program goal for hydrogen fermentation technology is to realize yields of 4 and 6 mol hydrogen per mole of glucose by 2013 and 2018, respectively, as well as to achieve 3 and 6 months of continuous operation for the same years. On this perspective, some integrated strategies have been investigated, such as the two-step fermentation process (acidogenic + photobiological or acidogenic +methanogenic processes) or the use of modified microbial fuel cells (de Vrije and

Claassen, 2003; Ueno et al., 2007). Through these coupled processes, more hydrogen or energy per mole of substrate can be achieved in the second stage.

Enhancement in hydrogen yield may be possible using a suitable microbial strain, process modification, efficient bioreactor design, and genetic and molecular engineering technique to redirect the metabolic pathway. Applying genetic and metabolic engineering techniques to improve the hydrogen yields of those microbial cultures with higher hydrogen production rates may also be another feasible option (Chittibabu et al., 2006; Nath and Das, 2006). Metabolic flux analysis (MFA) may serve as an important tool to guide a priori most suitable genetic modifications oriented to a hydrogen yield increase. MFA or flux balance analysis has been used to provide valuable information for the optimization and design of the fermentative hydrogen production process (Cai et al., 2010; Kim et al., 2009; Manish et al., 2007; Navarro et al., 2009; Niu et al., 2011; Oh et al., 2008; Pasika et al., 2011). It should be noted that the improvement of hydrogen production by gene manipulation or modification is focusing mainly on the disruption of endogenous genes and not introducing new activities in the microorganisms. New pathways need to be discovered to directly take full advantage of the 12 mol of hydrogen available in a mole of hexose.

Another challenge of hydrogen fermentation is maintaining stable production. Unstable hydrogen production is possibly attributed to the metabolic shift of hydrogen-producing bacteria, which could be minimized by an in-depth microbial study. A major technical prerequisite for stable and efficient hydrogen fermentations is the maintenance of low hydrogen partial pressures through continuous removal of hydrogen from the fermentation broth. An in-depth microbial study is also needed to shed light on the metabolic shift of hydrogen-producing bacteria causing unstable biochemical fermentation.

At the current technology development, biohydrogen production is more expensive than other fuel options. Biohydrogen would play a major role in the economy in the long run if technology improvements succeed in bringing down the costs. Cost of the feedstock is another main barrier for economically viable fermentation technology. Biohydrogen production employing a renewable biomass may be a potential answer to overcome some of the economics. There is scope to use sugarcane juice, molasses, or distillery effluent as substrates, as they contain sugar in significant quantities. Apart from that more easily degradable substance, green wastes will, most probably, to a large extent be the targeted feedstocks for hydrogen fermentation because of their abundance. Lignocellulose is the most abundant renewable natural resource and substrate available for conversion to fuels. There is great potential in the use of green waste biomass as a renewable source of energy via microbial breakdown to sugars that can then be converted to biohydrogen (Lo et al., 2008, 2009). With technology advancements, biohydrogen production may offer a sustainable alternative energy resource in the future. Rigorous technoeconomic analyses, however, are necessary to draw a cost-effective comparison between biologically produced hydrogen and the various other conventional fossil fuels.

CONCLUSIONS

Extensive work has indicated a promising prospect of biohydrogen production with substantial improvement in both the yield and the volumetric production rates of hydrogen fermentations. For realistic applications that make economic sense, hydrogen yields and production rates must be raised to a significant level, surpassing the current attainment. A technological breakthrough is needed to harvest most of the hydrogen available in the substrate. An intensified effort in scaling up of the process to full-scale systems would further refine the technology for practical application and commercialization. In order to accelerate and intensify technological advancement and to generate critical mass for development of a hydrogen-based economy, international knowledge exchange and cooperation need to be strengthened.

References

Adams, M.W.W., Mortenson, L.E., Chen, J.S., 1980. Hydrogenase. Biochim. Biophys. Acta 594, 105–176.

Ahn, Y., Park, E.J., Oh, Y.K., Park, S., Webster, G., Weightman, A.J., 2005. Biofilm microbial community of a thermophilic trickling biofilter used for continuous biohydrogen production. FEMS Microbiol. Lett. 249, 31–38.

Akkerman, I., Janssen, M., Rocha, J.M.S., Reith, J.H., Wijffels, R.H., 2003. Photobiological hydrogen production: Photochemical efficiency and bioreactor design. In: Reith, J.H., Wijffels, R.H., Barten, H. (Eds.), Biomethane and Biohydrogen: Status and Perspectives of Biological Methane and Hydrogen Production. Dutch Biological Hydrogen Foundation, Hague, The Netherlands.

Andersch, W., Bahl, H., Gottschalk, G., 1983. Level of enzymes involved in acetate, butyrate, acetone and butanol formation by Clostridium-Acetobutylicum. Eur. J. Appl. Microbiol. Biotechnol. 18, 327–332.

Basak, N., Das, D., 2007. The prospect of purple non-sulfur (PNS) photosynthetic bacteria for hydrogen production: The present state of the art. World J. Microbiol. Biotechnol. 23, 31–42.

Cai, G., Jin, B., Saint, C., Monis, P., 2010. Metabolic flux analysis of hydrogen production network by Clostridium butyricum W5: Effect of pH and glucose concentrations. Int. J. Hydrogen Energy 35, 6681–6690.

Call, D., Logan, B.E., 2008. Hydrogen production in a single chamber microbial electrolysis cell lacking a membrane. Environ. Sci. Technol. 42, 3401–3406.

Chang, F.Y., Lin, C.Y., 2004. Biohydrogen production using an up-flow anaerobic sludge blanket reactor. Int. J. Hydrogen Energy 29, 33–39.

Chen, C.K., Blaschek, H.P., 1999. Examination of physiological and molecular factors involved in enhanced solvent production by Clostridium beijerinckii BA101. Appl. Environ. Microbiol. 65, 2269–2271.

Chen, C.C., Lin, C.Y., 2003. Using sucrose as a substrate in an anaerobic hydrogen-producing reactor. Adv. Environ. Res. 7, 695–699.

Chen, W.M., Tseng, Z.J., Lee, K.S., Chang, J.S., 2005. Fermentative hydrogen production with Clostridium butyricum CGS5 isolated from anaerobic sewage sludge. Int. J. Hydrogen Energy 30, 1063–1070.

Cheng, S., Logan, B.E., 2007. Sustainable and efficient biohydrogen production via electrohydrogenesis. Proc. Natl. Acad. Sci. USA 104, 18871–18873.

Chittibabu, G., Nath, K., Das, D., 2006. Feasibility studies on the fermentative hydrogen production by recombinant Escherichia coli BL-21. Process Biochem. 41, 682–688.

Claassen, P.A.M., van Lier, J.B., Contreras, A.M.L., van Niel, E.W.J., Sijtsma, L., Stams, A.J.M., et al., 1999. Utilisation of biomass for the supply of energy carriers. Appl. Microbiol. Biotechnol. 52, 741–755.

Collet, C., Adler, N., Schwitzguebel, J.P., Peringer, P., 2004. Hydrogen production by Clostridium thermolacticum during continuous fermentation of lactose. Int. J. Hydrogen Energy 29, 1479–1485.

Dabrock, B., Bahl, H., Gottschalk, G., 1992. Parameters affecting solvent production by Clostridium pasteurianum. Appl. Environ. Microbiol. 58, 1233–1239.

Das, D., Veziroglu, T.N., 2001. Hydrogen production by biological processes: A survey of literature. Int. J. Hydrogen Energy 26, 13–28.

de Vrije, T., Claassen, P.A.M., 2003. Dark hydrogen fermentation. In: Reith, J.H., Wijffels, R.H., Barten, H. (Eds.), Bio-methane and Bio-hydrogen. Dutch Biological Hydrogen Foundation, Hague, The Netherlands, pp. 103–123.

Ditzig, J., Liu, H., Logan, B.E., 2007. Production of hydrogen from domestic wastewater using a bioelectrochemically assisted microbial reactor. Int. J. Hydrogen Energy 32, 2296–2304.

Evvyernie, D., Morimoto, K., Karita, S., Kimura, T., Sakka, K., Ohmiya, K., 2001. Conversion of chitinous wastes to hydrogen gas by *Clostridium paraputrificum* M-21. J. Biosci. Bioeng. 91, 339–343.

Fang, H.H.P., Liu, H., Zhang, T., 2002a. Characterization of a hydrogen-producing granular sludge. Biotechnol. Bioeng. 78, 44–52.

Fang, H.H.P., Zhang, T., Liu, H., 2002b. Microbial diversity of a mesophilic hydrogen-producing sludge. Appl. Microbiol. Biotechnol. 58, 112–118.

Ferchichi, M., Crabbe, E., Hintz, W., Gil, G.H., Almadidy, A., 2005. Influence of culture parameters on biological hydrogen production by *Clostridium saccharoperbutylacetonicum* ATCC 27021. World J. Microbiol. Biotechnol. 21, 855–862.

Forsberg, C.W., 2007. Future hydrogen markets for large-scale hydrogen production systems. Int. J. Hydrogen Energy 32, 431–439.

Garcia-Pena, E.I., Guerrero-Barajas, C., Ramirez, D., Arriaga-Hurtado, L.G., 2009. Semi-continuous biohydrogen production as an approach to generate electricity. Bioresour. Technol. 100, 6369–6377.

Hallenbeck, P.C., Benemann, J.R., 2002. Biological hydrogen production; fundamentals and limiting processes. Int. J. Hydrogen Energy 27, 1185–1193.

Hallenbeck, P.C., Ghosh, D., 2009. Advances in fermentative biohydrogen production: The way forward? Trends Biotechnol. 27, 287–297.

Hankamer, B., Lehr, F., Rupprecht, J., Mussgnug, J., Posten, C., Kruse, O., 2007. Photosynthetic biomass and H_2 production by green algae: From bioengineering to bioreactor scale-up. Physiol. Plant 131, 10–21.

Hawkes, F.R., Dinsdale, R., Hawkes, D.L., Hussy, I., 2002. Sustainable fermentative hydrogen production: Challenges for process optimisation. Int. J. Hydrogen Energy 27, 1339–1347.

Hawkes, F.R., Hussy, I., Kyazze, G., Dinsdale, R., Hawkes, D.L., 2007. Continuous dark fermentative hydrogen production by mesophilic microflora: Principles and progress. Int. J. Hydrogen Energy 32, 172–184.

Holladay, J.D., Hu, J., King, D.L., Wang, Y., 2009. An overview of hydrogen production technologies. Catalysis Today 139, 244–260.

Holt, R.A., Stephens, G.M., Morris, J.G., 1984. Production of solvents by Clostridium-Acetobutylicum cultures maintained at neutral Ph. Appl. Environ. Microbiol. 48, 1166–1170.

Horiuchi, J.I., Shimizu, T., Tada, K., Kanno, T., Kobayashi, M., 2002. Selective production of organic acids in anaerobic acid reactor by pH control. Bioresour. Technol. 82, 209–213.

Hwang, J.H., Choi, J.A., Abou-Shanab, R.A.I., Min, B., Song, H., Kim, Y., et al., 2011. Feasibility of hydrogen production from ripened fruits by a combined two-stage (dark/dark) fermentation system. Bioresour. Technol. 102, 1051–1058.

Janssen, M., 2002. Cultivation of Microalae: Effect of Light/Dark Cycles on Biomass Yield. Thesis Wageningen University, Wageningen, The Netherlands.

Kalia, V.C., Jain, S.R., Kumar, A., Joshi, A.P., 1994. Fermentation of biowaste to H_2 by *Bacillus licheniformis*. World J. Microbiol Biotechnol. 10, 224–227.

Kanai, T., Imanaka, H., Nakajima, A., Uwamori, K., Omori, Y., Fukui, T., et al., 2005. Continuous hydrogen production by the hyperthermophilic archaeon, *Thermococcus kodakaraensis* KOD1. J. Biotechnol. 116, 271–282.

Kapdan, I.K., Kargi, F., 2006. Bio-hydrogen production from waste materials. Enzyme Microbial Technol. 38, 569–582.

Kim, B.H., Bellows, P., Datta, R., Zeikus, J.G., 1984. Control of carbon and electron flow in clostridium-acetobutylicum fermentations: Utilization of carbon-monoxide to inhibit hydrogen-production and to enhance butanol yields. Appl. Environ. Microbiol. 48, 764–770.

Kim, J.O., Kim, Y.H., Ryu, J.Y., Song, B.K., Kim, I.H., Yeom, S.H., 2005. Immobilization methods for continuous hydrogen gas production biofilm formation versus granulation. Process Biochem. 40, 1331–1337.

Kim, S.Y., Seol, E.H., Oh, Y.K., Wang, G.Y., Park, S.H., 2009. Hydrogen production and metabolic flux analysis of metabolically engineered *Escherichia coli* strains. Int. J. Hydrogen Energy 34, 7417–7427.

Koutrouli, E.C., Kalfas, H., Gavala, H.N., Skiadas, I.V., Stamatelatou, K., Lyberatos, G., 2009. Hydrogen and methane production through two-stage mesophilic anaerobic digestion of olive pulp. Bioresour. Technol. 100, 3718–3723.

Kovács, K.L., Maróti, G., Rákhely, G., 2006. A novel approach for biohydrogen production. Int. J. Hydrogen Energy 31, 1460–1468.

Kuhn, M., Steinbüchel, A., Schlegel, H.G., 1984. H_2 evolution by strictly aerobic H_2 bacteria under anaerobic condition. J. Bacteriol. 159, 633–639.

Kumar, N., Das, D., 2000. Enhancement of hydrogen production by *Enterobacter cloacae* IIT-BT 08. Process Biochem. 35, 589–593.

Kumar, N., Das, D., 2001. Continuous hydrogen production by immobilized *Enterobacter cloacae* IIT-BT 08 using lignocellulosic materials as solid matrices. Enzyme Microb. Technol. 29, 280–287.

Kumar, A., Jain, S.R., Sharma, C.B., Joshi, A.P., Kalia, V.C., 1995. Increased H_2 production by immobilized microorganisms. World J. Microbiol. Biotechnol. 11, 156–159.

Kumar, N., Monga, P.S., Biswas, A.K., Das, D., 2000. Modeling and simulation of clean fuel production by *Enterobacter cloacae* IIT-BT 08. Int. J. Hydrogen Energy 25, 945–952.

Laurinavichene, T.V., Kosourov, S.N., Ghirardi, M.L., Seibert, M., Tsygankov, A.A., 2008. Prolongation of H_2 photoproduction by immobilized, sulfur-limited *Chlamydomonas reinhardtii* cultures. J. Biotechnol. 134, 275–277.

Lay, J.J., 2001. Biohydrogen generation by mesophilic anaerobic fermentation of microcrystalline cellulose. Biotechnol. Bioeng. 74, 280–287.

Lay, J.J., Lee, Y.J., Noike, T., 1999. Feasibility of biological hydrogen production from organic fraction of municipal solid waste. Water Res. 33, 2579–2586.

Lee, D.J., Show, K.Y., Su, A., 2011. Dark fermentation on biohydrogen production: Pure culture. Bioresour. Technol. 102, 8393–8402.

Lee, K.S., Lo, Y.C., Lin, P.J., Chang, J.S., 2006. Improving biohydrogen production in a carrier-induced granular sludge bed by altering physical configuration and agitation pattern of the bioreactor. Int. J. Hydrogen Energy 31, 1648–1657.

Levin, D.B., Pitt, L., Love, M., 2004. Biohydrogen production: Prospects and limitations to practical application. Int. J. Hydrogen Energy 29, 173–185.

Li, C.L., Fang, H.H.P., 2007. Fermentative hydrogen production from wastewater and solid wastes by mixed cultures. Crit. Rev. Environ. Sci. Technol. 37, 1–39.

Lin, C.Y., Chang, R.C., 1999. Hydrogen production during the anaerobic acidogenic conversion of glucose. J. Chem. Technol. Biotechnol. 74, 498–500.

Lin, C.Y., Chou, C.H., 2004. Anaerobic hydrogen production from sucrose using an acid-enriched sewage sludge microflora. Eng. Life Sci. 4, 66–70.

Liu, H., Zhang, T., Fang, H.H.P., 2003. Thermophilic H_2 production from a cellulose-containing wastewater. Biotechnol. Lett. 25, 365–369.

Lo, Y.C., Bai, M.D., Chen, W.M., Chang, J.S., 2008. Cellulosic hydrogen production with a sequencing bacterial hydrolysis and dark fermentation strategy. Bioresour. Technol. 99, 8299–8303.

Lo, Y.C., Su, Y.C., Chen, C.Y., Chen, W.M., Lee, K.S., Chang, J.S., 2009. Biohydrogen production from cellulosic hydrolysate produced via temperature-shift-enhanced bacterial cellulose hydrolysis. Bioresour. Technol. 100, 5802–5807.

Logan, B.E., 2004. Extracting hydrogen and electricity from renewable resources. Environ. Sci. Technol. 38, 160A–167A.

Logan, B.E., Oh, S.E., Kim, I.S., Van Ginkel, S., 2002. Biological hydrogen production measured in batch anaerobic respirometers. Environ. Sci. Technol. 37, 1055 1055.

Majizat, A., Mitsunori, Y., Mitsunori, W., Michimasa, N., Junichiro, M., 1997. Hydrogen gas production from glucose and its microbial kinetics in anaerobic systems. Water Sci. Technol. 36, 279–286.

Manis, S., Banerjee, R., 2008. Comparison of biohydrogen production processes. Int. J. Hydrogen Energy 33, 279–286.

Manish, S., Venkatesh, K., Banerjee, R., 2007. Metabolic flux analysis of biological hydrogen production *by Escherichia coli*. Int. J. Hydrogen Energy 32, 3820–3830.

Melis, A., 2002. Green alga hydrogen production: Progress, challenges and prospects. Int. J. Hydrogen Energy 27, 1217–1228.

Melis, A., Happe, T., 2001. Hydrogen production: Green algae as a source of energy. Plant Physiol. 127, 740–748.

Mizuno, O., Ohara, T., Shinya, M., Noike, T., 2000. Characteristics of hydrogen production from bean curd manufacturing waste by anaerobic microflora. Water Sci. Technol. 42, 345–350.

Monmoto, M., Atsuka, M., Atif, A.A.Y., Ngan, M.A., Fakhru'l-Razi, A., Iyuke, S.E., et al., 2004. Biological production of hydrogen from glucose by natural anaerobic microflora. Int. J. Hydrogen Energy 29, 709–713.

Mu, Y., Yu, H.Q., 2006. Biological hydrogen production in a UASB reactor with granules. I. Physicochemical characteristics of hydrogen-producing granules. Biotechnol. Bioeng. 94, 980–987.

Nandi, R., Sengupta, S., 1998. Microbial production of hydrogen: An overview. Crit. Rev. Microbiol. 24, 61–84.

Nath, K., Das, D., 2004. Improvement of fermentative hydrogen production: Various approaches. Appl. Microbiol. Biotechnol. 65, 520–529.

Nath, K., Das, D., 2006. Amelioration of biohydrogen production by a two-stage fermentation process. Indust. Biotechnol. 2, 44–47.

Navarro, E., Montagud, A., Fernández De Córdoba, P., Urchueguía, J.F., 2009. Metabolic flux analysis of the hydrogen production potential in *Synechocystis* sp. PCC6803. Int. J. Hydrogen Energy 34, 8828–8838.

Niu, K., Zhang, X., Tan, W.S., Zhu, M.L., 2011. Effect of culture conditions on producing and uptake hydrogen flux of biohydrogen fermentation by metabolic flux analysis method. Bioresour. Technol. 102, 4029–4033.

Ntaikou, I., Antonopoulou, G., Lyberatos, G., 2010. Biohydrogen production from biomass and wastes via dark fermentation: A review. Waste Biomass Valoriz. 1, 21–39.

Oh, Y.K., Park, M.S., Seol, E.H., Lee, S.J., Park, S., 2003. Isolation of hydrogen-producing bacteria from granular sludge of an upflow anaerobic sludge blanket reactor. Biotechnol. Bioprocess Eng. 8, 54–57.

Oh, S.E., Lyer, P., Bruns, M.A., Logan, B.E., 2004a. Biological hydrogen production using a membrane bioreactor. Biotechnol. Eng. 87, 119–127.

Oh, Y.K., Kim, S.H., Kim, M.S., Park, S., 2004b. Thermophilic biohydrogen production from glucose with trickling biofilter. Biotechnol. Eng. 88, 690–698.

Oh, Y.K., Kim, H.J., Park, S.H., Kim, M.S., Ryu, D.D., 2008. Metabolic-flux analysis of hydrogen production pathway in *Citrobacter amalonaticus* Y19. Int. J. Hydrogen Energy 33, 1471–1482.

Pasika, C., Zhang, J., Ferda, M., 2011. International Conference on Food Engineering and Biotechnology IPCBEE. Metabolic flux balance analysis for biological hydrogen production by purple pon-sulfur bacteria, vol. 9. IACSIT Press, Singapore.

Penfold, D.W., Forster, C.F., Macaskie, L.E., 2003. Increased hydrogen production by *Escherichia coli strain HD701* in comparison with the wild-type parent *strain MC4100*. Enzyme Microb. Technol. 33, 185–189.

Pinchukova, E.E., Varfolomeev, S.D., Kondrat'eva, E.N., 1979. Isolation, purification and study of the stability of the soluble hydrogenase from *Alcaligenes eutrophus* Z-1. Biokhimiya 44, 605–615.

Rachman, M.A., Furutani, Y., Nakashimada, Y., Kakizono, T., Nishio, N., 1997. Enhanced hydrogen production in altered mixed acid fermentation of glucose by *Enterobacter aerogenes*. J. Ferment. Bioeng. 83, 358–363.

Rachman, M.A., Nakashimada, Y., Kakizono, T., Nishio, N., 1998. Hydrogen production with high yield and high evolution rate by self-flocculated cells of *Enterobacter aerogenes* in a packed-bed reactor. Appl. Microbiol. Biotechnol. 49, 450–454.

Ravinder, T., Ramesh, B., Seenayya, G., Reddy, G., 2000. Fermentative production of acetic acid from various pure and natural cellulosic materials by *Clostridium lentocellum SG6*. World J. Microbiol. Biotechnol. 16, 507–512.

Schröder, C., Selig, M., Schönheit, P., 1994. Glucose fermentation to acetate, CO_2 and H_2 in the anaerobic hyperthermophilic Eubacterium Thermotoga-Maritima: Involvement of the Embden-Meyerhof pathway. Arch. Microbiol. 161, 460–470.

Shin, H.S., Youn, J.H., Kim, S.H., 2004. Hydrogen production from food waste in anaerobic mesophilic and thermophilic acidogenesis. Int. J. Hydrogen Energy 29, 1355–1363.

Show, K.Y., Zhang, Z.P., Tay, J.H., Liang, T.D., Lee, D.J., Jiang, W.J., 2007. Production of hydrogen in a granular sludge-based anaerobic continuous stirred tank reactor. Int. J. Hydrogen Energy 32, 4744–4753.

Show, K.Y., Zhang, Z., Lee, D.J., 2008. Design of bioreactors for biohydrogen production. J. Sci. Indust. Res. 67, 941–949.

Show, K.Y., Zhang, Z.P., Tay, J.H., Liang, T., Lee, D.J., Ren, N.Q., et al., 2010. Critical assessment of anaerobic processes for continuous biohydrogen production from organic wastewater. Int. J. Hydrogen Energy 35, 13350–13355.

Show, K.Y., Lee, D.J., Chang, J.S., 2011. Bioreactor and process design for biohydrogen production. Bioresour. Technol. 102, 8524–8533.

Sorensen, B., 2005. Hydrogen and Fuel Cells: Emerging Technologies and Applications. Elsevier Academic Press, New York.

Sparling, R., Risbey, D., Poggi-Varaldo, H.M., 1997. Hydrogen production from inhibited anaerobic composters. Int. J. Hydrogen Energy 22, 563–566.

Sridhar, J., Eiteman, M.A., 2001. Metabolic flux analysis of *Clostridium thermosuccinogenes*: Effects of pH and culture redox potential. Appl. Biochem. Biotechnol. 94, 51–69.

Tanisho, S., Ishiwata, Y., 1995. Continuous hydrogen production from molasses by fermentation using urethane foam as a support of flocks. Int. J. Hydrogen Energy 20, 541–545.

Tetali, S.D., Mitra, M., Melis, A., 2007. Development of the light-harvesting chlorophyll antenna in the green alga *Chlamydomonas reinhardtii* is regulated by the novel *Tla1* gene. Planta 225, 813–829.

Turner, J., Sverdrup, G., Mann, M.K., Maness, P.C., Kroposki, B., Ghirardi, M., et al., 2008. Renewable hydrogen production. Int. J. Energy Res. 32, 379–407.

Ueno, Y., Kawai, T., Sato, S., Otsuka, S., Morimoto, M., 1995. Biological production of hydrogen from cellulose by natural anaerobic microflora. J. Ferment. Bioeng. 79, 395–397.

Ueno, Y., Otsuka, S., Morimoto, M., 1996. Hydrogen production from industrial wastewater by anaerobic microflora in chemostat culture. J. Ferment. Bioeng. 82, 194–197.

Ueno, Y., Haruta, S., Ishii, M., Igarashi, Y., 2001. Microbial community in anaerobic hydrogen-producing microflora enriched from sludge compost. Appl. Microbiol. Biotechnol. 57, 555–562.

Ueno, Y., Fukui, H., Goto, M., 2007. Operation of a two-stage fermentation process producing hydrogen and methane from organic waste. Environ. Sci. Technol. 41, 1413–1419.

U.S. Department of Energy (DOE), 2007. Hydrogen, Fuel Cells and Infrastructure Technologies Program, Multi-Year Research, Development and Demonstration Plan. U.S. Department of Energy.

Valdez-Vazquez, I., Rios-Leal, E., Esparza-Garcia, F., Cecchi, F., Poggi-Varaldo, H.A., 2005. Semi-continuous solid substrate anaerobic reactors for H-2 production from organic waste: Mesophilic versus thermophilic regime. Int. J. Hydrogen Energy 30, 1383–1391.

Vanderhaegen, B., Ysebaert, E., Favere, K., Vanwambeke, M., Peeters, T., Panic, V., et al., 1992. Acidogenesis in relation to in-reactor granule yield. Water Sci. Technol. 25, 21–30.

van Niel, E.W.J., Budde, M.A.W., de Haas, G.G., van der Wal, F.J., Claassen, P.A.M., Stams, A.J.M., 2002. Distinctive properties of high hydrogen producing extreme thermophiles, *Caldicellulosiruptor saccharolyticus* and *Thermotoga elfii*. Int. J. Hydrogen Energy 27, 1391–1398.

Venkata Mohan, S., Veer Raghuvulu, S., Mohanakrishna, G., Srikanth, S., Sarma, P.N., 2009. Optimization and evaluation of fermentative hydrogen production process from mixed anaerobic culture employing data enveloping analysis (DEA) and Taguchi design of experimental (DOE) methodology. Int. J. Hydrogen Energy 34, 216–226.

Wang, J., Wan, W., 2009. Factors influencing fermentative hydrogen production: A review. Int. J. Hydrogen Energy 34, 799–811.

Wang, C.C., Chang, C.W., Chu, C.P., Lee, D.J., Chang, B.V., Liao, C.S., 2003a. Producing hydrogen from wastewater sludge by *Clostridium bifermentans*. J. Biotechnol. 102, 83–92.

Wang, C.C., Chang, C.W., Chu, C.P., Lee, D.J., Chang, B.V., Liao, C.S., et al., 2003b. Using filtrate of waste biosolids to effectively produce bio-hydrogen by anaerobic fermentation. Water Res. 37, 2789–2793.

Wang, A., Sun, D., Cao, G., Wang, H., Ren, N.Q., Wu, W.M., et al., 2011. Integrated hydrogen production process from cellulose by combining dark fermentation, microbial fuel cells, and a microbial electrolysis cell. Bioresour. Technol. 102, 4137–4143.

Wu, S.Y., Lin, C.N., Chang, J.S., Lee, K.S., Lin, P.J., 2002. Microbial hydrogen production with immobilized sewage sludge. Biotechnol. Progress 18, 921–926.

Wu, S.Y., Lin, C.N., Chang, J.S., 2003. Hydrogen production with immobilized sewage sludge in three-phase fluidized-bed bioreactors. Biotechnol. Progress 19, 828–832.

Wu, S.Y., Lin, C.N., Chang, J.S., 2005. Biohydrogen production with anaerobic sludge immobilized by ethylene-vinyl acetate copolymer. Int. J. Hydrogen Energy 30, 1375–1381.

Wu, S.Y., Hung, C.H., Lin, C.N., Chen, H.W., Lee, A.S., Chang, J.S., 2006. Fermentative hydrogen production and bacterial community structure in high-rate anaerobic bioreactors containing silicone-immobilized and self-flocculated sludge. Biotechnol. Bioeng. 93, 934–946.

Yokoi, H., Ohkawara, T., Hirose, J., Hayashi, S., Takasaki, Y., 1995. Characteristics of hydrogen production by aciduric *Enterobacter aerogenes* strain HO-39. J. Ferment. Bioeng. 80, 571–574.

Yokoi, H., Maeda, Y., Hirose, J., Hayashi, S., Takasaki, Y., 1997a. H_2 production by immobilized cells of Clostridium butyricum on porous glass beads. Biotechnol. Techniques 11, 431–433.

Yokoi, H., Tokushige, T., Hirose, J., Hayashi, S., Takahashi, Y., 1997b. Hydrogen production by immobilized cells of aciduric *Enterobacter aerogenes* strain HO-39. J. Ferment. Bioeng. 83, 481–484.

Yu, H.Q., Mu, Y., 2006. Biological hydrogen production in a UASB reactor with granules. II. Reactor performance in 3-year operation. Biotechnol. Bioeng. 94, 988–995.

Yu, H., Zhu, Z., Hu, W., Zhang, H., 2002. Hydrogen production from rice winery wastewater in an upflow anaerobic reactor by using mixed anaerobic cultures. Int. J. Hydrogen Energy 27, 1359–1365.

Zhang, L., Happe, T., Melis, A., 2002. Biochemical and morphological characterization of sulfur-deprived and H$_2$-producing *Chlamydomonas reinhardtii* (green alga). Planta 214, 552–561.

Zhang, T., Liu, H., Fang, H.H.P., 2003. Biohydrogen production from starch in wastewater under thermophilic condition. J. Environ. Manag. 69, 149–156.

Zhang, Y.F., Liu, G.Z., Shen, J.Q., 2005. Hydrogen production in batch culture of mixed bacteria with sucrose under different iron concentrations. Int. J. Hydrogen Energy 30, 855–860.

Zhang, H.S., Bruns, M.A., Logan, B.E., 2006a. Biological hydrogen production by *Clostridium acetobutylicum* in an unsaturated flow reactor. Water Res. 40, 728–734.

Zhang, Z.P., Show, K.Y., Tay, J.H., Liang, D.T., Lee, D.J., Jiang, W.J., 2006b. Effect of hydraulic retention time on biohydrogen production and anaerobic microbial community. Process Biochem. 41, 2118–2123.

Zhang, Z.P., Show, K.Y., Tay, J.H., Liang, D.T., Lee, D.J., 2007a. Biohydrogen production with anaerobic fluidized bed reactors-A comparison of biofilm-based and granule-based systems. Int. J. Hydrogen Energy 33, 1559–1564.

Zhang, Z.P., Show, K.Y., Tay, J.H., Liang, D.T., Lee, D.J., Jiang, W.J., 2007b. Rapid formation of hydrogen-producing granules in an anaerobic continuous stirred tank reactor induced by acid incubation. Biotechnol. Bioeng. 96, 1040–1050.

Zhang, Z.P., Tay, J.H., Show, K.Y., Yan, R., Liang, D.T., Lee, D.J., et al., 2007c. Biohydrogen production in a granular activated carbon anaerobic fluidized bed reactor. Int. J. Hydrogen Energy 32, 185–191.

Zhang, Z., Show, K.Y., Tay, J.H., Liang, T., Lee, D.J., 2008a. Biohydrogen production with anaerobic fluidized bed reactors- a comparison of biofilm-based and granule-based systems. Int. J. Hydrogen Energy 33, 1559–1564.

Zhang, Z.P., Adav, S.S., Show, K.Y., Tay, J.H., Liang, D.T., Lee, D.J., 2008b. Characteristics of rapidly formed hydrogen-producing granules and biofilms. Biotechnol. Bioeng. 101, 926–936.

Zhang, Z.P., Show, K.Y., Tay, J.H., Liang, T.D., Lee, D.J., 2008c. Enhanced continuous biohydrogen production by immobilized anaerobic microflora. Energy Fuels 22, 87–92.

Zhang, Z.P., Show, K.Y., Tay, J.H., Liang, T.D., Lee, D.J., Wang, J.Y., 2008d. The role of acid incubation in rapid immobilization of hydrogen-producing culture in anaerobic upflow column reactors. Int. J. Hydrogen Energy 33, 5151–5160.

Scale-up and Commercial Applications of Biohydrogen Production Processes

Biswarup Sen[*,†,‡], *Chen-Yeon Chu,*[†,‡,§] *Chiu-Yue Lin*[*,†,‡]

[*]Department of Environmental Engineering and Science, Feng Chia University, Taichung, Taiwan
[†]Green Energy Development Center, Feng Chia University, Taichung, Taiwan
[‡]Master Program of Green Energy Science and Technology, Feng Chia University, Taichung, Taiwan
[§]Department of Chemical Engineering, Feng Chia University, Taichung, Taiwan

INTRODUCTION

In recent years, the question has been asked repeatedly "Is the biohydrogen production process viable and have scalability?" The question is important and can be answered only if one considers its technoeconomic evaluation, net energy ratio, energy efficiency, greenhouse gas emissions, scale-up potential, and commercial feasibility. Extensive research in the past three decades has shown promising prospects of biohydrogen production, and there have been substantial improvements and developments in both yields and volumetric production rates of biohydrogen. However, for commercial-scale applications that make economic sense, hydrogen yields and production rates must surpass significantly the current achievements. In addition, a technological breakthrough is warranted to pull out most of the H_2 from the renewable feedstocks, if not all. Research addressing this challenge should be one of the focuses of future studies on biohydrogen production via biological pathways. Furthermore, scaling up of the process to a full-scale system will further refine the technology for future commercialization (Lee et al., 2011; Wang et al., 2010). At present, the biohydrogen production process is part of a tendency toward decentralized energy and power production on a small scale and is less energy-intensive than chemical and electrochemical processes (Brentner et al., 2010).

The global energy and environmental crises pose severe challenges to the development and sustainability of modern society (Nejat Veziroglu, 2007). The efficient use of biohydrogen depends on overcoming the problems of production, transportation, storage, and application efficiency. It is expected that with intensive investment in technology and infrastructure, a transition from a petroleum-based economy to a hydrogen-based economy could be achieved by 2050 (Lee and Chiu, 2012; Lee et al., 2011). In fact, "hydrogen roadmaps" have been proposed for the United States, the European Union, and other countries (Lee and Chiu, 2012). However, these roadmaps are based on H_2 from fossil fuels and renewable resources and not solely on H_2 produced via a biological route. Lee and Chiu (2012) studied the development of H_2 sectors in the United States, China, Japan, and India as examples of the four categories based on the human development index, using the refined input–output model, and found that China will have the largest biohydrogen market, followed by the United States, Japan, and India. In addition, India will see the most efficient investment in the biohydrogen sectors, whereas Japan will have the greatest potential to substitute fossil fuels for biohydrogen. These study results could provide a general guideline for decision makers in the development of biohydrogen in their own countries.

When it comes to establishing a hydrogen-based society and transportation, the major debate is the "chicken-and-egg" debate associated with the H_2 sector. To realize a H_2 economy, the major obstacle is the conflict between building technology or applications first or on building infrastructure first for promotion and commercialization. Kriston et al. (2010) argued "No hydrogen-powered car can be sold if it cannot be refueled, and nobody will invest in a H_2 refueling station if no one has a H_2-powered vehicle." Therefore, it is suggested that production technology and applications should go parallel or in tandem with the infrastructure.

Currently, biologically produced H_2 is more expensive than other fuels (Nath and Das, 2004), and the relatively low H_2 yield and production rate are two major challenges for biological H_2-producing systems. As a result, this prevents them from becoming a practical means of H_2 production. Despite profound fundamental knowledge on the biohydrogen fermentation technology and its operation, we are still far behind the commercial application of this process. Moreover, the process is still in its infancy to compete with fossil fuels in terms of cost, efficiency, and reliability. The present world economy strongly favors large-scale H_2 production systems, but is limited by the availability of a suitable biomass and the efficiency of converting that biomass into biohydrogen. Up until now, research has focused mainly on mesophilic and thermophilic biohydrogen fermentation from organic feedstock composed primarily of carbohydrates (starch, cellulose, and hemicellulose). Studies have also shown that biohydrogen can be produced from lipid-rich wastewaters and proteinaceous material as well (Antonopoulou et al., 2008; Yang et al., 2007; Zhou et al., 2006). Although it is now well realized that the biohydrogen production process is renewable, the process still needs to be promoted from a laboratory scale to a commercial scale. Only a few pilot-scale studies have been demonstrated so far. It is indeed very essential that technical issues need to be addressed in the laboratory for efficient biohydrogen production prior to scale-up and commercialization of the biohydrogen production process. This chapter discusses the strategies and process of scale-up and a description of the reported pilot-scale biohydrogen production units from different organic wastes as examples of commercial applications of the process.

TECHNOECONOMIC EVALUATION OF BIOHYDROGEN PRODUCTION PROCESS

At the moment, the biohydrogen production process is in the phase of commercialization; in fact, many companies engaged in biofuels production have set up their own in-house large-scale units that serve the dual benefit of waste utilization and energy generation. However, information on commercial-scale biohydrogen production process installations is property rights owned by the companies and not available to the public. As a universal rule, for commercialization of any product, it is most important to have done the technoeconomic evaluation of the new process, which in this case is the biohydrogen production process. Like most other biological processes involving microorganisms, the biohydrogen production process also suffers from serious issues of costs for maintenance and feedstock availability. Therefore, technoeconomic evaluation of the process is the first step toward assessing the viability and possibility of scaling up the laboratory-scale system to the commercial scale. Moreover, technoeconomic evaluation also serves as a guideline for further improvement of the process. The general steps (adapted from Kabir Kazi et al. 2010) involved in technoeconomic evaluation are as follow:

1. Search literature for biohydrogen production processes
2. Selection of a few case studies
3. Design process models using AspenPlus process simulation using available experimental data
4. Size and cost equipment using traditional methods such as literature references and trader quotations
5. Determine project investments and perform discounted cash flow analysis
6. Adjust sensitivity parameters and document results
7. Perform pioneer plant cost growth and performance analysis

It should be noted that the economic analyses are based on optimistic assumptions, are highly presumptive, and are intended primarily to ascertain the major cost factors for biohydrogen production. Currently, biologically produced H_2 is more costly than other fuel alternatives (Show et al., 2012). There is no doubt that many technical, engineering, and economic challenges have to be solved before scale-up and commercial application of the biohydrogen production process. The following section discusses the technoeconomic evaluation of the biohydrogen production process routed through dark and photofermentation modes of operation.

Dark Fermentation Process

The theoretical conversion of biomass to biohydrogen using dark fermentation is only one-third of the energy content of the sugars derived from starch or cellulose and the rest is in the form of organic acids. Therefore, the viability of the process relies mainly on harnessing the energy contained in the organic acids. Ljunggren and Zacchi (2010a) investigated three base cases reflecting the different strategies that can be used when performing dark fermentation: high productivity, high yield, and low productivity–low yield. Glucose concentrations, H_2 productivities, and yields are 4–20 g/liter, 7–45 $mmol_{H_2}$/(liter-h), and 1.4–3.5 $mmol_{H_2}$/$mole_{glucose}$, respectively. The authors also considered the production of pure methane as a reference case to investigate how the production of H_2 affects the production cost. This study found that cost estimates range from 50 to 340€/GJ for the three base cases and the reference case of process alternatives. It was also shown that capital costs and nutrients are the main contributors to the cost in all the base and reference cases. Surprisingly, by increasing the substrate concentration and decreasing the H_2 yield, there was a reduction in the production cost. Ljunggren et al. (2011) compared the H_2 process to a second-generation ethanol process with respect to cost and found that the H_2 production cost is about 20 times higher than that of ethanol production, that is, 421.7 €/GJ vs 19.5 €/GJ. The authors attributed low productivity, low energy efficiency, and the high cost of buffer and base required for pH control as the main drawbacks of the H_2 process.

Vatsala et al. (2008) evaluated the process economics of a pilot-scale (100 m^3) biohydrogen production unit fed with a sugar industry effluent in a batch dark fermentation mode. The authors reported a total estimated capital cost of \$44,444 and a total H_2 production cost of \$35.44/100 m^3. The total H_2 production cost included labor (\$8.88), chemicals (\$22.66), and power (\$3.88). The authors argued that their designed process could produce H_2 at the cost of \$0.9/gallon, where gasoline costs about \$2.5/gallon. However, in terms of energy content, gasoline has a fourfold higher energy content than H_2. Putting aside this comparison, the price estimated by the authors using their pilot-scale process is still promising and can certainly replace gasoline in the future.

Photofermentation Process

Benemann (1997) did a preliminary analysis of a two-stage process in which microalgae were cultivated in large open ponds to produce a high carbohydrate biomass that then produces H_2 in tubular photobioreactors. The author estimated an initial cost for an indirect microalgal biophotolysis system consisting of open ponds (140 ha) and photobioreactors (14 ha). Overall total H_2 production costs were estimated at US\$ 10/GJ with the algal ponds estimated at a cost of US\$ 6/m^2 and the photobioreactors with assumed costs of US\$ 100/m^2 as the major capital and operating cost factors. This cost analysis was further confirmed by another study on a large-scale (>100 ha), single-stage algal biophotolysis process in a near-horizontal tubular reactor system (Tredici and Zlttelli, 1998). Considering a 10% solar energy conversion efficiency, costs of the tubular photobioreactor were estimated at US\$ 50/m^2; however, the analysis did not include any costs for gas handling and assumed a relatively low annual capital charge at 17%. The H_2 production costs were estimated at US\$ 15/GJ.

Integrated Dark Photofermentation Process

An integrated biological process for the production of H_2 based on thermophilic and photoheterotrophic fermentation was evaluated from technical and economic standpoints (Ljunggren and Zacchi, 2010b). This study also included the pretreatment of raw material (potato steam peels) and purification of H_2 using amine absorption. The study analyzed the effects of different parameters (H_2 productivity and yield and substrate concentration) on the production costs of biohydrogen as a guideline for future improvements. It was found that photofermentation is the main contributor to the H_2 production cost—mainly because of the cost of plastic tubing for the photofermenters, which represents 40.5% of the H_2 production cost. Moreover, the costs of the capital investment and chemicals were also notable contributors to the H_2 production cost. The study suggested that major economic improvements could be achieved by increasing the productivity of the two fermentation steps on a medium- to long-term scale.

FUNDAMENTALS OF BIOHYDROGEN PRODUCTION PROCESS SCALE-UP

The main biohydrogen-producing bacteria are *Bacillus*, *Enterobacter*, and *Clostridium*. Many researches reveal that *Clostridium* outweighs other species. From fermentation, the best metabolic pathway of *Clostridium* is producing acetate and butyrate. When it comes to reactor design and scale-up, two fields are concerned with the realm of kinetics and hydrodynamic.

Kinetics

Batch, continuous, and pilot systems can be used for biohydrogen production. In application, each industrial or municipal has its individual requirements. Here the reactor design of how to scale up is introduced.

Components in a batch reactor with mixing (Fig. 1A) have the same concentration within the reactor at any time. The liquid volume, V, does not change with time. In addition, only component concentrations change with time or the rate of mass accumulation equals VdC/dt. In the system, bacteria and the substrate were selected as components, which are sufficiently high in concentration that imposes no organism growth rate. At time $(t)=0$, the batch reactor contains concentrations of a rate-limiting substrate (S_0, mg/liter) and microorganisms (X_0, mg/liter). While changes in substrate and microorganism concentrations are interdependent, the mass balance of each component is constructed. To start with, the mass balance in terms of mass changes in the working volume [Eq. (1)] is

$$\text{Rate of mass accumulation in working volume} = \text{input (rate of mass)}$$
$$- \text{ output (rate of mass)} + \text{generation (rate of mass)} \tag{1}$$

$$\text{Mass rate of accumulation} = 0 - 0 + \text{generation (rate of mass)} \tag{2}$$

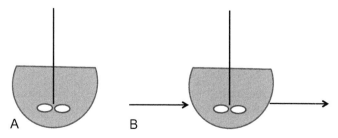

FIGURE 1 Basic reactor types: (A) batch reactor and (B) continuously stirred tank reactor.

While the microorganisms are consuming substrate, the system does not have input and output. Thus, the substrate is utilized or degraded by the microorganisms and generation has a negative value. In the mathematical form, Equation (2) becomes Equation (3):

$$V\frac{dS}{dt} = Vr_{ut} \tag{3}$$

The rate of substrate utilization is assumed to follow Monod kinetics, as given by Equation (4); in this situation, Equation (5) or Equation (5.1) is obtained:

$$r_{ut} = -\frac{\hat{q}S}{K+S}X_a \tag{4}$$

$$V\frac{dX_a}{dt} = V\left(-\frac{\hat{q}S}{K+S}\right)X_a \tag{5}$$

$$\frac{ds}{dt} = -\frac{\hat{q}S}{K+S}X_a \tag{5.1}$$

Equation (5) will be integrated to determine how S changes with time in the batch reactor, but X_a changes with time. For the sake of determining how X_a changes, the equation is constructed on the reactor microorganisms. Equation (1) can be converted into Equation (6):

$$\text{Mass rate of organism accumulation in working volume} \atop = 0 - 0 + \text{rate of mass generation} \tag{6}$$

With μ being the net-specific growth rate of organisms [Eq. (7)], the mathematical form is similar to Equation (3), and Equation (8) is obtained:

$$\mu = \frac{1}{X_a}\frac{dX_a}{dt} = \mu_{syn} + \mu_{dec} = \hat{\mu}\frac{S}{K+S} - b \tag{7}$$

$$V\frac{dX_a}{dt} = V(\mu X_a) \tag{8}$$

Assuming the organism growth rate follows the Monod kinetic and if decay, as well as growth, is considered, then combining Equation (8) with Equation (7) gives Equation (9) or Equation (9.1):

$$V\frac{dX_a}{dt} = V\left(\hat{\mu}\frac{S}{K+S} - b\right)X_a \tag{9}$$

$$\frac{dX_a}{dt} = \left(\hat{\mu}\frac{S}{K+S} - b\right)X_a \tag{9.1}$$

Hydrodynamic

Biohydrogen production fermenters, like three-phase reactors, include gas, liquid, and solid phases. However, the kinetics and hydrodynamics of biohydrogen production reactors are not so well described. The gas–liquid–solid systems display significant variation in the heterogeneous structure, especially when the hydraulic retention time (HRT) was varied. For gas–solid–liquid reactors, the bubble type will change from dispersed bubble to coalesced bubble and then to slugging; all the behaviors of bubble changing resulted in different hydrodynamic behavior (Chu et al., 2011a). Bubble movement, bursting, and size change are key factors affecting velocity distribution and sludge particle movement in the reactor (Wang et al., 2009). The coupled model also demonstrates a qualitative relationship between hydrodynamics and biohydrogen production, pointing out that controlling HRT is a key factor in biohydrogen production (Wang et al., 2010).

Flow Regimes

As mentioned earlier, flow regimes in a biohydrogen-producing reactor are affected strongly by the HRT at a fixed substrate concentration. The homogeneous flow regime was always found with a superficial gas velocity of 0.01 cm/s and a superficial liquid velocity of 0.0025 cm/s, when the HRT was higher than 8 h. The dispersed bubble regime is located with a superficial gas velocity of 0.04 cm/s. The transition state of dispersed bubbles and coalesced bubbles was always found between superficial gas velocities of 0.04 and 0.05 cm/s. As the substrate loading rate increases, and the chance of bubble coalescence is enhanced, a slugging regime is therefore formed. However, the counteracting effects of the increased mixing and working volume reduction make the biogas production linear to the liquid quantity that flowed into the bioreactor. A flow regime transition diagram is very useful for scaling up a dark fermentative hydrogen production reactor for further reactor design (Chu et al., 2011b).

Phase Holdups

Phase holdups play an important role in high-rate hydrogen production in an anaerobic fermentation reactor, especially in understanding the biomass content, biogas flow, and distribution that significantly affect the flow regime change in a reactor. The gas holdup (ε_g) and solid holdup (ε_s) increased, but the liquid holdup (ε_l) decreased when the HRT was decreased. In a biohydrogen production reactor, the three phase means biogas, liquid (substrate and nutrients), and solid (biomass, mostly bacteria). The gas and solid holdups were not significantly different in various configurations of bioreactors with the same operating conditions. Empirical correlations of the gas and liquid holdups were obtained by the regression of the bioreactors using the superficial velocities of gas (U_g) and liquid (U_l). The equations are shown in Equations (10) and (11). The relative confidence of the equations is above 86% (Chu et al., 2011b).

$$\varepsilon_g = 19.42 U_l^2 U_g^{-2} \tag{10}$$

$$\varepsilon_l = 6.70 \times 10^{-3} U_l^{-2} U_g^2 \tag{11}$$

where U_g is superficial gas velocity (m/s), coined by the biogas volumetric flow rate divided by the cross-section area of the reactors; and U_l is superficial liquid velocity (m/s), the liquid input volumetric flow rate divided by the cross-section area of the reactors.

Superficial liquid velocity has more influence on gas and liquid holdups than superficial gas velocity. In a traditional three-phase bioreactor, U_g is around 5–60 cm/s, but in anaerobic fermentative bioreactors, where gas is self-generated, U_g is around 0.01–0.12 cm/s. The superficial gas velocity in a traditional three-phase bioreactor is always far higher than that of a dark fermentation biohydrogen system. The empirical correlations obtained are satisfactory to predict the phase holdups in a dark fermentation biohydrogen system (Chu et al., 2011b).

COMMERCIAL APPLICATION OF BIOHYDROGEN PRODUCTION PROCESSES

Today hydrogen is mainly produced from fossil raw materials, such as oil and natural gas, which is not eco-friendly and clean. This hydrogen production process, therefore, does not present a solution to the problems of climate change and energy security. Biohydrogen, however, is an interesting alternative gaseous biofuel for the transportation sector, as well as an energy carrier for electricity generation. Moreover, biohydrogen is a clean energy carrier with a large potential of high conversion efficiency in fuel cells. In addition, biohydrogen can be added to biogas to improve the performance of combustion, thereby further lowering the CO_2 emissions and fuel consumption. Other applications of the biohydrogen production process are minimization of organic waste, which is generated as an enormous amount worldwide, and also simultaneous generation of other biofuels (ethanol and methane). In the past three decades, extensive research have been done to utilize the wastes from various streams, such as industrial waste, domestic waste, agricultural waste, and food waste. Despite a large set of laboratory-scale results, there are only a handful of studies on large-scale fermentors. In the following section of case studies, large-scale systems are discussed to demonstrate application of the biohydrogen production process in producing either hydrogen alone or in association with waste utilization.

Pilot-Scale Hydrogen Production from Sucrose

A pilot-scale, high-rate dark fermentative H_2 production plant has been established on the campus of Feng Chia University, Taiwan (Lin et al., 2011a). This pilot plant served to determine scale-up operation parameters for commercializing biohydrogen production technology. The working volume of this pilot plant system is 0.4 m^3 and is composed of two feedstock storage tanks (0.75 m^3 each), a nutrient storage tank (0.75 m^3), a mixing tank (0.6 m^3), an agitated granular sludge bed fermenter (working volume 0.4 m^3), a gas–liquid–solid separator (0.4 m^3), and a control panel (Fig. 2). The fermenter was seeded with mixed culture obtained from a laboratory-scale agitated granular sludge bed bioreactor and was operated for 67 days at 35 °C with sucrose feedstock. During the startup operation stage, the fermenter was fed on sucrose (20 g COD/liter) and operated at 35 °C in batch mode for 2 days and then switched to a continuous-feeding mode (HRT 12 h) for 1 month. The pH was found to markedly affect H_2 production efficiency and the bacterial community, and during the first 14 days of operation the H_2 production rate increased from 0.017 to 0.256 liter/liter-h (Lin et al., 2010). The organic loading rate (OLR) was then varied in the range of 40–240 kg$_{COD}$/m^3-day with influent sucrose concentrations of 20 and 40 kg $_{COD}$/m^3. Biogas and H_2 production rates

FIGURE 2 A pilot plant system (0.4 m^3) in operation at the Feng Chia University campus, Taiwan. (a) Side view of the plant, (b) fermenter and gas/liquid separator, (c) feedstock and medium storage tanks, and (d) automatic control panel.

increased with increasing OLR. The biomass concentration in the fermenter, in terms of volatile suspended solids (VSS), increased with an increasing OLR of 40–120 kg$_{COD}$/m^3-day. When the OLR reached 240 kg$_{COD}$/m^3-day, the biomass concentration decreased. This indicated that an OLR beyond 120 kg$_{COD}$/m^3 resulted in an inhibition of microbial growth. The biogas was mainly H$_2$ and CO$_2$ with a maximum H$_2$ content of 37% at OLR 240 kg$_{COD}$/m^3-day. This pilot plant fed on sucrose feedstock had a H$_2$ production rate of 15.6 m^3/m^3-day and a H$_2$ yield of 1.04 mol H$_2$/mol sucrose. The energy efficiency, that is, energy output to energy input (E_f), was calculated as 13–28. These E_f values were higher than the ones for ethanol production from corn, biodiesel, and sugarcane and are similar to those of ethanol production from cellulose.

Although the H$_2$ production rate was considerably high, the H$_2$ production performance of the pilot system was unsatisfactory when compared with that of laboratory-scale fermenters under the same operating conditions. Therefore, further studies have been conducted to improve the H_2 production performance of the pilot system (Lin et al., 2011b). Engineering strategies were applied to increase agitation and to promote mass transfer efficiency. In addition, the pilot system was operated under different OLR values with combinations of HRT and substrate concentration to improve biohydrogen production efficiency. The results of applying engineering strategies were very promising. It was found that with 25–30 rpm agitation and an OLR of 60 g_{COD}/liter-day, the H_2 production rate of the pilot system reached 0.55 mol_{H_2}/liter-day (13.4 m^3/m^3-day). This is threefold higher than the one obtained using a lower agitation of 10–15 rpm.

At the optimized operating conditions of HRT 6 h and sucrose concentration 30 gCOD/liter (120 gCOD/liter-day), the pilot system had the highest H_2 production rate, H_2 yield, and overall H_2 production efficiency of 1.18 mol_{H_2}/liter-day, 3.84 mol_{H_2}/$mol_{sucrose}$, and 47.2%, respectively. This study is a good example of how engineering strategies can improve H_2 production efficiency and performance in a pilot plant system to a level similar to a laboratory-scale system.

The structure of a microbial community and its variation during fermentor operation are of prime significance to a biohydrogen production process. Any notable change in the microbial community structure could lead to a significant variation in system performance. Therefore, it is suggested that microbial monitoring should be made while scaling-up and commercializing the biohydrogen production process. Cheng et al. (2011) demonstrated that selective primers for monitoring a specific microbial community dominated in anaerobic hydrogen fermentation can be designed and applied successfully to monitor the fluctuation of the community structure and diversity of the microorganisms. The findings of Cheng et al. (2011) are very helpful in revealing the correlation among biogas production, H_2 production rate, and percentage of *Clostridium* spp. Moreover, it was also shown that *C. pasteurianum* increased up to 90% of the total cell population at the maximum H_2 production.

Food Waste Utilization in a Pilot-Scale System

Food waste consists primarily of starch, protein, fat, and lipids and some amount of cellulose and hemicellulose. Food waste has been considered an excellent feedstock for biohydrogen production due to its abundant availability and suitability to fermentation (Kim et al., 2010). In fact, food waste could replace some of the fossil fuel energy that is currently consumed in homes, restaurants, and food/vegetable markets. Despite its high aptness for biohydrogen production, there is no commercial-scale technology based on food waste. Being a solid waste, food waste has several issues related to its continuous feeding operation. For example, OLR is considered a key parameter while evaluating pilot system performance fed with wastewaters; however, OLR may not be apt in the case of solid feedstock such as food waste. In cases of solid feedstock, the carbon/nitrogen (C/N) ratio is generally evaluated and monitored (Lay et al., 2012). A proper substrate C/N is essential for a successful pilot system operation, and the effect of C/N on production efficiency and performance should be taken into account while scaling up. A study utilizing food waste feedstock was carried out in a pilot-scale system (Kim et al., 2010). The working volume of the fermenter was 0.15 m^3 (total volume, 0.23 m^3) with a liquid depth of 770 mm and an inner diameter of 500 mm. This plant was an anaerobic sequencing batch reactor wherein the C/N ratio was adjusted from 10 to 30. At C/N 20, the H_2 yield was around 0.5 $mol_{H_2}/mol_{hexose\ added}$ but dropped at higher C/N values. With the applied operation conditions, the pilot plant could account for 2.3% of energy conversion efficiency contained in food waste to H_2. The performance was quite low and probably not suitable for scale up to a commercial scale. A close scrutiny of the metabolism revealed that several other products (lactate, propionate, and valerate) were formed, which probably pulled the reducing flux and was less available for hydrogen production. Further improvement in the H_2 yield was obtained (0.69 $mol_{H_2}/mol_{hexose\ added}$) by alkaline shock treatment (pH 12.5 for 1 day).

A pilot-scale hydrogen production plant has been established and operated at the solid municipal waste treatment facility in Korea (Lee and Chung, 2010). The plant consists of a hydrogen fermentation tank and a methane fermentation tank; the hydrogen fermentation tank is linked to a fuel cell system that recovers energy using the newly generated hydrogen. The hydrogen fermentation tank is stainless steel and has a working volume of 0.5 m^3. This pilot-scale plant produces 3.88 L_{H_2}/m^3-day at HRT 21 h and gives a hydrogen yield of 1.82 $mol_{H_2}/mol_{glucose}$ with 90% carbohydrate consumption. In addition to H_2, this pilot plant also generates methane. This study also evaluated the process economics and suggested that two-stage hydrogen/methane fermentation has a greater potential for recovering energy than single-stage methane fermentation.

Sugar Industry Effluents Utilization in a Pilot-Scale System

It has been well established that sugary substrates are an excellent source for a high rate of biohydrogen production (Lin et al., 2011a). Therefore, pilot-scale trials with feedstock containing high concentration of sugars would be an ideal start toward developing commercial-scale biohydrogen production technology. A pilot-scale biohydrogen production system was performed in a continuous flow mode (Ren et al., 2006). The plant was operated for more than 200 days at an OLR of 3.11 to 85.57 kg_{COD}/m^3-day with molasses substrate. Molasses is known to be degraded easily, despite its very high COD. Similar to the previous study (Lin et al., 2011a), the biogas and H_2 yields increased with OLR in the range of 3.11 to 68.21 kg_{COD}/m^3-day, but decreased at high OLR of above 68.21 kg_{COD}/m^3-day. The pilot plant was able to give a maximum H_2 production rate of 5.57 $m^3_{H_2}/m^3$-day with an H_2 yield of 26.13 $mol_{H_2}/kg_{CODremoved}$, at OLR values of 35 to 55 kg_{COD}/m^3-day.

Most of the aforementioned studies were conducted using sewage sludge as the seed incoculum. Although sewage sludge is an easily available source of microorganisms to start up a laboratory- or pilot-scale fermenter, it is most often the cause of low efficiency due to inconsistency of the microbial population and loading rate. Quite often it has been experienced that using same sewage sludge but leading to unsatisfactory results. Therefore, pretreated and enriched mixed cultures from sewage sludge should be induced as the seed inoculum. A mixed culture is often recommended for large-scale systems; however, some studies have also been carried out with a mixture of pure cultures from different genera. For example, a study (Vatsala et al., 2008) was done with a sugarcane distillery effluent using cocultures of *Citrobacter freundii* 01, *Enterobacter aerogenes* E10, and *Rhodopseudomonas palustris* P2 in a 100-m^3 fermentor. This pilot plant system is probably the largest scale-up reported in the literature up to now on biohydrogen production and simultaneous COD reduction of a sugar industry effluent. Results of a 100-m^3 fermentor revealed a maximum of 10,693 mol H_2 obtained through a batch operation for 40 h from 3862 mol glucose, which is an average yield of H_2 as 2.76 mol/$mol_{glucose}$. The H_2 production rate was 0.53 kg/100 m^3-h. This study clearly demonstrated the utility of a distillery effluent as a source of biohydrogen and simultaneous treatment of a high-strength industrial effluent.

SUMMARY

The technology evolution of developing a commercialized biohydrogen production system is shown in Figure 3. To develop commercial-scale technology, basic research in a batch reactor for kinetics should be carried out first. Then, stable operation at a continuous-flow reactor should be examined. The third period is enhancement of the performance of the biohydrogen production, including the rate and yield. After that, the performance of the biohydrogen production system should be verified in a pilot scale. If the test runs meet the predicted outcomes, then a commercial scale could be designed from the obtained data, including kinetics and hydrodynamics of the pilot plant.

A typical application of a biohydrogen fermenter in a conventional wastewater treatment process is shown in Figure 4. It does not need to change the original wastewater treatment process but only insert the biohydrogen production system into the conventional wastewater treatment process. This could easily recover the energy and useful bio-CO_2. Potential feedstocks for the commercial system are high organic wastewater, agricultural wastes, and fast-food wastes. Beverage and soft drink feedstocks have been well investigated in a pilot system at Feng Chia University, Taiwan. The main advantages of this system include minimizing COD impact, power generation, CO_2 reduction, and CO_2 reuse. Moreover, the CO_2 can be used to cultivate microalgae, which can produce H_2 by plasma, and produce biodiesel and bioethanol using appropriate devices and methods.

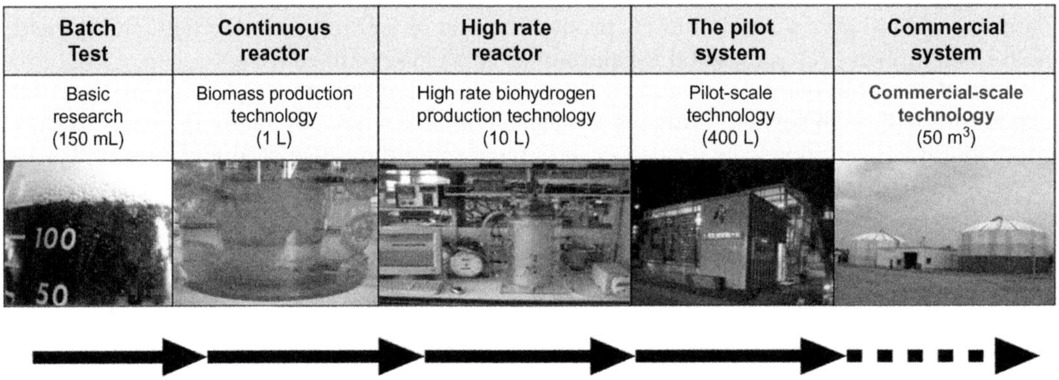

Batch Test	Continuous reactor	High rate reactor	The pilot system	Commercial system
Basic research (150 mL)	Biomass production technology (1 L)	High rate biohydrogen production technology (10 L)	Pilot-scale technology (400 L)	Commercial-scale technology (50 m^3)

FIGURE 3   Technology evolution of a biohydrogen production commercial system.

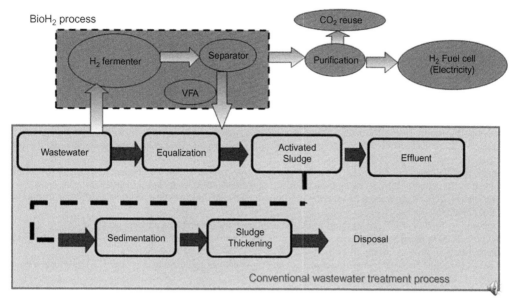

FIGURE 4 A typical application of a biohydrogen production system to a conventional wastewater treatment process.

References

Antonopoulou, G., et al., 2008. Biofuel generation from cheese whey in a two-stage anaerobic process. Indust. Eng. Chem. Res. 47 (15), 5227–5233.

Benemann, J.R., 1997. Feasibility analysis of photobiological hydrogen production. Int. J. Hydrogen Energy 22 (10–11), 979–987.

Brentner, L.B., et al., 2010. Challenges in developing biohydrogen as a sustainable energy source: Implications for a research agenda. Environ. Sci. Technol. 44 (7), 2243–2254.

Cheng, C.H., et al., 2011. Quantitative analysis of microorganism composition in a pilot-scale fermentative biohydrogen production system. Int. J. Hydrogen Energy 36 (21), 14153–14161.

Chu, C.Y., et al., 2011a. Phase holdups and microbial community in high-rate fermentative hydrogen bioreactors. Int. J. Hydrogen Energy 36 (1), 364–373.

Chu, C.Y., et al., 2011b. Hydrodynamic behaviors in fermentative hydrogen bioreactors by pressure fluctuation analysis. Bioresour. Technol 102 (18), 8669–8675.

Kabir Kazi, J.F., et al., 2010. Techno-Economic Analysis of Biochemical Scenarios for Production of Cellulosic Ethanol. T.R. NREL/TP-6A2-46588, Editor June.

Kim, D.H., et al., 2010. Experience of a pilot-scale hydrogen-producing anaerobic sequencing batch reactor (ASBR) treating food waste. Int. J. Hydrogen Energy 35 (4), 1590–1594.

Kriston, A., Szabó, Inzelt, G., 2010. The marriage of car sharing and hydrogen economy: A possible solution to the main problems of urban living. Int. J. Hydrogen Energy 35 (23), 12697–12708.

Lay, C.H., et al., 2012. Co-fermentation of water hyacinth and beverage wastewater in powder and pellet form for hydrogen production. Bioresour. Technol .

Lee, D.H., Chiu, L.H., 2012. Development of a biohydrogen economy in the United States, China, Japan, and India: With discussion of a chicken-and-egg debate. Int. J. Hydrogen Energy 1–10.

Lee, Y.W., Chung, J., 2010. Bioproduction of hydrogen from food waste by pilot-scale combined hydrogen/methane fermentation. Int. J. Hydrogen Energy 35 (21), 11746–11755.

Lee, D.H., Lee, D.J., Chiu, L.H., 2011. Biohydrogen development in United States and in China: An inpute-output model study. Int. J. Hydrogen Energy 1–7.

Lin, C.Y., et al., 2010. Pilot-scale hydrogen fermentation system start-up performance. Int. J. Hydrogen Energy 35 (24), 13452–13457.

Lin, C.Y., et al., 2011a. A pilot-scale high-rate biohydrogen production system with mixed microflora. Int. J. Hydrogen Energy 36 (14), 8758–8764.

Lin, P.J., et al., 2011b. Enhancing the performance of pilot-scale fermentative hydrogen production by proper combinations of HRT and substrate concentration. Int. J. Hydrogen Energy 36 (21), 14289–14294.

Ljunggren, M., Zacchi, G., 2010a. Techno-economic analysis of a two-step biological process producing hydrogen and methane. Bioresour. Technol. 101 (20), 7780–7788.

Ljunggren, M., Zacchi, G., 2010b. Techno-economic evaluation of a two-step biological process for hydrogen production. Biotechnol. Progress 26 (2), 496–504.

Ljunggren, M., Wallberg, O., Zacchi, G., 2011. Techno-economic comparison of a biological hydrogen process and a 2nd generation ethanol process using barley straw as feedstock. Bioresour. Technol. 102 (20), 9524–9531.

Nath, K., Das, D., 2004. Biohydrogen production as a potential energy resource: Present state-of-art. J. Sci. Indust. Res. 63 (9), 729–738.

Nejat Veziroglu, T., 2007. IJHE grows with hydrogen economy. Int. J. Hydrogen Energy 32 (1), 1–2.

Ren, N.Q., et al., 2006. Biohydrogen production from molasses by anaerobic fermentation with a pilot-scale bioreactor system. Int. J. Hydrogen Energy 31 (15), 2147–2157.

Show, K.Y., et al., 2012. Biohydrogen production: Current perspectives and the way forward. Int. J. Hydrogen Energy 37 (20), 15616–15631.

Tredici, M.R., Zlttelli, G.C., 1998. Efficiency of sunlight utilization: Tubular versus flat photobioreactors. Biotechnol. Bioeng. 57 (2), 187–197.

Vatsala, T.M., Raj, S.M., Manimaran, A., 2008. A pilot-scale study of biohydrogen production from distillery effluent using defined bacterial co-culture. Int. J. Hydrogen Energy 33 (20), 5404–5415.

Wang, X., et al., 2009. CFD simulation of an expanded granular sludge bed (EGSB) reactor for biohydrogen production. Int. J. Hydrogen Energy 34 (24), 9686–9695.

Wang, X., et al., 2010. Scale-up and optimization of biohydrogen production reactor from laboratory-scale to industrial-scale on the basis of computational fluid dynamics simulation. Int. J. Hydrogen Energy 35 (20), 10960–10966.

Yang, P., et al., 2007. Biohydrogen production from cheese processing wastewater by anaerobic fermentation using mixed microbial communities. Int. J. Hydrogen Energy 32 (18), 4761–4771.

Zhou, J.H., et al., 2006. Biohydrogen production from food wastes composed of carbohydrates, proteins and lipoids by fermentation. Zhejiang Daxue Xuebao (Gongxue Ban)/Journal of Zhejiang University (Engineering Science) 40 (11), 2007–2010.

Index

Note: Page numbers followed by *f* indicate figures and *t* indicate tables.

Printed and bound by CPI Group (UK) Ltd, Croydon, CR0 4YY

08/05/2025

01864826-0002